# College Algebra

**Second Edition**

## Louis Leithold

**Macmillan Publishing Co., Inc.**
*New York*

**Collier Macmillan Publishers**
*London*

Copyright © 1980, Interests International, Inc.

Printed in the United States of America

All rights reserved. No part of this book may be reproduced or transmitted in any form or by any means, electronic or mechanical, including photocopying, recording, or any information storage and retrieval system, without permission in writing from the Publisher.

Earlier edition copyright © 1975 by Interests International, Inc. Selected illustrations have been reprinted from *Intermediate Algebra for College Students*, Second edition, by Louis Leithold, © 1974 by Interests International, Inc.

Macmillan Publishing Co., Inc.
866 Third Avenue, New York, New York 10022

Collier Macmillan Canada, Ltd.

Library of Congress Cataloging in Publication Data

Leithold, Louis.
   College algebra.

   Includes index.
   1. Algebra.    I. Title.
QA154.2.L43         1980         512.9         79-15730

ISBN 0-02-369580-3

Cover art: "Keeta #2022"
Oil on canvas by Jerry Byrd
Courtesy of Tortue Gallery, Santa Monica
Chapter Opening art: Drawings by Jerry Byrd

Printing: 1 2 3 4 5 6 7 8        Year: 0 1 2 3 4 5 6

To L.T.

# Preface

In a course titled "College Algebra" a student should gain an appreciation of mathematics as a logical science, and the subject matter should be expounded in such a way that it conforms to the experience and maturity of the freshman mathematics student. Furthermore, college algebra is rarely a terminal course in mathematics, and thus another of its purposes is to develop skills that will enable a person to study effectively more advanced courses. With these objectives in mind, I have made every effort to write a textbook that students can read advantageously on their own.

In this second edition I have attempted to reflect the current consensus that mathematics should be more meaningful for students. The first edition's treatment of the real number system has been replaced by a less rigorous approach. While axioms, definitions, and theorems are still carefully stated, they are used to emphasize significant results instead of attempting to demonstrate the development of a mathematical system. It is assumed that the student has had a course in intermediate algebra, or the equivalent, and may need a review of some of the material but not a redevelopment. Consequently there is a much quicker pace for review and a reordering of topics in the hopes that a better classroom experience will be created.

Chapter 1 is concerned with real numbers and first-degree equations and inequalities. Second-degree equations and inequalities are topics in Chapter 2. In each of the first two chapters there is some analytic geometry. The straight line is treated in Chapter 1, and there is an introduction to conics in Chapter 2.

The notion of a function is introduced in Chapter 3, and it is used as a unifying concept throughout much of the remainder of the book. In Section 3.5, new to this edition, graphs of rational functions are given an extensive treatment as a preparation for their use in calculus. The inverse of a function is discussed in Section 3.6, and in Chapter 4 logarithmic functions are evolved as inverses of exponential functions. In the first two sections of Chapter 4, there is a review of properties of exponents and radicals from intermediate algebra. However, the emphasis in this chapter is upon the functional properties of the exponential and logarithmic functions.

In the first edition of this book, polynomial functions were presented in the final chapter. This topic now forms the basis of Chapter 5 in the hopes that it will give the student a positive feeling that he or she has now gone beyond intermediate algebra. There is a systematic method developed for finding the exact or approximate value of all real zeros of polynomial functions with real coefficients. Added to this edition is Section 5.8 that presents a complete discussion of graphs of polynomial functions.

A straight-forward coverage of systems of equations and inequalities is provided in Chapter 6. The "Gauss" technique is used to solve systems of linear equations in preparation for the treatment of matrices in Chapter 7. The material on matrices and determinants has been completely rewritten. Matrix arithmetic is given in Section 7.2 which precedes the discussion of determinants and their properties in Sections 7.3 and 7.4. The inverse of a square matrix is introduced in Section 7.5 and it is used to solve systems of equations.

The discussion of sequences and series in Chapter 8 is based on the function concept. Mathematical induction is the topic of Section 8.2 and it is used to prove properties of arithmetic and geometric sequences and series. Chapter 9 covers the topics of permutations, combinations, and the binomial theorem. In the discussion of the binomial theorem, the binomial coefficients are treated as combinations. Each of chapters 7, 8, and 9 is self-contained and may be omitted from a short course.

Many examples and illustrations that stress both theoretical and computational aspects of the subject are included. At the end of each section there is a generous list of exercises that are graded in difficulty, and for this edition there has been added a set of review exercises at the end of each chapter. The answers to the odd-numbered exercises are given in the back of the book, and the answers to the even-numbered ones are available in a separate booklet.

I am grateful for the suggestions for improving the manuscript that were made by the reviewers, Professors B. Patricia Barbalich of Jefferson Community College, Steven D. Kerr of Weber State College, John W. Milsom of Butler County Community College, Merlin C. Miller of Merritt College, Ethel Rogers, Donald G. Killian, and John S. Hutchinson of Wichita State University and William W. Mitchell of Phoenix College. I also appreciate the efforts of James Smith, mathematics editor; J. Edward Neve, production editor; and Andrew Zutis, designer. Thanks are also due to Jerry Byrd, whose works appear as cover and chapter-opening art.

Louis Leithold

# Contents

**Chapter 1**
Real Numbers, Equations, and Inequalities  1

1.1 The Set of Real Numbers, and Some of Its Subsets  2
1.2 Properties of Real Numbers  11
1.3 Some Algebraic Terminology  19
1.4 First-degree Equations in One Variable  28
1.5 Word Problems  37
1.6 First-degree Inequalities in One Variable  46
1.7 Equations and Inequalities Involving Absolute Value  55
1.8 Points in a Plane, Distance Formula, and Slope of a Line  63
1.9 First-degree Equations in Two Variables  75

**Chapter 2**
Second-degree Equations and Inequalities  89

2.1 Second-degree Equations in One Variable  90
2.2 Completing a Square and the Quadratic Formula  97
2.3 Other Equations in One Variable  107
2.4 Quadratic Equations in Two Variables  114
2.5 The Parabola  121
2.6 The Circle, the Ellipse, and the Hyperbola  126
2.7 Quadratic Inequalities  136
2.8 Inequalities in Two Variables  141

**Chapter 3**
Functions  151

3.1 Relations and Functions, and Their Graphs  152
3.2 Function Notation and Types of Functions  160
3.3 Operations on Functions  167
3.4 Quadratic Functions  171
3.5 Graphs of Rational Functions  177
3.6 The Inverse of a Function  187
3.7 Variation  196

## Chapter 4
Exponential and Logarithmic Functions 205

4.1 Properties of Exponents 206
4.2 Properties of Radicals 217
4.3 Exponential Functions 225
4.4 Logarithmic Functions 230
4.5 Properties of Logarithmic Functions 237
4.6 Common Logarithms 245
4.7 Exponential Equations 256

## Chapter 5
Polynomial Functions, the Theory of Polynomial Equations, and Complex Numbers 265

5.1 The Set of Complex Numbers 266
5.2 Geometric Representation of Complex Numbers 275
5.3 Equations Having Complex Roots 280
5.4 The Remainder Theorem and the Factor Theorem 286
5.5 Synthetic Division 290
5.6 Complex Zeros of Polynomial Functions 295
5.7 Rational Zeros of Polynomial Functions 303
5.8 Graphs of Polynomial Functions 312
5.9 Real Roots of Polynomial Equations 318

## Chapter 6
Systems of Equations and Inequalities 327

6.1 Systems of Linear Equations in Two Variables 328
6.2 Systems of Linear Equations in Three Variables 338
6.3 Word Problems 346
6.4 Systems Involving Quadratic Equations 355
6.5 Systems of Inequalities and Introduction to Linear Programming 363

## Chapter 7
Matrices and Determinants 375

7.1 Matrices 376
7.2 Properties of Matrices 383
7.3 Determinants 393
7.4 Properties of Determinants 403
7.5 The Inverse of a Square Matrix 412
7.6 Cramer's Rule 421

## Chapter 8
Sequences, Series, and Mathematical Induction 431

8.1 Sequences and Series 432
8.2 Mathematical Induction 438
8.3 Arithmetic Sequences and Series 444
8.4 Geometric Sequences and Series 451
8.5 Infinite Geometric Series 460

**Chapter 9**
Permutations, Combinations, and the Binomial Theorem   469

9.1   Counting   470
9.2   Permutations   474
9.3   Combinations   482
9.4   The Binomial Theorem   487

**Appendix**

Table 1   Powers and Roots   A-1
Table 2   Common Logarithms   A-2
Table 3   Exponential Functions   A-4
Table 4   Natural Logarithms   A-4

**Answers to Odd-numbered Exercises**   A-5

**Index**   A-21

# College Algebra

# 1

# Real Numbers, Equations, and Inequalities

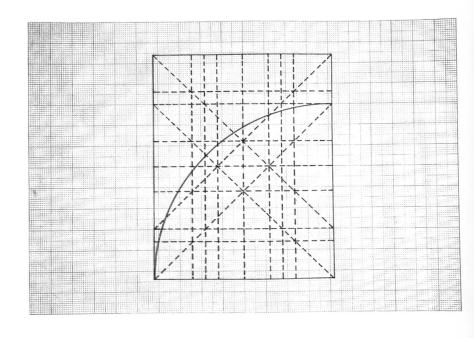

## 1.1 The Set of Real Numbers, and Some of Its Subsets

Probably the first numbers with which most people become concerned is the set of *natural numbers* (also called *counting numbers*), which may be listed as

$$1, 2, 3, \ldots$$

where the three dots are used to indicate that the list goes on and on with no last number. If we wish to list the set of natural numbers less than 100, we write

$$1, 2, 3, \ldots, 99$$

where here the three dots indicate the natural numbers between 3 and 99.

Note that in the paragraph above we use the word "set" and refer to the "set of natural numbers." The idea of "set" is used extensively in mathematics and is such a basic concept that it is not given a formal definition. We can say that a *set* is a collection of objects and we may speak of a set of books on a shelf, a set of students in a classroom, a set of trees in a park, and so on. Capital letters such as $A$, $B$, $C$, $R$, $S$, and $T$ are used to denote sets. The objects in a set are called the *elements* of a set. In algebra we are concerned mainly with sets whose elements are numbers.

We want every set to be *well defined;* that is, there should be some rule or property that enables one to decide whether a given object is or is not an element of a specific set. A pair of braces { } used with words or symbols can describe a set.

If $S$ is the set of natural numbers less than 8, we can write the set $S$ as

$$\{1, 2, 3, 4, 5, 6, 7\}$$

We can also write the set $S$ as

$$\{x, \text{such that } x \text{ is a natural number less than } 8\}$$

where the symbol "$x$" is called a "variable." A *variable* is a symbol used to represent any element of a given set, and the given set is called the *replacement set* of the variable. Another way of writing the above set $S$ is to use what is called *set-builder notation,* where a vertical bar is used in place of the words "such that." Thus using set-builder notation to describe the set $S$, we have

$$\{x \mid x \text{ is a natural number less than } 8\}$$

which is read "the set of all $x$ such that $x$ is a natural number less than 8."

The set of natural numbers is denoted by $N$. Hence we may write the set $N$ as $\{1, 2, 3, \ldots\}$ or, using set-builder notation, as $\{x \mid x \text{ is a natural number}\}$.

The symbol $\in$ is used to indicate that a specific element belongs to a set. Thus we may write $5 \in N$, which is read "5 is an element of $N$." The notation "$a, b \in R$" indicates that both $a$ and $b$ are elements of $R$. The symbol "$\notin$" is read "is not an element of." Hence we read $\frac{1}{2} \notin N$ as "$\frac{1}{2}$ is not an element of $N$."

Suppose that $S$ is the set denoted by $\{1, 2, 3, 4, 5, 6, 7\}$ and $T$ is the set denoted by $\{2, 4, 6\}$. Then every element of $T$ is also an element of $S$ and we say that $T$ is a "subset" of $S$.

**1.1.1 DEFINITION** The set $A$ is a *subset* of the set $B$, written $A \subseteq B$, if and only if every element of $A$ is also an element of $B$. If, in addition, there is at least one element of $B$ that is not an element of $A$, then $A$ is a *proper subset* of $B$, written $A \subset B$.

Observe from Definition 1.1.1 that every set is a subset of itself, but a set is not a proper subset of itself.

In Definition 1.1.1 the "if and only if" qualification is used to combine two statements: (1) "The set $A$ is a subset of the set $B$ *if* every element of $A$ is also an element of $B$"; and (2) "the set $A$ is a subset of set $B$ *only if* every element of $A$ is also an element of $B$," which is logically equivalent to the statement "every element of $A$ is also an element of $B$ if $A$ is a subset of $B$."

**ILLUSTRATION 1.** Let $X$ be the set denoted by $\{x \mid x$ is a natural number less than $8\}$ and $Y$ be the set denoted by $\{y \mid y$ is an even natural number less than $8\}$. Hence $X$ may also be denoted by $\{1, 2, 3, 4, 5, 6, 7\}$ and $Y$ may also be denoted by $\{2, 4, 6\}$. Because every element of $Y$ is also an element of $X$, $Y$ is a subset of $X$ and we write $Y \subseteq X$. Also, there is at least one element of $X$ that is not an element of $Y$, and therefore $Y$ is a proper subset of $X$ and we may write $Y \subset X$. Furthermore, since $\{5\}$ is the set consisting of the number 5, $\{5\} \subset X$, which states that the set consisting of the single element 5 is a proper subset of the set $X$. We may also write $5 \in X$, which states that the number 5 is an element of the set $X$. ●

The symbol $\not\subseteq$ is read "is not a subset of." Thus we may write $\{1, 2, 3, 4\} \not\subseteq \{1, 2, 3\}$.

Suppose that a set is described as one whose elements are the odd natural numbers that are divisible by 2. Because there are no such numbers having this property, this set contains no elements. Such a set is called the "empty set" or the "null set."

**1.1.2 DEFINITION** The *empty set* (or *null set*) is the set that contains no elements. It is denoted by $\emptyset$ or $\{\ \}$.

The empty set is considered to be a proper subset of every nonempty set. The justification of this can be seen by considering set $A$ to be a proper subset of set $B$ when there is no element in $A$ which is not in $B$ and there is at least one element in $B$ which is not in $A$. Therefore, if $B$ is any nonempty set, there is no element in $\emptyset$ which is not in $B$ and there is at least one element in $B$ which is not in $\emptyset$. Hence we may write $\emptyset \subset B$.

**EXAMPLE 1**
List all the subsets of $\{1, 3, 5\}$

**SOLUTION**

$$\{1, 3, 5\}, \{1, 3\}, \{1, 5\}, \{3, 5\}, \{1\}, \{3\}, \{5\}, \varnothing$$

In Example 1, note that all the subsets, other than the set itself, are proper subsets.

The concept of "subset" may be used to define what is meant by two sets being "equal."

**1.1.3 DEFINITION** Two sets $A$ and $B$ are said to be *equal*, written $A = B$, if and only if $A \subseteq B$ and $B \subseteq A$.

Essentially, Definition 1.1.3 states that the two sets $A$ and $B$ are equal if and only if every element of $A$ is an element of $B$ and every element of $B$ is an element of $A$, that is, if the sets $A$ and $B$ have exactly the same elements.

**ILLUSTRATION 2.** Let $A$ be the set denoted by $\{1, 2, 3, 4\}$, $B$ be the set denoted by $\{1, 3, 5, 7\}$, $C$ be the set denoted by $\{4, 3, 2, 1\}$, and $D$ be the set denoted by $\{x \mid x \text{ is an odd natural number less than } 8\}$. Then, because $A \subseteq C$ and $C \subseteq A$, we can write $A = C$. Similarly, $B \subseteq D$ and $D \subseteq B$ and therefore $B = D$. •

Changing the order of listing the elements of a set does not change the set. For instance, in Illustration 2, the elements of sets $A$ and $C$ are listed in a different order, but the two sets contain exactly the same elements; hence the two sets are equal.

The symbol $\neq$ is read "is not equal to." Thus $\{1, 2, 3\} \neq \{1, 2, 3, 4\}$.

Consider now the sets $S = \{a, b, c\}$ and $T = \{1, 2, 3\}$. The elements of set $S$ can be paired with the elements of set $T$ in such a way that each element in $S$ is associated with one and only one element in $T$ and each element in $T$ is associated with one and only one element in $S$. There are six possible pairings and they are shown in Figure 1.1.1. Any one of the six pairings is called a "one-to-one correspondence" between the two sets $S$ and $T$.

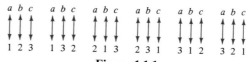

**Figure 1.1.1**

**1.1.4 DEFINITION** A *one-to-one correspondence* between two sets $A$ and $B$ is a pairing between the elements of $A$ with the elements of $B$ in such a way that each element of $A$ is paired with one and only one element of $B$ and each element of $B$ is paired with one and only one element of $A$.

The concept of "one-to-one correspondence" is used to define "equivalent sets."

**1.1.5 DEFINITION**  Set $A$ is said to be *equivalent* to set $B$, written $A \approx B$, if and only if there exists a one-to-one correspondence between sets $A$ and $B$.

Note that necessarily two sets are equivalent if they contain the same number of elements. If we compare Definitions 1.1.3 and 1.1.5, it follows that if two sets are equal, they are necessarily equivalent. However, if two sets are equivalent they are not necessarily equal as shown by the sets of Illustration 3.

**ILLUSTRATION 3.** If $S = \{a, b, c\}$ and $T = \{1, 2, 3\}$, then because of the discussion preceding Definition 1.1.4, $S$ is equivalent to $T$, and we write $S \approx T$. •

We shall define a "finite" set as one which is equivalent to a proper subset $\{1, 2, 3, \ldots, n\}$ of the set of natural numbers. For example, consider the set $V = \{a, e, i, o, u\}$. This set is equivalent to the set $\{1, 2, 3, 4, 5\}$. Hence we say that set $V$ is a "finite" set. The set of all natural numbers less than one hundred is another example of a "finite" set.

**1.1.6 DEFINITION**  A nonempty set is said to be *finite* if and only if it is equivalent to the set $\{1, 2, 3, \ldots, n\}$ for some fixed natural number $n$. The empty set is also said to be finite. A nonempty set that is not finite is said to be *infinite*.

In effect, Definition 1.1.6 states that a set is finite if the elements of the set can be arranged in some order and counted one by one until there is a last element. An example of an infinite set is the set of all the natural numbers, because no matter how many natural numbers are counted, there are always more. Other examples of infinite sets are the set of all even natural numbers and the set of all natural numbers divisible by five.

There are two operations on sets that we shall find useful as we proceed. One such operation consists of forming the "union" of two sets, which we now define.

**1.1.7 DEFINITION**  Let $A$ and $B$ be two sets. The *union* of $A$ and $B$, denoted by $A \cup B$ and read "$A$ union $B$," is the set of all elements which are in $A$ or in $B$ or in both $A$ and $B$.

**EXAMPLE 2**
Let $A = \{4, 6, 8, 10, 12\}$,
   $B = \{3, 6, 9, 12\}$, and
   $C = \{4, 10\}$.
Find
(a) $A \cup B$   (b) $A \cup C$
(c) $B \cup C$   (d) $A \cup A$

**SOLUTION**
(a) $A \cup B = \{3, 4, 6, 8, 9, 10, 12\}$
(b) $A \cup C = \{4, 6, 8, 10, 12\}$
(c) $B \cup C = \{3, 4, 6, 9, 10, 12\}$
(d) $A \cup A = \{4, 6, 8, 10, 12\}$
       $= A$

A second set operation, which will be useful later, consists of forming the "intersection" of two sets.

**1.1.8 DEFINITION** Let $A$ and $B$ be two sets. The *intersection* of $A$ and $B$, denoted by $A \cap B$ and read "$A$ intersection $B$," is the set of all elements that are in both $A$ and $B$.

**EXAMPLE 3**
If $A$, $B$, and $C$ are the sets defined in Example 2, find
(a) $A \cap B$   (b) $A \cap C$
(c) $B \cap C$   (d) $A \cap A$

**SOLUTION**
(a) $A \cap B = \{6, 12\}$
(b) $A \cap C = \{4, 10\}$
(c) $B \cap C = \emptyset$
(d) $A \cap A = \{4, 6, 8, 10, 12\}$
       $= A$

Note that in part (c) of Example 3 the intersection of sets $B$ and $C$ is the empty set. These two sets have no elements in common, and they are called "disjoint" sets.

**1.1.9 DEFINITION** Two sets $A$ and $B$ are said to be *disjoint* if and only if $A \cap B = \emptyset$.

In this chapter we are concerned with "the set of real numbers" and subsets of the real numbers. You have had previous experience with real numbers and are familiar with some of their properties, which have probably become familiar to you through informal means. However, the properties of the real numbers may be proved by deductive reasoning from a few basic assumptions called *axioms,* which we discuss in Section 1.2. We now give a brief intuitive discussion of sets of numbers that are subsets of the set of real numbers.

We have referred to the set of natural numbers, which we call $N$, so that

$$N = \{1, 2, 3, \ldots\}$$

The number "zero," denoted by the symbol 0 and formally defined in Section 1.2 (Axiom 1.2.5), is the number having the property that if it is added to any number the result is that number. The set of numbers whose elements are the natural numbers and zero is called the set of *whole numbers.*

Denoting this set by $W$, we write
$$W = \{0, 1, 2, 3, \ldots\}$$

The set of natural numbers is also called the set of *positive integers.* If $n$ is a natural number, it may also be referred to as the positive integer $n$. In particular, the natural number 14 is the same as the positive integer 14.

Corresponding to each positive integer $n$ there is a negative integer such that if the negative integer is added to $n$, the result is 0. For example, the negative integer $-5$, read "negative five," is the number which when added to 5 gives a result of 0.

The set of *negative integers* is denoted by $\bar{N}$ and we write
$$\bar{N} = \{-1, -2, -3, \ldots\}$$

The set of numbers whose elements are the positive integers, the negative integers, and zero is called the set of *integers.* Denoting this set by $J$, we have
$$J = \{\ldots, -3, -2, -1, 0, 1, 2, 3, \ldots\}$$

The set $J$ then is the union of the sets $\bar{N}$ and $W$, and symbolically we write
$$J = \bar{N} \cup W$$

Because $W = \{0\} \cup N$, we may write
$$J = \bar{N} \cup (\{0\} \cup N)$$

The set of integers then is the union of three *disjoint subsets:* the set of positive integers, the set of negative integers, and the set consisting of the single number 0. Note that the number 0 is an integer, but it is neither positive nor negative. Sometimes we refer to the set of *nonnegative integers,* which is the set consisting of the positive integers and 0 or, equivalently, the set of whole numbers. Similarly, the set of *nonpositive integers* is the set consisting of the negative integers and 0.

Consider now the set of numbers whose elements are those numbers which can be represented by the quotient of two integers $p$ and $q$, where $q$ is not 0, that is, the numbers that can be represented symbolically as

$$\frac{p}{q} \quad \text{where } q \text{ is not } 0$$

This set of numbers is called the set of *rational numbers,* which is denoted by $Q$. Symbolically, $Q$ is defined as

$$Q = \left\{ x \mid x \text{ can be represented by } \frac{p}{q}, p \in J, q \in J, q \text{ is not } 0 \right\}$$

Some numbers in the set $Q$ are $\frac{1}{2}$, $\frac{3}{4}$, $\frac{11}{5}$, $\frac{-2}{3}$, and $\frac{-31}{12}$. Every integer is a rational number because every integer can be represented as the quotient of itself and 1; that is, 8 can be represented by $\frac{8}{1}$, 0 can be represented by $\frac{0}{1}$, and $-15$ can be represented by $\frac{-15}{1}$. Hence the set $J$ is a proper subset of set $Q$ and we may write $J \subset Q$.

Any rational number can be written as a decimal. We are assuming that you are familiar with the process of using long division to do this. For example, $\frac{3}{10}$ can be written 0.3, $\frac{9}{4}$ can be written 2.25, and $\frac{83}{16}$ can be written 5.1875. These decimals are called *terminating decimals*. There are rational numbers whose decimal representation is nonterminating and repeating; for example, $\frac{1}{3}$ has the decimal representation 0.333. . . , where the digit 3 is repeated, and $\frac{47}{11}$ can be represented as 4.272727. . . , where the digits 2 and 7 are repeated in that order. It can be proved that the decimal representation of every rational number is either a terminating decimal or a nonterminating repeating decimal, although the proof is beyond the scope of this book. We shall, however, show in Section 8.5 that every nonterminating repeating decimal is a representation of a rational number.

The following question now arises: Are there numbers represented by nonterminating nonrepeating decimals? The answer is yes, and an example of such a number is the principal square root of 2, denoted symbolically by $\sqrt{2}$, and represented by a nonterminating nonrepeating decimal as 1.4142. . . . Another such number is $\pi$ (pi), which is the ratio of the circumference of a circle to its diameter and represented by a nonterminating nonrepeating decimal as 3.14159. . . . The numbers represented by nonterminating nonrepeating decimals cannot be expressed as the quotient of two integers (although we shall not prove this) and hence are not rational numbers. The set of numbers that are represented by nonterminating nonrepeating decimals is called the set of *irrational numbers*, which we shall denote by $H$. It may be defined symbolically by

$$H = \{x \mid \text{the decimal representation of } x \text{ is nonterminating nonrepeating}\}$$

We stated previously that any rational number has a decimal representation that is either terminating or nonterminating and repeating. We have just seen that any number whose decimal representation is nonterminating and nonrepeating is an irrational number. Hence the union of the set of rational numbers and the set of irrational numbers is the set of all the numbers that can be expressed as decimals, and this set is called the set of *real numbers*. Denoting the set of real numbers by $R$, we may define $R$ symbolically by

$$R = \{x \mid x \in (Q \cup H)\}$$

or, equivalently,

$$R = Q \cup H$$

The set of real numbers is also the union of three other disjoint subsets. One set is the set consisting of the single number 0. Another is the set consisting of the *positive real numbers* and is denoted by $R^+$. The third is the set of *negative real numbers*, which is denoted by $R^-$.

## EXAMPLE 4

The sets $N$, $W$, $\bar{N}$, $J$, $Q$, $H$, $R$, $R^+$, and $R^-$ are the sets of numbers defined in this section. Insert either $\subseteq$ or $\not\subseteq$ to make the statement correct.

(a) $N$ _____ $J$
(b) $Q$ _____ $W$
(c) $\{\sqrt{2}, \pi, 3.5\}$ _____ $H$
(d) $\{0\}$ _____ $W$
(e) $N$ _____ $R^+$
(f) $\bar{N}$ _____ $R$

### SOLUTION

(a) Because every natural number is an integer, $N \subseteq J$.
(b) Because every rational number is not a whole number, $Q \not\subseteq W$.
(c) $\sqrt{2}$ and $\pi$ are irrational numbers, but 3.5 is a rational number, and thus $\{\sqrt{2}, \pi, 3.5\} \not\subseteq H$.
(d) Zero is a whole number, and therefore $\{0\} \subseteq W$.
(e) Every natural number is a positive real number, and so $N \subseteq R^+$.
(f) Every negative integer is a real number, and thus $\bar{N} \subseteq R$.

## EXAMPLE 5

In each of the following, determine which one of the sets $N$, $W$, $J$, $Q$, $H$, $R$, $\{0\}$, and $\emptyset$ is equal to the given set

(a) $R^+ \cap J$    (b) $J \cup Q$
(c) $J \cap Q$     (d) $N \cap H$

### SOLUTION

(a) The intersection of $R^+$ and $J$ is the set of numbers that are both positive real numbers and integers, and this is the set of positive integers. Hence $R^+ \cap J = N$.
(b) The union of $J$ and $Q$ is the set of all numbers that are either integers or rational. Because the set of integers is a subset of the set of rational numbers, this union is the set of rational numbers. Therefore, $J \cup Q = Q$.
(c) The intersection of $J$ and $Q$ is the set of numbers that are both integers and rational numbers. This intersection is the set of integers, and therefore $J \cap Q = J$.
(d) Because the set of positive integers and the set of irrational numbers have no elements in common, $N \cap H = \emptyset$.

## EXERCISES 1.1

*In Exercises 1 through 6, describe the given set by listing the elements.*

1. The even natural numbers less than 10
2. The odd natural numbers between 6 and 18
3. The natural numbers less than 100 that are multiples of 8
4. The natural numbers between 9 and 29 that are multiples of 2 but not multiples of 4
5. The letters in the name of the day of the week following Tuesday
6. The letters in the name of the longest river in the United States

*In Exercises 7 through 10, use set-builder notation to describe the given set.*

7. $\{1, 2, 3, 4, 5, 6, 7, 8, 9\}$
8. $\{5, 10, 15, 20, 25, 30\}$
9. $\{a, b, c, d\}$
10. $\{x, y, z\}$

*In Exercises 11 through 14, indicate if the given set is finite or infinite.*

**11.** $\{x \mid x$ is a natural number greater than 1000$\}$
**12.** $\{x \mid x$ is a natural number less than 1000$\}$
**13.** $\{x \mid x$ is a grain of sand on the beach at Coney Island$\}$
**14.** $\{x \mid x$ is an even natural number divisible by 8$\}$

*In Exercises 15 and 16, list all the subsets of the given set.*

**15.** $\{0, 1, 2\}$
**16.** $\{r, s, t, u\}$

*In Exercises 17 through 22, insert either $\subseteq$ or $\not\subseteq$ in the blank to make the statement correct.* $S = \{1, 2, 3, 4, 5, 6, 7, 8, 9\}$, $T = \{1, 3, 5, 7, 9\}$, $U = \{2, 4, 6, 8\}$, and $V = \{4, 8\}$.

**17.** $V$ \_\_\_\_ $S$   **18.** $U$ \_\_\_\_ $V$   **19.** $S$ \_\_\_\_ $T$   **20.** $\emptyset$ \_\_\_\_ $S$   **21.** $V$ \_\_\_\_ $T$   **22.** $U$ \_\_\_\_ $S$

*In Exercises 23 through 42, list the elements of the given set if* $A = \{1, 3, 5, 7, 9\}$, $B = \{1, 2, 4, 8\}$, $C = \{2, 4, 6, 8\}$, $D = \{3, 6, 9\}$, and $E = \{1, 5, 9\}$.

**23.** $A \cup C$      **24.** $B \cup D$      **25.** $A \cap E$      **26.** $D \cap E$      **27.** $B \cup C$
**28.** $A \cup D$     **29.** $C \cap E$      **30.** $A \cup E$      **31.** $B \cap C$      **32.** $B \cup E$
**33.** $C \cap D$     **34.** $A \cap C$      **35.** $B \cup B$      **36.** $A \cap A$      **37.** $C \cap \emptyset$
**38.** $E \cup \emptyset$  **39.** $(C \cup E) \cap D$  **40.** $(A \cup B) \cap C$  **41.** $(B \cup C) \cup D$  **42.** $(A \cap E) \cap D$

*In Exercises 43 through 67, $N$ is the set of natural numbers, $W$ is the set of whole numbers, $\overline{N}$ is the set of negative integers, $J$ is the set of integers, $Q$ is the set of rational numbers, $H$ is the set of irrational numbers, $R$ is the set of real numbers, $R^+$ is the set of positive real numbers, and $R^-$ is the set of negative real numbers.*

*In Exercises 43 through 50, insert $\in$ or $\notin$ in the blank to make the statement correct.*

**43.** 15 \_\_\_\_ $N$       **44.** 2007 \_\_\_\_ $W$       **45.** 1.47 \_\_\_\_ $W$       **46.** $\frac{3}{7}$ \_\_\_\_ $R$
**47.** 0 \_\_\_\_ $Q$        **48.** $\pi$ \_\_\_\_ $Q$          **49.** $-5$ \_\_\_\_ $H$         **50.** $-3$ \_\_\_\_ $J$

*In Exercises 51 through 56, use the symbol $\subseteq$ to give a correct statement involving the two given sets.*

**51.** $N$ and $W$    **52.** $R$ and $W$    **53.** $\overline{N}$ and $R^-$    **54.** $N$ and $R^+$    **55.** $R$ and $H$    **56.** $J$ and $Q$

*In Exercises 57 through 62, determine which of the sets $N$, $W$, $J$, $Q$, $H$, $R$, $\{0\}$, and $\emptyset$ is equal to the given set.*

**57.** $Q \cap R$    **58.** $Q \cap H$    **59.** $Q \cup H$    **60.** $W \cup N$    **61.** $W \cap N$    **62.** $Q \cap J$

*In Exercises 63 through 67, list the elements of the given set if*

$$S = \{12, \tfrac{5}{3}, \sqrt{7}, 0, -38, -\sqrt{2}, 571, \pi, -\tfrac{1}{10}, 0.666\ldots, 16.34\}.$$

**63.** $S \cap J$      **64.** $S \cap Q$      **65.** $S \cap H$      **66.** $S \cap N$      **67.** $S \cap \overline{N}$

In Exercises 68 through 71, show that the given statement is true if $A = \{1, 2, 3, 4\}$, $B = \{0, 1, 2\}$, and $C = \{3, 4, 5\}$.

**68.** $A \cup (B \cup C) = (A \cup B) \cup C$
**69.** $A \cap (B \cap C) = (A \cap B) \cap C$
**70.** $A \cap (B \cup C) = (A \cap B) \cup (A \cap C)$
**71.** $A \cup (B \cap C) = (A \cup B) \cap (A \cup C)$

## 1.2 Properties of Real Numbers

The *real number system* consists of the set $R$ of real numbers and two operations called *addition* and *multiplication*. The operation of addition is denoted by the symbol "$+$," and the operation of multiplication is denoted by the symbol "$\cdot$" (or "$\times$"). If $a, b \in R$, $a + b$ denotes the *sum* of $a$ and $b$, and $a \cdot b$ (or $ab$) denotes their *product*.

We now present seven axioms that give laws governing the operations of addition and multiplication on the set $R$. The word *axiom* is used to indicate a formal statement about numbers, or properties of numbers, that is assumed to be true without proof.

**1.2.1 AXIOM** *Closure and Uniqueness Laws.* If $a, b \in R$, then $a + b$ is a unique real number, and $ab$ is a unique real number.

**1.2.2 AXIOM** *Commutative Laws.* If $a, b \in R$, then

$$a + b = b + a \quad \text{and} \quad ab = ba$$

**1.2.3 AXIOM** *Associative Laws.* If $a, b, c \in R$, then

$$a + (b + c) = (a + b) + c \quad \text{and} \quad a(bc) = (ab)c$$

**1.2.4 AXIOM** *Distributive Law.* If $a, b, c \in R$, then

$$a(b + c) = ab + ac$$

**1.2.5 AXIOM** *Existence of Identity Elements.* There exist two distinct numbers 0 and 1 such that for any real number $a$,

$$a + 0 = a \quad \text{and} \quad a \cdot 1 = a$$

**1.2.6 AXIOM** *Existence of Opposite or Additive Inverse.* For every real number $a$, there exists a real number called the *opposite of a* (or *additive inverse of a*), denoted by $-a$ (read "the opposite of $a$"), such that

$$a + (-a) = 0$$

**1.2.7 AXIOM**  *Existence of Reciprocal or Multiplicative Inverse.* For every real number $a$, except 0, there exists a real number called the *reciprocal of a* (or *multiplicative inverse of a*), denoted by $\frac{1}{a}$, such that

$$a \cdot \frac{1}{a} = 1$$

Axioms 1.2.1 through 1.2.7 are called *field axioms,* and if these axioms are satisfied by a set of elements, then the set is called a *field* under the two operations involved. Hence the set R is a field under addition and multiplication. For the set $J$ of integers, each of Axioms 1.2.1 through 1.2.6 is satisfied, but Axiom 1.2.7 is not satisfied (for instance, the integer 2 has no multiplicative inverse in $J$). Therefore, the set of integers is not a field under addition and multiplication.

**1.2.8 DEFINITION**  If $a, b \in R$, the operation of *subtraction* assigns to $a$ and $b$ a real number denoted by $a - b$ (read "$a$ minus $b$") called the *difference* of $a$ and $b$, where

$$a - b = a + (-b) \tag{1}$$

Equality (1) is read "$a$ minus $b$ equals $a$ plus the opposite of $b$."

**1.2.9 DEFINITION**  If $a, b \in R$, and $b \neq 0$, the operation of *division* assigns to $a$ and $b$ a real number, denoted by $a \div b$ (read "$a$ divided by $b$"), called the *quotient* of $a$ and $b$, where

$$a \div b = a \cdot \frac{1}{b}$$

The number $a$ is called the *dividend* and $b$ is called the *divisor.*

Other notations for the quotient of $a$ and $b$ are

$$\frac{a}{b} \quad \text{and} \quad a/b$$

The numerals $\frac{a}{b}$ and $a/b$ are called *fractions*. The number $a$ is called the *numerator* and $b$ is called the *denominator*.

An alternative definition of division is the statement

$$a \div b = q \quad \text{if and only if} \quad a = bq \quad \text{where } b \neq 0 \tag{2}$$

Observe in Definition 1.2.9 and statement (2) that, for the operation of division, the divisor is restricted to nonzero real numbers. This restriction is necessary because if $a \neq 0$ and $b = 0$, then if $a \div 0$ were equal to some real number $q$, it would follow that $0 \cdot q = a$, which is impossible because $0 \cdot q = 0$ and $a \neq 0$. Furthermore, if $a = 0$ and $b = 0$, then according to statement (2), $0 \div 0$ could equal any real number because $0 \cdot q = 0$ for any value of $q$. Hence *division by zero is not defined;* that is,

$$\frac{a}{0} \quad \text{is not defined}$$

By using the field axioms and Definitions 1.2.8 and 1.2.9, we can derive properties of the real numbers from which follow the familiar algebraic operations as well as the techniques of solving equations, factoring, and so forth. In this book we are not concerned with showing how such properties are derived from the axioms.

Properties that can be shown to be logical consequences of axioms are *theorems*. In the statement of most theorems there are two parts: the "if" part, called the *hypothesis,* and the "then" part, called the *conclusion*. The argument verifying a theorem is a *proof*. A proof consists of showing that the conclusion follows from the assumed truth of the hypothesis.

The concept of a real number being "positive" is given in the following axiom.

**1.2.10 AXIOM**    *Order Axiom.* In the set of real numbers there exists a subset called the *positive numbers* such that

(i) If $a \in R$, exactly one of the following three statements holds:

$$a = 0 \qquad a \text{ is positive} \qquad -a \text{ is positive}$$

(ii) The sum of two positive numbers is positive.
(iii) The product of two positive numbers is positive.

Axiom 1.2.10 is called the order axiom because it enables us to order the elements of the set $R$. In Definitions 1.2.12 and 1.2.13, we use the axiom to define the relations of "greater than" and "less than" on $R$.

The opposites of the elements of the set of positive numbers form the set of "negative" numbers, as given in the following definition.

**1.2.11 DEFINITION**    The real number $a$ is negative if and only if $-a$ is positive.

From Axiom 1.2.10 and Definition 1.2.11 it follows that a real number is either a positive number, a negative number, or zero.

The field axioms do not imply any order of the real numbers. That is, by means of the field axioms alone we cannot state that 2 is greater than 1, 3 is greater than 2, and so on. However, we introduced the order axiom and because the set $R$ of real numbers satisfies the order axiom and the field axioms, we say that $R$ is an *ordered field*. We use the concept of a positive number given in the order axiom to define what we mean by one real number being "greater than" another.

**1.2.12 DEFINITION** If $a, b \in R$, then *a is greater than b* (written $a > b$) if and only if $a - b$ is positive.

**ILLUSTRATION 1**

(a) $7 > 2$ because $7 - 2 = 5$, and 5 is positive.
(b) $-4 > -10$ because $-4 - (-10) = 6$, and 6 is positive. •

**1.2.13 DEFINITION** If $a, b \in R$, then *a is less than b* (written $a < b$) if an only if $b$ is greater than $a$; with symbols, we write

$$a < b \quad \text{if and only if} \quad b > a$$

**ILLUSTRATION 2**

(a) $2 < 7$ because $7 > 2$ \quad (b) $-10 < -4$ because $-4 > -10$ •

The statements "$a > b$" and "$a < b$" are called *inequalities*.

The proof of the following theorem is omitted but it involves the order axiom and Definitions 1.2.12 and 1.2.13.

**1.2.14 THEOREM** If $a \in R$, then

$$a \text{ is positive} \quad \text{if and only if} \quad a > 0 \tag{3}$$

and

$$a \text{ is negative} \quad \text{if and only if} \quad a < 0 \tag{4}$$

By using Theorem 1.2.14, the order axiom can be given in an alternative form, which is the next theorem.

**1.2.15 THEOREM** (i) If $a \in R$, exactly one of the following statements holds:

$$a = 0 \quad a > 0 \quad a < 0$$

(ii) If $a > 0$ and $b > 0$, then $a + b > 0$.
(iii) If $a > 0$ and $b > 0$, then $ab > 0$.

So far we have required the set $R$ of real numbers to satisfy the field axioms and the order axiom, and we stated that, because of this requirement, $R$ is an ordered field. There is one more condition that is imposed upon the set $R$. This condition is called the *completeness property*. We do not state the completeness property formally as an axiom because to state it precisely requires a more advanced approach than we wish to take in this book. However, we now give a geometric interpretation to the set of real numbers by associating with them the points on a horizontal line, called an *axis*. The completeness property guarantees that there is a one-to-one correspondence between the set $R$ and the set of points on an axis.

Refer to Figure 1.2.1. A point on the axis is chosen to represent the number 0. This point is called the *origin*. A unit of distance is selected arbitrarily. Then each positive integer $n$ is represented by the point at a distance of $n$ units to the right of the origin and each negative integer $-n$ is represented by the point at a distance of $n$ units to the left of the origin. We call these points *unit points*. They are labeled with the numbers with which they are associated. For example, 4 is represented by the unit point 4 units to the right of the origin and $-4$ is represented by the unit point 4 units to the left of the origin.

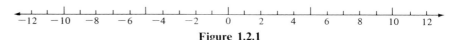

**Figure 1.2.1**

Figure 1.2.1 shows the unit points representing 0 and the first twelve positive integers and their corresponding negative integers. The unit point representing a particular integer can be obtained by going far enough along the axis to the right if the integer is positive, and to the left if the integer is negative. Because to each integer there corresponds a unique unit point on the axis, and with each unit point on the axis there is associated only one integer, there is a one-to-one correspondence between the set $J$ of integers and the set of unit points.

The elements of the set $Q$ of rational numbers may be associated with points on the axis of Figure 1.2.1. Recall that set $Q$ is defined by

$$Q = \left\{ x \mid x \text{ can be represented by } \frac{p}{q}, p \in J, q \in J, q \neq 0 \right\}$$

In particular, consider the number $\frac{1}{7}$ in $Q$. The segment of the axis from 0 to 1 may be divided into seven equal parts. The endpoint of the first such subdivision is then associated with the number $\frac{1}{7}$. Similarly, the endpoint of the second subdivision is associated with $\frac{2}{7}$, and so on. The point associated with the number $\frac{24}{7}$ is twenty-four sevenths units to the right of the origin, which is three sevenths of the distance from the unit point 3 to the unit point 4. A negative rational number, in a similar manner, is associated with a point to the left of the origin; that is, the point associated with the number $-\frac{24}{7}$ is the same distance to the left of the origin as the point associated with $\frac{24}{7}$ is to the right of the origin. Figure 1.2.2 shows some of the points associated with rational numbers.

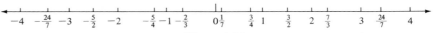

**Figure 1.2.2**

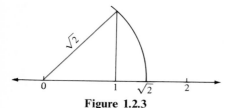

**Figure 1.2.3**

So far, then, we have shown how a rational number is associated with a point on the axis. It is not as easy to show the correspondence between irrational numbers and points on the axis. However, for certain irrational numbers we can find the point corresponding to it. In particular, the point corresponding to $\sqrt{2}$ may be found by a construction as indicated in Figure 1.2.3. From the point 1 a line segment of a length of one unit is drawn perpendicular to the axis. Then a right triangle is formed by connecting the endpoint of this segment with the origin. The length of the hypotenuse of this right triangle is $\sqrt{2}$. (This follows from the Pythagorean theorem, which states that $c^2$ has the same value as $a^2 + b^2$, where $c$ represents the length of the hypotenuse and $a$ and $b$ represent the lengths of the other two sides.) An arc of a circle having center at the origin and radius $\sqrt{2}$ is then drawn; the point of intersection of this arc with the axis will be the point associated with $\sqrt{2}$. We cannot find all points associated with irrational numbers in this way. For instance, the point corresponding to the number $\pi$ cannot be found in this manner; however, the position of the point can be approximated by using some of the digits in the decimal representation 3.14159.... At any rate, every irrational number can be associated with a unique point on the axis, and every point that does not correspond to a rational number can be associated with an irrational number. This indicates that a one-to-one correspondence between the set of real numbers and the points on the horizontal axis can be established. For this reason the horizontal axis is referred to as the *real number line,* which is a geometric representation of the set of real numbers. The real number that corresponds to a point on the real number line is called the *coordinate* of the point, and the point is called the *graph* of the real number.

Because of the completeness property, we can state that the set $R$ of real numbers is a *complete ordered field*. It is worth noting that the set $Q$ of rational numbers is an ordered field, but it is not a complete ordered field because the set of points on the axis that correspond to $Q$ is not the "complete" line; that is, there are points on the real number line that do not correspond to a number in $Q$.

Statements (3) and (4) of Theorem 1.2.14 are consistent with the fact that if $a$ is positive, the graph of $a$ on the real number line is to the right of the origin; and if $a$ is negative, the graph of $a$ is to the left of the origin. More generally, if $a > b$, then, on the real number line, the graph of $a$ is to the right of the graph of $b$. Because "$b < a$" is equivalent to "$a > b$," we can also state that if $b < a$, then on the real number line the graph of $b$ is to the left of the graph of $a$.

Sometimes the statements of equality and inequality are combined and written symbolically as follows:

$$a \geq b \quad \text{if and only if} \quad \text{either } a > b \text{ or } a = b$$
("$a \geq b$" is read "$a$ is greater than or equal to $b$")

$a \leq b$    if and only if    either $a < b$ or $a = b$
("$a \leq b$" is read "$a$ is less than or equal to $b$")

If we write

$$a \geq 0$$

it means that $a$ is a *nonnegative number,* and if we write

$$a \leq 0$$

it means that $a$ is a *nonpositive number.*

**ILLUSTRATION 3**
(a) Consider the set

$$\{x \mid -6 < x \leq 4, x \in R\} \quad (5)$$

(read "The set of all $x$ such that $x$ is greater than negative 6 and less than or equal to 4, and $x$ is an element of the set of real numbers"). The graph of this set is shown in Figure 1.2.4. The solid dot at 4 indicates that 4 is on the graph, and the open dot at $-6$ indicates that $-6$ is not on the graph.

(b) If we have the set

$$\{x \mid -6 < x \leq 4, x \in J\}$$

then we have the set of all integers greater than $-6$ and less than or equal to 4. This set is shown by the graph in Figure 1.2.5. It is a finite set. ●

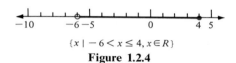

$\{x \mid -6 < x \leq 4, x \in R\}$
**Figure 1.2.4**

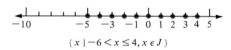

$\{x \mid -6 < x \leq 4, x \in J\}$
**Figure 1.2.5**

Usually, if it is understood that a variable $x$ is a real number, we omit the statement $x \in R$ in defining a set of numbers. In such a case set (5) can be written as

$$\{x \mid -6 < x \leq 4\}$$

**EXAMPLE 1**
Write in two ways each of the following statements, using one or more of the symbols $>$, $<$, $\geq$, and $\leq$.

(a) $x$ is between $-2$ and 2.
(b) $4t - 1$ is nonnegative.
(c) $y + 3$ is greater than 0 and less than or equal to 15.
(d) $2z$ is greater than or equal to $-5$ and less than or equal to $-1$.

**SOLUTION**
(a) $-2 < x < 2$ or, equivalently, $2 > x > -2$
(b) $4t - 1 \geq 0$ or, equivalently, $0 \leq 4t - 1$
(c) $0 < y + 3 \leq 15$ or, equivalently, $15 \geq y + 3 > 0$
(d) $-5 \leq 2z \leq -1$ or, equivalently, $-1 \geq 2z \geq -5$

## EXAMPLE 2

Show on the real number line the graph of each of the following sets (it is understood that $x \in R$).

(a) $\{x \mid -7 \leq x < -2\}$
(b) $\{x \mid x < 10 \text{ and } x > 1\}$
(c) $\{x \mid x \leq -5 \text{ or } x \geq 5\}$
(d) $\{x \mid x < 9\} \cap \{x \mid x \geq 2\}$
(e) $\{x \mid x \geq 3\} \cup \{x \mid x < 0\}$

## SOLUTION

The graphs are shown in Figures 1.2.6(a), (b), (c), (d), and (e), respectively.

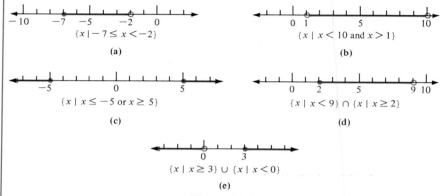

**Figure 1.2.6**

## EXERCISES 1.2

*In Exercises 1 through 24, the given equality follows immediately from one of the field axioms (Axioms 1.2.1 through 1.2.7). Indicate which axiom applies. Assume that each variable is a real number.*

1. $4 \cdot 5 = 5 \cdot 4$
2. $(6 + 2) + 4 = 6 + (2 + 4)$
3. $8 + 0 = 8$
4. $1 \cdot y = y$
5. $(5 + 2) + 4 = (2 + 5) + 4$
6. $17 + 41 = 41 + 17$
7. $3(xy) = (3x)y$
8. $(7a)b = b(7a)$
9. $\pi + (-\pi) = 0$
10. $x + (y + x) = (y + x) + x$
11. $7 + (8 + 11) = 7 + (11 + 8)$
12. $b + (-b) = 0$
13. $4 \cdot \dfrac{1}{4} = 1$
14. $x + 0 = x$
15. $3(a + b) = (a + b)3$
16. $11 + (y + 7) = (y + 7) + 11$
17. $a(b + 0) = ab$
18. $4(x + y) = 4x + 4y$
19. $0 \cdot 1 = 0$
20. $0 + 0 = 0$
21. $w + x(y + z) = w + (xy + xz)$
22. $(r + s)u + t = (s + r)u + t$
23. $(r + s) + (t + u) = r + [s + (t + u)]$
24. $(w + x) + (y + z) = [(w + x) + y] + z$

*In Exercises 25 and 26, arrange the elements of the given subset of R in the same order as their corresponding points from left to right on the real number line.*

25. $\{-2, 3, 21, 5, -7, \tfrac{2}{3}, \sqrt{2}, -\tfrac{7}{4}, -\sqrt{5}, -10, 0, \tfrac{3}{4}, -\tfrac{5}{3}, -1\}$
26. $\{\tfrac{11}{3}, \pi, -8, -\sqrt{2}, 3, -\sqrt{3}, 4, \tfrac{21}{4}, -\tfrac{3}{2}, 1.26, \tfrac{1}{2}\pi\}$

*In Exercises 27 through 38, write the statement in two ways, using one or more of the symbols* $>$, $<$, $\geq$, *and* $\leq$.

**27.** 8 is greater than $-9$.
**28.** $-5$ is less than 3.
**29.** $-12$ is less than $-3$.
**30.** $-16$ is greater than $-26$
**31.** $4x - 5$ is negative.
**32.** $2s + 3$ is nonpositive.
**33.** $3t + 7$ is nonnegative.
**34.** $8y - 4$ is positive.
**35.** $r$ is between 2 and 8.
**36.** $3x$ is greater than or equal to 10 and less than 20.
**37.** $a - 2$ is greater than $-5$ and less than or equal to 7.
**38.** $2z + 5$ is greater than or equal to $-1$ and less than or equal to 15.

*In Exercises 39 through 44, show on the real number line the graph of the given set. The set $N$ is the set of natural numbers and $J$ is the set of integers.*

**39.** $\{x \mid x < 12, x \in N\}$
**40.** $\{x \mid x \leq 7, x \in N\}$
**41.** $\{x \mid -5 \leq x < 5, x \in J\}$
**42.** $\{x \mid 2 < x < 10, x \in N\}$
**43.** $\{x \mid -6 \leq x \leq 0, x \in J\}$
**44.** $\{x \mid -11 \leq x < 3, x \in J\}$

*In Exercises 45 through 60, show on the real number line the graph of the given set if $x \in R$.*

**45.** $\{x \mid x > 2\}$
**46.** $\{x \mid x \leq 8\}$
**47.** $\{x \mid -4 < x \leq 4\}$
**48.** $\{x \mid 3 < x < 9\}$
**49.** $\{x \mid x < 12 \text{ and } x > 2\}$
**50.** $\{x \mid x \leq 5 \text{ and } x \geq -5\}$
**51.** $\{x \mid x \leq -4 \text{ or } x > 4\}$
**52.** $\{x \mid x < 3 \text{ or } x > 6\}$
**53.** $\{x \mid x < 12\} \cap \{x \mid x > 2\}$
**54.** $\{x \mid x \leq 5\} \cap \{x \mid x \geq -5\}$
**55.** $\{x \mid x \leq -4\} \cup \{x \mid x > 4\}$
**56.** $\{x \mid x < 3\} \cup \{x \mid x > 6\}$
**57.** $\{x \mid x \leq 0\} \cap \{x \mid x > -4\}$
**58.** $\{x \mid x > -8\} \cap \{x \mid x \leq 0\}$
**59.** $\{x \mid x \leq 7\} \cup \{x \mid x \leq 0\}$
**60.** $\{x \mid x > 10\} \cup \{x \mid x > 2\}$

## 1.3 Some Algebraic Terminology

To indicate a product, we have used the centered dot · or parentheses around one or more symbols, and sometimes we have omitted the symbol. For example, the product of the two factors $a$ and $b$ can be written in the following ways.

$$ab \quad a \cdot b \quad (a)(b) \quad a(b) \quad (a)b$$

Suppose that we have the product of two identical factors of $x$. We can use the notation $x^2$ to indicate this product, where the numeral 2 written to the upper right of the symbol $x$ is called an "exponent" and it indicates that the number represented by $x$ is to be used twice as a factor; that is,

$$x^2 = x \cdot x$$

In general, if $a$ is a real number and $n$ is a positive integer, then

$$a^n = a \cdot a \cdot a \cdot \ldots \cdot a \quad (n \text{ factors of } a)$$

where $n$ is called the *exponent,* $a$ is called the *base,* and $a^n$ is called the *nth power of a.* For example, $x^2$ is the second power of $x$ and $y^5$ is the fifth power of $y$, where

$$y^5 = y \cdot y \cdot y \cdot y \cdot y$$

The fourth power of 3 is denoted by $3^4$ and

$$3^4 = 3 \cdot 3 \cdot 3 \cdot 3$$

When a symbol is written without an exponent, the exponent is understood to be 1. Hence $x = x^1$.

In Section 1.1 we stated that a *variable* is a symbol used to represent any element of a given replacement set. If the replacement set is $R$, then the variable represents a real number. A *constant* is a symbol whose replacement set contains only one element. For example, if we write the sum

$$6x^2 + 2x + 5 \tag{1}$$

the symbol $x$ is a variable and the numerals 6, 2, and 5 are constants. The sum (1) is a particular "algebraic expression," which we now define.

**1.3.1 DEFINITION**  An *algebraic expression* is a constant, a variable, or indicated operations involving constants and variables.

**ILLUSTRATION 1.** The following are algebraic expressions.

$$7 \tag{2}$$

$$3x^2 y^5 \tag{3}$$

$$6x + 4 \tag{4}$$

$$8xy - 7x + y - 5 \tag{5}$$

$$(9r - 4s) \div 3 \tag{6}$$

$$\frac{5x^2 - 6xy + y^2}{x + 2y} \quad \bullet \tag{7}$$

When an algebraic expression is written as the sum of other algebraic expressions, each of the expressions is called a *term* of the given algebraic expression. For instance, algebraic expression (5) can be written as the sum $8xy + (-7x) + y + (-5)$ and so, $8xy, -7x, y,$ and $-5$ are the terms of (5). The algebraic expressions (1), (2), (3), and (4) have, respectively, three, one, one, and two terms.

An algebraic expression consisting of just one term that is either a constant or the product of a constant and positive integer powers of variables is called a *monomial.* For example, algebraic expressions (2) and (3) are monomials. Any sum of monomials is called a *polynomial.* A polynomial having two terms is called a *binomial*, and a polynomial having three terms is called a *trinomial*. Algebraic expression (4) is a binomial, (1) is a trinomial, and (5) is a polynomial with four terms. Algebraic expression (6) has one term whose numerator is a binomial, and (7) is an algebraic expression containing one term whose numerator is a trinomial and whose denominator is a binomial.

**EXAMPLE 1**
If $x = 3$, $y = -4$, and $z = 5$, find the value of each of the following algebraic expressions.

(a) $x^2 - 3yz + z^2$

(b) $\dfrac{z(x + 2y)^3}{x^3 - z^3}$

**SOLUTION**

(a) $x^2 - 3yz + z^2 = 3^2 - 3(-4)(5) + 5^2$
$= 9 + 60 + 25$
$= 94$

(b) $\dfrac{z(x + 2y)^3}{x^3 - z^3} = \dfrac{5[3 + 2(-4)]^3}{3^3 - 5^3}$

$= \dfrac{5[3 + (-8)]^3}{27 - 125}$

$= \dfrac{5(-5)^3}{-98}$

$= \dfrac{5(-125)}{-98}$

$= \dfrac{-625}{-98}$

$= \dfrac{625}{98}$

In a product, any factors that are constants are called *numerical factors*, and factors that are variables are called *literal factors*. For example, in the monomial $3x^2y^5$, 3 is a numerical factor and $x^2$ and $y^5$ are literal factors. If a term is a product of two factors, each factor is said to be the *coefficient* of the other factor. For instance, in the term $5xyz$, the coefficient of $yz$ is $5x$, the coefficient of $x$ is $5yz$, the coefficient of $5z$ is $xy$, and so on. If a coefficient is a numerical factor, then it is called a *numerical coefficient*. Hence in the term $5xyz$, 5 is the numerical coefficient of $xyz$.

By the *degree* of a monomial in one variable, we mean the exponent of that variable appearing in the monomial. In particular, the monomial $5x^3$ has a degree of 3. If a monomial has more than one variable, the *degree* of the monomial in any one of its variables is the exponent of that variable, and the *degree* of the monomial is the sum of the exponents of all the variables that appear. For example, the monomial $3x^2y^5$ has a degree of 2 in $x$ and a degree of 5 in $y$, and furthermore, we say that the degree of the monomial is 7. The monomial $-2xyz$ has a degree of 1 in each of its variables, $x, y,$ and $z$, and the degree of the monomial is 3. The degree of a nonzero-constant monomial is zero. The constant 0 has no degree.

The *degree* of a polynomial is the same as the degree of the term of highest degree appearing in the polynomial. Therefore, $7x^2 - 4x + 2$ is a second-degree polynomial, and $3x + 6$ is a first-degree polynomial. The polynomial $6x^2y^2 - 4x^3 + 2y$ has a degree of 3 in $x$ and a degree of 2 in $y$, and the degree of the polynomial is 4.

**EXAMPLE 2**

For each of the following algebraic expressions, determine if it is a monomial, a binomial, or a trinomial. Also, for each expression state its degree in each of the variables appearing and the degree of the expression.

(a) $5xy^2 + xy - 6y$
(b) $7x^2y^3z + 2xy^4$
(c) $4u - 8t$
(d) $11rs + 2s^3 - r^2s^2$

**SOLUTION**

(a) The expression $5xy^2 + xy - 6y$ is a trinomial; it has a degree of 1 in $x$ and a degree of 2 in $y$; its degree is 3.
(b) The expression $7x^2y^3z + 2xy^4$ is a binomial; it has a degree of 2 in $x$, a degree of 4 in $y$, and a degree of 1 in $z$; its degree is 6.
(c) The expression $4u - 8t$ is a binomial; it has a degree of 1 in $u$ and a degree of 1 in $t$; its degree is 1.
(d) The expression $11rs + 2s^3 - r^2s^2$ is a trinomial; it has a degree of 2 in $r$ and a degree of 3 in $s$; its degree is 4.

Because a polynomial is a sum of monomials and a monomial is a symbol representing a real number, we can apply to polynomials the definitions, axioms, and theorems involving operations of real numbers. We are assuming that you are familiar with addition, subtraction, multiplication, and division of polynomials and rational expressions as well as factoring polynomials. If you need a review of this material, you may consult a text in Intermediate Algebra [in particular, *Intermediate Algebra for College Students*, by Louis Leithold (New York: Macmillan Publishing Co., Inc.)].

We discussed positive integer powers of real numbers. We are now concerned with starting with a positive integer power of a real number $b$ and finding $b$.

**1.3.2 DEFINITION**  If $a, b \in R$ and $n$ is a positive integer such that

$$b^n = a \tag{8}$$

then $b$ is called an *nth root of a*.

**ILLUSTRATION 2**

(a) 2 is a square root of 4 because $2^2 = 4$; furthermore, $-2$ is also a square root of 4 because $(-2)^2 = 4$.
(b) 3 is a fourth root of 81 because $3^4 = 81$. Also, $-3$ is a fourth root of 81 because $(-3)^4 = 81$.
(c) $-4$ is a cube root of $-64$ because $(-4)^3 = -64$.
(d) 4 is a cube root of 64 because $4^3 = 64$.

In Illustration 2(a) we see that there are two real square roots of 4, and in Illustration 2(b) we see that there are two real fourth roots of 81. To distinguish between the two roots in such cases, we introduce the concept of principal $n$th root.

**1.3.3 DEFINITION**  If $n$ is a positive integer greater than 1, and $\sqrt[n]{a}$ denotes the *principal $n$th root of a*, then

(i) If $a > 0$, $\sqrt[n]{a}$ is the positive $n$th root of $a$.
(ii) If $a < 0$, and $n$ is odd, $\sqrt[n]{a}$ is the negative $n$th root of $a$.
(iii) $\sqrt[n]{0} = 0$.

**ILLUSTRATION 3**

(a) From Definition 1.3.3(i),

$$\sqrt[4]{81} = 3 \quad \text{(read "The principal fourth root of 81 equals 3")}$$

Note that $-3$ is also a fourth root of 81 but it is not the principal fourth root of 81. However, we can write

$$-\sqrt[4]{81} = -3$$

(b) From Definition 1.3.3(ii),

$$\sqrt[3]{-1000} = -10$$

(read "The principal cube root of $-1000$ equals $-10$"). ●

The symbol $\sqrt{\phantom{a}}$ is called a *radical sign*, and the entire expression $\sqrt[n]{a}$ is called a *radical*, where the number $a$ is the *radicand* and the number $n$ is the *index*. The *order* of a radical is the same as its index. If no index appears, the index is understood to be 2. Thus $\sqrt{36} = 6$ (read "The principal square root of 36 equals 6").

The principal $n$th root of a real number $b$ is a rational number if and only if $b$ is the $n$th power of a rational number. For instance,

$$\sqrt{9} = 3 \text{ because } 3^2 = 9$$

$$\sqrt[3]{-\frac{1}{27}} = -\frac{1}{3} \text{ because } \left(-\frac{1}{3}\right)^3 = -\frac{1}{27}$$

$$\sqrt[4]{625} = 5 \text{ because } 5^4 = 625$$

Recall from Section 1.1 that a real number that is not rational is called an irrational number and an irrational number cannot be represented by a terminating decimal or a nonterminating repeating decimal. Because 3 is not the square of a rational number, $\sqrt{3}$ is an irrational number. Other examples of irrational numbers are

$$\sqrt{2} \qquad \sqrt[3]{4} \qquad \sqrt[3]{-5} \qquad \sqrt[4]{15}$$

Table 1 in the Appendix gives decimal approximations for some irrational numbers. In particular, from the table, we have

$$\sqrt{3} \approx 1.732$$

where the symbol $\approx$ is read "approximately equals."

A review of operations involving radicals along with properties of rational exponents is given in Sections 4.1 and 4.2.

If $a < 0$ and $n$ is a positive even integer, there is no real $n$th root of $a$ because an even power of any real number is a nonnegative number. For instance, suppose that in equality (8) $n$ is 2 and $a$ is $-25$. Then equality (8) becomes

$$b^2 = -25 \tag{9}$$

Because the square of any real number is a nonnegative number, there is no real number that can be substituted for $b$ in equality (9). Therefore, there is no real square root of $-25$; in a similar manner it follows that there is no real square root of any negative number.

We see then that in order to consider square roots of negative numbers, we must deal with numbers other than real numbers. Thus we develop a set of numbers that contains the set $R$ of real numbers as a subset and also square roots of negative numbers. We denote such a set of numbers by $C$ and refer to it as the set of complex numbers. In this section our discussion of $C$ is a very brief one. A complete discussion is given in Chapter 5, where we define operations of *addition* and *multiplication* on $C$ so that the axioms of addition and multiplication for the set $R$ are satisfied. We use the same symbols ($+$ and $\cdot$) for these operations on $C$ as we do when the operations are performed on real numbers. We first require that the set $C$ is such that the real number $-1$ has a square root. Let $i$ be a symbol for a number in $C$ whose square is $-1$; that is, we define $i$ as a number such that

$$i^2 = -1$$

Because every real number is to be an element of $C$, it follows that if $b \in R$, then $b \in C$. In order for the closure law for multiplication to hold, the number $b \cdot i$, abbreviated $bi$, must be an element of $C$. Furthermore, if $a \in R$, then $a \in C$, and if the closure law for addition is to hold, the number $a + bi$ must be an element of $C$. We now have a set $C$, which we define formally.

**1.3.4 DEFINITION**  The set of all numbers of the form

$$a + bi \qquad \text{where } a, b \in R \quad \text{and} \quad i^2 = -1$$

is called the *set of complex numbers* and is denoted by $C$; that is,

$$\{a + bi \mid a, b \in R, i^2 = -1\} = C$$

For the complex number $a + bi$, the number $a$ is called the *real part* and the number $b$ is called the *imaginary part*.

**ILLUSTRATION 4**

(a) The number $-3 + 6i$ is a complex number whose real part is $-3$ and whose imaginary part is 6.
(b) The number $7 + (-4)i$ is a complex number whose real part is 7 and whose imaginary part is $-4$. •

If $-p$ is a negative number, then the complex number $a + (-p)i$ can be written as $a - pi$. Hence

$$7 + (-4)i = 7 - 4i$$

The real number $a$ is a complex number and it can be written in the form $a + 0i$; that is,

$$a = a + 0i$$

Hence a real number is a complex number whose imaginary part is 0. The number $0 + bi$ can be written more simply as $bi$; that is,

$$bi = 0 + bi$$

The number $bi$, $b \neq 0$, is called a *pure imaginary number*. More generally, any complex number $a + bi$, where $b \neq 0$, is called an *imaginary number*.

**ILLUSTRATION 5**

(a) The complex number $-5 + 2i$ is an imaginary number.
(b) The complex number $8i$ is a pure imaginary number.
(c) The real number $-2$ is a complex number, and it can be written as $-2 + 0i$.
(d) The real number 0 is a complex number, and it can be written as $0 + 0i$. •

We now state a definition that allows every real number (positive, negative, or zero) to have a square root.

**1.3.5 DEFINITION** A number $s$ is said to be a square root of a real number $r$ if and only if

$$s^2 = r$$

We have learned that any positive number has two square roots, one positive and one negative, and the number 0 has only one square root, 0.

Now, let $-p$ be any negative number; then $p$ is a positive number and
$$(i\sqrt{p})^2 = i^2(\sqrt{p})^2$$
$$= (-1)p$$
$$= -p$$

Therefore, because $(i\sqrt{p})^2 = -p$, it follows from Definition 1.3.5 that $i\sqrt{p}$ is a square root of $-p$. In a similar way, $(-i\sqrt{p})^2 = -p$, and therefore $-i\sqrt{p}$ is a square root of $-p$. As we did with square roots of positive numbers, we distinguish between the two square roots by using the concept of principal square root.

**1.3.6 DEFINITION** If $p$ is a positive number, then the *principal square root of* $-p$, denoted by $\sqrt{-p}$, is defined by

$$\boxed{\sqrt{-p} = i\sqrt{p}}$$

The two square roots of $-p$ are written as $\sqrt{-p}$ and $-\sqrt{-p}$, or as $i\sqrt{p}$ and $-i\sqrt{p}$.

**ILLUSTRATION 6**

(a) $\sqrt{-5} = i\sqrt{5}$
The two square roots of $-5$ are $i\sqrt{5}$ and $-i\sqrt{5}$.

(b) $\sqrt{-16} = i\sqrt{16}$
$= 4i$
The two square roots of $-16$ are $4i$ and $-4i$.

(c) $\sqrt{-1} = i\sqrt{1}$
$= i$
The two square roots of $-1$ are $i$ and $-i$. ●

**EXAMPLE 3**
Write the given complex number in the form

$$a + bi$$

(a) $\sqrt{-9}$
(b) $5 - 6\sqrt{-4}$
(c) $\sqrt{25} + \sqrt{-25}$
(d) $-\sqrt{\dfrac{16}{49}} + 3\sqrt{-\dfrac{16}{49}}$
(e) $\sqrt{24} + 5\sqrt{-27}$

**SOLUTION**

(a) $\sqrt{-9} = i\sqrt{9}$
$= 3i$
$= 0 + 3i$

(b) $5 - 6\sqrt{-4} = 5 - 6(i\sqrt{4})$
$= 5 - 6(2i)$
$= 5 + (-12i)$

(c) $\sqrt{25} + \sqrt{-25} = 5 + i\sqrt{25}$
$= 5 + 5i$

(d) $-\sqrt{\dfrac{16}{49}} + 3\sqrt{-\dfrac{16}{49}} = -\dfrac{4}{7} + 3\left(i\sqrt{\dfrac{16}{49}}\right)$
$= -\dfrac{4}{7} + 3\left(\dfrac{4}{7}i\right)$
$= -\dfrac{4}{7} + \dfrac{12}{7}i$

(e) $\sqrt{24} + 5\sqrt{-27} = \sqrt{4}\sqrt{6} + 5(i\sqrt{9}\sqrt{3})$
$= 2\sqrt{6} + 5(i \cdot 3\sqrt{3})$
$= 2\sqrt{6} + 15\sqrt{3}i$

## EXERCISES 1.3

In Exercises 1 through 8, determine if the given algebraic expression is a monomial, a binomial, or a trinomial. Also, for each expression state its degree in each of the variables appearing and the degree of the expression.

1. $7x^2 + 4y$
2. $3x^2y - 3xy - 1$
3. $4xy^3z^2$
4. $r^3s^2t + rs^3$
5. $5uv - v^3 + 2u^2v^2$
6. $-8xyz^2$
7. $xy^2z^3 + y^3z$
8. $6xy - x^3 + x^2y^2$

In Exercises 9 through 16, find the value of the given expression if $x = -4$, $y = 3$, and $z = -2$.

9. $2x^3 - 5z^2$
10. $xyz + 3x^2$
11. $(y - z)(x + z)$
12. $4z^2 - 2yz - y^3$
13. $\dfrac{x^2 - yz}{2xz - y^3}$
14. $\dfrac{3y^2 + 5xy + x^2}{z - x^2z}$
15. $\dfrac{(3x + 2y)^2}{z^3 - x^2}$
16. $\dfrac{3x^2 - y^2 - z^2 + 1}{(x + y^2 + z^3)^3}$

In Exercises 17 through 30, find the indicated root.

17. $\sqrt{81}$
18. $\sqrt[4]{81}$
19. $\sqrt[3]{-64}$
20. $-\sqrt[6]{64}$
21. $\sqrt[3]{-0.001}$
22. $\sqrt[3]{-0.027}$
23. $\sqrt[4]{\frac{16}{625}}$
24. $\sqrt{\frac{16}{625}}$
25. $-\sqrt[5]{-32}$
26. $-\sqrt[7]{-128}$
27. $\sqrt[3]{\frac{216}{125}}$
28. $\sqrt[5]{\frac{243}{100,000}}$
29. $\sqrt{(-3)^2}$
30. $\sqrt{(-2)^4}$

In Exercises 31 through 44, write the given complex number in the form $a + bi$. Simplify each radical.

31. $\sqrt{-49}$
32. $\sqrt{-144}$
33. $3\sqrt{-12}$
34. $5\sqrt{-75}$
35. $4 + 3\sqrt{-1}$
36. $8 - 5\sqrt{-1}$
37. $-4 + \sqrt{-4}$
38. $\sqrt{36} + \sqrt{-36}$
39. $\sqrt{48} - \sqrt{-48}$
40. $\dfrac{1}{3} + \dfrac{1}{5}\sqrt{-45}$
41. $2 - \sqrt{-\dfrac{25}{16}}$
42. $\dfrac{2}{3} - \sqrt{-\dfrac{16}{25}}$
43. $\sqrt{50} + \sqrt{-200}$
44. $\sqrt{54} + \sqrt{-162}$

## 1.4 First-degree Equations in One Variable

We have used the equals sign to denote an equality, where the equality $a = b$ means that the numerals $a$ and $b$ represent the same number. We now introduce the concept of "algebraic equation" and use the equals sign in a different sense.

**1.4.1 DEFINITION** An *algebraic equation* in the variable $x$ is a statement of the form

$$E = F$$

where $E$ and $F$ are algebraic expressions in $x$.

The replacement set of a variable in an equation is the set of numbers for which the algebraic expressions in the equation are defined. Unless otherwise stated, we consider the replacement set to be a set of real numbers. A variable in an equation is sometimes called an *unknown*.

**ILLUSTRATION 1.** The following statements are algebraic equations in $x$.

$$x - 7 = 0 \tag{1}$$

$$x^2 + 12 = 7x \tag{2}$$

$$x + 5 = 5 + x \tag{3}$$

$$x + 2 = x + 3 \tag{4}$$

$$\frac{x}{x} = 1 \tag{5}$$

$$\frac{2}{x + 1} = \frac{3}{2x - 1} \tag{6}$$

For equations (1) through (4) the replacement set is $R$. Because the left member of equation (5) is not defined if $x = 0$, the replacement set of $x$ in equation (5) is the set of all real numbers except zero. The left member of equation (6) is not defined if $x = -1$, and the right member is not defined if $x = \frac{1}{2}$; therefore, the replacement set of $x$ in equation (6) is the set of all real numbers except $-1$ and $\frac{1}{2}$. ●

When a variable in an equation is replaced by a specific numeral, the resulting statement may be either true or false. For instance, in equation (1), if $x$ is replaced by 7, the resulting statement is true, but if $x$ is replaced by a numeral representing a number other than seven, the resulting statement is false. In equation (2), if $x$ is replaced by either 3 or 4, we obtain a true statement, and if $x$ is replaced by a numeral representing a number other than three or four, we obtain a false statement.

**1.4.2 DEFINITION** Suppose that we have an algebraic equation in the variable $x$. If, when a specific numeral is substituted for $x$, the resulting statement is true, then the number represented by that numeral is called a *solution* (or *root*) of the equation. The set of all solutions of an equation is called the *solution set* of the equation.

A number that is a solution of an equation is said to *satisfy the equation*.

**ILLUSTRATION 2.** The only solution of equation (1) is 7, so the solution set of equation (1) is $\{7\}$.

Equation (2) has two solutions, 3 and 4; hence the solution set of equation (2) is $\{3, 4\}$.

If, in equation (3), any numeral representing a real number is substituted for $x$, a true statement is obtained. Therefore, the solution set of equation (3) is $R$.

When any numeral representing a real number is substituted for $x$ in equation (4), we get a false statement. Hence the solution set of equation (4) is $\emptyset$.

In equation (5), if any numeral other than 0 is substituted for $x$, a true statement is obtained. Therefore, the solution set of equation (5) is $\{x \mid x \in R, x \neq 0\}$. •

If the solution set of an equation in one variable is the same as the replacement set of the variable, the equation is called an *identity*. Equations (3) and (5) are examples of identities. If there is at least one element in the replacement set of the variable that is not in the solution set of the equation, the equation is called a *conditional equation*. Equations (1), (2), (4), and (6) are conditional equations.

An important type of equation is the *polynomial equation* in one variable, which can be written in the form $E = 0$, where $E$ is a polynomial in one variable. The degree of the polynomial is the *degree of the equation*.

**ILLUSTRATION 3.** Particular examples of polynomial equations are the following ones.

$$7x - 21 = 0 \tag{7}$$

$$4y^3 - 8y^2 - y + 2 = 0 \tag{8}$$

$$2w^2 - 3w - 5 = 0 \tag{9}$$

$$9z^4 - 13z^2 + 4 = 0 \tag{10}$$

Equation (7) is of the first degree in $x$, equation (8) is of the third degree in $y$, equation (9) is of the second degree in $w$, and equation (10) is of the fourth degree in $z$. •

Determining the solution set of a conditional equation is called *solving* the equation. As shown by equation (1) in Illustration 1, it is sometimes possible to solve an equation by inspection. In general, however, we must first learn the concept of equivalent equations.

**1.4.3 DEFINITION**  Equations that have the same solution set are called *equivalent equations*.

We can often solve an equation by replacing it by a succession of equivalent equations, each one in some way simpler than the preceding one so that we eventually obtain an equation for which the solution set is apparent. The following theorem, which follows from the addition and multiplication properties for real numbers, is used to replace an equation by an equivalent one.

**1.4.4 THEOREM**  Suppose that

$$E = F \tag{11}$$

is an algebraic equation in the variable $x$. Then, if $G$ is an algebraic expression in $x$, equation (11) is equivalent to each of the following equations.

$$E + G = F + G \tag{12}$$

$$E - G = F - G \tag{13}$$

$$E \cdot G = F \cdot G \quad (G \text{ is never zero}) \tag{14}$$

$$\frac{E}{G} = \frac{F}{G} \quad (G \text{ is never zero}) \tag{15}$$

Because equations (12) and (13) are equivalent to equation (11), it follows that the solution set of an equation is not changed if the same algebraic expression is added to or subtracted from both members of an equation. Because equation (14) is equivalent to equation (11), the solution set of an equation is not changed if both members of an equation are multiplied by the same algebraic expression, where the condition that $G$ is never zero is necessary so that equation (11) can be obtained from equation (14) by dividing on both sides by $G$. Furthermore, because equation (15) is equivalent to equation (11), the solution set of an equation is not changed if both members of the equation are divided by the same algebraic expression provided that the algebraic expression is never zero.

**ILLUSTRATION 4.**  To solve the equation

$$3x - 5 = 7 \tag{i}$$

we first add 5 to both members of the equation, and we obtain

$$(3x - 5) + 5 = 7 + 5 \qquad \text{(ii)}$$
$$3x = 12 \qquad \text{(iii)}$$

Dividing each member of equation (iii) by 3, we have

$$\frac{3x}{3} = \frac{12}{3} \qquad \text{(iv)}$$
$$x = 4 \qquad \text{(v)}$$

Equations (i) through (v) are all equivalent. The solution set of equation (v) is obviously $\{4\}$, so the solution set of the given equation (i) is $\{4\}$.

We check the work for possible errors in arithmetic by substituting the solution in the original equation.

$$3(4) - 5 = 12 - 5$$
$$= 7$$

Thus the solution checks. ●

To solve an equation involving rational expressions, we obtain an equivalent equation by multiplying both members of the equation by the least common denominator (LCD) of the fractions.

**ILLUSTRATION 5.** To solve the equation

$$\frac{2x}{3} - \frac{x}{2} + 6 = \frac{5x}{6} - 2 \qquad \text{(i)}$$

we multiply both members by 6, and we obtain

$$(6)\frac{2x}{3} - (6)\frac{x}{2} + (6)6 = (6)\frac{5x}{6} - (6)2 \qquad \text{(ii)}$$
$$4x - 3x + 36 = 5x - 12 \qquad \text{(iii)}$$

We now subtract $5x$ and 36 from each member of equation (iii), and we get

$$(x + 36) - 5x - 36 = (5x - 12) - 5x - 36 \qquad \text{(iv)}$$
$$-4x = -48 \qquad \text{(v)}$$

Dividing both members of equation (v) by $-4$, we have

$$\frac{-4x}{-4} = \frac{-48}{-4} \qquad \text{(vi)}$$
$$x = 12 \qquad \text{(vii)}$$

Equations (i) through (vii) are all equivalent. Because the solution set of equation (vii) is $\{12\}$, it follows that the solution set of the given equation (i) is $\{12\}$. We check the solution.

$$\text{Does } \frac{2(12)}{3} - \frac{12}{2} + 6 = \frac{5(12)}{6} - 2?$$

$$\frac{2(12)}{3} - \frac{12}{2} + 6 = 8 - 6 + 6 \quad \text{and} \quad \frac{5(12)}{6} - 2 = 10 - 2$$
$$= 8 \qquad\qquad\qquad\qquad = 8$$

Hence the solution checks. ●

**EXAMPLE 1**

Find the solution set of the equation

$$4[3x - (5x - 1)] = 3 - 4x$$

**SOLUTION**

We simplify the algebraic expression in the left member and obtain a succession of equivalent equations.

$$4[3x - 5x + 1] = 3 - 4x$$
$$4[-2x + 1] = 3 - 4x$$
$$-8x + 4 = 3 - 4x$$
$$-8x + 4 + 4x - 4 = 3 - 4x + 4x - 4$$
$$-4x = -1$$
$$\frac{-4x}{-4} = \frac{-1}{-4}$$
$$x = \frac{1}{4}$$

Therefore, the solution set is $\{\frac{1}{4}\}$.

Each of the equations in Illustrations 4 and 5 and Example 1 is a "first-degree equation in one variable."

**1.4.5 DEFINITION** An equation of the form

$$ax + b = 0$$

where $a, b \in R$ and $a \neq 0$, or any equation equivalent to this equation, is called a *first-degree equation in one variable*.

If an equation involving rational expressions contains fractions having a variable in the denominator, then the LCD contains the variable. In such a case it is possible that, when multiplying both members by the LCD, the resulting equation will not be equivalent to the given one. This can be seen in the following illustration.

## 1.4] First-degree Equations in One Variable

**ILLUSTRATION 6.** If we have the equation

$$\frac{2x}{x-3} + 1 = \frac{6}{x-3} \qquad (16)$$

we see that the LCD is $(x-3)$; furthermore, $x - 3 \neq 0$ because division by zero is undefined. Multiplying each member by $(x-3)$, we get

$$(x-3)\frac{2x}{x-3} + (x-3)(1) = (x-3)\frac{6}{x-3} \qquad (17)$$

$$2x + x - 3 = 6$$
$$3x = 9$$
$$x = 3$$

However, $x \neq 3$, and so there is no value of $x$ that satisfies the equation. Thus there is no solution for equation (16), so its solution set is $\emptyset$.

**EXAMPLE 2**
Find the solution set of the equation

$$\frac{3}{2x+1} = \frac{4}{5x-3}$$

**SOLUTION**
The LCD is $(2x+1)(5x-3)$, and $x \neq -\frac{1}{2}$, $x \neq \frac{3}{5}$. Multiplying both members of the equation by the LCD, we get

$$(2x+1)(5x-3)\frac{3}{2x+1} = (2x+1)(5x-3)\frac{4}{5x-3}$$

$$3(5x-3) = 4(2x+1)$$
$$15x - 9 = 8x + 4$$
$$7x = 13$$
$$x = \frac{13}{7}$$

Therefore, the solution set is $\{\frac{13}{7}\}$.

**EXAMPLE 3**
Find the solution set of each equation.

(a) $\frac{x}{x} = 0$

(b) $\frac{x}{x} = 1$

(c) $\frac{x}{x} = 2$

**SOLUTION**
In each equation the LCD is $x$, and $x \neq 0$.
(a) We multiply on each side of the equation by $x$ and obtain

$$x = 0$$

But $x \neq 0$; therefore, the solution set is $\emptyset$.
(b) Multiplying on each side of the equation by $x$, we get

$$x = x$$

This equation is true for any value of $x$, but $x \neq 0$. Therefore, the solution set is $\{x \mid x \in R, x \neq 0\}$.
(c) When multiplying on each side of the equation by $x$, we have

$$x = 2x$$

This equation is only true if $x = 0$; however, $x \neq 0$. Therefore, the solution set is $\emptyset$.

**EXAMPLE 4**
Find the solution set of the equation
$$\frac{1}{2x+5} - \frac{4}{2x-1} = \frac{4x+4}{4x^2+8x-5}$$

**SOLUTION**
We factor the denominator in the rational expression occurring in the right member, and we have
$$\frac{1}{2x+5} - \frac{4}{2x-1} = \frac{4x+4}{(2x+5)(2x-1)}$$

The LCD is $(2x+5)(2x-1)$ and $x \neq -\frac{5}{2}$, $x \neq \frac{1}{2}$. We multiply each member of the equation by the LCD, and obtain
$$2x - 1 - 4(2x+5) = 4x + 4$$
$$2x - 1 - 8x - 20 = 4x + 4$$
$$-6x - 21 = 4x + 4$$
$$-10x = 25$$
$$x = -\frac{5}{2}$$

But $x \neq -\frac{5}{2}$ because when $x = -\frac{5}{2}$, the denominator in the first fraction in the left member of the given equation and the denominator in the fraction in the right member are both zero. Hence, $-\frac{5}{2}$ is not a solution. Therefore, the solution set is $\varnothing$.

An equation may contain more than one variable or it may contain symbols, such as *a* and *b*, representing constants. An equation of this type is sometimes called a *literal equation,* and often we wish to solve for one of the variables in terms of the other variables or symbols. The method of solution consists of treating the variable for which we are solving as the unknown and the other variables and symbols as known.

**ILLUSTRATION 7.** If *F* is the number of degrees in the Fahrenheit temperature reading and *C* is the number of degrees in the Celsius temperature reading, then
$$F = \frac{9}{5}C + 32$$

To solve this equation for *C*, we first multiply each member by 5 and obtain
$$5F = 9C + 160$$
Adding $-160$ to both members, we get
$$5F - 160 = 9C$$
Dividing both members by 9, we have
$$C = \frac{1}{9}(5F - 160)$$

## EXAMPLE 5

If $p$ dollars is invested at the rate of $100r$ per cent at simple interest for $t$ years, and $A$ dollars is the amount of the investment at $t$ years, then the formula for finding $A$ is

$$A = p(1 + rt)$$

(a) Solve for $p$.
(b) Solve for $t$.

## SOLUTION

(a) We divide both members of the given equation by $(1 + rt)$, and we get

$$p = \frac{A}{1 + rt}$$

(b) Using the distributive law in the right member of the given equation, we have

$$A = p + prt$$

Adding $-p$ to both members, we get

$$A - p = prt$$

Dividing both members by $pr$, we have

$$t = \frac{A - p}{pr}$$

## EXAMPLE 6

Solve the following equation for $a$ and $b$.

$$b = act + abc$$

## SOLUTION

To solve for $a$, we first factor the right member.

$$b = a(ct + bc)$$

We now divide both members by $(ct + bc)$ and obtain

$$a = \frac{b}{ct + bc} \qquad (ct + bc \neq 0)$$

To solve for $b$, we add $-abc$ to each member of the given equation and obtain

$$b - abc = act$$

Factoring the left member of the equation, we get

$$b(1 - ac) = act$$

Dividing both members of the equation by $(1 - ac)$, we have

$$b = \frac{act}{1 - ac} \qquad (ac \neq 1)$$

## EXERCISES 1.4

*In Exercises 1 through 28, find the solution set of the given equation.*

1. $7x + 4 = 25$
2. $7 = y + 10$
3. $4w - 3 = 11 - 3w$
4. $x - 9 = 3x + 3$
5. $2(t - 5) = 3 - (4 + t)$
6. $1 - 3(2x - 4) = 4(6 - x) - 8$
7. $x = x + 1$
8. $x + 3 = 1 + x + 2$

9. $3(4y + 9) = 7(2 - 5y) - 2y$
10. $-2[s - (5 - 4s)] + 4 = -3s$
11. $x - 6[2x - 2(x + 3)] + 5 = 8$
12. $10 - 2[3x - 3(7 - x)] = x$
13. $\dfrac{3x - 2}{3} + \dfrac{x - 3}{2} = \dfrac{5}{6}$
14. $\dfrac{3}{8} + \dfrac{1}{2x} = \dfrac{2}{x}$
15. $\dfrac{5}{2y} - \dfrac{1}{y} = \dfrac{3}{4}$
16. $\dfrac{1}{4 - t} + \dfrac{3}{6 + t} = 0$
17. $\dfrac{2}{3x - 4} = \dfrac{5}{6x - 7}$
18. $\dfrac{3}{x^2 - 9} - \dfrac{7}{x - 3} = -\dfrac{4}{x + 3}$
19. $\dfrac{x - 2}{x - 2} = 1$
20. $\dfrac{x - 2}{x - 2} = 0$
21. $\dfrac{4}{25w^2 - 1} + \dfrac{3}{5w - 1} = \dfrac{2}{5w - 1}$
22. $\dfrac{3}{x^2 - x - 6} = \dfrac{4}{2x^2 + x - 6}$
23. $\dfrac{5}{x^2 + 6x - 7} = \dfrac{2}{x^2 - 1}$
24. $\dfrac{2}{y + 1} - \dfrac{3}{1 - y} = \dfrac{5}{y}$
25. $\dfrac{t + 3}{t - 2} - \dfrac{t - 3}{t + 2} = \dfrac{5}{t^2 - 4}$
26. $\dfrac{x + 17}{x^2 - 6x + 8} + \dfrac{x - 2}{x - 4} = \dfrac{x - 4}{x - 2}$
27. $\dfrac{3x^2 + 4}{x^3 + 8} = \dfrac{3}{x + 2}$
28. $\dfrac{w}{3w^2 - 8w + 4} = \dfrac{w + 2}{3w^2 + w - 2}$

*In Exercises 29 through 34, solve for x or y in terms of the other symbols.*

29. $3ax + 6ab = 7ax + 3ab$
30. $\dfrac{a + 3x}{b} = \dfrac{c}{2}$
31. $a(y - a) - 2b(y - 3b) = ab$
32. $5a(5a + x) = 2a(2a - x)$
33. $\dfrac{x + b}{3a - 4b} = \dfrac{x - a}{2a - 5b}$
34. $\dfrac{1}{c - y} + \dfrac{2}{c + y} = \dfrac{1}{y}$

*In Exercises 35 through 42, solve the given formula for the indicated symbol.*

35. $A = \dfrac{1}{2}(a + b)h$; for $h$
36. $A = \dfrac{1}{2}(a + b)h$; for $b$
37. $E = I(R + r)$; for $r$
38. $A = P\left(1 + \dfrac{r}{n}\right)$; for $r$
39. $\dfrac{1}{f} = \dfrac{1}{p} + \dfrac{1}{q}$; for $p$
40. $E = I\left(R + \dfrac{r}{n}\right)$; for $n$
41. $S = \dfrac{a - rl}{1 - r}$; for $r$
42. $S = \dfrac{a - rl}{1 - r}$; for $l$

## 1.5 Word Problems

In many applications of algebra, the problems are stated in words and are called *word problems*. They give relationships between known numbers and unknown numbers to be determined. In this section we solve word problems by using a first-degree equation in one variable. Thus after determining the unknown numbers from the statement of the problem, it is necessary to represent these numbers by symbols using only one variable. After this is done, two algebraic expressions for the same number are obtained and an equation is formed from them. The procedure is shown in the following examples.

**EXAMPLE 1**
If a rectangle has a length that is three centimeters less than four times its width, and its perimeter is 19 centimeters, what are the dimensions of the rectangle?

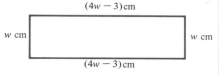

**Figure 1.5.1**

**SOLUTION**
Let $w$ centimeters represent the width of the rectangle. Then $(4w - 3)$ centimeters represents the length of the rectangle. Refer to Figure 1.5.1.

The perimeter of a rectangle is the total distance around the rectangle. Therefore, the number of centimeters in the perimeter can be represented by either $[w + (4w - 3) + w + (4w - 3)]$ or 19; thus we have the equation

$$w + (4w - 3) + w + (4w - 3) = 19$$

Solving the equation, we have

$$10w - 6 = 19$$
$$10w = 25$$
$$w = 2\tfrac{1}{2}$$
$$4w - 3 = 7$$

Hence the width of the rectangle is $2\tfrac{1}{2}$ centimeters and the length of the rectangle is 7 centimeters.

**CHECK:** The perimeter is $2\tfrac{1}{2}$ centimeters + 7 centimeters + $2\tfrac{1}{2}$ centimeters + 7 centimeters, which equals 19 centimeters.

Certain word problems can be classified according to type and the solution of a particular type often utilizes a specific formula or procedure. Some common types of problems are now considered.

The next example can be classified as an *investment problem*, because it is one involving income from an investment. The income in an investment problem can be in the form of interest and in such a case we use the formula

$$I = P \cdot R$$

where $I$ dollars is the annual interest earned when $P$ dollars is invested at a rate of $100R$ per cent per year.

## EXAMPLE 2

A man invested part of $15,000 at 12 per cent and the remainder at 8 per cent. If his annual income from the two investments is $1456, how much does he have invested at each rate?

### SOLUTION

Let $x$ dollars represent the amount he has invested at 12 per cent and $(15,000 - x)$ dollars represent the amount he has invested at 8 per cent. We use the formula $I = P \cdot R$ and make Table 1.5.1.

Table 1.5.1

|  | Number of Dollars Invested | × Rate = | Number of Dollars in Interest |
|---|---|---|---|
| 12 per cent investment | $x$ | 0.12 | $0.12x$ |
| 8 per cent investment | $15,000 - x$ | 0.08 | $0.08(15,000 - x)$ |

Therefore, we have the following equation.

$$0.12x + 0.08(15,000 - x) = 1456$$
$$0.12x + 1200 - 0.08x = 1456$$
$$0.04x = 256$$
$$x = 6400$$
$$15,000 - x = 8600$$

Thus the man has $6400 invested at 12 per cent and $8600 at 8 per cent.

**CHECK:** The annual interest from the $6400 invested at 12 per cent is $768 and from the $8600 invested at 8 per cent is $688; and $768 + $688 = $1456.

●

A *mixture problem* can involve mixing solutions, containing different per cents of a substance, in order to obtain a solution containing a certain per cent of the substance. For instance, one may wish to obtain 50 liters of a 20 per cent acid solution by mixing a 35 per cent acid solution with a 14 per cent acid solution (Example 3). Another kind of mixture problem, for which the method of solving is similar, involves mixing commodities of different values in order to obtain a combination worth a specific amount (Example 4).

## EXAMPLE 3

How many liters of a 35 per cent acid solution and how many liters of a 14 per cent acid solution should be combined to obtain 50 liters of a 20 per cent acid solution?

### SOLUTION

Let $x$ represent the number of liters of the 35 per cent acid solution to be used. Then $(50 - x)$ represents the number of liters of the 14 per cent acid solution to be used. To obtain an equation, we use the data in Table 1.5.2.

From the last column in the table we see that the total number of liters of acid in the mixture can be represented by either 10 or $[0.35x + 0.14(50 - x)]$. Thus we have the following equation

$$0.35x + 0.14(50 - x) = 10$$

Multiplying each member of the equation by 100, we obtain

$$35x + 14(50 - x) = 1000$$

## Word Problems

**Table 1.5.2**

|  | Per Cent of Acid | × | Number of Liters of Solution | = | Number of Liters of Acid |
|---|---|---|---|---|---|
| 35 per cent acid solution | 35% |  | $x$ |  | $0.35x$ |
| 14 per cent acid solution | 14% |  | $50 - x$ |  | $0.14(50 - x)$ |
| mixture | 20% |  | 50 |  | 10 |

We solve this equation.

$$35x + 700 - 14x = 1000$$
$$21x = 300$$
$$x = \frac{300}{21}$$
$$x = 14\frac{2}{7}$$
$$50 - x = 35\frac{5}{7}$$

Therefore, $14\frac{2}{7}$ liters of the 35 per cent acid solution and $35\frac{5}{7}$ liters of the 14 per cent acid solution should be used.

**CHECK:** The $14\frac{2}{7}$ liters of the 35 per cent acid solution gives 5 liters of acid and $35\frac{5}{7}$ liters of the 14 per cent acid solution also gives 5 liters of acid; and $5 + 5 = 10$.

### EXAMPLE 4

A merchant has 50 pounds of peanuts worth $1.60 per pound and 60 pounds of walnuts worth $1.90 per pound. How many pounds of almonds worth $2.40 per pound should be mixed with these nuts to obtain a mixture to sell for $2 per pound?

### SOLUTION

Let $a$ represent the number of pounds of almonds to be used. Then $(110 + a)$ represents the number of pounds of the mixture. We refer to Table 1.5.3 to obtain an equation.

**Table 1.5.3**

|  | Number of Dollars in the Price per Pound | × | Number of Pounds | = | Number of Dollars in the Total Value |
|---|---|---|---|---|---|
| Peanuts | 1.60 |  | 50 |  | 80 |
| Walnuts | 1.90 |  | 60 |  | 114 |
| Almonds | 2.40 |  | $a$ |  | $2.4a$ |
| Mixture | 2 |  | $110 + a$ |  | $2(110 + a)$ |

From the last column in the table we see that the number of dollars in the total value of the mixture can be represented by either $2(110 + a)$ or $(80 + 114 + 2.4a)$, and we get the following equation.

$$2(110 + a) = 80 + 114 + 2.4a$$
$$220 + 2a = 194 + 2.4a$$
$$2a - 2.4a = 194 - 220$$
$$-0.4a = -26$$
$$a = 65$$

Therefore, 65 pounds of almonds should be added.

**CHECK:** If 65 pounds of almonds are added, then the number of dollars in the total value of the almonds is $2.4 \cdot 65 = 156$ and the number of dollars in the total value of the mixture is $2 \cdot 175 = 350$; and $80 + 114 + 156 = 350$.

If an object travels at a uniform rate of $r$ miles per hour for a time of $t$ hours, then, if $d$ miles is the distance traveled,

$$t \cdot r = d \tag{1}$$

A problem involving the use of formula (1) is called a *uniform-motion problem*. In applying formula (1), the units of measurement of the rate, time, and distance must be consistent. In Example 5 the rate is measured in meters per second, and therefore the time is measured in seconds and the distance is measured in meters. In Example 6 the distance is measured in kilometers and the time is measured in hours, and therefore the rate is measured in kilometers per hour.

**EXAMPLE 5**
One runner took 3 minutes and 45 seconds to complete a race and another runner required 4 minutes to run the same race. If the rate of the faster runner is 0.4 meter per second faster than the rate of the slower runner, find their rates.

**SOLUTION**
We let the measurement of the time be in seconds. Let $r$ represent the number of meters per second in the rate of the slower runner. Then $(r + 0.4)$ represents the number of meters per second in the rate of the faster runner. We apply formula (1) and make Table 1.5.4.

The number of meters in the distance traveled by each runner is the same

Table 1.5.4

|  | Number of Seconds in Time | × | Number of Meters per Second in Rate | = | Number of Meters in Distance |
|---|---|---|---|---|---|
| Slower runner | 240 |  | $r$ |  | $240r$ |
| Faster runner | 225 |  | $r + 0.4$ |  | $225(r + 0.4)$ |

because each is running in the same race. From the last column in the table, we note that the number of meters in the distance traveled can be represented by either $240r$ or $225(r + 0.4)$. Therefore, we have the following equation.

$$240r = 225(r + 0.4)$$
$$240r = 225r + 90$$
$$15r = 90$$
$$r = 6$$
$$r + 0.4 = 6.4$$

Thus the rate of the slower runner is 6 meters per second and the rate of the faster runner is 6.4 meters per second.

**CHECK:** In 4 minutes the slower runner travels 1440 meters ($240 \cdot 6 = 1440$) and in $3\frac{3}{4}$ minutes the faster runner travels 1440 meters ($225 \cdot 6.4 = 1440$).

If the formula given by equation (1) is solved for $r$, we obtain

$$r = \frac{d}{t}$$

and if it is solved for $t$, we get

$$t = \frac{d}{r} \tag{2}$$

The next example involves a uniform-motion problem that leads to an equation containing rational expressions.

**EXAMPLE 6**

A father and daughter leave home at the same time in separate automobiles. The father drives to his office, a distance of 24 kilometers, and the daughter drives to school, a distance of 28 kilometers. They arrive at their destinations at the same time. Find their average speeds if the daughter's average speed is 12 kilometers per hour more than her father's.

**SOLUTION**

Let $r$ represent the number of kilometers per hour in the average speed of the daughter. Then $(r - 12)$ is the number of kilometers per hour in the average speed of the father. We apply formula (2) and make Table 1.5.5.

**Table 1.5.5**

|          | Number of Miles in Distance | ÷ | Number of Miles per Hour in Rate | = | Number of Hours in Time |
|----------|---|---|---|---|---|
| Daughter | 28 |   | $r$ |   | $\dfrac{28}{r}$ |
| Father   | 24 |   | $r - 12$ |   | $\dfrac{24}{r - 12}$ |

The number of hours of driving time of both the daughter and the father is the same. From the last column in the table we see that the number of hours can be represented by either $\dfrac{28}{r}$ or $\dfrac{24}{r-12}$. Thus we have the following equation.

$$\dfrac{28}{r} = \dfrac{24}{r-12}$$

We solve the equation.

$$r(r-12)\dfrac{28}{r} = r(r-12)\dfrac{24}{r-12}$$
$$(r-12)28 = (r)24$$
$$(r-12)7 = r(6)$$
$$7r - 84 = 6r$$
$$r = 84$$
$$r - 12 = 72$$

Therefore, the daughter's average speed is 84 kilometers per hour and the father's average speed is 72 kilometers per hour.

**CHECK:** The time for the daughter to travel 28 kilometers is 20 minutes ($28 \div 84 = \tfrac{1}{3}$, and $\tfrac{1}{3}$ hour is 20 minutes). The time for the father to travel 24 kilometers is 20 minutes ($24 \div 72 = \tfrac{1}{3}$).

A *work problem* is one in which a specific job is done in a certain amount of time when a uniform rate of work is assumed. For instance, if it takes a painter 10 hours to paint a room, then his rate of work is $\tfrac{1}{10}$ of the room per hour. In solving a work problem, we multiply the rate of work by the time in order to obtain the fractional part of the work completed. In particular, if the painter works for 7 hours, then the fractional part of the work completed is $\tfrac{7}{10}$. Furthermore, if a faucet can fill a tub in 15 minutes, then the faucet's rate of work is $\tfrac{1}{15}$ of the tub per minute and in $x$ minutes the fractional part of the tub filled by the faucet is $\dfrac{x}{15}$.

**EXAMPLE 7**
One painter can paint a room in 12 hours and another can paint the same room in 10 hours. How long will it take to paint the room if they work together?

**SOLUTION**
Let $x$ represent the number of hours it takes the two painters to paint the room when they are working together.

The rate of work of the first painter is $\tfrac{1}{12}$ of the room per hour, and the rate of work of the second painter is $\tfrac{1}{10}$ of the room per hour. We use Table 1.5.6 to obtain an equation.

Because the two painters complete the work together (they paint the room), the sum of the entries in the last column of the table is 1; that is, the fractional part of the work done by the first painter plus the fractional part of

Table 1.5.6

|  | Rate of Work | × | Number of Hours Worked | = | Fractional Part of Work Done |
|---|---|---|---|---|---|
| First painter | $\frac{1}{12}$ | | $x$ | | $\frac{x}{12}$ |
| Second painter | $\frac{1}{10}$ | | $x$ | | $\frac{x}{10}$ |

the work done by the second painter equals 1. We then have the following equation.

$$\frac{x}{12} + \frac{x}{10} = 1$$

We solve the equation.

$$60 \cdot \frac{x}{12} + 60 \cdot \frac{x}{10} = 60 \cdot 1$$
$$5x + 6x = 60$$
$$11x = 60$$
$$x = 5\frac{5}{11}$$

Hence it takes the painters $5\frac{5}{11}$ hours to paint the room together.

**CHECK:** The fractional part of the work done by the first painter is $5\frac{5}{11} \div 12$ or $\frac{5}{11}$, and the fractional part of the work done by the second painter is $5\frac{5}{11} \div 10$ or $\frac{6}{11}$; and $\frac{5}{11} + \frac{6}{11} = 1$.

**EXAMPLE 8**
One pipe takes 30 minutes to fill a tank and, after it has been running for 10 minutes, it is shut off. A second pipe is then opened and it finishes filling the tank in 15 minutes. How long would it take the second pipe alone to fill the tank?

**SOLUTION**
Let $x$ represent the number of minutes it would take the second pipe alone to fill the tank.

The rate of work of the second pipe is $\frac{1}{x}$ of the tank per minute and the rate of work of the first pipe is $\frac{1}{30}$ of the tank per minute. We have Table 1.5.7.

Table 1.5.7

|  | Rate of Work | × | Number of Minutes Worked | = | Fractional Part of the Tank Filled |
|---|---|---|---|---|---|
| First pipe | $\frac{1}{30}$ | | 10 | | $\frac{1}{3}$ |
| Second pipe | $\frac{1}{x}$ | | 15 | | $\frac{15}{x}$ |

Because the entire tank is filled, the sum of the entries in the last column of the table is 1. Therefore, we have the following equation.

$$\frac{1}{3} + \frac{15}{x} = 1$$

$$3x \cdot \frac{1}{3} + 3x \cdot \frac{15}{x} = 3x \cdot 1$$

$$x + 45 = 3x$$

$$-2x = -45$$

$$x = 22\frac{1}{2}$$

Thus it would take the second pipe alone $22\frac{1}{2}$ minutes to fill the tank.

**CHECK:** The first pipe fills $\frac{1}{3}$ of the tank and the fractional part of the tank filled by the second pipe is $15 \div 22\frac{1}{2}$ or $\frac{2}{3}$; and $\frac{1}{3} + \frac{2}{3} = 1$.

## EXERCISES 1.5

1. If the width of a rectangle is 2 centimeters more than one-half its length and its perimeter is 40 centimeters, what are the dimensions of the rectangle?
2. Admission tickets to a motion picture theatre cost $2 for adults and $1.50 for students. If 810 tickets were sold and the total receipts was $1426.50, how many of each type of tickets were sold?
3. The amount invested in two real estate developments is $30,000. One investment yields 14 per cent, and the annual income from this investment is $120 less than the annual income from the other investment, which yields 10 per cent. Find the amount invested at each rate.
4. Determine the number of grams of pure silver that should be added to 110 grams of an alloy that is 60 per cent silver in order to obtain an alloy that contains 75 per cent silver.
5. How many liters of a solution containing 55 per cent glycerine should be added to 25 liters of a solution that contains 28 per cent glycerine to give a solution that contains 35 per cent glycerine?
6. A woman invested $25,000 in two business ventures. Last year she made a profit of 15 per cent from the first venture but she lost 5 per cent from the second venture. If last year's income from the two investments was equivalent to a return of 8 per cent on the entire amount invested, how much had she invested in each venture?
7. Perfume to sell for $20 an ounce is to be a blend of perfume selling for $26 an ounce and perfume selling for $12.50 an ounce. If 270 ounces of the blend is desired, how much of each kind of perfume should be used?
8. A gardener has 26 pounds of a mixture of fertilizer and weed killer. If 1 pound of the mixture is replaced by weed killer, the result is a mixture that is 5 per cent weed killer. What per cent of the original mixture was weed killer?
9. A man leaves his home in an automobile on a business trip. Twenty minutes later his wife discovers that he left his briefcase behind and decides to overtake him in another car. If the wife knows that her husband averages 75 kilometers per hour and she averages 100 kilometers per hour, how long will it take her to overtake her husband?
10. At 2 P.M., a train traveling 76 kilometers per hour, leaves Denver for Kansas City, a distance of 1000

kilometers. At 3 p.m., a train traveling 100 kilometers per hour leaves Kansas City for Denver. At what time do the two trains pass each other?

11. One runner ran 840 meters in the same time that another runner ran 810 meters. If the average speed of the faster runner was one-third of a meter per second more than the average speed of the slower runner, what was the average speed of each runner?

12. A pipe can fill a swimming pool in 10 hours. If a second pipe is open, the two pipes together can fill the pool in 4 hours. How long would it take the second pipe alone to fill the pool?

13. Each of two brothers can wash a car in 1 hour; however, their sister can wash the car in 45 minutes. If all three work together, how long will it take to wash the car?

14. In an automobile race the average speed of one car was 240 kilometers per hour and the average speed of another car was 210 kilometers per hour. If the faster car finished the race 20 minutes before the slower car, what was the distance of the race?

15. Two printing presses are available to print the daily college newspaper. If only one press is used, it takes the older press twice as long to print an edition as it takes the newer press. If the two presses together can print an edition in three hours, how long would it take each press alone to print an edition?

16. An investor wishes to realize a return of 9 per cent on his total investments. If he has $10,000 invested at 7.5 per cent, how much additional money should be invested at 12 per cent?

17. Every freshman student at a particular college is required to take an English aptitude test. A student who passes the examination enrolls in English Composition, and a student who fails the test must enroll in English Fundamentals. In a freshman class of 1240 students there are more students enrolled in English Fundamentals than in English Composition. However, if 30 more students had passed the test, each course would have the same enrollment. How many students are taking each course?

18. The annual sophomore class picnic is planned by a committee consisting of 17 members. A vote, to determine if the picnic should be held at the beach or in the mountains, resulted in a victory for the beach location. However, if two committee members had changed their vote from favoring the beach to favoring the mountains, then the mountain site would have won by one vote. How many votes did each picnic location receive?

19. A retail merchant invested $6500 in three kinds of cameras. His profit on the sales of camera A was 25 per cent, on the sales of camera B his profit was 12 per cent, and he lost 1 per cent on the sales of camera C. If he invested an equal amount in cameras A and B, and his overall profit on the total investment was 14 per cent, how much did he invest in each kind of camera?

20. Determine how much water is required to dilute 15 liters of a solution containing 12 per cent dye so that a 5 per cent dye solution is obtained.

21. How much water must be evaporated from the 15 liters of the 12 per cent dye solution in Exercise 20 in order to obtain a solution containing 20 per cent dye? Assume that the total amount of dye is not affected by the process of evaporation.

22. A car overtakes a truck that is traveling in the same direction at the rate of 75 kilometers per hour. If the car is 5 meters long and the truck is 10 meters long, what must be the speed of the car so that the car passes the truck in 3 seconds?

23. A freight train, $\frac{1}{2}$ kilometer long, is moving at the rate of 43 kilometers per hour. The train passes a girl who is walking in the opposite direction beside the track at a rate of 7 kilometers per hour. How long will it take the train to pass the girl?

24. A woman can do a certain job in 10 hours and her younger daughter can do it in 12 hours. After the woman and her younger daughter have been working for 1 hour, they are joined by the older daughter, and the three complete the job in 3 more hours. How long will it take the older daughter to do the job by herself?

25. One pipe can fill a tank in 45 minutes and another pipe can fill it in 30 minutes. If these two pipes are

open and a third pipe is draining water from the tank, it takes 27 minutes to fill the tank. How long will it take the third pipe alone to empty a full tank?

26. Fifty boxes of cookies each weighing 1 pound and selling for $2.30 per box are to be made up of shortbread worth $2.40 per pound, tea biscuits worth $1.80 per pound, and macaroons worth $2.60 per pound. If 12 pounds of macaroons are used, how many pounds of shortbread and how many pounds of tea biscuits should be used?

## 1.6 First-degree Inequalities in One Variable

In Section 1.2 we mentioned that the statements

$$a > b \quad a \geq b \quad a < b \quad a \leq b$$

are called inequalities. Before discussing inequalities involving a variable, we state some properties of inequalities. One such property is the transitive property of order, which we now state and prove.

**1.6.1 THEOREM** *Transitive Property of Order.* If $a, b, c \in R$, and

if $a > b$ and $b > c$, then $a > c$

**Proof.** Because $a > b$, it follows from the definition of order (1.2.12) and Theorem 1.2.14 that

$$a - b > 0 \quad \text{(i)}$$

Similarly, because $b > c$, it follows that

$$b - c > 0 \quad \text{(ii)}$$

From Theorem 1.2.15(ii), the sum of two positive numbers is positive, and so

$$(a - b) + (b - c) > 0 \quad \text{(iii)}$$

Applying the definition of subtraction (1.2.8) in each set of parentheses in the left member of inequality (iii), we have

$$[a + (-b)] + [b + (-c)] > 0 \quad \text{(iv)}$$

By repeated applications of the commutative and associative laws of addition to the left member of inequality (iv), we have

$$[a + (-c)] + [(-b) + b] > 0 \quad \text{(v)}$$

From the definition of subtraction, Axiom 1.2.6 (existence of opposite), and inequality (v), we have

$$(a - c) + 0 > 0 \quad \text{(vi)}$$

Because 0 is the additive identity, we have from inequality (vi),

$$a - c > 0$$

and therefore from the definition of order,

$$a > c$$

which is what we wished to prove.

**ILLUSTRATION 1.** If $x > 5$ and $5 > y$, then by the transitive property of order, it follows that $x > y$. •

Three other important properties of inequalities are given in the next theorem.

**1.6.2 THEOREM**  Suppose that $a, b, c \in R$.

(i) If $a > b$, then $a + c > b + c$.
(ii) If $a > b$ and $c > 0$, then $ac > bc$. •
(iii) If $a > b$ and $c < 0$, then $ac < bc$.

Before proving this theorem, we give an illustration of its content.

**ILLUSTRATION 2**

(a) Because $9 > 4$, it follows from Theorem 1.6.2(i) that $9 + 3 > 4 + 3$ or, equivalently, $12 > 7$.
   Furthermore, because $9 > 4$, it follows that $9 - 11 > 4 - 11$ or, equivalently, $-2 > -7$.
(b) Because $7 > 5$, it follows from Theorem 1.6.2(ii) that $7 \cdot 2 > 5 \cdot 2$ or, equivalently, $14 > 10$.
(c) Because $7 > 5$, it follows from Theorem 1.6.2(iii) that $7(-2) < 5(-2)$ or, equivalently, $-14 < -10$. •

***Proof of Theorem 1.6.2.*** Because $a > b$, it follows from the definition of order that

$$a - b > 0 \quad (1)$$

But
$$\begin{aligned}
a - b &= (a - b) + 0 \\
&= (a - b) + (c - c) \\
&= [a + (-b)] + [c + (-c)] \\
&= (a + c) + [(-b) + (-c)] \\
&= (a + c) - (b + c)
\end{aligned} \qquad (2)$$

Therefore, from inequality (1) and equality (2), we have

$$(a + c) - (b + c) > 0 \qquad (3)$$

From inequality (3) and the definition of order, it follows that

$$a + c > b + c$$

which proves part (i) of the theorem.

We now prove part (ii). Because $a > b$, we have inequality (1), and from this inequality and the fact that $c > 0$, we have from Theorem 1.2.15(iii),

$$\begin{aligned}
(a - b)c &> 0 \\
ac - bc &> 0 \\
ac &> bc
\end{aligned}$$

which proves part (ii).

In the hypothesis of part (iii), $c$ is negative and so from Definition 1.2.11, $-c$ is positive. Therefore,

$$-c > 0 \qquad (4)$$

Because $a > b$, we have inequality (1), and from inequalities (1) and (4) and Theorem 1.2.15(iii), it follows that

$$\begin{aligned}
-c(a - b) &> 0 \\
bc - ac &> 0
\end{aligned}$$

Therefore, from the definition of order,

$$bc > ac$$

or, equivalently,

$$ac < bc$$

which proves part (iii).

The following theorem is similar to Theorem 1.6.2 except that the direction of the inequality is reversed.

**1.6.3 THEOREM**  Suppose that $a, b, c \in R$.

(i) If $a < b$, then $a + c < b + c$.
(ii) If $a < b$, and $c > 0$, then $ac < bc$.
(iii) If $a < b$, and $c < 0$, then $ac > bc$.

**ILLUSTRATION 3**

(a) Because $-8 < -2$, from Theorem 1.6.3(i) we have $-8 + 5 < -2 + 5$ or, equivalently, $-3 < 3$.
    Furthermore, because $-8 < -2$, it follows that $-8 - 4 < -2 - 4$ or, equivalently, $-12 < -6$.
(b) Because $-5 < 3$, it follows from Theorem 1.6.3(ii) that $(-5)(4) < (3)(4)$ or, equivalently, $-20 < 12$.
(c) Because $-5 < 3$, from Theorem 1.6.3(iii) we have $(-5)(-4) > (3)(-4)$ or, equivalently, $20 > -12$. ●

Theorems 1.6.2 and 1.6.3 also are valid if in each theorem the symbol $>$ is replaced by $\geq$ and $<$ is replaced by $\leq$.

Parts (ii) of Theorems 1.6.2 and 1.6.3 state that if both members of an inequality are multiplied by a positive number, the direction of the inequality remains unchanged, whereas parts (iii) state that if both members of an inequality are multiplied by a negative number, the direction of the inequality is reversed. Parts (ii) and (iii) also hold for division because dividing both members of an inequality by a number $d(d \neq 0)$ is equivalent to multiplying both members by $\frac{1}{d}$.

We now discuss first-degree (or linear) inequalities involving a single variable. Examples of such inequalities are

$$3x - 8 > 7 \tag{5}$$

$$\frac{x - 7}{4} \leq x \tag{6}$$

$$\frac{3x}{x + 2} < 5 \tag{7}$$

As with equations, the *replacement set* of a variable in an inequality is a set of numbers, and, unless otherwise stated, the replacement set is the set of real numbers for which the members of the inequality are defined. For inequalities (5) and (6) the replacement set is the set $R$ of real numbers, but because the left member of inequality (7) is not defined when $x$ is $-2$, its replacement set is $\{x \mid x \in R \text{ and } x \neq -2\}$. Any number in the replacement set for which the inequality is true is called a *solution* of the inequality. The set of all solutions is called the *solution set* of the inequality.

An *absolute inequality* is one that is true for every number in the replacement set. For instance, if $x \in R$, the inequalities

$$x + 2 > x + 1$$

and

$$x^2 \geq 0$$

are absolute inequalities. A *conditional inequality* is one that is true for only the numbers in a proper subset of the replacement set. Inequalities (5), (6), and (7) are examples of conditional inequalities. To find the solution set of a conditional inequality, we proceed in a manner similar to that used to solve an equation; that is, we obtain equivalent inequalities (inequalities having the same solution set) until we obtain one whose solution set is apparent. The following theorem, which is similar to Theorem 1.4.4, follows from Theorems 1.6.2 and 1.6.3. It is used to replace an inequality by an equivalent one.

**1.6.4 THEOREM**  Suppose that

$$E > F \tag{8}$$

is an inequality, where $E$ and $F$ are algebraic expressions in the variable $x$. Then if $G$ is an algebraic expression in $x$, inequality (8) is equivalent to each of the following inequalities.

(i) $E + G > F + G$
(ii) $E - G > F - G$
(iii) $E \cdot G > F \cdot G$, for all $x$ for which $G > 0$
(iv) $E \cdot G < F \cdot G$, for all $x$ for which $G < 0$
(v) $\dfrac{E}{G} > \dfrac{F}{G}$, for all $x$ for which $G > 0$
(vi) $\dfrac{E}{G} < \dfrac{F}{G}$, for all $x$ for which $G < 0$

**ILLUSTRATION 4.**  Inequality (5) is

$$3x - 8 > 7$$

or, equivalently (by applying Theorem 1.6.4(i)),

$$3x - 8 + 8 > 7 + 8$$

or, equivalently,

$$3x > 15$$

or, equivalently (by applying Theorem 1.6.4(v)),

$$\frac{3x}{3} > \frac{15}{3}$$

or, equivalently,

$$x > 5$$

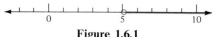

Figure 1.6.1

Hence the solution set of the given inequality is $\{x \mid x > 5\}$ (read "The set of all elements $x$ such that $x$ is greater than 5"). This solution set can be represented by a graph on the real number line. Figure 1.6.1 shows a sketch of such a graph where the portion of the line in color consists of the points whose coordinates are in the solution set. The open dot at 5 indicates that 5 is not in the solution set. •

**ILLUSTRATION 5.** Inequality (6) is

$$\frac{x-7}{4} \leq x$$

or, equivalently (by applying Theorem 1.6.4(iii)),

$$(4)\frac{x-7}{4} \leq (4)(x)$$

$$x - 7 \leq 4x$$

or, equivalently (by applying Theorem 1.6.4(i) and (ii)),

$$x - 7 - 4x + 7 \leq 4x - 4x + 7$$

$$-3x \leq 7$$

or, equivalently (by applying Theorem 1.6.4(vi)),

$$\frac{-3x}{-3} \geq \frac{7}{-3}$$

$$x \geq -\frac{7}{3}$$

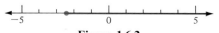

Figure 1.6.2

Therefore, the solution set of the given inequality is $\{x \mid x \geq -\frac{7}{3}\}$. A sketch of the graph of the solution set is shown in Figure 1.6.2. The solid dot at $-\frac{7}{3}$ indicates that $-\frac{7}{3}$ is in the solution set. •

**EXAMPLE 1**
Find and draw a sketch of the graph of the solution set of the inequality

$$8x + 14 \leq 3x - 6$$

**SOLUTION**

$$8x + 14 \leq 3x - 6$$
$$8x + 14 - 3x - 14 \leq 3x - 6 - 3x - 14$$
$$5x \leq -20$$
$$x \leq -4$$

Thus the solution set of the given inequality is $\{x \mid x \leq -4\}$. A sketch of the graph of the solution set is shown in Figure 1.6.3.

Figure 1.6.3

## EXAMPLE 2
Find and draw a sketch of the graph of the solution set of the inequality

$$\frac{4x-3}{3} > 2x + 4$$

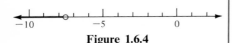

**Figure 1.6.4**

### SOLUTION

$$\frac{4x-3}{3} > 2x + 4$$

$$3\left(\frac{4x-3}{3}\right) > 3(2x+4)$$

$$4x - 3 > 6x + 12$$
$$4x - 3 - 6x + 3 > 6x + 12 - 6x + 3$$
$$-2x > 15$$
$$\frac{-2x}{-2} < \frac{15}{-2}$$
$$x < -\frac{15}{2}$$

Hence the solution set of the given inequality is $\{x \mid x < -\frac{15}{2}\}$. A sketch of the graph of the solution set is shown in Figure 1.6.4.

The next example involves a continued inequality, and the procedure for finding the solution set of such an inequality is identical with the method used in Examples 1 and 2.

## EXAMPLE 3
Find and draw a sketch of the graph of the solution set of the inequality

$$3 < 4x + 7 \leq 15$$

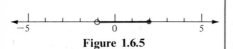

**Figure 1.6.5**

### SOLUTION

$$3 < 4x + 7 \leq 15$$
$$3 - 7 < 4x + 7 - 7 \leq 15 - 7$$
$$-4 < 4x \leq 8$$
$$\frac{-4}{4} < \frac{4x}{4} \leq \frac{8}{4}$$
$$-1 < x \leq 2$$

Therefore, the solution set of the given inequality is $\{x \mid -1 < x \leq 2\}$. A sketch of the graph of the solution set is shown in Figure 1.6.5.

## EXAMPLE 4
If the temperature using the Fahrenheit scale is $F$ degrees and using the Celsius scale is $C$ degrees, then

$$C = \frac{5}{9}(F - 32)$$

Find the range of values of $F$ if $C$ is to be between 10 and 20.

### SOLUTION
In the continued inequality $10 < C < 20$, we replace $C$ by $\frac{5}{9}(F - 32)$ and obtain

$$10 < \frac{5}{9}(F - 32) < 20$$
$$90 < 5F - 160 < 180$$
$$90 + 160 < 5F < 180 + 160$$
$$250 < 5F < 340$$
$$50 < F < 68$$

## EXAMPLE 5
Find the solution set of the inequality

$$\frac{5}{x} > 2$$

**SOLUTION**
We wish to multiply both members of the inequality by $x$. However, we must consider two cases because the direction of the inequality that results will depend upon whether $x$ is positive or negative.

**CASE 1:** $x$ is positive; that is, $x > 0$.
If we multiply on both sides of the given equality by $x$, we obtain

$$5 > 2x$$

$$\frac{5}{2} > x$$

$$x < \frac{5}{2}$$

Because for Case 1, $x > 0$, the solution set of the given inequality is

$$\{x \mid x > 0\} \cap \left\{x \mid x < \frac{5}{2}\right\} \quad \text{or, equivalently,} \quad \left\{x \mid 0 < x < \frac{5}{2}\right\}.$$

**CASE 2:** $x$ is negative; that is, $x < 0$.
Multiplying on both sides of the given inequality by $x$ and reversing the direction of the inequality, we get

$$5 < 2x$$

$$\frac{5}{2} < x$$

$$x > \frac{5}{2}$$

The solution set of the given inequality for Case 2 is

$$\{x \mid x < 0\} \cap \left\{x \mid x > \frac{5}{2}\right\}$$

which is the empty set $\emptyset$.
The solution set of the given inequality is the union of the solution sets of Cases 1 and 2, which is $\{x \mid 0 < x < \frac{5}{2}\}$.

## EXAMPLE 6
Find the solution set of the inequality

$$\frac{3x}{x+2} < 5$$

and draw a sketch of its graph.

**SOLUTION**
When multiplying both sides of the inequality by $x + 2$, we must consider two cases, as in Example 5.

**CASE 1:** $x + 2 > 0$; that is, $x > -2$.
Multiplying on both sides of the inequality by $x + 2$, we obtain

$$3x < 5x + 10$$
$$-2x < 10$$
$$x > -5$$

The solution set for Case 1 is $\{x \mid x > -2\} \cap \{x \mid x > -5\}$ or, equivalently, $\{x \mid x > -2\}$.

**CASE 2:** $x + 2 < 0$; that is, $x < -2$.
Multiplying on both sides of the inequality by $x + 2$ and reversing the direction of the inequality, we have

$$3x > 5x + 10$$
$$-2x > 10$$
$$x < -5$$

The solution set for Case 2 is $\{x \mid x < -2\} \cap \{x \mid x < -5\}$ or, equivalently, $\{x \mid x < -5\}$.

The solution set of the given inequality is the union of the solution sets of Cases 1 and 2 which is $\{x \mid x < -5\} \cup \{x \mid x > -2\}$. A sketch of the graph of the solution set is shown in Figure 1.6.6.

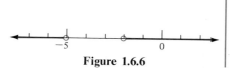

**Figure 1.6.6**

**EXAMPLE 7**
A company that builds and sells desks has a weekly overhead (including salaries and plant cost) of $3400. The cost of materials for each desk is $40 and it is sold for $200. How many desks must be built and sold each week so that the company is guaranteed a profit?

**SOLUTION**
Let $x$ be the number of desks built and sold each week. Then the number of dollars in the total revenue received each week is $200x$, and the number of dollars in the total cost each week is $3400 + 40x$. If $P$ dollars is the weekly profit, then because profit equals revenue minus cost, we have

$$P = 200x - (3400 + 40x)$$
$$= 160x - 3400$$

For a profit we must have $P > 0$; that is,

$$160x - 3400 > 0$$
$$160x > 3400$$
$$x > 21\tfrac{1}{4}$$

Because $x$ must be a positive integer, we conclude that the company must build and sell at least 22 desks each week to have a profit.

## EXERCISES 1.6

*In Exercises 1 through 28, find the solution set of the given inequality. Draw a sketch of the graph of the solution set.*

1. $4x \geq 20$
2. $3x - 5 < 7$
3. $2x - 1 < 6$
4. $3x + 1 \geq 4x - 3$
5. $2x + 1 > x - 4$
6. $5x + 6 \leq x - 2$
7. $\dfrac{2x - 5}{4} \leq 3$
8. $\dfrac{2x - 9}{7} > 0$
9. $-3 > \dfrac{3x + 5}{4}$
10. $\dfrac{x - 5}{4} \leq x$
11. $\dfrac{2x - 5}{3} < x + 1$
12. $6x - 1 < \dfrac{5x - 1}{3}$
13. $5 \leq 2x - 3 < 13$
14. $11 \geq 3x - 5 > 2$
15. $-7 < 2x + 1 < 3$
16. $1 \leq 3x - 2 \leq 16$
17. $8 \geq 2 - x \geq 6$
18. $19 > 4 - 3x > 10$
19. $2 > -3 - 3x \geq -7$
20. $-1 < 2 - 2x \leq 3$

21. $\dfrac{5}{x} < \dfrac{3}{4}$  22. $\dfrac{1}{x} > \dfrac{2}{3}$  23. $\dfrac{4}{x} - 3 \geq \dfrac{2}{x} - 7$  24. $\dfrac{2}{1-x} < 1$

25. $\dfrac{5x}{x-4} \geq 6$  26. $\dfrac{3x-8}{2x+3} < 4$  27. $\dfrac{3x-11}{3x+5} < 7$  28. $\dfrac{4x-7}{5x+1} \geq 2$

*In Exercises 29 through 36, draw a sketch of the graph of the given set.*

29. $\{x \mid x > -4\} \cap \{x \mid x < 3\}$
30. $\{x \mid x \leq 10\} \cap \{x \mid x \geq 3\}$
31. $\{x \mid 5 - 2x \geq 1\} \cap \{x \mid -(x+2) \leq 5\}$
32. $\{x \mid 3(x-2) < 4\} \cap \{x \mid -2(x-3) < 5\}$
33. $\{x \mid x < -2\} \cup \{x \mid x > 4\}$
34. $\{x \mid x < 1\} \cup \{x \mid x \geq 9\}$
35. $\{x \mid 2x + 6 \leq 3\} \cup \{x \mid -(4x-1) \leq -4\}$
36. $\{x \mid -(x-4) > 5\} \cup \{x \mid 3x - 7 > 3\}$

37. A silversmith wants to obtain an alloy containing at least 72 per cent silver and at most 75 per cent silver. Determine the greatest and least amounts of an 80 per cent silver alloy that should be combined with a 65 per cent silver alloy in order to have 30 grams of the required alloy.

38. If in a particular course, a student has an average score of less than 90 and not below 80 on four examinations, the student will receive a grade of B. If the student's grades on the first three examinations are 87, 94, and 73, what score on the fourth examination will result in a B grade in the course?

39. An investor has $8000 invested at 9 per cent and wishes to invest some additional money at 16 per cent in order to realize a return of at least 12 per cent on the total investment. What is the least amount of money that should be invested?

40. Part of $20,000 is to be invested at 6 per cent and the remainder is to be invested at 8 per cent. What is the least amount that can be invested at 8 per cent in order to have a yearly income of at least $1500 from the two investments?

41. A lamp manufacturer sells only to wholesalers through its showroom. The weekly overhead (including salaries, plant cost, and showroom rental) is $3000. If each lamp sells for $84 and the material used in its production costs $22, how many lamps must be made and sold each week so that the manufacturer realizes a profit?

42. What is the minimum amount of pure alcohol that must be added to 24 liters of a 20 per cent alcohol solution to obtain a mixture that is at least 30 per cent alcohol?

## 1.7 Equations and Inequalities Involving Absolute Value

Associated with each real number is a nonnegative number, called its absolute value.

**1.7.1 DEFINITION** If $a \in R$, the *absolute value of a*, denoted by $|a|$, is $a$ if $a$ is nonnegative and is the opposite of $a$ if $a$ is negative. With symbols, we write

$$|a| = \begin{cases} a, & \text{if } a \geq 0 \\ -a, & \text{if } a < 0 \end{cases}$$

**56** Real Numbers, Equations, and Inequalities  [Ch. 1

**ILLUSTRATION 1.** If in Definition 1.7.1 we take $a$ as 6, 0, and $-6$, we have

$$|6| = 6 \qquad |0| = 0 \qquad |-6| = -(-6) = 6$$

The absolute value of a number $a$ can be considered as the distance (without regard to direction, left or right) of the graph of $a$ from the origin. In particular, the graphs of both 6 and $-6$ are 6 units from the origin.

### EXAMPLE 1
Find the value of each of the following expressions if $x = 3$ and $y = -6$.

(a) $|3x + 2y|$
(b) $|3x| + |2y|$
(c) $|xy|$
(d) $|x| \cdot |y|$
(e) $|x - y|$
(f) $|x| - |y|$
(g) $\left|\dfrac{x}{y}\right|$
(h) $\dfrac{|x|}{|y|}$

### SOLUTION

(a) $|3x + 2y| = |3(3) + 2(-6)|$
$= |9 + (-12)|$
$= |-3|$
$= 3$

(b) $|3x| + |2y| = |3(3)| + |2(-6)|$
$= |9| + |-12|$
$= 9 + 12$
$= 21$

(c) $|xy| = |3(-6)|$
$= |-18|$
$= 18$

(d) $|x| \cdot |y| = |3| \cdot |6|$
$= 3 \cdot 6$
$= 18$

(e) $|x - y| = |3 - (-6)|$
$= |3 + 6|$
$= |9|$
$= 9$

(f) $|x| - |y| = |3| - |-6|$
$= 3 - 6$
$= -3$

(g) $\left|\dfrac{x}{y}\right| = \left|\dfrac{3}{-6}\right|$
$= \left|-\dfrac{1}{2}\right|$
$= \dfrac{1}{2}$

(h) $\dfrac{|x|}{|y|} = \dfrac{|3|}{|-6|}$
$= \dfrac{3}{6}$
$= \dfrac{1}{2}$

### EXAMPLE 2
Express each of the following without absolute value bars.

(a) $|2 - \sqrt{2}|$   (b) $|\sqrt{2} - 2|$

### SOLUTION

(a) Because $2 > \sqrt{2}$, $(2 - \sqrt{2}) > 0$. Hence
$$|2 - \sqrt{2}| = 2 - \sqrt{2}$$

(b) Because $(\sqrt{2} - 2) < 0$,
$$|\sqrt{2} - 2| = -(\sqrt{2} - 2)$$
$$= 2 - \sqrt{2}$$

From Definition 1.7.1, we have

$$|x - a| = \begin{cases} x - a, & \text{if } x - a \geq 0 \\ -(x - a), & \text{if } x - a < 0 \end{cases}$$

or, equivalently,

$$|x - a| = \begin{cases} x - a, & \text{if } x \geq a \\ -(x - a), & \text{if } x < a \end{cases}$$

As a distance on the real number line, we can interpret $|x - a|$ as the distance between the graphs of $x$ and $a$ without regard to direction. Refer to Figure 1.7.1.

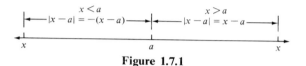

**Figure 1.7.1**

**ILLUSTRATION 2.** Suppose that we have the equation

$$|x - 4| = 7 \tag{1}$$

This equation will be satisfied if either

$$x - 4 = 7 \quad \text{or} \quad -(x - 4) = 7$$

We solve each equation.

$$\begin{aligned} x - 4 &= 7 & -(x - 4) &= 7 \\ x &= 11 & -x + 4 &= 7 \\ & & -x &= 3 \\ & & x &= -3 \end{aligned}$$

Therefore, the solution set of equation (1) is $\{11, -3\}$. ●

**EXAMPLE 3**

Find the solution set of the equation

$$|3x + 5| = 9$$

**SOLUTION**

The given equation will be satisfied if either

$$\begin{aligned} 3x + 5 &= 9 & -(3x + 5) &= 9 \\ 3x &= 4 & -3x - 5 &= 9 \\ x &= \tfrac{4}{3} & -3x &= 14 \\ & & x &= -\tfrac{14}{3} \end{aligned}$$

The solution set is $\{\tfrac{4}{3}, -\tfrac{14}{3}\}$.

We now consider inequalities involving absolute values.

**ILLUSTRATION 3.** Suppose that we have the inequality

$$|x| < 7 \tag{2}$$

Because $|x| = x$, if $x \geq 0$, and $|x| = -x$, if $x < 0$, it follows that the solution set of inequality (2) is the union of the sets

$$\{x \mid x < 7 \text{ and } x \geq 0\} \tag{3}$$

and

$$\{x \mid -x < 7 \text{ and } x < 0\} \tag{4}$$

Set (3) is equivalent to

$$\{x \mid 0 \leq x < 7\} \tag{5}$$

Because $-x < 7$ is equivalent to $x > -7$ (from Theorem 1.6.4(iv)), then set (4) is equivalent to

$$\{x \mid -7 < x < 0\} \tag{6}$$

The solution set of inequality (2) is therefore, the union of the sets (5) and (6), which is

$$\{x \mid 0 \leq x < 7\} \cup \{x \mid -7 < x < 0\}$$

or, equivalently,

$$\{x \mid -7 < x < 7\} \tag{7}$$

**Figure 1.7.2**

A sketch of the graph of this set is shown in Figure 1.7.2. ●

Comparing inequality (2) and its solution set (7), we conclude that the inequality

$|x| < 7 \quad$ is equivalent to $\quad -7 < x < 7$

More generally, if $b > 0$,

$$\boxed{|x| < b \quad \text{is equivalent to} \quad -b < x < b} \tag{8}$$

The next theorem follows immediately from statement (8).

**1.7.2 THEOREM**    If $E$ is an algebraic expression in the variable $x$, and $b$ is a positive number, then the inequality

$$\boxed{|E| < b \quad \text{is equivalent to} \quad -b < E < b}$$

**ILLUSTRATION 4.** From Theorem 1.7.2, the inequality

$$|x - 5| < 8 \tag{9}$$

is equivalent to

$$-8 < x - 5 < 8$$

or, equivalently,
$$-8+5 < x < 8+5$$
$$-3 < x < 13$$

Therefore, the solution set of inequality (9) is $\{x \mid -3 < x < 13\}$. ●

**EXAMPLE 4**
Find the solution set of the inequality
$$|2x - 7| \leq 9$$
Draw a sketch of the graph of the solution set.

**SOLUTION**
The given inequality is equivalent to
$$-9 \leq 2x - 7 \leq 9$$
$$-9 + 7 \leq 2x \leq 9 + 7$$
$$-2 \leq 2x \leq 16$$
$$-1 \leq x \leq 8$$

Therefore, the solution set is $\{x \mid -1 \leq x \leq 8\}$, and a sketch of its graph is shown in Figure 1.7.3.

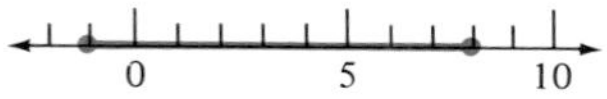

**Figure 1.7.3**

**ILLUSTRATION 5.** Consider the inequality
$$|x| > 4 \qquad (10)$$
From the definition of $|x|$, the solution set of this inequality is the union of the sets
$$\{x \mid x > 4 \text{ and } x \geq 0\} \qquad (11)$$
and
$$\{x \mid -x > 4 \text{ and } x < 0\} \qquad (12)$$
Set (11) is equivalent to
$$\{x \mid x > 4\} \qquad (13)$$
Because $-x > 4$ is equivalent to $x < -4$, set (12) is equivalent to
$$\{x \mid x < -4\} \qquad (14)$$
The solution set of inequality (10) is then the union of the sets (13) and (14), which is
$$\{x \mid x > 4\} \cup \{x \mid x < -4\} \qquad (15)$$
A sketch of the graph of this set is shown in Figure 1.7.4. ●

**Figure 1.7.4**

Comparing inequality (10) and its solution set (15), we conclude that the inequality
$$|x| > 4 \quad \text{is equivalent to} \quad \text{``}x > 4 \text{ or } x < -4\text{''}$$

More generally, if $b > 0$,

$$|x| > b \text{ is equivalent to } \text{``}x > b \text{ or } x < -b\text{''} \tag{16}$$

The following theorem is a direct consequence of statement (16).

**1.7.3 THEOREM** If $E$ is an algebraic expression in the variable $x$ and $b$ is a positive number, then the solution set of the inequality

$$|E| > b$$

is the union of the solution sets of the inequalities

$$E > b \text{ and } E < -b$$

**ILLUSTRATION 6.** By Theorem 1.7.3, the solution set of the inequality

$$|x + 4| > 2$$

is the set of all $x$ such that

$$x + 4 > 2 \quad \text{or} \quad x + 4 < -2$$

or, equivalently,

$$x > -2 \quad \text{or} \quad x < -6$$

Therefore, the solution set of the given inequality is

$$\{x \mid x < -6\} \cup \{x \mid x > -2\}$$

A sketch of the graph of the solution set is shown in Figure 1.7.5. ●

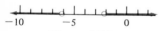

Figure 1.7.5

**EXAMPLE 5**
Find and draw a sketch of the graph of the solution set of the inequality

$$|\tfrac{2}{3}x - 5| \geq 3$$

**SOLUTION**
The solution set of the given inequality is the union of the solution sets of the inequalities

$$\begin{array}{ll} \tfrac{2}{3}x - 5 \geq 3 & \text{and} \quad \tfrac{2}{3}x - 5 \leq -3 \\ 2x - 15 \geq 9 & \phantom{\text{and}} \quad 2x - 15 \leq -9 \\ 2x \geq 24 & \phantom{\text{and}} \quad 2x \leq 6 \\ x \geq 12 & \phantom{\text{and}} \quad x \leq 3 \end{array}$$

Thus the solution set is $\{x \mid x \leq 3\} \cup \{x \mid x \geq 12\}$. A sketch of the graph of the solution set is shown in Figure 1.7.6.

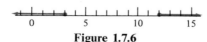

Figure 1.7.6

Theorems 1.7.2 and 1.7.3 are also valid if in each theorem the symbol $<$ is replaced by $\leq$ and $>$ is replaced by $\geq$.

**EXAMPLE 6**
Find the solution set of the inequality
$$\left|\frac{x+4}{x-3}\right| \leq 2$$

**SOLUTION**
The given inequality is equivalent to
$$-2 \leq \frac{x+4}{x-3} \leq 2$$

If we multiply by $x - 3$, we must consider two cases, depending upon whether $x - 3$ is positive or negative.

**CASE 1:** $x - 3 > 0$ or, equivalently, $x > 3$.
Then we have
$$-2(x-3) \leq x+4 \leq 2(x-3)$$
$$-2x+6 \leq x+4 \leq 2x-6$$

Thus, if $x > 3$, then also $-2x + 6 \leq x + 4$ and $x + 4 \leq 2x - 6$. We solve these two inequalities. The first inequality is
$$-2x + 6 \leq x + 4$$

Adding $-x - 6$ to both members gives
$$-3x \leq -2$$

Dividing both members by $-3$ and reversing the inequality sign, we obtain
$$x \geq \tfrac{2}{3}$$

The second inequality is
$$x + 4 \leq 2x - 6$$

Adding $-2x - 4$ to both members gives
$$-x \leq -10$$

Multiplying both members by $-1$ and reversing the inequality sign, we obtain
$$x \geq 10$$

Therefore, if $x > 3$, then the original inequality holds if and only if $x \geq \tfrac{2}{3}$ and $x \geq 10$. Because all three inequalities, $x > 3$, $x \geq \tfrac{2}{3}$, and $x \geq 10$ must be satisfied by the same values of $x$, we have as the solution set for Case 1 $\{x \mid x \geq 10\}$.

**CASE 2:** $x - 3 < 0$ or, equivalently, $x < 3$.
Thus, we have
$$-2(x - 3) \geq x + 4 \geq 2(x - 3)$$
$$-2x + 6 \geq x + 4 \geq 2x - 6$$

Considering the left inequality, we have
$$-2x + 6 \geq x + 4$$
$$-3x \geq -2$$
$$x \leq \frac{2}{3}$$

From the right inequality, we have
$$x + 4 \geq 2x - 6$$
$$-x \geq -10$$
$$x \leq 10$$

Therefore, if $x < 3$, the original inequality holds if and only if $x \leq \frac{2}{3}$ and $x \leq 10$. Because all three inequalities must be satisfied by the same values of $x$, we have as the solution set for Case 2 $\{x \mid x \leq \frac{2}{3}\}$.
 The solution set of the given inequality is the union of the solution sets of Cases 1 and 2, which is $\{x \mid x \leq \frac{2}{3}\} \cup \{x \mid x \geq 10\}$.

## EXERCISES 1.7

*In Exercises 1 through 6, find the value of the given expression if $x = -4$, $y = 2$, and $z = -3$.*

1. (a) $|2x - y|$ (b) $2|x| - |y|$ (c) $|x - 2y|$ (d) $|x| - 2|y|$
2. (a) $|x + 3y|$ (b) $|x| + 3|y|$ (c) $|x - 3y|$ (d) $|x| - 3|y|$
3. (a) $|xyz|$ (b) $|xz|y$ (c) $x|zy|$ (d) $|xy|z$ (e) $|-xyz|$
4. (a) $\left|\dfrac{xz}{y}\right|$ (b) $\left|\dfrac{x}{y}\right|z$ (c) $x\left|\dfrac{z}{y}\right|$ (d) $\dfrac{|xz|}{y}$
5. (a) $\dfrac{|x - y|}{z}$ (b) $\dfrac{|x + 2z|}{y}$ (c) $\dfrac{|x - y|}{|3y + z|}$ (d) $\dfrac{|x| - |y|}{3|y| + |z|}$
6. (a) $\dfrac{|xy|}{|x| + |y|}$ (b) $\dfrac{|x| - |y|}{|x||y|}$ (c) $\dfrac{|x| + |y| + |z|}{x + y + z}$

In Exercises 7 through 18, find the solution set of the given equation.

7. $|x - 5| = 4$
8. $|2x + 3| = 7$
9. $|3x - 8| = 4$
10. $|4x - 9| = 11$
11. $|4x + 5| = 15$
12. $|8 - x| = 4$
13. $|7 - 2x| = 9$
14. $|3x - 2| = |2x + 3|$
15. $|x - 4| = |5 - 2x|$
16. $|5x| = 6 - x$
17. $\left|\dfrac{x + 3}{x - 3}\right| = 7$
18. $\left|\dfrac{2x + 1}{x - 1}\right| = 3$

In Exercises 19 through 33, find the solution set of the given inequality. Draw a sketch of the graph of the solution set.

19. $|x| \le 5$
20. $|x| > 6$
21. $|x - 1| > 7$
22. $|x + 1| < 5$
23. $|x - 5| \le 3$
24. $|3x - 4| \le 2$
25. $|2x - 7| < 9$
26. $|3x - 4| \ge 2$
27. $|2x - 7| > 9$
28. $6 < |4x + 7|$
29. $4 < |3x + 12|$
30. $|3x - 5| \ge 10$
31. $\left|\dfrac{x - 5}{x + 3}\right| \le 7$
32. $\left|\dfrac{x + 2}{2x - 3}\right| \ge 4$
33. $\left|\dfrac{6 - 5x}{3 + x}\right| > \dfrac{1}{2}$
34. $\left|\dfrac{x + 1}{x - 1}\right| < 4$

## 1.8 Points in a Plane, Distance Formula, and Slope of a Line

We now consider "ordered pairs" of real numbers. Any two real numbers form a *pair*, and when an order is designated for the pair of real numbers we call it an *ordered pair of real numbers*. If the first real number is represented by $x$ and the second real number is represented by $y$, then this ordered pair of real numbers is represented by writing $x$ and $y$ in parentheses with a comma separating them as $(x, y)$. For the ordered pair $(x, y)$, the number $x$ is called the *first component* and the number $y$ is called the *second component*. The ordered pair $(3, 7)$ having 3 as the first component and 7 as the second component is different from the ordered pair $(7, 3)$ for which 7 is the first component and 3 is the second component.

A set of ordered pairs is obtained from two given sets by considering the "Cartesian product" of the two sets.

**1.8.1 DEFINITION** If $S$ and $T$ are two given sets, then the *Cartesian product* of $S$ and $T$, denoted by $S \times T$ (read "$S$ cross $T$"), is the set of all possible ordered pairs for which the first component is an element of $S$ and the second component is an element of $T$. Symbolically, we write

$$S \times T = \{(x, y) \mid x \in S \text{ and } y \in T\}$$

**ILLUSTRATION 1.** If $S = \{5, -1\}$ and $T = \{1, 2, 3\}$, then

$$S \times T = \{(5, 1), (5, 2), (5, 3), (-1, 1), (-1, 2), (-1, 3)\} \quad \bullet$$

In this book we are concerned with the Cartesian product of the set $R$ of real numbers with itself, that is, $R \times R$, and subsets of $R \times R$.

$$R \times R = \{(x,y) \mid x \in R \text{ and } y \in R\}$$

The set $R \times R$ is an infinite set of ordered pairs of real numbers.

Just as the set $R$ of real numbers is represented geometrically by a line (the real number line), the set $R \times R$ is represented geometrically by a plane called the *real plane*. The method we use with $R \times R$ is the one attributed to the French mathematician René Descartes (1596–1650), who is credited with the invention of analytic geometry in 1637.

We select a horizontal line in the plane. This line, extending indefinitely to the left and to the right is called the $x$ *axis*. A vertical line is chosen, extending indefinitely up and down; it is called the $y$ *axis*. The point of intersection of the $x$ axis and the $y$ axis is called the *origin* and is denoted by the letter $O$. We establish the positive direction on the $x$ axis to the right of the origin and the positive direction on the $y$ axis above the origin. See Figure 1.8.1.

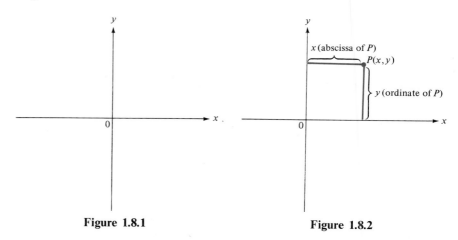

**Figure 1.8.1**   **Figure 1.8.2**

We now associate an ordered pair of real numbers $(x, y)$ with a point $P$ in the real plane. Refer to Figure 1.8.2. The distance of $P$ from the $y$ axis (considered as positive if $P$ is to the right of the $y$ axis and negative if $P$ is to the left of the $y$ axis) is called the *abscissa* (or $x$ *coordinate*) of $P$ and is denoted by $x$. The distance of $P$ from the $x$ axis (considered as positive if $P$ is above the $x$ axis and negative if $P$ is below the $x$ axis) is called the *ordinate* (or $y$ *coordinate*) of $P$ and is denoted by $y$. The abscissa and the ordinate of a point are called the *rectangular Cartesian coordinates* of the point. There is a one-to-one correspondence between the points in the real plane and the rectangular Cartesian coordinates. That is, with each point there corresponds a unique ordered pair $(x, y)$ and with each ordered pair $(x, y)$ there is associated only one point. This one-to-one correspondence is called a *rectangular Cartesian coordinate system.*

When we use the terminology "the point $(x, y)$," we mean the point having abscissa $x$ and ordinate $y$. When we use the notation $P(x, y)$, we mean the point $P$ whose coordinates are $x$ and $y$. To locate on a rectangular Cartesian coordinate system a point $(x, y)$ is to *plot* the point, and the point is represented by a dot (the dot is not the point, but it is used to visualize the point). The point associated with a given ordered pair $(x, y)$ is called the graph of the ordered pair.

EXAMPLE 1
Plot the points $(-4, 0)$, $(-3, -7)$, $(-1, 5)$, $(0, -6)$, $(2, 8)$, $(3, 0)$, $(5, -3)$, and $(8, 4)$.

SOLUTION
Figure 1.8.3 shows a rectangular Cartesian coordinate system with the given points plotted.

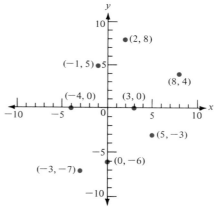

**Figure 1.8.3**

The $x$ and $y$ axes are called the *coordinate axes,* and they divide the plane into four parts, called *quadrants* (the quadrants do not include the coordinate axes). The first quadrant is the one in which the abscissa and the ordinate are both positive, that is, the upper right quadrant. The other quadrants are numbered in the counterclockwise direction, with the fourth, for example, being the lower right quadrant. See Figure 1.8.4.

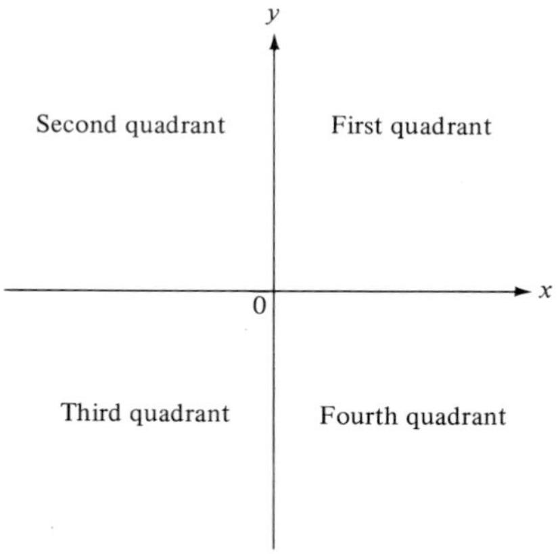

Suppose that $A$ and $B$ denote two points having the same ordinate but different abscissas, and let $A$ be the point $(x_1, y)$ and $B$ be the point $(x_2, y)$. The subscripts with the letter $x$ are used to indicate that $x_1$ and $x_2$ are different values; hence $(x_1, y)$ and $(x_2, y)$ are two ordered pairs having the same second component and different first components. Then the *directed distance* from $A$ to $B$ is denoted by $\overline{AB}$, and we define

$$\overline{AB} = x_2 - x_1$$

**Figure 1.8.4**

The directed distance from $B$ to $A$ is denoted by $\overline{BA}$, and

$$\overline{BA} = x_1 - x_2$$

**ILLUSTRATION 2**
(a) Figure 1.8.5 shows points $A(3, 4)$ and $B(8, 4)$, where

$$\overline{AB} = 8 - 3 \qquad \overline{BA} = 3 - 8$$
$$\phantom{\overline{AB}} = 5 \qquad \phantom{\overline{BA}} = -5$$

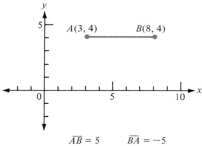

$\overline{AB} = 5 \qquad \overline{BA} = -5$
**Figure 1.8.5**

(b) Figure 1.8.6 shows points $A(-5, -2)$ and $B(6, -2)$, where

$$\overline{AB} = 6 - (-5) \qquad \overline{BA} = -5 - 6$$
$$\phantom{\overline{AB}} = 11 \qquad \phantom{\overline{BA}} = -11$$

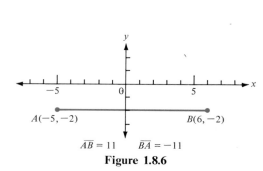

$\overline{AB} = 11 \quad \overline{BA} = -11$ $\qquad\qquad$ $\overline{AB} = -6 \quad \overline{BA} = 6$
**Figure 1.8.6** $\qquad\qquad\qquad\qquad$ **Figure 1.8.7**

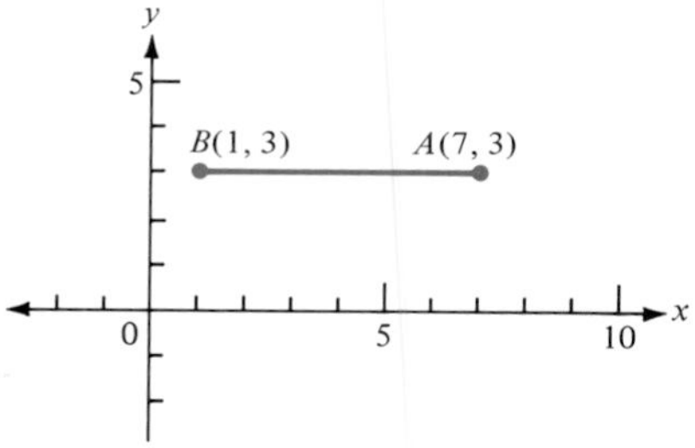

(c) Figure 1.8.7 shows points $A(7, 3)$ and $B(1, 3)$, where

$$\overline{AB} = 1 - 7 \qquad \overline{BA} = 7 - 1$$
$$\phantom{\overline{AB}} = -6 \qquad \phantom{\overline{BA}} = 6$$

In Illustration 2 we see that the number $\overline{AB}$ is positive if $B$ is to the right of $A$ and $\overline{AB}$ is negative if $B$ is to the left of $A$.

If $C$ is the point $(x, y_1)$ and $D$ is the point $(x, y_2)$, then the *directed distance* from $C$ to $D$, denoted by $\overline{CD}$, is defined by

$$\overline{CD} = y_2 - y_1$$

The directed distance from $D$ to $C$ is denoted by $\overline{DC}$, and

$$\overline{DC} = y_1 - y_2$$

**ILLUSTRATION 3**
(a) Figure 1.8.8 shows points $C(-2, 3)$ and $D(-2, -5)$, where

$$\begin{aligned} \overline{CD} &= -5 - 3 & \overline{DC} &= 3 - (-5) \\ &= -8 & &= 8 \end{aligned}$$

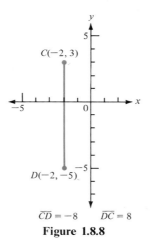
$\overline{CD} = -8 \quad \overline{DC} = 8$
**Figure 1.8.8**

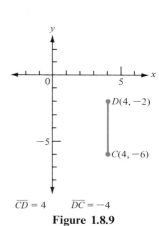
$\overline{CD} = 4 \quad \overline{DC} = -4$
**Figure 1.8.9**

(b) Figure 1.8.9 shows points $C(4, -6)$ and $D(4, -2)$, where

$$\begin{aligned} \overline{CD} &= -2 - (-6) & \overline{DC} &= -6 - (-2) \\ &= 4 & &= -4 \end{aligned}$$

In Illustration 3 note that the number $\overline{CD}$ is positive if $D$ is above $C$ and $\overline{CD}$ is negative if $D$ is below $C$.

Observe that the terminology "directed distance" indicates both a distance and a direction (positive or negative). If we are concerned only with the length of the line segment between two points $P_1$ and $P_2$ (that is, the distance between the points $P_1$ and $P_2$ without regard to direction), then we use the terminology "undirected distance." We denote the *undirected distance* from $P_1$ to $P_2$ by $|P_1P_2|$, which is a nonnegative number.

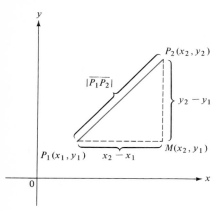

**Figure 1.8.10**

We now wish to obtain a formula for computing $\overline{P_1P_2}$ if $P_1(x_1, y_1)$ and $P_2(x_2, y_2)$ are any two points in the plane. We use the Pythagorean theorem from plane geometry, which is as follows.

*In a right triangle, the sum of the squares of the lengths of the perpendicular sides is equal to the square of the length of the hypotenuse.*

Figure 1.8.10 shows $P_1$ and $P_2$ in the first quadrant and the point $M(x_2, y_1)$. Note that $\overline{|P_1P_2|}$ is the length of the hypotenuse of right triangle $P_1MP_2$. Using the Pythagorean theorem, we have

$$\overline{|P_1P_2|}^2 = \overline{|P_1M|}^2 + \overline{|MP_2|}^2$$

Hence

$$\overline{|P_1P_2|} = \sqrt{\overline{|P_1M|}^2 + \overline{|MP_2|}^2}$$

or, equivalently,

$$\overline{|P_1P_2|} = \sqrt{(x_2 - x_1)^2 + (y_2 - y_1)^2} \qquad (1)$$

Note that in formula (1) we do not have a $\pm$ symbol in front of the radical in the right member because $\overline{|P_1P_2|}$ is a nonnegative number. Formula (1) holds for all possible positions of $P_1$ and $P_2$ in all four quadrants. The length of the hypotenuse is always $\overline{|P_1P_2|}$, and the lengths of the legs are always $\overline{|P_1M|}$ and $\overline{|MP_2|}$. The result is stated as a theorem.

**1.8.2 THEOREM** The undirected distance between two points, $P_1(x_1, y_1)$ and $P_2(x_2, y_2)$, is given by

$$\overline{|P_1P_2|} = \sqrt{(x_2 - x_1)^2 + (y_2 - y_1)^2}$$

Observe that if $P_1$ and $P_2$ are on the same horizontal line, then $y_1 = y_2$ and

$$\overline{|P_1P_2|} = \sqrt{(x_2 - x_1)^2 + 0^2}$$

or, equivalently (because $\sqrt{a^2} = |a|$),

$$\overline{|P_1P_2|} = |x_2 - x_1|$$

Furthermore, if $P_1$ and $P_2$ are on the same vertical line, then $x_1 = x_2$ and

$$\overline{|P_1P_2|} = \sqrt{0^2 + (y_2 - y_1)^2}$$

or, equivalently,

$$\overline{|P_1P_2|} = |y_2 - y_1|$$

## EXAMPLE 2
Find the lengths of the sides of the triangle having vertices at $A(2, -1)$, $B(-2, -4)$, and $C(5, 3)$. Show that the triangle is isosceles.

### SOLUTION
The triangle is shown in Figure 1.8.11.

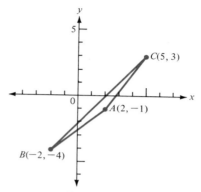

**Figure 1.8.11**

Using the formula of Theorem 1.8.2, we have

$$|\overline{AB}| = \sqrt{(-2-2)^2 + (-4+1)^2}$$
$$= \sqrt{16+9}$$
$$= 5$$

$$|\overline{AC}| = \sqrt{(5-2)^2 + (3+1)^2}$$
$$= \sqrt{9+16}$$
$$= 5$$

$$|\overline{BC}| = \sqrt{(5+2)^2 + (3+4)^2}$$
$$= \sqrt{49+49}$$
$$= 7\sqrt{2}$$

An isosceles triangle is one for which two of the sides have equal lengths. Because $|\overline{AB}| = |\overline{AC}|$, we conclude that triangle $ABC$ is isosceles.

Suppose that $L$ is a nonvertical line and $P_1(x_1, y_1)$ and $P_2(x_2, y_2)$ are any two distinct points on $L$. Figure 1.8.12 shows such a line. $R$ is the point $(x_2, y_1)$ and the points $P_1$, $P_2$, and $R$ are vertices of a right triangle; furthermore, $\overline{P_1R} = x_2 - x_1$ and $\overline{RP_2} = y_2 - y_1$. The number $y_2 - y_1$ is called the *rise* from $P_1$ to $P_2$; it gives the measure of the change in the ordinate from $P_1$ to $P_2$, and it may be positive, negative, or zero. The number $x_2 - x_1$ is called the *run* from $P_1$ to $P_2$; it gives the measure of the change in the abscissa from $P_1$ to $P_2$, and it may be positive or negative. The run may not be zero because $x_2 \neq x_1$, since the line $L$ is not vertical. For all choices of the points $P_1$ and $P_2$ on $L$, the quotient

$$\frac{\text{rise from } P_1 \text{ to } P_2}{\text{run from } P_1 \text{ to } P_2}$$

is constant, and this quotient is called the "slope" of the line. Following is the formal definition.

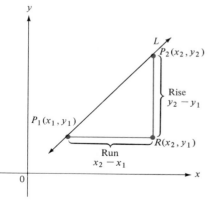

**Figure 1.8.12**

**1.8.3 DEFINITION**  If $P_1(x_1, y_1)$ and $P_2(x_2, y_2)$ are any two distinct points on $L$, which is not parallel to the $y$ axis, then the *slope* of $L$, denoted by $m$, is given by

$$m = \frac{y_2 - y_1}{x_2 - x_1} \qquad (2)$$

**ILLUSTRATION 4.** Let $A$ be the point $(-5, 4)$ and $B$ be the point $(3, -6)$. See Figure 1.8.13. To find the slope of the line through $A$ and $B$, we can use formula (2) with $A$ as $P_1$ and $B$ as $P_2$, and we have

$$m = \frac{-6 - 4}{3 - (-5)}$$

$$= \frac{-10}{8}$$

$$= -\frac{5}{4}$$

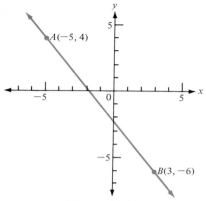

Figure 1.8.13

If we use formula (2) with $B$ as $P_1$ and $A$ as $P_2$, we have

$$m = \frac{4 - (-6)}{-5 - 3}$$

$$= \frac{10}{-8}$$

$$= -\frac{5}{4}$$

When using formula (2), it makes no difference which point is taken as $P_1$ and which point is taken as $P_2$ because

$$\frac{y_2 - y_1}{x_2 - x_1} = \frac{y_1 - y_2}{x_1 - x_2}$$

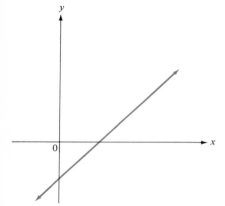

Figure 1.8.14

Thus we can assume that the points $P_1$ and $P_2$ are chosen so that $x_1 < x_2$, that is $x_2 - x_1 > 0$. Hence if $y_2 - y_1 > 0$, the slope $m$ (computed from formula (1)) is positive, and if $y_2 - y_1 < 0$, the slope $m$ is negative. A line having a positive slope is shown in Figure 1.8.14, and a line having a negative slope is shown in Figure 1.8.15.

If a line is parallel to the $x$ axis, then $y_2 = y_1$, and thus $m = 0$. If a line is parallel to the $y$ axis, $x_2 = x_1$; hence formula (2) is meaningless because we cannot divide by zero. This is the reason that lines parallel to the $y$ axis, or vertical lines, are excluded in Definition 1.8.3. We say that a vertical line does not have a slope.

To show that the value of $m$ computed from formula (2) is independent of

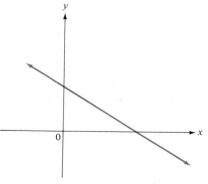

Figure 1.8.15

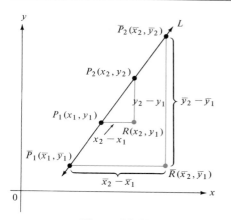

**Figure 1.8.16**

the choice of the two points $P_1$ and $P_2$, we choose two other points $\bar{P}_1(\bar{x}_1, \bar{y}_1)$ and $\bar{P}_2(\bar{x}_2, \bar{y}_2)$ and compute a number $\bar{m}$ from formula (2).

$$\bar{m} = \frac{\bar{y}_2 - \bar{y}_1}{\bar{x}_2 - \bar{x}_1}$$

Refer to Figure 1.8.16. Triangles $\bar{P}_1 \bar{R} \bar{P}_2$ and $P_1 R P_2$ are similar, and hence from a theorem of plane geometry it follows that

$$\frac{\bar{y}_2 - \bar{y}_1}{\bar{x}_2 - \bar{x}_1} = \frac{y_2 - y_1}{x_2 - x_1}$$

or, equivalently,

$$\bar{m} = m$$

Thus the value of $m$ computed from formula (2) is the same number, no matter what two points on $L$ are selected.

If we multiply on both sides of equation (2) by $x_2 - x_1$, we obtain

$$y_2 - y_1 = m(x_2 - x_1) \qquad (3)$$

It follows from equation (3) that if we consider a particle moving along a line $L$, the change in the ordinate of the particle is equal to the product of the slope and the change in the abscissa.

**ILLUSTRATION 5.** If $L$ is the line through the points $A(-1, 2)$ and $B(1, 8)$, and $m$ is the slope of $L$, then

$$\begin{aligned} m &= \frac{8 - 2}{1 - (-1)} \\ &= \frac{6}{2} \\ &= 3 \end{aligned}$$

Hence if a particle is moving along the line $L$, the change in the ordinate is three times the change in the abscissa. Refer to Figure 1.8.17. If a particle is at $B(1, 8)$ and the abscissa is increased by one unit, then the ordinate is increased by three units, and the particle is at the point $C(2, 11)$. Similarly, if the particle is at $A(-1, 2)$ and the abscissa is decreased by two units, then the ordinate is decreased by six units, and the particle is at the point $D(-3, -4)$. ●

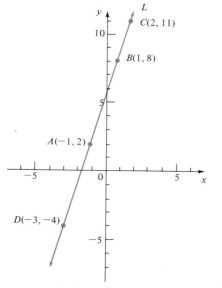

**Figure 1.8.17**

It is proved in analytic geometry that two distinct lines are parallel if and only if they have the same slope.

**ILLUSTRATION 6.** Let $L_1$ be the line through the points $A(1, 2)$ and $B(3, -6)$, and $m_1$ be the slope of $L_1$; and let $L_2$ be the line through the points $C(2, -5)$ and $D(-1, 7)$, and $m_2$ be the slope of $L_2$. Then

$$m_1 = \frac{-6 - 2}{3 - 1} \qquad m_2 = \frac{7 - (-5)}{-1 - 2}$$

$$= \frac{-8}{2} \qquad\qquad = \frac{12}{-3}$$

$$= -4 \qquad\qquad = -4$$

Because $m_1 = m_2$, it follows that $L_1$ and $L_2$ are parallel. See Figure 1.8.18. ●

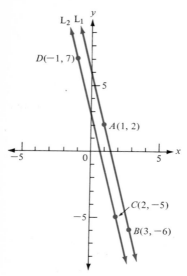

**Figure 1.8.18**

Any two distinct points determine a line. Three distinct points may or may not lie on the same line. If three or more points lie on the same line, they are said to be *collinear*. Hence three points $A$, $B$, and $C$ are collinear if and only if the line through the points $A$ and $B$ is the same as the line through the points $B$ and $C$. Because the line through $A$ and $B$ and the line through $B$ and $C$ both contain the point $B$, they are the same line if and only if their slopes are equal.

**EXAMPLE 3**
Prove that the points $A(-3, -4)$, $B(2, -1)$, and $C(7, 2)$ are collinear.

**SOLUTION**
If $m_1$ is the slope of the line through $A$ and $B$, and $m_2$ is the slope of the line through $B$ and $C$, then

$$m_1 = \frac{-1 - (-4)}{2 - (-3)} \qquad m_2 = \frac{2 - (-1)}{7 - 2}$$

$$= \frac{3}{5} \qquad\qquad\qquad = \frac{3}{5}$$

Hence $m_1 = m_2$. Therefore, the line through $A$ and $B$ and the line through $B$ and $C$ have the same slope and contain the common point $B$. Thus they are the same line, and therefore $A$, $B$, and $C$ are collinear.

Another theorem from analytic geometry states that two lines $L_1$ and $L_2$, neither of which is vertical, are perpendicular if and only if the product of their slopes is $-1$. That is, if $m_1$ is the slope of $L_1$ and $m_2$ is the slope of $L_2$, then $L_1$ and $L_2$ are perpendicular if and only if

$$m_1 m_2 = -1$$

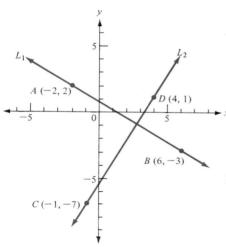

**Figure 1.8.19**

**ILLUSTRATION 7.** Refer to Figure 1.8.19. Let $L_1$ be the line through $A(-2, 2)$ and $B(6, -3)$, and $m_1$ be the slope of $L_1$. Let $L_2$ be the line through $C(-1, -7)$ and $D(4, 1)$, and $m_2$ be the slope of $L_2$. Then

$$m_1 = \frac{-3-2}{6-(-2)} \qquad m_2 = \frac{1-(-7)}{4-(-1)}$$

$$= \frac{-5}{8} \qquad\qquad = \frac{8}{5}$$

$$= -\frac{5}{8}$$

Because

$$m_1 m_2 = \left(-\frac{5}{8}\right)\frac{8}{5}$$

$$= -1$$

it follows that $L_1$ and $L_2$ are perpendicular. ●

**EXAMPLE 4**

Prove by means of slopes that the four points $A(4, 2)$, $B(2, 6)$, $C(6, 8)$, and $D(8, 4)$ are the vertices of a rectangle.

**SOLUTION**

Refer to Figure 1.8.20. Let $m_1$ be the slope of the line through $A$ and $B$, $m_2$ be the slope of the line through $A$ and $D$, $m_3$ be the slope of the line through $B$ and $C$, and $m_4$ be the slope of the line through $D$ and $C$. Then

$$m_1 = \frac{6-2}{2-4} \qquad m_2 = \frac{4-2}{8-4} \qquad m_3 = \frac{8-6}{6-2} \qquad m_4 = \frac{8-4}{6-8}$$

$$= \frac{4}{-2} \qquad\qquad = \frac{2}{4} \qquad\qquad = \frac{2}{4} \qquad\qquad = \frac{4}{-2}$$

$$= -2 \qquad\qquad = \frac{1}{2} \qquad\qquad = \frac{1}{2} \qquad\qquad = -2$$

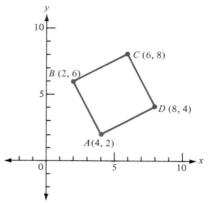

**Figure 1.8.20**

Because $m_1 = m_4$, it follows that the line through $A$ and $B$ is parallel to the line through $D$ and $C$. Because $m_2 = m_3$, it follows that the line through $A$ and $D$ is parallel to the line through $B$ and $C$. Because $m_1 m_2 = -1$, we conclude that the line through $A$ and $B$ is perpendicular to the line through $A$ and $D$. Therefore, the quadrilateral $ABCD$ has its opposite sides parallel, and a pair of adjacent sides are perpendicular. Thus the quadrilateral is a rectangle.

## EXERCISES 1.8

In Exercises 1 through 8, draw the graph of the given set of points.

1. $\{(-5, -3), (0, 3), (5, 0), (5, 3)\}$
2. $\{(-2, 4), (-4, 2), (2, -4), (4, -2)\}$
3. $\{(-6, -6), (0, 0), (1, 6), (6, -6)\}$
4. $\{(-5, 0), (0, -3), (6, -1), (8, 3)\}$
5. $\{(-9, 4), (-4, 0), (0, 9), (9, -4)\}$
6. $\{(-3, -3), (0, 0), (3, 5), (5, -3)\}$
7. $\{(-1, 1), (2, 2), (4, 0), (5, -1), (7, 7)\}$
8. $\{(-10, -5), (-5, -10), (0, -5), (4, 10), (10, 5)\}$

9. For the points $A(-1, 7)$ and $B(6, 7)$, find (a) $\overline{AB}$; (b) $\overline{BA}$.
10. For the points $A(-2, 3)$ and $B(-4, 3)$, find (a) $\overline{AB}$; (b) $\overline{BA}$.
11. For the points $A(3, -4)$ and $B(3, -8)$, find (a) $\overline{AB}$; (b) $\overline{BA}$.
12. For the points $A(-4, -5)$ and $B(-4, 6)$, find (a) $\overline{AB}$; (b) $\overline{BA}$.
13. If $A$ is the point $(-2, 3)$ and $B$ is the point $(x, 3)$, find $x$ such that (a) $\overline{AB} = -8$; (b) $\overline{BA} = -8$.
14. If $A$ is the point $(-4, y)$ and $B$ is the point $(-4, 3)$, find $y$ such that (a) $\overline{AB} = -3$; (b) $\overline{BA} = -3$.

In Exercises 15 through 22, draw a sketch of the line segment through the two given points and find the length of the line segment.

15. $(1, 3), (-2, 7)$
16. $(-4, -1), (4, 5)$
17. $(8, 5), (3, -7)$
18. $(6, -5), (2, -2)$
19. $(-1, -3), (-4, 0)$
20. $(0, -2), (2, 0)$
21. $(-4, 7), (1, -3)$
22. $(-3, -4), (-4, 3)$

In Exercises 23 and 24, find the lengths of the sides of the triangle having vertices at the three given points. Draw a sketch of the triangle.

23. $A(4, -5), B(-2, 3), C(-1, 7)$
24. $A(2, 3), B(3, -3), C(-1, -1)$

In Exercises 25 through 30, draw a sketch of the line through the two given points and find the slope of the line.

25. $(3, 2), (5, 8)$
26. $(-4, -7), (-1, 2)$
27. $(6, -2), (2, 1)$
28. $(5, -3), (3, 7)$
29. $(8, -2), (-7, -2)$
30. $(0, -3), (-3, 0)$

31. By showing that two sides have the same length, prove that the three points $A(-8, 1)$, $B(-1, -6)$, and $C(2, 4)$ are the vertices of an isosceles triangle. Draw a sketch of the triangle.

32. By showing that the three sides have the same length, prove that the three points $A(-5, 0)$, $B(3, 0)$, and $C(-1, 4\sqrt{3})$ are the vertices of an equilateral triangle. Draw a sketch of the triangle.

33. By showing that each pair of opposite sides have the same length, prove that the four points $A(-2, -3)$, $B(1, 6)$, $C(3, -4)$, and $D(6, 5)$ are the vertices of a parallelogram. Draw a sketch of the parallelogram.
34. Use the converse of the Pythagorean theorem to prove that the points $A(-3, -4)$, $B(-1, 6)$, and $C(3, 2)$ are the vertices of a right triangle. Draw a sketch of the triangle.
35. By showing that opposite sides are parallel, prove that the four points $A(0, 0)$, $B(-2, 1)$, $C(3, 4)$, and $D(5, 3)$ are the vertices of a parallelogram. Draw a sketch of the parallelogram.
36. Prove that the four points of Exercise 33 are the vertices of a parallelogram by showing that opposite sides are parallel.
37. By showing that two opposite sides are parallel, prove that the four points $A(-1, -3)$, $B(8, 3)$, $C(3, 4)$, and $D(0, 2)$ are the vertices of a trapezoid. Draw a sketch of the trapezoid.
38. Prove that the points $A(-13, 6)$, $B(-5, 21)$, $C(2, -2)$, and $D(10, 13)$ are the vertices of a square. Draw a sketch of the square.
39. Prove that the triangle with vertices at $A(1, 1)$, $B(2, 3)$, and $C(5, -1)$ is a right triangle by two methods: (a) use the converse of the Pythagorean theorem; (b) show that two sides are perpendicular by using slopes. Draw a sketch of the right triangle.
40. Prove that the three points $A(-2, -3)$, $B(-1, 0)$, and $C(1, 6)$ are collinear by two methods: (a) use slopes; (b) use the distance formula.
41. Find the ordinate of the point whose abscissa is $-3$ and which is collinear with the points $(3, 2)$ and $(0, 5)$.
42. Find the ordinate of the point whose abscissa is 4 and for which the line through it and the point $(-2, 5)$ is perpendicular to the line through the points $(8, -4)$ and $(-1, 2)$.

## 1.9 First-degree Equations in Two Variables

If $S$ is a subset of $R \times R$, then $S$ is a set of ordered pairs of real numbers. We now define what we mean by the "graph" of such a set.

### 1.9.1 DEFINITION

Let $S$ be a subset of $R \times R$. Then the *graph* of $S$ is the set of all points $(x, y)$ in a rectangular Cartesian coordinate system for which $(x, y)$ is an ordered pair in $S$.

**ILLUSTRATION 1.** If $S = \{(-5, -4), (0, -2), (3, 0), (5, 6)\}$, the graph of $S$ consists of four points, as shown in Figure 1.9.1.

An equation in two variables, $x$ and $y$, defines a set of ordered pairs. For example, consider the equation

$$4x + 3y = 12 \tag{1}$$

for which the replacement set of $x$ is $R$ and the replacement set of $y$ is $R$. An ordered pair of numbers $(u, v)$ is said to *satisfy* equation (1) if, when $u$ is substituted for $x$ and $v$ is substituted for $y$, the resulting statement is true. An ordered pair that satisfies an equation is called a *solution* of the equation, and the set of all such ordered pairs is the *solution set* of the equation.

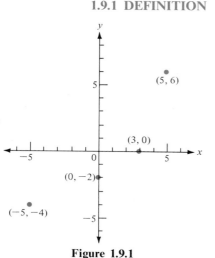

**Figure 1.9.1**

The solution set of equation (1) is an infinite set. Some of the ordered pairs in this solution set are $(0, 4)$, $(1, \frac{8}{3})$, $(2, \frac{4}{3})$, $(3, 0)$, $(4, -\frac{4}{3})$, $(-1, \frac{16}{3})$, $(-2, \frac{20}{3})$, and $(-3, 8)$. These solutions can be verified by substituting into equation (1). For instance, $(4, -\frac{4}{3})$ is a solution because

$$4 \cdot 4 + 3 \cdot \left(-\frac{4}{3}\right) = 12$$

When discussing equations in one variable, we stated (Definition 1.4.3) that equations which have the same solution set are called *equivalent equations*. This definition applies to equations in two or more variables, and as with equations in one variable, an equation equivalent to a given equation can be obtained by applying definitions, axioms, and theorems involving operations of real numbers.

The solution set of an equation in two variables can be described using set notation and the given equation. For example, the solution set of equation (1) can be expressed as

$$\{(x, y) \mid 4x + 3y = 12\}$$

However, sometimes when using set notation the given equation is replaced by an equivalent equation in which one of the variables is expressed in terms of the other. Equation (1) is equivalent to

$$y = 4 - \frac{4}{3}x \qquad (2)$$

Thus the solution set of equation (1) can also be expressed as

$$\left\{(x, y) \mid y = 4 - \frac{4}{3}x\right\}$$

**1.9.2 DEFINITION** The *graph of an equation* in two variables is the graph of its solution set.

The graph of an equation in two variables is also called a *curve*.

Because the solution set of equation (1) is an infinite set, its graph consists of infinitely many points. We can determine some of the points by substituting into the equation values for one of the variables and computing the corresponding values of the other variable. In Table 1.9.1, we list these representative solutions, which are obtained by substituting values of $x$ into equivalent equation (2).

Table 1.9.1

| $x$ | $-4$ | $-3$ | $-2$ | $-1$ | $0$ | $1$ | $2$ | $3$ | $4$ |
|---|---|---|---|---|---|---|---|---|---|
| $y$ | $\frac{28}{3}$ | $8$ | $\frac{20}{3}$ | $\frac{16}{3}$ | $4$ | $\frac{8}{3}$ | $\frac{4}{3}$ | $0$ | $-\frac{4}{3}$ |

The points $(-4, \frac{28}{3})$, $(-3, 8)$, $(-2, \frac{20}{3})$, $(-1, \frac{16}{3})$, $(0, 4)$, $(1, \frac{8}{3})$, $(2, \frac{4}{3})$ $(3, 0)$, and $(4, -\frac{4}{3})$ are plotted in Figure 1.9.2. These points appear to lie on a straight line. In fact, it can be shown that every solution of equation (1) corresponds to a point on the line, and, conversely, the coordinates of each point on the line satisfy equation (1). Thus the line is the graph of equation (1). It is impossible to draw the complete graph because $x$ and $y$ both have values which are numerically as large as desired. However, we can draw what we call a *sketch of the graph,* as shown in Figure 1.9.3. This sketch is understood to be only an approximation of the complete graph; that is, the line does not terminate as shown but continues on in both directions. This continuation is indicated by the arrows at the extremities of the portion of the line shown.

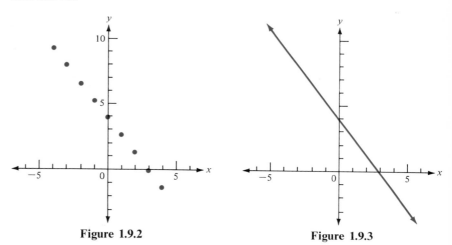

**Figure 1.9.2**  **Figure 1.9.3**

Consider the general first-degree equation in two variables $x$ and $y$, which is
$$Ax + By + C = 0 \qquad (3)$$
where $A$, $B$, and $C$ are constants and not both $A$ and $B$ are zero. Equation (1) is equivalent to the equation
$$4x + 3y - 12 = 0$$
which is in the form of equation (3), where $A = 4$, $B = 3$, and $C = -12$. Later in this section (Theorem 1.9.4) we prove that any equation equivalent to one of the form of equation (3) has a graph that is a straight line. Therefore, such an equation is called a *linear equation.*

Because a line is determined by any two distinct points on the line, it is only necessary to find two solutions of a linear equation in order to determine its graph. Generally, the solution for which the first component is zero and the solution for which the second component is zero are the easiest solutions to obtain.

### EXAMPLE 1
Draw a sketch of the graph of the equation
$$5x - 2y = 10$$

### SOLUTION
Because the equation is a first-degree equation, it is a linear equation and its graph is a straight line. When $x$ is 0, we obtain $-2y = 10$ or, equivalently, $y = -5$. Thus the point $(0, -5)$ is on the line. When $y$ is 0, we obtain $5x = 10$ or, equivalently, $x = 2$. Hence the point $(2, 0)$ is on the line. We plot these two points and draw the line through them as shown in Figure 1.9.4.

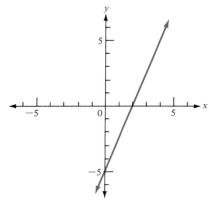

**Figure 1.9.4**

The abscissa ($x$ coordinate) of the point at which a line intersects the $x$ axis is called the $x$ *intercept* of the line and the ordinate ($y$ coordinate) of the point at which a line intersects the $y$ axis is called the $y$ *intercept* of the line. The $x$ and $y$ intercepts of a line are usually denoted by $a$ and $b$, respectively. For the line of Example 1, $a$ is 2 and $b$ is $-5$.

### EXAMPLE 2
Draw a sketch of the graph of each of the following equations.

(a) $3x - 8y + 9 = 0$
(b) $4x + 5y = 0$

### SOLUTION
Each of the equations is a linear equation and therefore each graph is a straight line.

(a) For the line having the equation $3x - 8y + 9 = 0$, the $x$ intercept is $-3$ and the $y$ intercept is $\tfrac{9}{8}$. Therefore, we plot the points $(-3, 0)$ and $(0, \tfrac{9}{8})$ and draw the line through them. See Figure 1.9.5.
(b) For the line having the equation $4x + 5y = 0$, both the $x$ intercept and the $y$ intercept are zero. Thus the line contains the origin. We need another point on the line; for instance, when $x$ is 5, $y$ is $-4$. We plot the point $(5, -4)$ and draw the line through this point and the origin as shown in Figure 1.9.6.

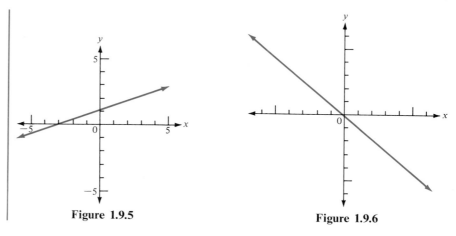

Figure 1.9.5

Figure 1.9.6

The next example gives two special cases of linear equations in two variables. In each case the coefficient of one of the variables is zero.

EXAMPLE 3
Draw a sketch of the graph of each of the following sets.
(a) $\{(x,y) \mid x = -6\}$
(b) $\{(x,y) \mid y = 4\}$

SOLUTION
For each set, the graph is given by the solution set of the equation, and each of the equations is a linear equation.
(a) For the equation $x = -6$, the coefficient of $y$ is 0; that is, the equation can be written in the form $x + 0y = -6$. Therefore, any ordered pair of the form $(-6, y)$ is a solution of the equation. For instance, the ordered pairs $(-6, -4)$, $(-6, -2)$, $(-6, 0)$, $(-6, 1)$, $(-6, 5)$, and $(-6, 10)$ are solutions of the equation. If we plot the graph of two of these ordered pairs and draw a line through the two points, we have the graph of the given set. It is a line parallel to the $y$ axis and six units to the left of the $y$ axis. See Figure 1.9.7.

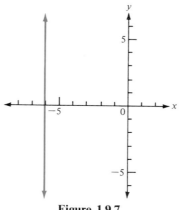

Figure 1.9.7

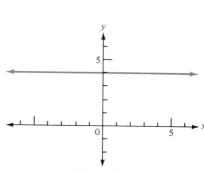

Figure 1.9.8

(b) The equation $y = 4$ can be written in the form $0x + y = 4$. Thus any ordered pair of the form $(x, 4)$ is a solution of the equation. Two such ordered pairs are $(-3, 4)$ and $(3, 4)$, and if we plot the corresponding two points and draw a line through them, we have the graph of the given set. The graph of the given set is a line parallel to the $x$ axis and four units above the $x$ axis; it is shown in Figure 1.9.8.

We have discussed the graph of a first-degree equation in two variables. We now define what is meant by an "equation of a graph."

**1.9.3 DEFINITION** An *equation of a graph* is an equation that is satisfied by the coordinates of those, and only those, points on the graph.

From Definition 1.9.3 it follows that an equation of a graph has the following properties.

1. If a point $P$ is on the graph, then its coordinates satisfy the equation.
2. If a point $P$ is not on the graph, then its coordinates do not satisfy the equation.

We apply Definition 1.9.3 to find an equation of the line through two given points $P_1(x_1, y_1)$ and $P_2(x_2, y_2)$. Let the coordinates of any point $P$ on the line be represented by $(x, y)$. Then we want an equation that is satisfied by $x$ and $y$ if and only if $P(x, y)$ is on the line through $P_1(x_1, y_1)$ and $P_2(x_2, y_2)$. We consider two cases, $x_2 = x_1$ and $x_2 \neq x_1$.

**CASE 1:** $x_2 = x_1$
In this case the line through $P_1$ and $P_2$ is parallel to the $y$ axis, and all points on this line have the same abscissa. Therefore, $P(x, y)$ is any point on the line if and only if

$$x = x_1 \qquad (4)$$

Equation (4) is an equation of a line parallel to the $y$ axis. Note that this equation is independent of $y$; that is, the ordinate may have any value whatsoever, and the point $P(x, y)$ is on the line whenever the abscissa is $x_1$.

**CASE 2:** $x_2 \neq x_1$
The slope of the line through $P_1$ and $P_2$ is given by

$$m = \frac{y_2 - y_1}{x_2 - x_1} \qquad (5)$$

If $P(x, y)$ is any point on the line except $(x_1, y_1)$, the slope is also given by

$$m = \frac{y - y_1}{x - x_1} \qquad (6)$$

The point $P$ is on the line through $P_1$ and $P_2$ if and only if the value of $m$ from equation (5) is the same as the value of $m$ from equation (6), that is, if and only if

$$\frac{y - y_1}{x - x_1} = \frac{y_2 - y_1}{x_2 - x_1}$$

Multiplying on both sides of this equation by $x - x_1$, we obtain

$$y - y_1 = \frac{y_2 - y_1}{x_2 - x_1}(x - x_1) \qquad (7)$$

Equation (7) is satisfied by the coordinates of $P_1$ as well as by the coordinates of any other point on the line through $P_1$ and $P_2$.

Equation (7) is called the *two-point* form of an equation of the *line*. It gives an equation of the line if two points on the line are known.

If in equation (7) we replace $\frac{y_2 - y_1}{x_2 - x_1}$ by $m$, we get

$$y - y_1 = m(x - x_1) \qquad (8)$$

Equation (8) is called the *point-slope* form of an equation of the line. It gives an equation of the line if a point $P_1(x_1, y_1)$ on the line and the slope $m$ of the line are known. It is recommended that you use the point-slope form even when two points are given, as shown in the following illustration.

**ILLUSTRATION 2.** To find an equation of the line through the two points $A(-1, 3)$ and $B(5, 2)$, we first compute $m$.

$$m = \frac{2 - 3}{5 - (-1)}$$
$$= \frac{-1}{6}$$
$$= -\frac{1}{6}$$

Using the point-slope form of an equation of the line, with $A$ as $P_1$, we have

$$y - 3 = -\frac{1}{6}[x - (-1)]$$
$$6y - 18 = -x - 1$$
$$x + 6y - 17 = 0 \qquad \bullet \quad (9)$$

Observe that equation (9) is written in the form of equation (3), where $A$ is 1, $B$ is 6, and $C$ is $-17$. When a first-degree equation in two variables is written in the form of equation (3), it is said to be the *standard form* of a linear equation.

If in equation (8) the point $(x_1, y_1)$ is $(0, b)$, we have

$$y - b = m(x - 0)$$

$$\boxed{y = mx + b} \qquad (10)$$

Because $b$ is the $y$ intercept of the line, equation (10) is called the *slope-intercept* form of an equation of the line. This form is especially important because it enables us to find the slope of a line from its equation.

**ILLUSTRATION 3.** To find the slope of the line having the equation $6x + 5y - 7 = 0$, we solve the equation for $y$, and we have

$$5y = -6x + 7$$

$$y = -\frac{6}{5}x + \frac{7}{5}$$

Comparing this equation with equation (10), we see that $m = -\frac{6}{5}$ and $b = \frac{7}{5}$. Hence the slope is $-\frac{6}{5}$. ●

EXAMPLE 4
Find the slope-intercept form of an equation of the line through the points $(-7, -5)$ and $(-3, -2)$. Draw a sketch of the line.

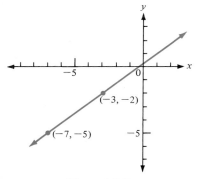

**Figure 1.9.9**

SOLUTION
If $m$ is the slope of the line, then

$$m = \frac{-2 - (-5)}{-3 - (-7)}$$

$$= \frac{3}{4}$$

Using the point-slope form of an equation of the line with $(-7, -5)$ as $P_1$, we have

$$y - (-5) = \frac{3}{4}[x - (-7)]$$

$$4(y + 5) = 3(x + 7)$$
$$4y + 20 = 3x + 21$$
$$4y = 3x + 1$$

$$y = \frac{3}{4}x + \frac{1}{4}$$

A sketch of the line is shown in Figure 1.9.9.

Earlier in this section we stated that the graph of any first-degree equation in two variables is a straight line. We are now in a position to prove this statement.

**1.9.4 THEOREM** The graph of the equation
$$Ax + By + C = 0$$
where $A$, $B$, and $C$ are constants and not both $A$ and $B$ are zero, is a straight line.

*Proof.* We consider two cases: $B \neq 0$ and $B = 0$.

**CASE 1:** $B \neq 0$
Because $B \neq 0$, we divide on both sides of the given equation by $B$ and obtain
$$y = -\frac{A}{B}x - \frac{C}{B}$$
This is an equation of a straight line because it is the slope-intercept form, where $m = -\frac{A}{B}$ and $b = -\frac{C}{B}$.

**CASE 2:** $B = 0$
Because $B = 0$, we may conclude that $A \neq 0$ and thus have
$$Ax + C = 0$$
$$x = -\frac{C}{A}$$
This equation is in the form of equation (4), and hence the graph is a straight line parallel to the $y$ axis.

**EXAMPLE 5**
Find an equation of the line that contains the point $(-3, 5)$ and that is perpendicular to the line having the equation $2x - 7y = 4$. Draw a sketch of each line on the same coordinate system.

**SOLUTION**
First we write the given equation in the slope-intercept form.
$$2x - 7y = 4$$
$$-7y = -2x + 4$$
$$y = \frac{2}{7}x - \frac{4}{7}$$

Thus the slope of the given line is $\frac{2}{7}$. Let $m$ be the slope of the required line. Because the two lines are perpendicular, the product of their slopes is $-1$. Therefore,
$$\frac{2}{7}m = -1$$
$$m = -\frac{7}{2}$$

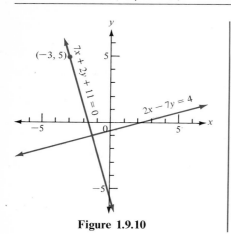

**Figure 1.9.10**

Because the required line has a slope $-\frac{7}{2}$ and contains the point $(-3, 5)$ we have, from the point-slope form of an equation of the line,

$$y - 5 = -\frac{7}{2}[x - (-3)]$$
$$2y - 10 = -7(x + 3)$$
$$2y - 10 = -7x - 21$$
$$7x + 2y + 11 = 0$$

In Figure 1.9.10 we have a sketch of each of the lines.

## EXERCISES 1.9

In Exercises 1 through 8, draw a sketch of the graph of the given equation.

1. $y = 2x + 5$
2. $y = 6 - 3x$
3. $3x - 7y = 21$
4. $4x + 5y = 20$
5. $6y - x = 6$
6. $3x + 9y = 6$
7. $6x - y = 9$
8. $2y - 9x = 18$

In Exercises 9 through 16, draw a sketch of the graph of the given set.

9. $\{(x, y) \,|\, 3x + 4y = 0\}$
10. $\{(x, y) \,|\, y = 4x - 2\}$
11. $\{(x, y) \,|\, y = 8 - 3x\}$
12. $\{(x, y) \,|\, 3x - 4y = 7\}$
13. $\{(x, y) \,|\, y = -7\}$
14. $\{(x, y) \,|\, 2x = 5\}$
15. $\{(x, y) \,|\, 3x + 10y = 15\}$
16. $\{(x, y) \,|\, 2x - 5y = 0\}$

In Exercises 17 through 22, find an equation of the line through the two given points. Draw a sketch of the line.

17. $(0, 0), (7, 3)$
18. $(3, 2), (-2, 3)$
19. $(5, 6), (-6, -5)$
20. $(-4, -5), (-3, -1)$
21. $(4, -4), (4, 2)$
22. $(-2, -3), (8, -3)$

In Exercises 23 through 28, find an equation of the line through the given point and whose slope is m. Draw a sketch of the line.

23. $(3, -4), m = 2$
24. $(-4, -1), m = -3$
25. $(-2, 5), m = \frac{2}{3}$
26. $(6, 2), m = -\frac{1}{4}$
27. $(1, -3), m = 0$
28. $(7, 4)$, there is no slope

In Exercises 29 through 34, write the equation of the line in the slope-intercept form and determine the slope and the y intercept. Draw a sketch of the line.

29. $2x - y + 5 = 0$
30. $x = -3y + 4$
31. $6x - 3y - 2 = 0$
32. $4x - 7y = 0$
33. $3y + 8 = 0$
34. $3x - 9y + 4 = 0$

35. Find an equation of the line through the point $(3, -1)$ and parallel to the line whose equation is $4x - 3y + 12 = 0$. Draw a sketch of each line on the same coordinate system.
36. Find an equation of the line through the origin and perpendicular to the line whose equation is $2x - 5y + 6 = 0$. Draw a sketch of each line on the same coordinate system.
37. Find an equation of the line through the point $(2, 4)$ and perpendicular to the line whose equation is $x - 5y + 10 = 0$. Draw a sketch of each line on the same coordinate system.
38. Find an equation of the line through the point $(-5, -1)$ and parallel to the line whose equation is $3x - y + 9 = 0$. Draw a sketch of each line on the same coordinate system.
39. Find an equation of the line whose $x$ intercept is 5 and whose slope is $-6$. Draw a sketch of the line.
40. Find an equation of the line whose $x$ intercept is $-2$ and whose $y$ intercept is 6. Draw a sketch of the line.
41. (a) Write an equation whose graph is the $x$ axis.
    (b) Write an equation whose graph is the $y$ axis.
42. (a) Write an equation whose graph consists of all points having an abscissa of $-7$.
    (b) Write an equation whose graph consists of all points having an ordinate of 3.

## REVIEW EXERCISES (CHAPTER 1)

In Exercises 1 through 9, list the elements of the given set (or sets) if $N$ is the set of natural numbers, $J$ is the set of integers, $Q$ is the set of rational numbers, and $H$ is the set of irrational numbers.

$$A = \{1, 2, 4, 9\} \quad B = \{2, 4, 6, 8\} \quad C = \{4, 8\} \quad D = \{-4, \sqrt{2}, 17, \tfrac{3}{4}, -5, 0, -\tfrac{1}{3}, 2, -\sqrt{3}\}$$

1. $A \cup B$    2. $A \cap B$    3. $A \cap C$    4. $B \cup C$    5. $D \cap N$    6. $D \cap J$    7. $D \cap Q$    8. $D \cap H$
9. The subsets of $B$ that have $C$ as a subset.

In Exercises 10 through 18, the given equality follows immediately from one of the field axioms (Axioms 1.2.1 through 1.2.7). Indicate which axiom applies. Assume that each variable is a real number.

10. $(2 + 3) + 5 = 2 + (3 + 5)$
11. $3 \cdot 7 = 7 \cdot 3$
12. $5 \cdot \dfrac{1}{5} = 1$
13. $-2 + 0 = -2$
14. $4 + (x + 3) = (x + 3) + 4$
15. $3 \cdot (4 \cdot 5) = (3 \cdot 4) \cdot 5$
16. $5(a + b) = 5a + 5b$
17. $8 + (-8) = 0$
18. $(xy)(zw) = x[y(zw)]$

In Exercises 19 and 20, find the value of the given expression if $a = -4$, $b = 3$, and $c = -2$.

19. $c^3 - 2b^2 c + c^2 - a^2$
20. $\dfrac{(2a - 3c)^2}{a^3 + 2b^3}$

In Exercises 21 through 23, find the indicated root.

21. $\sqrt{\dfrac{25}{81}}$
22. $\sqrt[4]{0.0016}$
23. $-\sqrt[3]{-27}$

*In Exercises 24 through 26, write the given complex number in the form a + bi.*

**24.** $-\sqrt{4} + \sqrt{-9}$   **25.** $\sqrt{64} - \sqrt{-64}$   **26.** $-\dfrac{4}{25} - \sqrt{-\dfrac{4}{25}}$

*In Exercises 27 through 32, find the solution set of the given equation.*

**27.** $2(5x - 4) = 11 - (3 + 2x)$

**28.** $4x - (3x - 5) = x - 7$

**29.** $\dfrac{x}{2x - 2} = \dfrac{x - 4}{2x - 4}$

**30.** $\dfrac{t^2 - 3}{t^2 - 6t + 5} = \dfrac{2t + 3}{t - 5} - \dfrac{t + 3}{t - 1}$

**31.** $\left|\dfrac{x + 1}{x - 1}\right| = 3$

**32.** $|4x - 1| = |x + 5|$

*In Exercises 33 and 34, solve for x in terms of the other symbols.*

**33.** $a^2 x - ab = b^2 x + a^2$

**34.** $\dfrac{d}{10x} - \dfrac{d}{5} = \dfrac{1}{x}$

*In Exercises 35 and 36, solve the given formula for the indicated symbol.*

**35.** $Ax + By + C = 0$, for $y$

**36.** $F = G\dfrac{m_1 m_2}{d^2}$, for $m_1$

*In Exercises 37 through 44, find the solution set of the given inequality. Draw a sketch of the graph of the solution set.*

**37.** $3x - 1 \leq 11$

**38.** $5x + 7 \geq 2x - 2$

**39.** $\dfrac{x + 1}{2} - 3 \geq \dfrac{x + 2}{4}$

**40.** $-14 < 1 - 5x < 11$

**41.** $\dfrac{2x}{x - 3} \geq 5$

**42.** $\dfrac{3x + 4}{x - 2} < 4$

**43.** $|2x - 5| < 7$

**44.** $|3x + 7| > 11$

**45.** Find the distance between the points $(7, -4)$ and $(3, 8)$.

**46.** Find the slope of the line through the points in Exercise 45.

**47.** Show that the line through the points $A(-1, 3)$ and $B(2, -5)$ is parallel to the line through the points $C(-1, 12)$ and $D(2, 4)$. Draw a sketch of the two lines.

**48.** By showing that two sides have the same length, prove that the three points $A(-1, 3)$, $B(2, 6)$, and $C(3, 2)$ are the vertices of an isosceles triangle. Draw a sketch of the triangle.

**49.** Prove that the three points $A(0, -3)$, $B(1, 4)$, and $C(2, 11)$ are collinear by two methods: (a) use slopes; (b) use the distance formula.

**50.** Show that the line through the points $A(-8, -1)$ and $B(-3, 5)$ is perpendicular to the line through the points $C(1, 3)$ and $D(7, -2)$. Draw a sketch of the two lines.

In Exercises 51 through 53, the points A, B, C, and D are the vertices of a quadrilateral. Use slopes to determine if the quadrilateral is a rectangle, a parallelogram, or a trapezoid. Draw a sketch of the quadrilateral.

51. $A(3, 1)$, $B(5, 2)$, $C(15, 5)$, $D(17, 6)$
52. $A(-8, 0)$, $B(-3, -5)$, $C(1, 4)$, $D(3, 2)$
53. $A(3, 1)$, $B(2, -2)$, $C(-1, -1)$, $D(0, 2)$
54. Find an equation of the line through the point $(5, -2)$ and having a slope equal to $-\frac{3}{2}$. Draw a sketch of the line.
55. Find an equation of the line through the two points $(-7, -2)$ and $(4, 1)$. Draw a sketch of the line.
56. Find an equation of the line through the point $(-3, -2)$ and parallel to the line having the equation $7x - 3y - 4 = 0$. Draw a sketch of each line on the same coordinate system.
57. Find an equation of the line through the point $(-2, 3)$ and perpendicular to the line having the equation $3x + 2y + 4 = 0$. Draw a sketch of each line on the same coordinate system.
58. At a post office, a woman purchased some stamps having denominations of 10, 15, and 31 cents. She purchased three times as many 10-cent stamps as 15-cent stamps and five more 15-cent stamps than 31-cent stamps. If she paid $5.29 for the stamps, how many of each kind did she buy?
59. A company obtained two loans totaling $30,000. The interest rate for one loan was 8 per cent, and for the other loan it was 11 per cent. The total annual interest for the two loans was equal to the interest the company would have paid if the entire amount had been borrowed at 10 per cent. What was the amount of each loan?
60. How many liters of a solution that is 55 per cent glycerine should be added to 25 liters of a solution that is 28 per cent glycerine to give a solution that is 35 per cent glycerine?
61. A group of 4 students decide to hire a tutor for a review session prior to an examination, and the tutor's fee is to be shared by each student. If two additional students join the group, the cost to each student is reduced by $3. What is the cost per student if 4 are in the group, and what is it if 6 are in the group?
62. A man can paint a room by himself in 8 hours, and it takes his son 12 hours to do the same job. The man works by himself for 3 hours. Then his son joins him and the two together complete the job. How long does the son work?
63. In a long-distance race around a 400-meter track, the winner finished the race one lap ahead of the loser. If the average speed of the winner was 6 meters per second and the average speed of the loser was 5.75 meters per second, how soon after the start did the winner complete the race?
64. A man leaves home at 8 A.M. and walks to his office at the rate of 8 kilometers per hour. At 8:15 A.M., the man's son leaves home and rides his bicycle at a rate of 20 kilometers per hour along the same route to school. At what time does the son overtake his father?
65. An automobile radiator contains 8 liters of a solution that is 10 per cent antifreeze and 90 per cent water. How much of the solution should be drained and replaced with pure antifreeze to obtain a solution that is 25 per cent antifreeze?
66. The perimeter of a rectangle must not be greater than 30 centimeters and the length must be 8 centimeters What is the range of values for the width?
67. A student must receive an average score of at least 90 on five examinations in order to receive a grade of A in a particular course. If the student's scores on the first four examinations are 93, 95, 79, and 88, what must be the score on the fifth examination in order for the course grade to be A?

# 2
# Second-degree Equations and Inequalities

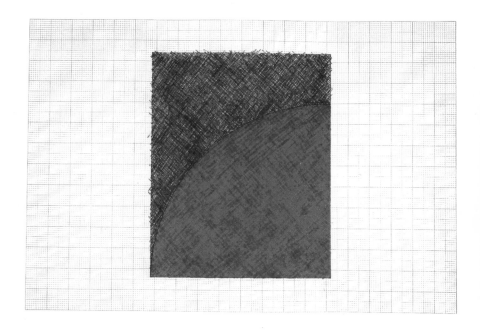

## 2.1 Second-degree Equations in One Variable

An equation equivalent to an equation of the form
$$ax^2 + bx + c = 0 \tag{1}$$
where $a$, $b$, and $c$ are constants representing real numbers and $a \neq 0$, is called a second-degree equation, or *quadratic equation,* in the variable $x$. The equations

$$6x^2 + 7x - 3 = 0 \tag{2}$$
$$x + 10 = 2x^2 \tag{3}$$
$$7x = 3 - 4x^2 \tag{4}$$
$$x^2 - 7 = 0 \tag{5}$$
$$3x^2 - 4x = 0 \tag{6}$$

are examples of quadratic equations in $x$. A quadratic equation is said to be in *standard form* when it is written in the form of equation (1). Equation (2) is in standard form, where $a$ is 6, $b$ is 7, and $c$ is $-3$. Equation (5) is in standard form, where $a$ is 1, $b$ is 0 and $c$ is $-7$, and equation (6) is in standard form, where $a$ is 3, $b$ is $-4$, and $c$ is 0. Note that it is possible for either $b$ or $c$ to be 0 as in equations (5) and (6), respectively. However, the restriction that $a \neq 0$ is necessary in order to have a second-degree equation.

Equation (3) is equivalent to the equation
$$2x^2 - x - 10 = 0 \tag{7}$$
which is obtained from equation (3) by adding $-2x^2$ to both members and then multiplying both members by $-1$. Equation (7) is a standard form of equation (3). By adding $4x^2$ and $-3$ to both members of equation (4), we obtain
$$4x^2 + 7x - 3 = 0$$
which is a standard form of equation (4).

Recall that the replacement set of a variable in an equation is the set of numbers for which the algebraic expressions in the equation are defined. For quadratic equations in standard form we consider the replacement set to be the set $C$ (the set of complex numbers). We wish now to find the solution set of a given quadratic equation, that is, all numbers in the replacement set which satisfy the equation.

One method of finding the solution set of a quadratic equation involves the following theorem, the proof of which is based on properties of the set $C$ similar to those for $R$ given in Section 1.2.

**2.1.1 THEOREM** If $r, s \in C$, then

$$rs = 0 \quad \text{if and only if} \quad r = 0 \text{ or } s = 0$$

Theorem 2.1.1 can be extended to a product of more than two factors. For instance, if $r, s, t, u \in C$, then $rstu = 0$ if and only if at least one of the numbers $r$, $s$, $t$, or $u$ is 0.

**ILLUSTRATION 1.** To find the solution set of the equation

$$x^2 + 3x - 10 = 0 \qquad (8)$$

we factor the left member and obtain

$$(x + 5)(x - 2) = 0 \qquad (9)$$

By applying Theorem 2.1.1, it follows that equation (9) gives a true statement if and only if

$$x + 5 = 0 \qquad (10)$$

or

$$x - 2 = 0 \qquad (11)$$

Hence both equation (10) and equation (11) give a solution for equation (8). The solution of equation (10) is $-5$ and the solution of equation (11) is 2. Therefore, the solution set of equation (8) is $\{-5, 2\}$.

The solutions can be checked by substituting $-5$ and 2 into the original equation (8), as follows.

Does $(-5)^2 + 3(-5) - 10 = 0?$      Does $2^2 + 3(2) - 10 = 0?$
$(-5)^2 + 3(-5) - 10 = 25 - 15 - 10$      $2^2 + 3(2) - 10 = 4 + 6 - 10$
$\qquad\qquad\qquad\qquad = 0$                                $= 0$

Therefore, both of the solutions check.    ●

In Illustration 1 we see that the solution set of equation (8) is the union of the solution sets of equations (10) and (11). The method used in this illustration can be applied to any quadratic equation in standard form (the right member is zero) for which the left member can be factored. That is, after the left member is factored, each factor is set equal to zero, and the solutions of these first-degree equations are found. Then the solution set of the given quadratic equation is the union of the solution sets of the two first-degree equations.

**ILLUSTRATION 2.** To solve the equation

$$16x^2 - 56x + 49 = 0 \qquad (12)$$

we factor the left member and obtain

$$(4x - 7)^2 = 0$$
$$(4x - 7)(4x - 7) = 0 \qquad (13)$$

Each of the factors in equation (13) gives the same value of $x$ as a solution. Therefore, equation (12) has *two equal roots* of $\frac{7}{4}$. The solution set is $\{\frac{7}{4}\}$. •

In Illustration 2 the solution set of the quadratic equation consists of just one number; however, this number is called a *double root* (or a *root of multiplicity two*) of the given equation. Every quadratic equation has two roots; in Illustration 1 the roots are unequal, and in Illustration 2 the roots are equal.

### EXAMPLE 1
Find the solution set of each of the following equations.

(a) $1 + \dfrac{5x}{6} = \dfrac{2x^2}{3}$

(b) $\dfrac{3x}{x+2} = \dfrac{x+4}{x}$

### SOLUTION
We first write the given equation in standard form. Then we factor the left member, set each factor equal to zero, and solve the equations.

(a)
$$1 + \frac{5x}{6} = \frac{2x^2}{3}$$
$$(6)(1) + (6)\frac{5x}{6} = (6)\frac{2x^2}{3}$$
$$6 + 5x = 4x^2$$
$$-4x^2 + 5x + 6 = 0$$
$$4x^2 - 5x - 6 = 0$$
$$(4x + 3)(x - 2) = 0$$
$$4x + 3 = 0 \qquad x - 2 = 0$$
$$4x = -3 \qquad\qquad x = 2$$
$$x = -\frac{3}{4}$$

The solution set is $\{-\frac{3}{4}, 2\}$.

(b)
$$\frac{3x}{x+2} = \frac{x+4}{x}$$
$$x(x+2)\frac{3x}{x+2} = x(x+2)\frac{x+4}{x}$$
$$3x^2 = (x+2)(x+4)$$
$$3x^2 = x^2 + 6x + 8$$
$$2x^2 - 6x - 8 = 0$$
$$x^2 - 3x - 4 = 0$$
$$(x-4)(x+1) = 0$$
$$x - 4 = 0 \qquad x + 1 = 0$$
$$x = 4 \qquad\qquad x = -1$$

The solution set is $\{-1, 4\}$.

If we have a quadratic equation of the form $x^2 = d$ (that is, there is no first-degree term), then an equivalent equation is

$$x^2 - d = 0$$

and factoring the left member, we obtain

$$(x + \sqrt{d})(x - \sqrt{d}) = 0$$

We set each factor equal to zero and solve the equations.

$$x + \sqrt{d} = 0 \qquad\qquad x - \sqrt{d} = 0$$
$$x = -\sqrt{d} \qquad\qquad x = \sqrt{d}$$

Therefore, the solution set of the equation $x^2 = d$ is $\{-\sqrt{d}, \sqrt{d}\}$.

Another approach to solving the equation $x^2 = d$ is to apply the definition of a square root of a number. That is,

$$x^2 = d$$

## 2.1] Second-degree Equations in One Variable 93

if and only if

$$x = \sqrt{d} \quad \text{or} \quad x = -\sqrt{d}$$

A more concise way of writing the two numbers $\sqrt{d}$ and $-\sqrt{d}$ is to write $\pm\sqrt{d}$. Hence

$$x^2 = d \quad \text{if and only if} \quad x = \pm\sqrt{d} \tag{14}$$

where $x = \pm\sqrt{d}$ stands for the two equations $x = \sqrt{d}$ and $x = -\sqrt{d}$.

**EXAMPLE 2**
Find the solution set of each of the following equations.

(a) $x^2 - 25 = 0$
(b) $3x^2 + 7 = 0$

**SOLUTION**

(a) $x^2 - 25 = 0$
$x^2 = 25$
$x = \pm\sqrt{25}$
$x = \pm 5$
The solution set is $\{-5, 5\}$.

(b) $3x^2 + 7 = 0$
$3x^2 = -7$
$x^2 = -\dfrac{7}{3}$

$$x = \pm\sqrt{-\dfrac{7}{3}}$$

$$x = \pm i\sqrt{\dfrac{7 \cdot 3}{3^2}}$$

$$x = \pm i\dfrac{\sqrt{21}}{3}$$

The solution set is $\left\{-i\dfrac{\sqrt{21}}{3}, i\dfrac{\sqrt{21}}{3}\right\}$.

To demonstrate the check of complex roots, we verify the solutions found in Example 2(b).

Does $3\left(-i\dfrac{\sqrt{21}}{3}\right)^2 + 7 = 0$?

$3\left(-i\dfrac{\sqrt{21}}{3}\right)^2 + 7 = 3i^2\left(\dfrac{21}{9}\right) + 7$
$= (-1)(7) + 7$
$= 0$

Does $3\left(i\dfrac{\sqrt{21}}{3}\right)^2 + 7 = 0$?

$3\left(\dfrac{i\sqrt{21}}{3}\right)^2 + 7 = 3i^2\left(\dfrac{21}{9}\right) + 7$
$= (-1)(7) + 7$
$= 0$

Therefore, both of the solutions check.

**ILLUSTRATION 3.** To find the solution set of the equation

$$5x^2 = 4x \tag{15}$$

we first write the equation in standard form as

$$5x^2 - 4x = 0$$

Then we factor the left member and obtain

$$x(5x - 4) = 0$$

By setting each factor equal to zero, we have

$$x = 0 \quad 5x - 4 = 0$$
$$x = \frac{4}{5}$$

Therefore, the solution set is $\{0, \frac{4}{5}\}$.

Note that it is incorrect if in equation (15) both members are divided by $x$ because if this is done we obtain the equation

$$5x = 4$$

which has only the solution $\frac{4}{5}$, and the solution 0 is lost. •

Illustration 3 indicates why you should avoid dividing both members of an equation by an algebraic expression containing the unknown.

Consider now the problem of finding a quadratic equation having two given numbers as solutions. Suppose, for instance, that $r$ and $s$ are the given roots. We know that the quadratic equation

$$(x - r)(x - s) = 0 \tag{16}$$

has the solution set $\{r, s\}$. Hence by multiplying the factors in the left member of equation (16), we have a quadratic equation in standard form for which $r$ and $s$ are solutions.

**ILLUSTRATION 4.** A quadratic equation having the solutions 4 and $-3$ is

$$(x - 4)[x - (-3)] = 0$$
$$(x - 4)(x + 3) = 0$$
$$x^2 - x - 12 = 0$$ •

EXAMPLE 3

Find a quadratic equation in standard form having the given pair of roots.

(a) $-\frac{1}{7}$ and $\frac{2}{5}$  (b) $\frac{3}{4}i$ and $-\frac{3}{4}i$

SOLUTION

(a) The following equations are equivalent, having $-\frac{1}{7}$ and $\frac{2}{5}$ as roots.

$$\left[x - \left(-\frac{1}{7}\right)\right]\left(x - \frac{2}{5}\right) = 0$$

$$\left(x + \frac{1}{7}\right)\left(x - \frac{2}{5}\right) = 0$$

$$x^2 - \frac{2}{5}x + \frac{1}{7}x - \frac{2}{35} = 0$$

Multiplying each member of the equation by 35, we obtain

$$35x^2 - 14x + 5x - 2 = 0$$
$$35x^2 - 9x - 2 = 0$$

(b) The following equations are equivalent equations having $\frac{3}{4}i$ and $-\frac{3}{4}i$ as roots.

$$\left(x - \frac{3}{4}i\right)\left[x - \left(-\frac{3}{4}i\right)\right] = 0$$

$$\left(x - \frac{3}{4}i\right)\left(x + \frac{3}{4}i\right) = 0$$

$$x^2 - \frac{9}{16}i^2 = 0$$

$$x^2 - \frac{9}{16}(-1) = 0$$

$$16x^2 + 9 = 0$$

If a quadratic equation is obtained when solving a word problem, it is possible that one of the two roots of the quadratic equation does not satisfy the conditions of the problem. For instance, if a quadratic equation has both a positive number and a negative number as roots and the variable in the equation represents the measure of the length of a room, we reject the negative number as a solution to the word problem. Of course, it can happen that both roots of a quadratic equation are acceptable solutions to a word problem, and furthermore it is possible that neither of the two roots of a quadratic equation fits the conditions of a problem.

**EXAMPLE 4**

A girl can row a boat 12 kilometers downstream and return a distance of 12 kilometers upstream in 4 hours. If the rate of the current is 4 kilometers per hour, what is the girl's rate of rowing in still water?

**SOLUTION**

Let $r$ represent the number of kilometers per hour in the rate of the boat in still water. Then, because the rate of the current is 4 kilometers per hour, the effective rate of the boat going downstream is $(r + 4)$ kilometers per hour, and the effective rate of the boat going upstream is $(r - 4)$ kilometers per hour. Table 2.1.1 gives the number of hours it takes to travel each way.

Table 2.1.1

|  | Number of Kilometers in Distance Traveled | ÷ | Number of Kilometers per Hour in Effective Rate | = | Number of Hours in Time |
|---|---|---|---|---|---|
| Downstream | 12 |  | $r + 4$ |  | $\dfrac{12}{r + 4}$ |
| Upstream | 12 |  | $r - 4$ |  | $\dfrac{12}{r - 4}$ |

The total number of hours of travel time can be represented by either 4 or the sum of the entries in the last column of the table. Thus we have the following equation.

$$\frac{12}{r + 4} + \frac{12}{r - 4} = 4 \qquad (17)$$

We solve the equation

$$(r+4)(r-4)\frac{12}{r+4} + (r+4)(r-4)\frac{12}{r-4} = (r+4)(r-4)(4)$$

$$12(r-4) + 12(r+4) = 4(r^2 - 16)$$
$$12r - 48 + 12r + 48 = 4r^2 - 64$$
$$-4r^2 + 24r + 64 = 0$$
$$r^2 - 6r - 16 = 0$$
$$(r-8)(r+2) = 0$$
$$r - 8 = 0 \qquad r + 2 = 0$$
$$r = 8 \qquad r = -2$$

Both 8 and $-2$ satisfy equation (17) and therefore the solution set of equation (17) is $\{-2, 8\}$. We reject the negative root because $r$ must be a positive number.

Hence the rate of the boat in still water is 8 kilometers per hour.

**CHECK:** The effective rate of the boat going downstream is 12 kilometers per hour, so it takes 1 hour for the boat to go 12 kilometers downstream; the effective rate of the boat going upstream is 4 kilometers per hour, so it takes 3 hours for the boat to go 12 kilometers upstream; and $1 + 3 = 4$.

## EXERCISES 2.1

*In Exercises 1 through 28, find the solution set of the given equation. If more than one variable appears in the equation, solve for x in terms of the other variables.*

1. $x^2 = 49$
2. $25x^2 - 16 = 0$
3. $x^2 + 9 = 0$
4. $3y^2 - 5 = 0$
5. $5t^2 - 12 = 0$
6. $3x^2 + 7 = 0$
7. $4x^2 = x$
8. $\frac{1}{6}x^2 + x = 0$
9. $x^2 = 8x - 15$
10. $y^2 - 11y + 28 = 0$
11. $8w^2 + 10w - 3 = 0$
12. $14x^2 - x - 3 = 0$
13. $2t + 15 = 24t^2$
14. $35 - 32s^2 = 12s$
15. $49x^2 + 84x + 36 = 0$
16. $25x^2 - 85x + 30 = 0$
17. $18y^2 + 57y + 45 = 0$
18. $64y^2 - 80y + 25 = 0$
19. $9x^2 - 9ax + 2a^2 = 0$
20. $3x^2 + 7bx - 6b^2 = 0$
21. $10a^2x^2 - 9abx - 7b^2 = 0$
22. $(x-5)^2 = 4(x-2)$
23. $(x-6)(x+2) = 9$
24. $\dfrac{y+4}{y} = \dfrac{3y}{y+2}$
25. $\dfrac{3t}{3t+4} + \dfrac{2}{5} = \dfrac{t}{3t-4}$
26. $\dfrac{32}{x^2+3x+2} - 3 = \dfrac{x-3}{x+1}$
27. $\dfrac{70}{x^2-4x+3} = \dfrac{23}{1-x} - 3$
28. $\dfrac{6}{x^2-1} + \dfrac{1}{x-1} = \dfrac{1}{2}$

In Exercises 29 through 38, find a quadratic equation in standard form having the given pair of roots.

29. 5 and $-2$
30. $-4$ and $-7$
31. $\frac{2}{3}$ and $\frac{3}{4}$
32. $\frac{2}{5}$ and $-\frac{1}{8}$
33. 0 and $-\frac{1}{9}$
34. $6a$ and $\frac{1}{3}a$
35. $\dfrac{b}{a}$ and $\dfrac{a}{b}$
36. $i$ and $-i$
37. $4i$ and $-4i$
38. $\frac{3}{5}i$ and $-\frac{3}{5}i$

39. The sum of the reciprocals of two consecutive even integers is $\frac{9}{40}$. What are the integers?
40. It takes a boy 15 minutes longer to mow the lawn than it takes his sister, and if they both work together it takes them 56 minutes. How long does it take the boy to mow the lawn by himself?
41. A page of print containing 144 square centimeters of printed area has a margin of $4\frac{1}{2}$ centimeters at the top and bottom, and a margin of 2 centimeters at the sides. What are the dimensions of the page if the width (across the page) is four-ninths of the length?
42. It is desired to have an open bin with a square bottom, rectangular sides, and a height of 3 meters. If the material for the bottom costs $5.40 per square meter and the material for the sides costs $2.40 per square meter, what is the volume of the bin that can be constructed for $63 worth of material?
43. It took a faster runner 10 seconds longer to run a distance of 1500 feet than it took a slower runner to run a distance of 1000 feet. If the speed of the faster runner was 5 feet per second more than the speed of the slower runner, what was the speed of each runner?
44. A motorboat takes 50 minutes longer to go 40 kilometers up a river than to return the same distance. If the rate of the boat in still water is 20 kilometers per hour, what is the rate of the current?
45. A college rowing team that can row 24 kilometers per hour in still water rowed a distance of 15 kilometers downstream. If the rate of the current had been double its actual rate, the team would have covered the distance in 5 minutes less time. Find the actual rate of the current.
46. Are there two consecutive even integers, the sum of whose reciprocals is $\frac{8}{45}$? If your answer is yes, find them. If your answer is no, prove it.

## 2.2 Completing a Square and the Quadratic Formula

In Section 2.1 we learned that the solution set of an equation of the form

$$x^2 = d$$

is the union of the solution sets of the two equations

$$x = \sqrt{d} \quad \text{and} \quad x = -\sqrt{d}$$

In a similar way, if $E$ as an algebraic expression, then the solution set of the equation

$$E^2 = d \tag{1}$$

is the union of the solution sets of the equations

$$E = \sqrt{d} \quad \text{and} \quad E = -\sqrt{d}$$

**ILLUSTRATION 1.** The solution set of the equation

$$(x - 6)^2 = 10 \tag{2}$$

is the union of the solution sets of the equations

$$x - 6 = \sqrt{10} \quad \text{and} \quad x - 6 = -\sqrt{10} \tag{3}$$
$$x = 6 + \sqrt{10} \quad \text{and} \quad x = 6 - \sqrt{10} \tag{4}$$

Hence the solution set of equation (2) is $\{6 - \sqrt{10}, 6 + \sqrt{10}\}$. ●

Note that equations (3) can be written more concisely as

$$x - 6 = \pm\sqrt{10}$$

and equations (4) can be written more concisely as

$$x = 6 \pm \sqrt{10}$$

The method used in Illustration 1 can be applied to find the solution set of any quadratic equation. The first step is to write the equation in a form similar to equation (1); that is, the left member is a square of an algebraic expression containing the variable and the right member is a constant.

**ILLUSTRATION 2.** To find the solution set of the equation

$$2x^2 + 6x - 5 = 0 \tag{5}$$

we first add 5 to each member and then divide each member by 2 (the coefficient of $x^2$).

$$x^2 + 3x = \frac{5}{2} \tag{6}$$

We now add to each member the square of one-half of the coefficient of the first-degree term; that is, we add $(\frac{3}{2})^2$ to each member. We obtain

$$x^2 + 3x + \frac{9}{4} = \frac{5}{2} + \frac{9}{4} \tag{7}$$

The left member is now the square of $(x + \frac{3}{2})$, so we have

$$\left(x + \frac{3}{2}\right)^2 = \frac{19}{4} \tag{8}$$

Equations (5), (6), (7), and (8) are all equivalent, and equation (8) is in the form of equation (1). The solution set of equation (8) is the union of the solution sets of the equations

$$x + \frac{3}{2} = \pm\sqrt{\frac{19}{4}}$$

$$x = -\frac{3}{2} \pm \frac{\sqrt{19}}{2}$$

Hence the solution set of equation (5) is $\left\{\dfrac{-3 - \sqrt{19}}{2}, \dfrac{-3 + \sqrt{19}}{2}\right\}$. •

The method used to obtain equation (8) in Illustration 2 is called *completing a square*. The important step is obtaining equation (7) equivalent to equation (6). Note that in equation (6) the left member is $x^2 + 3x$ (the coefficient of $x^2$ is 1 and the coefficient of $x$ is 3). We added the square of one-half of 3 to each member of equation (6) which gives the equivalent equation (7), in which the left member is $(x + \frac{3}{2})^2$. More generally, if we have

$$x^2 + kx \qquad (9)$$

we add $\left(\dfrac{k}{2}\right)^2$ to complete the square; that is,

$$x^2 + kx + \dfrac{k^2}{4} = \left(x + \dfrac{k}{2}\right)^2$$

Observe in binomial (9) that the coefficient of $x^2$ is 1.

### EXAMPLE 1
Add a term to each of the following algebraic expressions in order to make it a square of a binomial; also write the resulting expression as a square of a binomial.

(a) $x^2 + 6x$
(b) $x^2 - 5x$
(c) $x^2 + \dfrac{3}{4}x$

### SOLUTION
The coefficient of $x^2$ in each of the given expressions is 1. Hence, to complete a square, we add the square of one half of the coefficient of $x$.

(a) $x^2 + 6x + \left(\dfrac{6}{2}\right)^2 = x^2 + 6x + 9$
$\qquad\qquad\qquad\quad = (x + 3)^2$

(b) $x^2 - 5x + \left(-\dfrac{5}{2}\right)^2 = x^2 - 5x + \dfrac{25}{4}$
$\qquad\qquad\qquad\quad = \left(x - \dfrac{5}{2}\right)^2$

(c) $x^2 + \dfrac{3}{4}x + \left(\dfrac{3}{8}\right)^2 = x^2 + \dfrac{3}{4}x + \dfrac{9}{64}$
$\qquad\qquad\qquad\quad = \left(x + \dfrac{3}{8}\right)^2$

### EXAMPLE 2
Find the solution set of the following equation by completing a square.

$$5x^2 - 6x - 1 = 0$$

### SOLUTION
For the given equation we first find an equivalent equation in which the left member is of the form $x^2 + kx$; hence we add 1 to each member and then divide each member by 5. Doing this, we have

$$x^2 - \dfrac{6}{5}x = \dfrac{1}{5} \qquad (10)$$

The coefficient of $x$ is $-\frac{6}{5}$ and one half of $-\frac{6}{5}$ is $-\frac{3}{5}$. We therefore add $(-\frac{3}{5})^2$ to each member of equation (10) and obtain

$$x^2 - \frac{6}{5}x + \frac{9}{25} = \frac{1}{5} + \frac{9}{25}$$

$$\left(x - \frac{3}{5}\right)^2 = \frac{14}{25} \tag{11}$$

The square roots of $\frac{14}{25}$ are $\frac{\sqrt{14}}{5}$ and $-\frac{\sqrt{14}}{5}$, and therefore the solution set of equation (11) is the union of the solution sets of the equations

$$x - \frac{3}{5} = \frac{\sqrt{14}}{5} \quad \text{and} \quad x - \frac{3}{5} = -\frac{\sqrt{14}}{5}$$

We solve each of these equations.

$$x = \frac{3}{5} + \frac{\sqrt{14}}{5} \qquad x = \frac{3}{5} - \frac{\sqrt{14}}{5}$$

$$x = \frac{3 + \sqrt{14}}{5} \qquad x = \frac{3 - \sqrt{14}}{5}$$

The solution set is then $\left\{\dfrac{3 - \sqrt{14}}{5}, \dfrac{3 + \sqrt{14}}{5}\right\}$.

Consider now the general quadratic equation in standard form

$$ax^2 + bx + c = 0 \tag{12}$$

We solve this equation for $x$ in terms of $a$, $b$, and $c$ by completing a square. We first add $-c$ to both members of the equation and then we divide both members by $a$ (remember $a \neq 0$), and we have

$$x^2 + \frac{b}{a}x + \left(\frac{b}{2a}\right)^2 = -\frac{c}{a} + \left(\frac{b}{2a}\right)^2$$

$$\left(x + \frac{b}{2a}\right)^2 = \frac{b^2}{4a^2} - \frac{c}{a}$$

$$\left(x + \frac{b}{2a}\right)^2 = \frac{b^2 - 4ac}{4a^2} \tag{13}$$

We solve equation (13) by equating $x + \dfrac{b}{2a}$ to the two square roots of the right member, and we have

$$x + \frac{b}{2a} = \pm \frac{\sqrt{b^2 - 4ac}}{2a}$$

$$x = -\frac{b}{2a} \pm \frac{\sqrt{b^2 - 4ac}}{2a}$$

$$x = \frac{-b \pm \sqrt{b^2 - 4ac}}{2a} \tag{14}$$

The two values of $x$ given by equations (14) are the solutions of equation (12). Thus the solution set of equation (12) is

$$\left\{ \frac{-b - \sqrt{b^2 - 4ac}}{2a}, \frac{-b + \sqrt{b^2 - 4ac}}{2a} \right\}$$

Equations (14) are known as the *quadratic formula*. The quadratic formula can be used to find the solution set of any quadratic equation by substituting into the formula the values of $a$, $b$, and $c$ given by the equation.

**EXAMPLE 3**
Use the quadratic formula to find the solution set of each of the following equations.
(a) $6x^2 = 10 + 11x$
(b) $x^2 + \frac{5}{3}x + 1 = 0$

**SOLUTION**
(a) We write the given equation in standard form as

$$6x^2 - 11x - 10 = 0$$

Using the quadratic formula (14) where $a$ is 6, $b$ is $-11$, and $c$ is $-10$, we have

$$x = \frac{-b \pm \sqrt{b^2 - 4ac}}{2a}$$
$$= \frac{-(-11) \pm \sqrt{(-11)^2 - 4(6)(-10)}}{2(6)}$$
$$= \frac{11 \pm \sqrt{121 + 240}}{12}$$
$$= \frac{11 \pm \sqrt{361}}{12}$$
$$= \frac{11 \pm 19}{12}$$

Therefore,

$$x = \frac{11 + 19}{12} \qquad x = \frac{11 - 19}{12}$$
$$x = \frac{30}{12} \qquad x = \frac{-8}{12}$$
$$x = \frac{5}{2} \qquad x = -\frac{2}{3}$$

The solution set is $\{-\frac{2}{3}, \frac{5}{2}\}$.

(b) Writing the given equation in standard form, we have
$$3x^2 + 5x + 3 = 0$$
We now use the quadratic formula, where $a$ is 3, $b$ is 5, and $c$ is 3, and obtain

$$\begin{aligned}
x &= \frac{-b \pm \sqrt{b^2 - 4ac}}{2a} \\
&= \frac{-5 \pm \sqrt{(5)^2 - 4(3)(3)}}{2(3)} \\
&= \frac{-5 \pm \sqrt{25 - 36}}{6} \\
&= \frac{-5 \pm \sqrt{-11}}{6} \\
&= \frac{-5 \pm i\sqrt{11}}{6}
\end{aligned}$$

The solution set is $\left\{\dfrac{-5 - i\sqrt{11}}{6}, \dfrac{-5 + i\sqrt{11}}{6}\right\}$.

Let $r$ and $s$ denote the roots of the general quadratic equation
$$ax^2 + bx + c = 0$$
where $a$, $b$, and $c$ are real numbers. Let
$$r = \frac{-b + \sqrt{b^2 - 4ac}}{2a} \quad \text{and} \quad s = \frac{-b - \sqrt{b^2 - 4ac}}{2a}$$

The number represented by $b^2 - 4ac$ is called the *discriminant of the quadratic equation*. By finding the value of the discriminant, we obtain information about the character of the roots without actually solving the equation.

1. If $b^2 - 4ac = 0$, then
$$r = \frac{-b}{2a} \quad \text{and} \quad s = \frac{-b}{2a}$$
and therefore $r$ and $s$ are equal real numbers. In such a case the root $-\dfrac{b}{2a}$ is said to be of multiplicity two.
2. If $b^2 - 4ac < 0$, then $b^2 - 4ac$ is a negative number, and therefore $r$ and $s$ are imaginary numbers.
3. If $b^2 - 4ac > 0$, then $b^2 - 4ac$ is a positive real number, and therefore $r$ and $s$ are unequal real numbers. If, furthermore, $a$, $b$, and $c$ are rational numbers, then $b^2 - 4ac$ is a rational number because when the operations

of addition, multiplication, and subtraction are performed on rational numbers the results are rational numbers. In such a case
(i) $r$ and $s$ are rational numbers if and only if $b^2 - 4ac$ is the square of a rational number;
(ii) $r$ and $s$ are irrational numbers if and only if $b^2 - 4ac$ is not the square of a rational number.

**ILLUSTRATION 3.** To determine the character of the roots of the equation
$$3x^2 - 2x - 6 = 0$$
we compute the value of the discriminant.
$$b^2 - 4ac = (-2)^2 - 4(3)(-6)$$
$$= 76$$

The discriminant is positive and not the square of a rational number. Hence the roots are unequal irrational numbers. •

**EXAMPLE 4**
Determine the character of the roots of each of the following equations.
(a) $2x^2 + 6x + 7 = 0$
(b) $5x^2 - 11x + 2 = 0$
(c) $11x^2 + \sqrt{5}x - 1 = 0$

**SOLUTION**
(a) For the given equation $a$ is 2, $b$ is 6, and $c$ is 7. Hence
$$b^2 - 4ac = 6^2 - 4(2)(7)$$
$$= 36 - 56$$
$$= -20$$
The discriminant is negative and therefore the roots are imaginary numbers.
(b) Because $a$ is 5, $b$ is $-11$, and $c$ is 2,
$$b^2 - 4ac = (-11)^2 - 4(5)(2)$$
$$= 121 - 40$$
$$= 81$$
The discriminant is positive and is the square of a rational number, where $a$, $b$, and $c$ are rational. Hence the roots are unequal rational numbers.
(c) For the given equation, $a$ is 11, $b$ is $\sqrt{5}$, and $c$ is $-1$. Hence
$$b^2 - 4ac = (\sqrt{5})^2 - 4(11)(-1)$$
$$= 5 + 44$$
$$= 49$$
The discriminant is positive. The roots are unequal irrational numbers. Note that even though the discriminant is 49, which is the square of a rational number, the roots are not rational because the coefficient of $x$ in the equation is $\sqrt{5}$ (an irrational number).

## EXAMPLE 5

Find the values of $k$ for which the equation $9x^2 + kx + 16 = 0$ has equal roots.

## SOLUTION

The equation has equal roots if and only if the discriminant is zero. Here, $a$ is 9, $b$ is $k$, and $c$ is 16. Thus

$$b^2 - 4ac = k^2 - 4(9)(16)$$

Equating the discriminant to zero and solving the quadratic equation, we have

$$k^2 - 576 = 0$$
$$k^2 = 576$$
$$k = \pm 24$$

**CHECK:** When $k$ is 24, the given equation is

$$9x^2 + 24x + 16 = 0$$
$$(3x + 4)^2 = 0$$

and the roots are $-\tfrac{4}{3}$ and $-\tfrac{4}{3}$.

When $k$ is $-24$, the given equation is

$$9x^2 - 24x + 16 = 0$$
$$(3x - 4)^2 = 0$$

and the roots are $\tfrac{4}{3}$ and $\tfrac{4}{3}$.

As discussed in Section 1.4, if an equation contains more than one variable and we wish to solve for one of the variables in terms of the other variables, then we treat the variable for which we are solving as the unknown and the other variables as known.

**ILLUSTRATION 4.** To solve the equation

$$3x^2 + xy + x - 3y = 5$$

for $x$ in terms of $y$, we first write the equation as a standard form of a quadratic equation in $x$.

$$3x^2 + (y + 1)x + (-3y - 5) = 0$$

We now use the quadratic formula, where $a$ is 3, $b$ is $y + 1$, and $c$ is $-3y - 5$.

$$x = \frac{-b \pm \sqrt{b^2 - 4ac}}{2a}$$
$$= \frac{-(y + 1) \pm \sqrt{(y + 1)^2 - 4(3)(-3y - 5)}}{2(3)}$$
$$= \frac{-y - 1 \pm \sqrt{y^2 + 2y + 1 + 36y + 60}}{6}$$
$$= \frac{-y - 1 \pm \sqrt{y^2 + 38y + 61}}{6}$$

●

## EXAMPLE 6

If a projectile is shot vertically upward with an initial velocity of $V$ feet per second, and $s$ feet is the distance of the projectile from the starting point at $t$ seconds, then

$$s = Vt - 16t^2$$

Solve this formula for $t$.

### SOLUTION

Writing the formula as a standard form of a quadratic equation in $t$, we have

$$16t^2 - Vt + s = 0$$

We use the quadratic formula, where $a$ is 16, $b$ is $-V$, and $c$ is $s$, and we have

$$t = \frac{-b \pm \sqrt{b^2 - 4ac}}{2a}$$

$$= \frac{-(-V) \pm \sqrt{(-V)^2 - 4(16)(s)}}{2(16)}$$

$$= \frac{V \pm \sqrt{V^2 - 64s}}{32}$$

## EXAMPLE 7

A park contains a flower garden, 50 meters long and 30 meters wide, and a path of uniform width around it. If the area of the path is 600 square meters, find its width.

### SOLUTION

Let $w$ represent the number of meters in the width of the path. Refer to Figure 2.2.1. The area of the park minus the area of the garden is equal to the area of the path; thus we have the equation

$$(50 + 2w)(30 + 2w) - 50 \cdot 30 = 600$$
$$1500 + 160w + 4w^2 - 1500 = 600$$
$$4w^2 + 160w - 600 = 0$$
$$w^2 + 40w - 150 = 0 \qquad (15)$$

We solve equation (15) by using the quadratic formula, where $a$ is 1, $b$ is 40, and $c$ is $-150$.

$$w = \frac{-b \pm \sqrt{b^2 - 4ac}}{2a}$$

$$= \frac{-40 \pm \sqrt{(40)^2 - 4(1)(-150)}}{2(1)}$$

$$= \frac{-40 \pm \sqrt{2200}}{2}$$

$$= \frac{-40 \pm 10\sqrt{22}}{2}$$

$$= -20 \pm 5\sqrt{22}$$

Therefore, the solution set of equation (15) is $\{-20 - 5\sqrt{22}, -20 + 5\sqrt{22}\}$. Because $w$ must be a positive number, we reject the negative root. Because $\sqrt{22} \approx 4.69$, an approximate value for $w$ is $-20 + 5(4.69)$; that is $w \approx 3.45$. Therefore, the width of the path is 3.45 meters.

**CHECK:** The park is 56.90 meters long and 36.90 meters wide; hence the area of the park is 2100 square meters. The area of the garden is 1500 square meters and the area of the path is 600 square meters; and $2100 - 1500 = 600$.

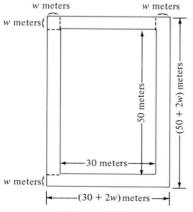

**Figure 2.2.1**

## EXERCISES 2.2

In Exercises 1 through 8, add a term to the given algebraic expression in order to make it a square of a binomial; also write the resulting expression as a square of a binomial.

1. $x^2 + 6x$
2. $x^2 - 4x$
3. $x^2 - 11x$
4. $y^2 + y$
5. $w^2 + \frac{1}{2}w$
6. $x^2 - \frac{3}{5}x$
7. $x^2 - \frac{4}{3}ax$
8. $t^2 + \frac{1}{3}st$

In Exercises 9 through 16, find the solution set of the given equation by completing a square.

9. $x^2 + 7x + 10 = 0$
10. $x^2 - 5x + 6 = 0$
11. $3t^2 = 12 - 5t$
12. $2y^2 = 11y - 12$
13. $x^2 - x - 1 = 0$
14. $3x^2 + x - 1 = 0$
15. $3y^2 + 4y + 2 = 0$
16. $w = 4w^2 + 2$

In Exercises 17 through 26, find the solution set of the given equation by using the quadratic formula.

17. $x^2 - 3x - 4 = 0$
18. $x^2 + 2x - 3 = 0$
19. $2x^2 + x - 15 = 0$
20. $4y^2 + 4y - 3 = 0$
21. $2x + 2 = x^2$
22. $x^2 + 1 = 6x$
23. $t^2 - 4t + 7 = 0$
24. $4s^2 - 12s + 13 = 0$
25. $25y^2 - 20y + 7 = 0$
26. $9x^2 + 12x + 52 = 0$

In Exercises 27 through 36, find the discriminant and determine the character of the roots of the given quadratic equation; do not solve the equation.

27. $6x^2 - 11x - 10 = 0$
28. $4x^2 + 12x + 9 = 0$
29. $3x^2 - 4x = 3$
30. $4t^2 + 2t + 1 = 0$
31. $3y = 2y^2 + 5$
32. $4x - 4 = -5x^2$
33. $25x^2 - 40x + 16 = 0$
34. $5x^2 = 4x - 8$
35. $5w^2 + \sqrt{5}w - 11 = 0$
36. $14y^2 + 11y - 15 = 0$

In Exercises 37 through 42, find the values (or value) of k for which the given equation has equal roots. Verify your results.

37. $25x^2 + kx + 36 = 0$
38. $4x^2 + 2kx + 9 = 0$
39. $x^2 + 2kx - 1 = 2k$
40. $2y^2 - ky + k = 0$
41. $2(ky + 1)(3y - 4) + 9 = 0$
42. $7x^2 - 1 = 8kx^2 - 2kx$

In Exercises 43 through 52, solve for x in terms of the other symbols.

43. $5ax^2 - 3x - 2a = 0$
44. $6dx^2 - 3dx + 5 = 0$
45. $x^2 + b^2 = 2bx + a^2x^2$
46. $x^2 + 2ax + a^2 = b^2x^2$
47. $2x^2 + 3xy - 7x = 2y^2 - 11y + 15$
48. $x^2 + xy + 4x = 2y^2 - 17y + 21$
49. $2x^2 + 2xy - y^2 = 0$
50. $x^2 + xy + 2x - 1 = 0$
51. $x^2 - 2xy - 4x - 3y^2 = 0$
52. $9x^2 - 6xy + y^2 - 3y = 0$

**53.** If $r_1$ and $r_2$ are the roots of the quadratic equation $ax^2 + bx + c = 0$, show that

$$r_1 + r_2 = -\frac{b}{a} \quad \text{and} \quad r_1 r_2 = \frac{c}{a}$$

*In Exercises 54 through 57, use the formulas of Exercise 51 to find the sum and product of the roots of the given equation.*

**54.** $4x^2 + 5 = 6x$
**55.** $6x^2 = 3 + 2x$
**56.** $rx^2 - sx = t$
**57.** $mx^2 = n - 3x$

**58.** If a regular polygon of ten sides is inscribed in a circle of radius $r$ units, then, if $s$ units is the length of a side,

$$\frac{r}{s} = \frac{s}{r-s}$$

Solve this formula for $s$ in terms of $r$.

**59.** The standard form of an equation of a parabola having a vertical axis is $y = ax^2 + bx + c$. Solve for $x$ in terms of $y$, $a$, $b$, and $c$.

**60.** If $r$ units is the radius of a circular arch of height $h$ units and width $w$ units, then

$$r = \frac{4h^2 + w^2}{8h}$$

Solve this formula for $h$ in terms of $r$ and $w$.

**61.** What is the width of a strip that must be plowed around a rectangular field, 100 meters long by 60 meters wide, so that the field will be two-thirds plowed?

**62.** A park in the shape of a rectangle has dimensions 60 meters by 100 meters. If the park contains a rectangular garden enclosed by a concrete terrace, how wide is the terrace if the area of the garden is one-half the area of the park?

## 2.3 Other Equations in One Variable

Suppose that $E = F$ is an algebraic equation in $x$. If this equation contains radicals or rational exponents, we can solve it by raising both members of the equation to the same positive integer power. However, when we do this we must apply the following theorem.

**2.3.1 THEOREM** If

$$E = F$$

is an algebraic equation in $x$, its solution set is a subset of the solution set of the equation

$$E^n = F^n$$

where $n$ is any positive integer.

The theorem follows immediately from the fact that if $a, b \in C$ and $a = b$, then $a^n = b^n$, where $n$ is any positive integer.

**ILLUSTRATION 1.** If we square each member of the equation

$$x = 5 \tag{1}$$

we obtain

$$x^2 = 25 \tag{2}$$

The solution set of equation (1) is $\{5\}$ and the solution set of equation (2) is $\{-5, 5\}$. Thus the solution set of equation (1) is a subset of the solution set of equation (2). This result agrees with Theorem 2.3.1, •

**ILLUSTRATION 2.** To solve the equation

$$\sqrt{2x + 5} = 5 - x \tag{3}$$

we first square each member to obtain

$$(\sqrt{2x + 5})^2 = (5 - x)^2$$
$$2x + 5 = 25 - 10x + x^2 \tag{4}$$
$$x^2 - 12x + 20 = 0 \tag{5}$$

We solve equation (5).

$$(x - 10)(x - 2) = 0$$
$$x - 10 = 0 \quad\quad x - 2 = 0$$
$$x = 10 \quad\quad\quad x = 2$$

Thus the solution set of equation (4) is $\{2, 10\}$. According to Theorem 2.3.1 the solution set of equation (3) is a subset of $\{2, 10\}$; that is, both of the numbers 2 and 10 may be solutions of equation (3), only one of these numbers may be a solution, or neither of these numbers may be a solution. To determine which case applies, we substitute each of the numbers into equation (3) to see if the equation is satisfied.

$$\sqrt{2(2) + 5} \stackrel{?}{=} 5 - 2 \quad\quad \sqrt{2(10) + 5} \stackrel{?}{=} 5 - 10$$
$$\sqrt{9} \stackrel{?}{=} 3 \quad\quad\quad\quad \sqrt{25} \stackrel{?}{=} -5$$
$$3 = 3 \quad\quad\quad\quad\quad 5 \neq -5$$

Hence 2 is a solution of equation (3) and 10 is not. Therefore, the solution set of equation (3) is $\{2\}$. •

In Illustration 2 the number 10 is called an *extraneous solution* of equation (3); it was introduced when both members of equation (3) were squared. The reason that this extraneous solution was introduced should be apparent after reading Illustration 3.

**Illustration 3.** If both members of the equation

$$-\sqrt{2x+5} = 5 - x \qquad (6)$$

are squared, we obtain

$$(-\sqrt{2x+5})^2 = (5-x)^2$$
$$2x + 5 = 25 - 10x + x^2$$

which is equation (4). In Illustration 2 we showed that the solution set of equation (4) is $\{2, 10\}$. Hence, by Theorem 2.3.1, the solution set of equation (6) is a subset of $\{2, 10\}$. Substituting each of these numbers into equation (6), we have

$$-\sqrt{2(2)+5} \stackrel{?}{=} 5 - 2 \qquad\qquad -\sqrt{2(10)+5} \stackrel{?}{=} 5 - 10$$
$$-\sqrt{9} \stackrel{?}{=} 3 \qquad\qquad\qquad -\sqrt{25} \stackrel{?}{=} -5$$
$$-3 \neq 3 \qquad\qquad\qquad\qquad -5 = -5$$

Hence 10 is a solution of equation (6) and 2 is an extraneous solution. Therefore, the solution set of equation (6) is $\{10\}$. •

Observe in Illustration 2 that equation (3) states that the principal square root of $(2x + 5)$ is $5 - x$, and in Illustration 3 equation (6) states that the negative square root of $(2x + 5)$ is $5 - x$. Hence, when squaring both members of either equation (3) or (6), we obtain equation (4) and so in each case an extraneous solution is introduced. This discussion should convince you that when Theorem 2.3.1 is used to solve an equation, all solutions obtained must be checked in the original equation. The check is for possible extraneous solutions and is not just a check for computational errors.

When solving an equation having terms involving one or more radicals, the first step is to write the equation so that the term involving the most complicated radical belongs to one member and all the other terms belong to the other member. Then apply Theorem 2.3.1 and raise both members of the equation to the power corresponding to the index of the radical.

**EXAMPLE 1**

Find the solution set of the equation

$$\sqrt{x}\sqrt{x-8} = 3$$

**SOLUTION**

We square both members of the equation.

$$(\sqrt{x}\sqrt{x-8})^2 = 3^2$$
$$x(x-8) = 9$$
$$x^2 - 8x = 9$$
$$x^2 - 8x - 9 = 0$$
$$(x-9)(x+1) = 0$$
$$x - 9 = 0 \qquad x + 1 = 0$$
$$x = 9 \qquad\quad x = -1$$

Hence the solution set of the given equation is a subset of $\{-1, 9\}$.

We substitute these numbers into the given equation.

$$\sqrt{9}\sqrt{9-8} \stackrel{?}{=} 3 \qquad \sqrt{-1}\sqrt{-1-8} \stackrel{?}{=} 3$$
$$3 \cdot 1 \stackrel{?}{=} 3 \qquad i \cdot 3i \stackrel{?}{=} 3$$
$$3 = 3 \qquad 3i^2 \stackrel{?}{=} 3$$
$$3(-1) \stackrel{?}{=} 3$$
$$-3 \neq 3$$

Therefore, 9 is a solution and $-1$ is an extraneous solution. Thus the solution set of the given equation is $\{9\}$.

Sometimes, if an equation contains more than one radical, it is necessary to apply Theorem 2.3.1 more than once before obtaining an equation free of radicals.

EXAMPLE 2
Find the solution set of each of the following equations.
(a) $\sqrt{3-3x} - \sqrt{3x+2} = 3$
(b) $\sqrt{2x+3} - \sqrt{x-2} - 2 = 0$

SOLUTION
(a) $\sqrt{3-3x} - \sqrt{3x+2} = 3$
$$\sqrt{3-3x} = 3 + \sqrt{3x+2}$$
$$(\sqrt{3-3x})^2 = (3 + \sqrt{3x+2})^2$$
$$3 - 3x = 9 + 6\sqrt{3x+2} + 3x + 2$$
$$-6\sqrt{3x+2} = 6x + 8$$
$$-3\sqrt{3x+2} = 3x + 4$$
$$9(\sqrt{3x+2})^2 = (3x+4)^2$$
$$9(3x+2) = 9x^2 + 24x + 16$$
$$27x + 18 = 9x^2 + 24x + 16$$
$$-9x^2 + 3x + 2 = 0$$
$$9x^2 - 3x - 2 = 0$$
$$(3x-2)(3x+1) = 0$$
$$3x - 2 = 0 \qquad 3x + 1 = 0$$
$$3x = 2 \qquad 3x = -1$$
$$x = \frac{2}{3} \qquad x = -\frac{1}{3}$$

Therefore, the solution set of the given equation is a subset of $\{-\frac{1}{3}, \frac{2}{3}\}$. We substitute these numbers into the given equation.

$$\sqrt{3-3(-\frac{1}{3})} - \sqrt{3(-\frac{1}{3})+2} \stackrel{?}{=} 3 \qquad \sqrt{3-3(\frac{2}{3})} - \sqrt{3(\frac{2}{3})+2} \stackrel{?}{=} 3$$
$$\sqrt{4} - \sqrt{1} \stackrel{?}{=} 3 \qquad \sqrt{1} - \sqrt{4} \stackrel{?}{=} 3$$
$$2 - 1 \stackrel{?}{=} 3 \qquad 1 - 2 \stackrel{?}{=} 3$$
$$1 \neq 3 \qquad -1 \neq 3$$

We see that each of the numbers is an extraneous solution of the given equation. Thus there is no solution to the given equation, and so its solution set is ∅.

(b) $\sqrt{2x+3} - \sqrt{x-2} - 2 = 0$

$$\sqrt{2x+3} = \sqrt{x-2} + 2$$
$$(\sqrt{2x+3})^2 = (\sqrt{x-2} + 2)^2$$
$$2x + 3 = x - 2 + 4\sqrt{x-2} + 4$$
$$x + 1 = 4\sqrt{x-2}$$
$$(x+1)^2 = 16(\sqrt{x-2})^2$$
$$x^2 + 2x + 1 = 16(x-2)$$
$$x^2 + 2x + 1 = 16x - 32$$
$$x^2 - 14x + 33 = 0$$
$$(x-3)(x-11) = 0$$
$$x - 3 = 0 \quad\quad x - 11 = 0$$
$$x = 3 \quad\quad x = 11$$

Therefore, the solution set of the given equation is a subset of $\{3, 11\}$. We substitute these numbers into the given equation.

$$\sqrt{2(3)+3} - \sqrt{3-2} - 2 \stackrel{?}{=} 0 \quad\quad \sqrt{2(11)+3} - \sqrt{11-2} - 2 \stackrel{?}{=} 0$$
$$\sqrt{9} - \sqrt{1} - 2 \stackrel{?}{=} 0 \quad\quad \sqrt{25} - \sqrt{9} - 2 \stackrel{?}{=} 0$$
$$3 - 1 - 2 \stackrel{?}{=} 0 \quad\quad 5 - 3 - 2 \stackrel{?}{=} 0$$
$$0 = 0 \quad\quad 0 = 0$$

Each of the numbers, 3 and 11, is a solution of the given equation, so its solution set is $\{3, 11\}$.

In the next example we have a polynomial equation of the fourth degree, which is solved by factoring and applying Theorem 2.1.1.

**EXAMPLE 3**
Find the solution set of the equation
$$x^4 - 2x^2 - 15 = 0$$

**SOLUTION**
We factor the left member and obtain

$$(x^2 - 5)(x^2 + 3) = 0 \tag{7}$$

From Theorem 2.1.1, equation (7) gives a true statement if and only if $x^2 - 5 = 0$ or $x^2 + 3 = 0$. Therefore, we set each of the factors in equation (7) equal to zero and solve the equations.

$$x^2 - 5 = 0 \quad\quad x^2 + 3 = 0$$
$$x^2 = 5 \quad\quad x^2 = -3$$
$$x = \pm\sqrt{5} \quad\quad x = \pm i\sqrt{3}$$

The solution set of equation (7) is therefore $\{-i\sqrt{3}, i\sqrt{3}, -\sqrt{5}, \sqrt{5}\}$.

**2.3.2 DEFINITION**  An equation in a single variable $x$ is said to be *an equation quadratic in form* if it can be written as

$$au^2 + bu + c = 0 \quad (a \neq 0) \tag{8}$$

where $u$ is an algebraic expression in $x$.

**ILLUSTRATION 4.** The following equations are quadratic in form.

(a) $x^4 - 2x^2 - 15 = 0$ is quadratic in $x^2$

(b) $3\left(4x - \dfrac{1}{x}\right)^2 - 4\left(4x - \dfrac{1}{x}\right) - 15 = 0$ is quadratic in $\left(4x - \dfrac{1}{x}\right)$ ●

The solution set of an equation quadratic in form can be found by solving equation (8), and then solving the equations obtained by replacing $u$ with the algebraic expression in $x$.

**ILLUSTRATION 5.** The equation of Example 3 is quadratic in form because if we make the substitution $u = x^2$, the equation becomes

$$u^2 - 2u - 15 = 0$$

We solve this equation for $u$.

$$(u - 5)(u + 3) = 0$$
$$u - 5 = 0 \qquad u + 3 = 0$$
$$u = 5 \qquad u = -3$$

Now we replace $u$ with $x^2$ and solve the resulting equations.

$$x^2 = 5 \qquad x^2 = -3$$
$$x = \pm\sqrt{5} \qquad x = \pm i\sqrt{3}$$

The solution set is that of Example 3. ●

It is apparent that the method used in Illustration 5 is an alternative to that used in Example 3.

**EXAMPLE 4**
Find the solution set of the equation
$$3\left(4x - \dfrac{1}{x}\right)^2 - 4\left(4x - \dfrac{1}{x}\right) = 15$$

**SOLUTION**
The given equation is quadratic in $\left(4x - \dfrac{1}{x}\right)$, so we substitute $u = 4x - \dfrac{1}{x}$ and the equation becomes

$$3u^2 - 4u - 15 = 0$$
$$(3u + 5)(u - 3) = 0$$
$$3u + 5 = 0 \qquad u - 3 = 0$$
$$u = -\dfrac{5}{3} \qquad u = 3$$

We replace $u$ with $\left(4x - \dfrac{1}{x}\right)$ and solve for $x$.

$$4x - \frac{1}{x} = -\frac{5}{3} \qquad\qquad 4x - \frac{1}{x} = 3$$

$$3x(4x) - 3x\left(\frac{1}{x}\right) = 3x\left(-\frac{5}{3}\right) \qquad\qquad x(4x) - x\left(\frac{1}{x}\right) = x(3)$$

$$12x^2 - 3 = -5x \qquad\qquad 4x^2 - 1 = 3x$$

$$12x^2 + 5x - 3 = 0 \qquad\qquad 4x^2 - 3x - 1 = 0$$

$$(4x + 3)(3x - 1) = 0 \qquad\qquad (4x + 1)(x - 1) = 0$$

$$4x + 3 = 0 \quad 3x - 1 = 0 \qquad\qquad 4x + 1 = 0 \quad x - 1 = 0$$

$$x = -\frac{3}{4} \quad x = \frac{1}{3} \qquad\qquad x = -\frac{1}{4} \quad x = 1$$

Therefore, the solution set is $\{-\tfrac{3}{4}, -\tfrac{1}{4}, \tfrac{1}{3}, 1\}$.

## EXERCISES 2.3

*In Exercises 1 through 42, find the solution set of the given equation.*

1. $\sqrt{x} - 5 = 3$
2. $\sqrt{x} + 4 = 7$
3. $\sqrt{x + 5} = 3$
4. $\sqrt{x - 4} = 7$
5. $\sqrt{2x - 3} = 2$
6. $\sqrt{2x + 5} = 3$
7. $\sqrt{y} + y = 6$
8. $\sqrt{t} + 6 = t$
9. $\sqrt{3x + 7} = x + 1$
10. $2\sqrt{x + 4} - 1 = x$
11. $\sqrt{x + 5} - \sqrt{x} = 1$
12. $\sqrt{2t + 1} - 2\sqrt{t} - 1 = 0$
13. $\sqrt{3x - 4} + 8 = 0$
14. $\sqrt{5x + 1} + \sqrt{3x + 1} = 0$
15. $\sqrt{5w + 1} - \sqrt{3w} - 1 = 0$
16. $\sqrt{2 + 4y} + \sqrt{3 - 4y} = 3$
17. $\sqrt{2x + 11} + 1 = \sqrt{5x + 1}$
18. $7 - \sqrt{8 - x} = \sqrt{2x + 25}$
19. $\sqrt{t}\sqrt{t - 6} + 4 = 0$
20. $\sqrt{x}\sqrt{x - 5} - 4 = 0$
21. $\sqrt{4 - 3x} + \sqrt{3x - 9} = \sqrt{3x - 14}$
22. $\sqrt{2x - 1} + \sqrt{x + 4} = 3\sqrt{x - 1}$
23. $\sqrt{y - 2\sqrt{y} + 3} + 6 = 0$
24. $\sqrt{w + 2\sqrt{w - 1}} + 2 = 0$
25. $\sqrt{\sqrt{5x - 1} - x} = 1$
26. $\sqrt{2x - 1} - \sqrt{x - 7} - 4 = 0$
27. $\sqrt[3]{3x + 1} = x + 1$
28. $\sqrt[3]{x^2 - 1} = x - 1$
29. $x^4 - 5x^2 + 4 = 0$
30. $9x^4 - 8x^2 - 1 = 0$
31. $t^4 - 5t^2 + 6 = 0$
32. $6w^4 - 17w^2 + 12 = 0$
33. $8x^4 - 6x^2 - 5 = 0$
34. $8x^4 + 6x^2 - 9 = 0$
35. $y^6 - 35y^3 + 216 = 0$
36. $27z^6 - 35z^3 + 8 = 0$
37. $\sqrt[4]{x} + 2\sqrt{x} = 3$
38. $\sqrt{x} - 5\sqrt[4]{x} + 4 = 0$
39. $(x^2 + 2x)^2 - 14(x^2 + 2x) - 15 = 0$
40. $(2x^2 + 7x)^2 - 12(2x^2 + 7x) - 45 = 0$
41. $\sqrt[4]{3x^2 - 3x + 1} = \sqrt{x - 2}$
42. $\sqrt[4]{2x^2 + 5} = \sqrt{x + 2}$

## 2.4 Quadratic Equations in Two Variables

Our discussion, in Section 1.9, of equations in two variables was confined to first-degree equations. Now consider the equation

$$y = x^2 - 3 \qquad (1)$$

which is a second-degree equation (or quadratic equation). As with linear equations in two variables, the solution set of equation (1) is the infinite set of all ordered pairs $(x, y)$ which satisfy the equation, and the graph of the equation is the graph of its solution set. Table 2.4.1 gives a few representative solutions of equation (1); these solutions are obtained by substituting arbitrary numbers for $x$ in the right side of equation (1) and computing the corresponding values for $y$.

**Table 2.4.1**

| $x$ | 0 | 1 | 2 | 3 | 4 | $-1$ | $-2$ | $-3$ | $-4$ |
|---|---|---|---|---|---|---|---|---|---|
| $y = x^2 - 3$ | $-3$ | $-2$ | 1 | 6 | 13 | $-2$ | 1 | 6 | 13 |

In Figure 2.4.1 we have plotted the points having as coordinates the number pairs $(x, y)$ given in Table 2.4.1. These points are on the graph of equation (1). We can get a better idea of the appearance of the graph by plotting additional points between these points. In particular, the $x$ intercepts of the graph are found by substituting 0 for $y$ and solving for $x$. Doing this, we have

$$x^2 - 3 = 0$$
$$x^2 = 3$$
$$x = \pm\sqrt{3}$$

Therefore, the points $(\sqrt{3}, 0)$ and $(-\sqrt{3}, 0)$ are on the graph. More points are obtained from the solutions of equation (1) that appear in Table 2.4.2.

**Table 2.4.2**

| $x$ | $\frac{1}{2}$ | $\frac{3}{2}$ | $\frac{5}{2}$ | $\frac{7}{2}$ | $-\frac{1}{2}$ | $-\frac{3}{2}$ | $-\frac{5}{2}$ | $-\frac{7}{2}$ |
|---|---|---|---|---|---|---|---|---|
| $y = x^2 - 3$ | $-\frac{11}{4}$ | $-\frac{3}{4}$ | $\frac{13}{4}$ | $\frac{37}{4}$ | $-\frac{11}{4}$ | $-\frac{3}{4}$ | $\frac{13}{4}$ | $\frac{37}{4}$ |

Figure 2.4.2 shows the points obtained from the solutions in both Tables 2.4.1 and 2.4.2 as well as the points of intersection of the graph with the $x$ axis. If we connect these points with a smooth curve, the graph has the appearance shown in Figure 2.4.3. As with the line, it is impossible to show the complete graph. The curve in Figure 2.4.3 is an approximation to the graph of equation (1), and we call it a sketch of the graph.

The graph of equation (1) is called a *parabola;* thus Figure 2.4.3 shows a sketch of a parabola. Parabolas and some of their properties are discussed further in Section 2.5.

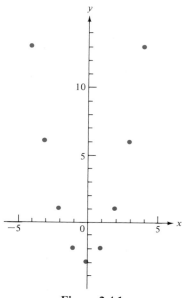

Figure 2.4.1

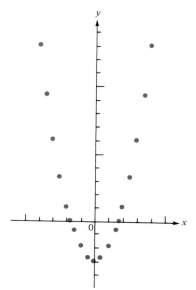

Figure 2.4.2

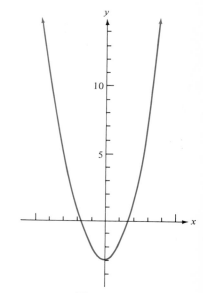

Figure 2.4.3

EXAMPLE 1

Draw a sketch of the graph of the equation

$$y^2 - x - 1 = 0 \qquad (2)$$

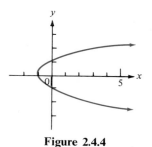

Figure 2.4.4

SOLUTION

Solving equation (2) for $y$, we have

$$y = \pm\sqrt{x + 1} \qquad (3)$$

Equations (3) are equivalent to the two equations

$$y = \sqrt{x + 1} \qquad (4)$$

and

$$y = -\sqrt{x + 1} \qquad (5)$$

The coordinates of all points that satisfy equation (2) satisfy either equation (4) or (5), and the coordinates of any point that satisfy equation (4) or (5) satisfy equation (2). Table 2.4.3 gives some of these values of $x$ and $y$.

Table 2.4.3

| $x$ | 0 | 0  | 1          | 1           | 2          | 2           | 3 | 3  | 4          | 4           | $-1$ |
|-----|---|----|------------|-------------|------------|-------------|---|----|------------|-------------|------|
| $y$ | 1 | $-1$ | $\sqrt{2}$ | $-\sqrt{2}$ | $\sqrt{3}$ | $-\sqrt{3}$ | 2 | $-2$ | $\sqrt{5}$ | $-\sqrt{5}$ | 0    |

Observe that for any value of $x < -1$ there is no real value for $y$. Also, for each value of $x > -1$, there are two values for $y$. A sketch of the graph of equation (2) is shown in Figure 2.4.4.

The graph of equation (2) is also a parabola, and hence Figure 2.4.4 shows a sketch of a parabola.

EXAMPLE 2
Draw sketches of the graphs of the equations

$$y = \sqrt{x+1} \qquad (6)$$

and

$$y = -\sqrt{x+1} \qquad (7)$$

SOLUTION
Equation (6) is the same as equation (4). The value of $y$ is nonnegative; thus the graph of equation (6) is the upper half of the graph of equation (2). A sketch of this graph is shown in Figure 2.4.5.

Equation (7) is the same as equation (5). Hence the graph of equation (7), a sketch of which is shown in Figure 2.4.6, is the lower half of the parabola of Figure 2.4.4.

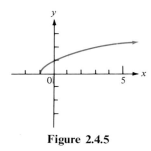

Figure 2.4.5

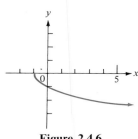

Figure 2.4.6

EXAMPLE 3
Draw a sketch of the graph of the equation

$$16x^2 - 25y^2 = 0$$

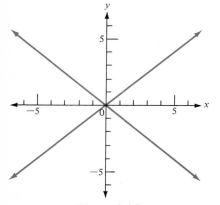

Figure 2.4.7

SOLUTION
Factoring the left member of the equation, we obtain

$$(4x - 5y)(4x + 5y) = 0 \qquad (8)$$

By the property of real numbers, that $ab = 0$ if and only if $a = 0$ or $b = 0$, it follows that equation (8) is satisfied by an ordered pair $(x, y)$ if and only if

$$4x - 5y = 0 \qquad (9)$$

or

$$4x + 5y = 0 \qquad (10)$$

Thus the solution set of equation (8) consists of all of the solutions of the equation $4x - 5y = 0$ and all of the solutions of the equation $4x + 5y = 0$; that is, the solution set of equation (8) is the union of the solution set of equation (9) and the solution set of equation (10). Therefore, the graph of equation (8) consists of the two lines having equations (9) and (10). A sketch of the graph is shown in Figure 2.4.7.

**EXAMPLE 4**
Discuss the graph of each of the following equations.
(a) $2x^2 + 7y^2 = 0$
(b) $x^2 + y^2 = -4$

**SOLUTION**
(a) Both $2x^2$ and $7y^2$ are nonnegative real numbers for any ordered pair of real numbers $(x, y)$. Because the sum of two nonnegative real numbers is zero if and only if the two numbers are zero, it follows that the only ordered pair of real numbers satisfying the equation $2x^2 + 7y^2 = 0$ is the ordered pair $(0, 0)$. Therefore, the graph of the given equation is a single point, the origin.

(b) Because both $x^2$ and $y^2$ are nonnegative real numbers for any ordered pair of real numbers $(x, y)$, the equation $x^2 + y^2 = -4$ has no solution. That is, the solution set of the given equation is the empty set $\emptyset$. Hence the equation has no graph.

When drawing a sketch of the graph of an equation, it is often helpful to consider properties of symmetry of a graph.

**2.4.1 DEFINITION** Two points $P$ and $Q$ are said to be *symmetric with respect to a line* if and only if the line is the perpendicular bisector of the line segment $PQ$.

**ILLUSTRATION 1**

(a) The points $(6, 3)$ and $(6, -3)$ are symmetric with respect to the $x$ axis because the $x$ axis is the perpendicular bisector of the line segment joining $(6, 3)$ and $(6, -3)$. See Figure 2.4.8.

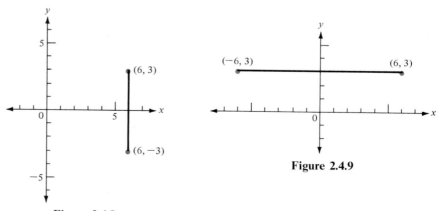

**Figure 2.4.8**

**Figure 2.4.9**

(b) The points $(6, 3)$ and $(-6, 3)$ are symmetric with respect to the $y$ axis because the $y$ axis is the perpendicular bisector of the line segment joining $(6, 3)$ and $(-6, 3)$. See Figure 2.4.9. ●

In general, the points $(x, y)$ and $(x, -y)$ are symmetric with respect to the $x$ axis and the points $(x, y)$ and $(-x, y)$ are symmetric with respect to the $y$ axis.

**2.4.2 DEFINITION** The graph of an equation is symmetric with respect to a line $L$ if and only if for every point $P$ on the graph there is a point $Q$, also on the graph, such that $P$ and $Q$ are symmetric with respect to $L$.

**ILLUSTRATION 2.** The graph of equation (2) in Example 1 is symmetric with respect to the $x$ axis, as can be seen by the sketch of the graph shown in Figure 2.4.4. Equation (2) is

$$y^2 - x - 1 = 0$$

If $y$ is replaced by $-y$ in this equation, we obtain

$$(-y)^2 - x - 1 = 0$$

or, equivalently,

$$y^2 - x - 1 = 0$$

which is equation (2). That is, if in equation (2), $y$ is replaced by $-y$, we obtain an equivalent equation. •

Illustration 2 gives a particular case of part (i) of the following theorem.

**2.4.3 THEOREM** The graph of an equation in $x$ and $y$ is

(i) symmetric with respect to the $x$ axis if and only if an equivalent equation is obtained when $y$ is replaced by $-y$ in the equation;
(ii) symmetric with respect to the $y$ axis if and only if an equivalent equation is obtained when $x$ is replaced by $-x$ in the equation.

**PROOF OF PART (i):** From Definition 2.4.2 it follows that if a point $(x, y)$ is on a graph that is symmetric with respect to the $x$ axis then the point $(x, -y)$ also must be on the graph. Furthermore, if both the points $(x, y)$ and $(x, -y)$ are on the graph, then the graph is symmetric with respect to the $x$ axis. Therefore, the coordinates of the point $(x, -y)$ as well as $(x, y)$ must satisfy an equation of the graph. Hence we may conclude that the graph of an equation in $x$ and $y$ is symmetric with respect to the $x$ axis if and only if an equivalent equation is obtained when $y$ is replaced by $-y$ in the equation.

The proof of part (ii) is similar to the proof of part (i).

**ILLUSTRATION 3.** Equation (1) is

$$y = x^2 - 3$$

If $x$ is replaced by $-x$ in this equation, we obtain

$$y = (-x)^2 - 3$$

which is equivalent to the original equation. Therefore, by Theorem 2.4.3(ii) the graph of equation (1) is symmetric with respect to the $y$ axis. See Figure 2.4.3 for a sketch of the graph of equation (1). •

**2.4.4 DEFINITION**  Two points $P$ and $Q$ are said to be *symmetric with respect to a third point* if and only if the third point is the midpoint of the line segment $PQ$.

**ILLUSTRATION 4.** The points $(6, 3)$ and $(-6, -3)$ are symmetric with respect to the origin because the origin is the midpoint of the line segment joining $(6, 3)$ and $(-6, -3)$. See Figure 2.4.10. ●

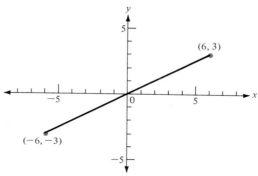

**Figure 2.4.10**

**2.4.5 DEFINITION**  The graph of an equation is symmetric with respect to a point $R$ if and only if for every point $P$ on the graph there is a point $S$, also on the graph, such that $P$ and $S$ are symmetric with respect to $R$.

The following example gives a graph that is symmetric with respect to the origin.

**EXAMPLE 5**

Draw a sketch of the graph of the equation

$$xy = 4 \qquad (11)$$

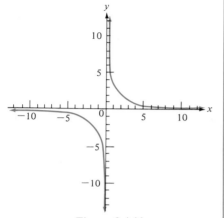

**Figure 2.4.11**

**SOLUTION**

Table 2.4.4 gives some values of $x$ and $y$ satisfying equation (11).

**Table 2.4.4**

| $x$ | $-12$ | $-8$ | $-4$ | $-3$ | $-2$ | $-1$ | $-\frac{1}{2}$ | $-\frac{1}{3}$ | $\frac{1}{3}$ | $\frac{1}{2}$ | $1$ | $2$ | $3$ | $4$ | $8$ | $12$ |
|---|---|---|---|---|---|---|---|---|---|---|---|---|---|---|---|---|
| $y$ | $-\frac{1}{3}$ | $-\frac{1}{2}$ | $-1$ | $-\frac{4}{3}$ | $-2$ | $-4$ | $-8$ | $-12$ | $12$ | $8$ | $4$ | $2$ | $\frac{4}{3}$ | $1$ | $\frac{1}{2}$ | $\frac{1}{3}$ |

From Equation (11), we obtain

$$y = \frac{4}{x}$$

We see that as $x$ increases through positive values, $y$ decreases through positive values and gets closer and closer to zero. As $x$ decreases through positive values, $y$ increases through positive values and gets larger and larger. As $x$ increases through negative values (that is, $x$ takes on the values $-12$, $-8$, $-4$, $-3$, $-2$, $-1$, $-\frac{1}{2}$, $-\frac{1}{3}$, and so on), $y$ takes on negative values having larger and larger absolute values. A sketch of the graph is shown in Figure 2.4.11.

The curve of Example 5 is called a *hyperbola* and the $x$ and $y$ axes are called *asymptotes* of this hyperbola. Other hyperbolas are discussed in Section 2.6.

**2.4.6 THEOREM** The graph of an equation in $x$ and $y$ is symmetric with respect to the origin if and only if an equivalent equation is obtained when $x$ is replaced by $-x$ and $y$ is replaced by $-y$ in the equation.

The proof of Theorem 2.4.6 is similar to the proof of Theorem 2.4.3(i) and is omitted.

**ILLUSTRATION 5.** Equation (11) is

$$xy = 4$$

If $x$ is replaced by $-x$ and $y$ is replaced by $-y$ in this equation, we have

$$(-x)(-y) = 4$$

which is equivalent to the original equation. Therefore, by Theorem 2.4.6, the graph of equation (11) is symmetric with respect to the origin. See Figure 2.4.11 for a sketch of the graph of equation (11). ●

## EXERCISES 2.4

*In Exercises 1 through 26, draw a sketch of the graph of the given equation.*

1. $y = x^2 - 6$
2. $y = x^2 + 3$
3. $y = x^2 + 1$
4. $y = x^2 - 2$
5. $y = -x^2 + 2$
6. $y = -x^2 + 4$
7. $y = \sqrt{x - 4}$
8. $y = -\sqrt{x - 4}$
9. $y^2 = x - 4$
10. $y^2 = x + 3$
11. $x = y^2 + 5$
12. $2y^2 + x = 4$
13. $y = \sqrt{3x + 4}$
14. $y = -\sqrt{2x - 5}$
15. $9x^2 - 4y^2 = 0$
16. $25x^2 + y^2 = 0$
17. $9x^2 + 4y^2 = 0$
18. $25x^2 - y^2 = 0$
19. $x^2 + 2y^2 + 1 = 0$
20. $(2x + y)(3x - y + 6) = 0$
21. $(x + y + 7)(4x + 5y - 20) = 0$
22. $(x - y)(x^2 - y) = 0$
23. $(x^2 + y)(2x + y - 1) = 0$
24. $4 + 4x^2 + y^2 = 0$
25. $xy = 9$
26. $4xy = 1$

27. Draw a sketch of the graph of each of the following equations on a different coordinate system:
    (a) $y = \sqrt{x}$  (b) $y = -\sqrt{x}$
    (c) $y^2 = x$
28. Draw a sketch of the graph of each of the following equations on a different coordinate system:
    (a) $y = \sqrt{-x}$  (b) $y = -\sqrt{-x}$
    (c) $y^2 = -x$
29. Draw a sketch of the graph of each of the following equations on a different coordinate system:
    (a) $y = \sqrt{9 - x}$  (b) $y = -\sqrt{9 - x}$
    (c) $y^2 = 9 - x$
30. Draw a sketch of the graph of each of the following equations on a different coordinate system:
    (a) $y = \sqrt{2x + 1}$  (b) $y = -\sqrt{2x + 1}$
    (c) $y^2 = 2x + 1$
31. Write an equation whose graph is the set of all points on either the $x$ axis or the $y$ axis.
32. Draw a sketch of the graph of each of the following equations on a different coordinate system:
    (a) $y^2 = 9 - x^2$  (b) $y = \sqrt{9 - x^2}$
    (c) $y = -\sqrt{9 - x^2}$

## 2.5 The Parabola

A complete treatment of the parabola and its properties belongs to a course in analytic geometry. However, because of the importance of the parabola in our study of quadratic functions (Section 3.4) we give a brief discussion here.

**2.5.1 DEFINITION**  A parabola is the set of all points in a plane equidistant from a fixed point and a fixed line. The fixed point is called the *focus*, and the fixed line is called the *directrix*.

We use Definition 2.5.1 to derive an equation of a parabola where the rectangular Cartesian coordinate system is chosen so that the $y$ axis contains the focus and is perpendicular to the directrix. Furthermore, we choose the origin as the point on the $y$ axis midway between the focus and the directrix. Observe that we are choosing the coordinate axes (not the parabola) in a special way. See Figure 2.5.1.

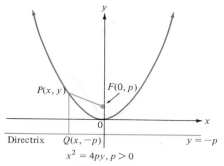

Figure 2.5.1

Let $p$ be the directed distance $\overline{OF}$. The focus is the point $F(0, p)$, and the directrix is the line having the equation $y = -p$. A point $P(x, y)$ is on the parabola if and only if $P$ is equidistant from $F$ and the directrix. That is, if $Q(x, -p)$ is the foot of the perpendicular line from $P$ to the directrix, then $P$ is on the parabola if and only if

$$|\overline{FP}| = |\overline{QP}|$$

Because

$$|\overline{FP}| = \sqrt{x^2 + (y - p)^2}$$

and

$$|\overline{QP}| = \sqrt{(x - x)^2 + (y + p)^2}$$

the point $P$ is on the parabola if and only if

$$\sqrt{x^2 + (y - p)^2} = \sqrt{(y + p)^2}$$

By squaring on both sides of the equation, we obtain

$$x^2 + y^2 - 2py + p^2 = y^2 + 2py + p^2$$
$$x^2 = 4py$$

This result is stated as a theorem.

**2.5.2 THEOREM**  An equation of the parabola having its focus at $(0, p)$ and having as its directrix the line $y = -p$ is

$$x^2 = 4py \tag{1}$$

In Figure 2.5.1, $p$ is positive; $p$ may be negative, however, because it is the directed distance $\overline{OF}$. Figure 2.5.2 shows a parabola for $p < 0$.

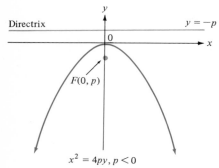

$x^2 = 4py, p < 0$

**Figure 2.5.2**

From Figures 2.5.1 and 2.5.2 we see that for the equation $x^2 = 4py$ the parabola opens upward if $p > 0$ and downward if $p < 0$. The point midway between the focus and the directrix on the parabola is called the *vertex*. The vertex of the parabolas in Figures 2.5.1 and 2.5.2 is the origin. The line through the vertex and the focus is called the *axis* of the parabola. A parabola is symmetric with respect to its axis. The axis of the parabolas of Figures 2.5.1 and 2.5.2 is the $y$ axis.

**ILLUSTRATION 1.** The graph of the equation

$$x^2 = 10y$$

is a parabola whose vertex is at the origin and whose axis is the $y$ axis. Because $4p = 10$, $p = \frac{5}{2} > 0$, and therefore the parabola opens upward. The focus is at the point $F(0, \frac{5}{2})$ and an equation of the directrix is $y = -\frac{5}{2}$. Two points on the parabola are the points $(5, \frac{5}{2})$ and $(-5, \frac{5}{2})$. Figure 2.5.3 shows a sketch of the parabola, the focus, and the directrix. ●

The points $(5, \frac{5}{2})$ and $(-5, \frac{5}{2})$ on the parabola of Illustration 1 are the endpoints of the chord through the focus, perpendicular to the axis of the parabola. This chord is called the *latus rectum* of the parabola. When drawing a sketch of the graph of a parabola, it is helpful to plot the points that are the endpoints of the latus rectum.

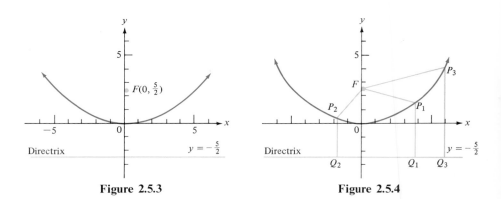

**Figure 2.5.3**     **Figure 2.5.4**

In Figure 2.5.4 the parabola of Figure 2.5.3 is shown and three points $P_1$, $P_2$, and $P_3$ on the parabola are chosen. The definition of a parabola states that any point on the parabola is equidistant from the focus and the directrix; therefore,

$$|FP_1| = |Q_1P_1| \qquad |FP_2| = |Q_2P_2| \qquad |FP_3| = |Q_3P_3|$$

## EXAMPLE 1

Draw a sketch of the graph of the parabola having the equation

$$x^2 = -9y$$

Find the focus and an equation of the directrix.

## SOLUTION

The graph is a parabola whose vertex is at the origin and whose axis is the $y$ axis. Because $4p = -9$, $p = -\frac{9}{4} < 0$, and so the parabola opens downward. The focus is at the point $F(0, -\frac{9}{4})$, and an equation of the directrix is $y = \frac{9}{4}$. The endpoints of the latus rectum are $(\frac{9}{2}, -\frac{9}{4})$ and $(-\frac{9}{2}, -\frac{9}{4})$. Figure 2.5.5 shows a sketch of the parabola, the focus, and the directrix.

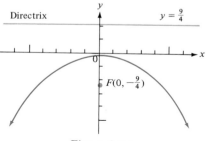

**Figure 2.5.5**

Our work with parabolas has been concerned so far with those having the origin as the vertex and having the $y$ axis as the axis. We now consider a more general situation and discuss parabolas for which the axis is parallel to the $y$ axis.

If a parabola has its vertex at the point $(h, k)$ and the axis of the parabola is the line $x = h$, then an equation of the parabola is

$$(x - h)^2 = 4p(y - k) \tag{2}$$

If $p > 0$, the parabola opens upward, and a sketch of such a parabola is shown in Figure 2.5.6. A sketch of a parabola having an equation of the form (2) where $p < 0$ is shown in Figure 2.5.7; the parabola opens downward.

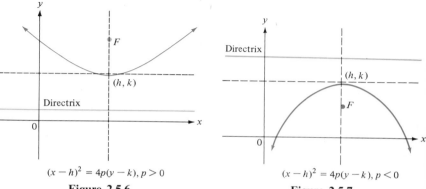

$(x - h)^2 = 4p(y - k), p > 0$
**Figure 2.5.6**

$(x - h)^2 = 4p(y - k), p < 0$
**Figure 2.5.7**

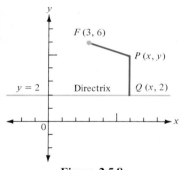

Figure 2.5.8

We omit the proof that the graph of equation (2) is the indicated parabola. However, the next illustration shows the derivation of an equation of a parabola whose axis is a line parallel to the $y$ axis, and the result is an equation of the form (2).

**ILLUSTRATION 2.** We use Definition 2.5.1 to find an equation of the parabola whose focus is at the point $(3, 6)$ and whose directrix has the equation $y = 2$. See Figure 2.5.8.

Let $P(x, y)$ be any point on the parabola and $Q(x, 2)$ be the foot of the perpendicular line from $P$ to the directrix. Then, from Definition 2.5.1, $P$ is on the parabola if and only if

$$|\overline{FP}| = |\overline{QP}|$$

Because

$$|\overline{FP}| = \sqrt{(x-3)^2 + (y-6)^2}$$

and

$$|\overline{QP}| = \sqrt{(x-x)^2 + (y-2)^2}$$

the point $P$ is on the parabola if and only if

$$\sqrt{(x-3)^2 + (y-6)^2} = \sqrt{(y-2)^2}$$

By squaring on both sides of the equation, we obtain

$$(x-3)^2 + (y-6)^2 = (y-2)^2$$

Simplifying the squares of the binomials involving $y$, we have

$$(x-3)^2 + y^2 - 12y + 36 = y^2 - 4y + 4$$
$$(x-3)^2 = 8(y-4) \qquad \bullet \quad (3)$$

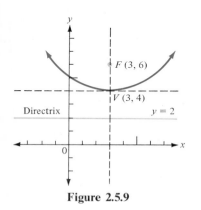

Figure 2.5.9

Equation (3) is of the form of equation (2). Note that we are given that the focus of the parabola is at the point $(3, 6)$ and the directrix has the equation $y = 2$. Because the axis of the parabola is the line through the focus and perpendicular to the directrix, it follows that the axis has the equation $x = 3$, that is, $x = k$, where $k$ is 3. Furthermore, the vertex is the point on the axis midway between the directrix and the focus; hence the vertex is at the point $(3, 4)$, and in equation (3) $(h, k)$ is $(3, 4)$. The focus is above the vertex and the measure of the distance from the vertex to the focus is 2; in equation (3) $p$ is 2. We have therefore verified equation (2) for this specific parabola; a sketch of its graph is shown in Figure 2.5.9.

The graph of any quadratic equation of the form

$$y = Ax^2 + Bx + C \qquad (4)$$

where $A$, $B$, and $C$ are constants and $A \neq 0$, is a parabola whose axis is parallel to the $y$ axis. This statement can be proved by showing that equation (4) is equivalent to an equation of the form of equation (2). We omit this proof, but the next illustration shows such a proof for a particular case.

**ILLUSTRATION 3.** The equation

$$y = \frac{1}{4}x^2 + x + 6 \tag{5}$$

is in the form of equation (4) and is equivalent to the equation

$$4y = x^2 + 4x + 24$$
$$x^2 + 4x = 4y - 24$$

Completing the square in the left member by adding 4 to each member, we have

$$x^2 + 4x + 4 = 4y - 24 + 4$$
$$(x + 2)^2 = 4(y - 5) \tag{6}$$

which is of the form of equation (2). The given equation (5) then has a parabola as its graph, and because equation (6) is equivalent to equation (5) we see that the vertex of the parabola is at $(-2, 5)$ and the axis has the equation $x = -2$. ●

**EXAMPLE 2**

Show that the graph of the equation

$$y = 2x^2 + 4x - 1$$

is a parabola. Find the vertex, the axis, the focus, and the directrix. Also, draw a sketch of the parabola.

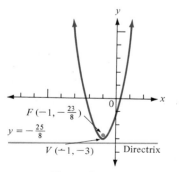

**Figure 2.5.10**

**SOLUTION**

The given equation is equivalent to

$$2(x^2 + 2x) = y + 1$$

We complete the square of the binomial within the parentheses in the left member by adding 2 to each member (note that when we add 1 within the parentheses, we are actually adding 2 because each term within the parentheses is multiplied by 2), and we have

$$2(x^2 + 2x + 1) = y + 1 + 2$$
$$(x + 1)^2 = \frac{1}{2}(y + 3) \tag{7}$$

Equation (7) is of the form of equation (2), where $(h, k)$ is $(-1, -3)$ and $4p = \frac{1}{2}$; therefore, $p = \frac{1}{8}$. The vertex is at $(-1, -3)$ and the axis is the line having the equation $x = -1$. Because $p > 0$, the parabola opens upward. The measure of the distance from the vertex to the focus is $p = \frac{1}{8}$. Thus the focus is at the point $(-1, -\frac{23}{8})$. Because the vertex is midway between the focus and the directrix, an equation of the directrix is $y = -\frac{25}{8}$. Figure 2.5.10 shows a sketch of the parabola, the focus, and the directrix.

## EXERCISES 2.5

*For each of the parabolas in Exercises 1 through 8, draw a sketch, and find the focus and an equation of the directrix.*

1. $y = x^2$
2. $y = -4x^2$
3. $x^2 - 8y = 0$
4. $x^2 + 10y = 0$
5. $x^2 = -16y$
6. $x^2 = 5y$
7. $x^2 + 2y = 0$
8. $5x^2 + 9y = 0$

*In Exercises 9 through 18, show that the graph of the given equation is a parabola. Find the vertex, the axis, the focus, and the directrix. Also, draw a sketch of the parabola.*

9. $y = -x^2 + 3$
10. $y = x^2 - 5$
11. $y = x^2 - 6x$
12. $y = x^2 + 4x$
13. $y = x^2 + 4x - 2$
14. $y = -x^2 + 6x - 5$
15. $y = -2x^2 - 8x - 5$
16. $y = 2x^2 + 10x + 3$
17. $y = \frac{1}{8}x^2 - \frac{1}{2}x - \frac{3}{2}$
18. $y = -\frac{1}{4}x^2 - \frac{3}{2}x - 2$

*In Exercises 19 through 24, use Definition 2.5.1 to find an equation of the parabola having the given focus and directrix.*

19. Focus at $(0, -3)$; directrix, $y = 3$
20. Focus at $(0, 4)$; directrix, $y = -4$
21. Focus at $(0, \frac{3}{4})$; directrix, $4y + 3 = 0$
22. Focus at $(0, -\frac{1}{2})$; directrix, $2y - 1 = 0$
23. Focus at $(-2, 9)$; directrix, $y = 5$
24. Focus at $(4, -\frac{3}{4})$; directrix, $y = -\frac{5}{4}$

25. A parabola has its focus at $F(p, 0)$ and its directrix is the line having the equation $x = -p$. Prove that an equation of the parabola is $y^2 = 4px$.

*In Exercises 26 through 29, use the result of Exercise 25 to find the focus and an equation of the directrix of the given parabola. Draw a sketch of the parabola.*

26. $x = y^2$
27. $x = -8y^2$
28. $y^2 + 6x = 0$
29. $2y^2 - 5x = 0$

## 2.6 The Circle, the Ellipse, and the Hyperbola

Curves, other than the parabola, that are graphs of quadratic equations in two variables are the circle, the ellipse, and the hyperbola. As with the parabola, a complete discussion of these curves belongs to a course in analytic geometry. However, the curves occur so often that a brief discussion is given here. The simplest of the three curves is the circle.

**2.6.1 DEFINITION**  A *circle* is the set of all points in a plane equidistant from a fixed point. The fixed point is called the *center* of the circle, and the measure of the constant equal distance is called the *radius* of the circle.

**2.6.2 THEOREM** The circle with center at the point $C(h, k)$ and radius $r$ has as an equation

$$(x - h)^2 + (y - k)^2 = r^2 \tag{1}$$

**Proof.** The point $P(x, y)$ lies on the circle if and only if

$$|\overline{PC}| = r$$

that is, if and only if

$$\sqrt{(x - h)^2 + (y - k)^2} = r$$

Squaring on both sides of the equation, we obtain the equivalent equation (remember that $r$ is a positive number)

$$(x - h)^2 + (y - k)^2 = r^2$$

which is equation (1). Equation (1) is satisfied by the coordinates of those and only those points that lie on the given circle; hence (1) is an equation of the circle.

Figure 2.6.1 shows the circle with center at $(h, k)$ and radius $r$. If the center of the circle is at the origin, then $h = k = 0$; therefore, its equation is

$$x^2 + y^2 = r^2 \tag{2}$$

Such a circle is shown in Figure 2.6.2.

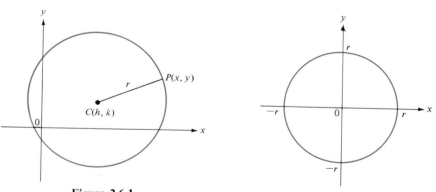

**Figure 2.6.1**    **Figure 2.6.2**

If the center and the radius of a circle are known, the circle can be drawn by using a compass.

### EXAMPLE 1

Find an equation of the circle with center at $(5, -3)$ and radius equal to 4. Draw the circle.

### SOLUTION

The circle is shown in Figure 2.6.3. An equation of the circle is of the form of equation (1), where $h = 5$, $k = -3$, and $r = 4$. Substituting into equation (1), we have

$$(x - 5)^2 + [y - (-3)]^2 = 4^2$$
$$(x - 5)^2 + (y + 3)^2 = 16$$
$$x^2 - 10x + 25 + y^2 + 6y + 9 = 16$$
$$x^2 + y^2 - 10x + 6y + 18 = 0$$

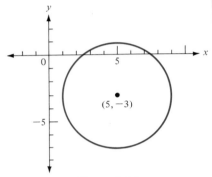

**Figure 2.6.3**

### EXAMPLE 2

Show that the graph of the equation

$$x^2 + y^2 + 4x - 8y - 5 = 0$$

is a circle and find its center and radius.

### SOLUTION

The given equation may be written as

$$(x^2 + 4x) + (y^2 - 8y) = 5$$

Completing the squares of the terms in parentheses by adding 4 and 16 on both sides of the equation, we have

$$(x^2 + 4x + 4) + (y^2 - 8y + 16) = 5 + 4 + 16$$
$$(x + 2)^2 + (y - 4)^2 = 25$$

Comparing this equation with equation (1), we see that this is an equation of a circle with center at $(-2, 4)$ and a radius of 5.

Consider now the equation

$$Ax^2 + By^2 = C \qquad (3)$$

where $A$, $B$, and $C$ are nonzero constants having the same sign. A special case of this equation occurs when $A = B$. Then, we have

$$Ax^2 + Ay^2 = C \qquad (4)$$

Dividing on both sides of this equation by $A$, we obtain

$$x^2 + y^2 = \frac{C}{A} \qquad (5)$$

Because $C$ and $A$ are nonzero numbers having the same sign, it follows that $\frac{C}{A} > 0$, and hence equation (5) is of the form of equation (2). Therefore, the graph of equation (4) is a circle having its center at the origin and radius $\sqrt{\frac{C}{A}}$.

**ILLUSTRATION 1.** The equation

$$4x^2 + 4y^2 = 1$$

is equivalent to the equation

$$x^2 + y^2 = \frac{1}{4}$$

which is of the form of equation (2). Hence the graph of the equation is a circle having its center at the origin and radius $\frac{1}{2}$. ●

The next illustration is a special case of equation (3), where $A$ and $B$ are not equal.

**ILLUSTRATION 2.** To draw the graph of the equation

$$4x^2 + 9y^2 = 36 \tag{6}$$

we first find the points of intersection of the graph with the $x$ and $y$ axes. We find the $x$ intercepts by replacing $y$ by 0. Doing this, we have

$$4x^2 = 36$$
$$x^2 = 9$$
$$x = \pm 3$$

Therefore, the graph intersects the $x$ axis at the points $(-3, 0)$ and $(3, 0)$. We find the $y$ intercepts by replacing $x$ by 0, and we have

$$9y^2 = 36$$
$$y^2 = 4$$
$$y = \pm 2$$

Thus the graph intersects the $y$ axis at the points $(0, -2)$ and $(0, 2)$. We solve equation (6) for $y$, and we have

$$9y^2 = 36 - 4x^2$$
$$y^2 = \frac{4}{9}(9 - x^2)$$
$$y = \pm \frac{2}{3}\sqrt{9 - x^2} \tag{7}$$

Recalling Theorems 2.4.3 and 2.4.6, we see that the graph of equation (6) is symmetric with respect to the $x$ axis, the $y$ axis, and the origin. Therefore, if we have the part of the graph that is in the first quadrant, we can use the properties of symmetry to complete the graph in the other three quadrants. To obtain representative points in the first quadrant, we assign positive values to $x$ and find the corresponding positive value of $y$ by substituting into equation (7) with the plus sign. Two such points are $(2, \frac{2}{3}\sqrt{5})$ and $(1, \frac{4}{3}\sqrt{2})$. If we plot these two points and the points $(0, 2)$ and $(3, 0)$ and connect the points by a smooth curve, we have the part of the graph in the first quadrant as shown in Figure 2.6.4. The complete graph is shown in Figure 2.6.5.

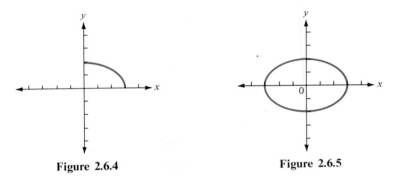

Figure 2.6.4      Figure 2.6.5

In Illustration 2 observe that because $y$ is a real number, it follows from equations (7) that we are concerned only with values of $x$ for which $9 - x^2 \geq 0$ (otherwise, the radicand is a negative number); that is, it is not necessary to assign any values to $x$ for which $|x| > 3$.

The curve shown in Figure 2.6.5 is an *ellipse*. Our discussion of the ellipse, as well as that of the hyperbola, is less formal than that of the parabola and the circle. Our treatment is restricted to those ellipses and hyperbolas having fairly simple equations.

Refer again to equation (3). It can be proved that any equation of the form

$$Ax^2 + By^2 = C$$

where $A$, $B$, and $C$ are nonzero real numbers having the same sign, has a graph that is an ellipse. In Illustration 2, we have such an equation where $A$ is 4, $B$ is 9, and $C$ is 36; the graph is an ellipse. Because equation (4) is a special case of equation (3) (that is, $A = B$), and because the graph of equation (4) is a circle, it follows that a circle is a special case of an ellipse.

Suppose now that we have an equation of the form

$$Ax^2 + By^2 = C$$

where $A$ and $B$ are real numbers having opposite signs, and $C$ is a nonzero real number. In the next illustration we have such an equation.

**ILLUSTRATION 3.** We wish to determine the graph of the equation

$$25x^2 - 9y^2 = 225 \tag{8}$$

If we replace $y$ by 0, we have

$$25x^2 = 225$$
$$x^2 = 9$$
$$x = \pm 3$$

Hence the graph intersects the $x$ axis at the points $(-3, 0)$ and $(3, 0)$. If we replace $x$ by 0, we have

$$-9y^2 = 225$$
$$y^2 = -25$$

Because the only values of $y$ satisfying this equation are $5i$ and $-5i$, it follows that the graph does not intersect the $y$ axis.

The graph is symmetric with respect to the $y$ axis because we obtain an equivalent equation when, in equation (8), $x$ is replaced by $-x$. Also, the graph is symmetric with respect to the $x$ axis, because we obtain an equivalent equation when $y$ is replaced by $-y$. Furthermore, when both $x$ is replaced by $-x$ and $y$ is replaced by $-y$, we also obtain an equation equivalent to equation (8); thus the graph is symmetric with respect to the origin. Therefore, as in Illustration 2, we need only to find the part of the graph that is in the first quadrant, and then we complete the graph by using the properties of symmetry.

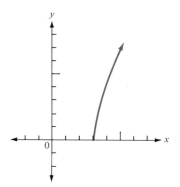

**Figure 2.6.6**

Solving equation (8) for $y$, we have

$$-9y^2 = -25x^2 + 225$$
$$y^2 = \frac{25}{9}(x^2 - 9)$$
$$y = \pm \frac{5}{3}\sqrt{x^2 - 9} \tag{9}$$

Some representative points in the first quadrant can be obtained by taking values of $x > 3$ and finding the corresponding positive value of $y$ by substituting into (9) with the plus sign. Doing this, we see that the points $(5, \frac{20}{3})$ and $(7, \frac{10}{3}\sqrt{10})$ are on the graph. If we plot these two points and the point $(3, 0)$, and connect the points by a smooth curve, we have a sketch of the part of the graph in the first quadrant as shown in Figure 2.6.6. A sketch of the complete graph is shown in Figure 2.6.7. ●

The curve of Figure 2.6.7 is a *hyperbola*. It can be proved that the graph of any equation of the form

$$Ax^2 + By^2 = C \tag{10}$$

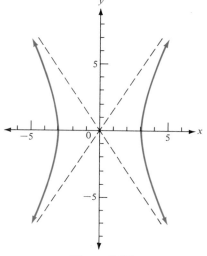

**Figure 2.6.7**

where $A$ and $B$ are real numbers having opposite signs and $C$ is a nonzero real number, is a hyperbola.

The dashed lines in Figure 2.6.7 are called *asymptotes* of the hyperbola. They are shown as dashed lines because they are not part of the graph, but they are used as guides in drawing a sketch of the graph. A rigorous definition of an asymptote of a graph requires the concept of "limit" which is discussed in the calculus. However, the following statement can be considered as an intuitive idea of an asymptote: If the distance between a graph and a line gets smaller and smaller as either $|x|$ or $|y|$ gets larger and larger, then the line is an asymptote of the graph. Every hyperbola has two asymptotes. If an equation of a hyperbola is in the form of equation (10), then the equations of the asymptotes can be obtained by replacing $C$ by zero. For instance, if in equation (8) we replace 225 by 0, we have

$$25x^2 - 9y^2 = 0$$

The left member of this equation can be factored as the difference of two squares, and we have

$$(5x - 3y)(5x + 3y) = 0$$

The solution set of this equation is the union of the solution sets of the equations

$$5x - 3y = 0 \tag{11}$$

and

$$5x + 3y = 0 \tag{12}$$

Equations (11) and (12) are equations of the asymptotes of the hyperbola having equation (8). These lines are the ones shown as dashed lines in Figure 2.6.7.

For other equations of the form

$$Ax^2 + By^2 = C$$

we refer back to Section 2.4. In Example 3 of Section 2.4 we have the equation

$$16x^2 - 25y^2 = 0$$

which is of the form $Ax^2 + By^2 = C$, where $A$ and $B$ have opposite signs and $C = 0$. The graph consists of two lines shown in Figure 2.4.7. In Example 4(a) of Section 2.4 we have the equation

$$2x^2 + 7y^2 = 0$$

which is of the form $Ax^2 + By^2 = C$, where $A$ and $B$ have the same sign and $C = 0$. The graph consists of one point, the origin. In Example 4(b) of Section 2.4 we have the equation

$$x^2 + y^2 = -4$$

which is of the form $Ax^2 + By^2 = C$, where $A > 0$, $B > 0$, and $C < 0$. The equation has no graph.

In Table 2.6.1 we have a summary of our discussion of the graphs of the various types of equations of the form

$$Ax^2 + By^2 = C$$

Table 2.6.1  Graphs of equations of the form $Ax^2 + By^2 = C$
($A$, $B$, and $C$ are real)

| Conditions on $A$, $B$, and $C$ | Conclusion | Particular Example |
|---|---|---|
| $A$, $B$, and $C$ have the same sign | Graph is an ellipse | $4x^2 + 9y^2 = 36$ |
| $A$, $B$, and $C$ have the same sign and $A = B$ | Graph is a circle | $4x^2 + 4y^2 = 1$ |
| $A$ and $B$ have opposite signs and $C \neq 0$ | Graph is a hyperbola | $25x^2 - 9y^2 = 225$ |
| $A$ and $B$ have opposite signs and $C = 0$ | Graph is two distinct lines through the origin | $16x^2 - 25y^2 = 0$ |
| $A$ and $B$ have the same sign and $C = 0$ | Graph is the origin | $2x^2 + 7y^2 = 0$ |
| $A$ and $B$ have the same sign and $C$ has the opposite sign | No graph | $x^2 + y^2 = -4$ |

**EXAMPLE 3**

Name the graph of the equation

$$4y^2 - x^2 = 16 \qquad (13)$$

Draw a sketch of the graph.

**SOLUTION**

The equation is of the form $Ax^2 + By^2 = C$, where $A = -1$, $B = 4$, and $C = 16$. Because $A$ and $B$ have opposite signs and $C \neq 0$, the graph is a hyperbola. When $x = 0$, we have $4y^2 = 16$, and so $y = \pm 2$. Therefore, the graph intersects the $y$ axis at the points $(0, -2)$ and $(0, 2)$. When $y = 0$, we have $-x^2 = 16$, which has no real value solutions. Therefore, the graph does not intersect the $x$ axis. If in the given equation we replace 16 by 0, we get

$$4y^2 - x^2 = 0$$
$$(2y - x)(2y + x) = 0$$

The lines having the equations

$$2y - x = 0$$

and

$$2y + x = 0$$

are asymptotes of the hyperbola. These lines are drawn as dashed lines in Figure 2.6.8.

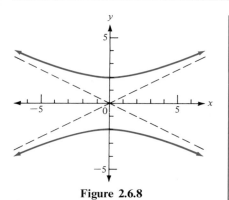

Figure 2.6.8

The graph is symmetric with respect to the $x$ axis, the $y$ axis, and the origin. Therefore, we need only to plot points in the first quadrant, and then apply properties of symmetry to get the complete graph.

If we solve equation (13) for $x$, we obtain

$$-x^2 = -4y^2 + 16$$
$$x^2 = 4(y^2 - 4)$$
$$x = \pm 2\sqrt{y^2 - 4} \qquad (14)$$

We see from (14) that if $|y| < 2$, the radicand is negative and hence $x$ is not a real number. Thus we are only concerned with values of $y$ for which $|y| \geq 2$. When $y = 3$ and $x > 0$, $x = 2\sqrt{5}$. When $y = 4$ and $x > 0$, $x = 4\sqrt{3}$. We plot the points $(0, 2)$, $(2\sqrt{5}, 3)$, and $(4\sqrt{3}, 4)$ and connect them with a smooth curve. This gives us a sketch of the graph in the first quadrant. A sketch of the complete graph is drawn from the properties of symmetry. The asymptotes are used as guides. See Figure 2.6.8.

A *conic section* (or *conic*) is a curve of intersection of a plane with a right-circular cone of two nappes. Three types of curves of intersection that occur are the parabola, the ellipse (including the circle as a special case), and the hyperbola. The Greek mathematician Apollonius studied conic sections, in terms of geometry, by using this concept.

In consideration of the geometry of conic sections, a cone is regarded as having two nappes, extending indefinitely far in both directions. A portion of a right-circular cone of two nappes is shown in Figure 2.6.9. A *generator* (or element) of the cone is a line lying in the cone, and all the generators of a cone contain the point $V$, called the *vertex* of the cone.

In Figure 2.6.10 we have a cone and a cutting plane which is parallel to

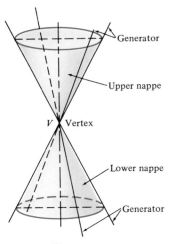

Figure 2.6.9

Figure 2.6.10

Figure 2.6.11

Figure 2.6.12

Figure 2.6.13

one and only one generator of the cone. This conic is a parabola. If the cutting plane is parallel to two generators, it intersects both nappes of the cone and we have a hyperbola (see Figure 2.6.11). An ellipse is obtained if the cutting plane is parallel to no generator, in which case the cutting plane intersects each generator, as in Figure 2.6.12. A circle, a special case of an ellipse, is formed if the cutting plane, which intersects each generator, is also perpendicular to the axis of the cone. See Figure 2.6.13.

There are many applications of conic sections, and we mention a few of them. The orbits of planets and satellites are ellipses. Ellipses are used in making machine gears. Arches of bridges are sometimes elliptical or parabolic in shape. The path of a projectile is a parabola if motion is considered to be in a plane and air resistance is neglected. Parabolas are used in the design of parabolic mirrors, searchlights, and automobile headlights. Hyperbolas are used in combat in "sound ranging" to locate the position of enemy guns by the sound of the firing of those guns.

## EXERCISES 2.6

*In Exercises 1 through 18, name and draw a sketch of the graph of the given equation.*

1. $x^2 + y^2 = 49$
2. $x^2 + y^2 = 1$
3. $9x^2 + 9y^2 = 1$
4. $7x^2 + 7y^2 = 1$
5. $25x^2 + y^2 = 25$
6. $9x^2 + 4y^2 = 36$
7. $36x^2 - 25y^2 = 900$
8. $x^2 + 8y^2 = 0$
9. $64x^2 - y^2 = 0$
10. $9y^2 - 16x^2 = 144$
11. $4x^2 + 9y^2 = 0$
12. $9x^2 - 49y^2 = 0$
13. $y^2 - 16x^2 = 64$
14. $25x^2 + 100y^2 = 1$
15. $8x^2 + 5y^2 = 40$
16. $12x^2 - y^2 = 27$
17. $2x^2 + 3y^2 + 4 = 0$
18. $10y^2 + 5x^2 + 1 = 0$

*In Exercises 19 through 22, name and draw a sketch of the graph of each of the four given equations.*

19. (a) $4x^2 + 25y^2 = 100$ (b) $4x^2 - 25y^2 = 100$ (c) $4x^2 + 25y^2 = 0$ (d) $4x^2 - 25y^2 = 0$
20. (a) $16x^2 + 9y^2 = 25$ (b) $16x^2 - 9y^2 = 25$ (c) $16x^2 + 9y^2 = 0$ (d) $16x^2 - 9y^2 = 0$
21. (a) $25x^2 + 4y^2 = 100$ (b) $25x^2 - 4y^2 = -100$ (c) $25x^2 + 4y^2 = -100$ (d) $25x^2 - 4y^2 = 0$
22. (a) $9x^2 + 16y^2 = 25$ (b) $9x^2 - 16y^2 = 25$ (c) $9x^2 + 16y^2 = -25$ (d) $9x^2 - 16y^2 = 0$

23. Find an equation of the circle with center at $(3, -5)$ and radius equal to 4.

24. Find an equation of the circle with center at $(-2, 2)$ and radius equal to 5.

*In Exercises 25 and 26, show that the graph of the given equation is a circle and find its center and radius.*

25. $x^2 + y^2 + 10x - 12y + 12 = 0$
26. $x^2 + y^2 - 8x + 6y - 24 = 0$

27. Show that the general form of an equation of an ellipse, $Ax^2 + By^2 = C$, where $A$, $B$, and $C$ are positive, can be written in the form

$$\frac{x^2}{a^2} + \frac{y^2}{b^2} = 1$$

where $a^2$ and $b^2$ are the squares of the $x$ and $y$ intercepts, respectively, of the ellipse.

28. Show that the general form of an equation of a hyperbola, $Ax^2 - By^2 = C$, where $A$, $B$, and $C$ are positive, can be written in the form

$$\frac{x^2}{a^2} - \frac{y^2}{b^2} = 1$$

where $a^2$ is the square of the $x$ intercepts of the hyperbola.

## 2.7 Quadratic Inequalities

An inequality equivalent to one of the form $E > 0$, $E < 0$, $E \geq 0$, or $E \leq 0$, where $E$ is a polynomial of the second degree is called a *quadratic inequality*. A graphical method of solving quadratic inequalities in one variable is shown in the following illustration.

ILLUSTRATION 1. To find the solution set of the inequality

$$x^2 - 2x - 3 < 0 \tag{1}$$

we consider the graph of the equation

$$y = x^2 - 2x - 3$$

This equation is equivalent to

$$x^2 - 2x = y + 3$$

We complete the square in the left member by adding 1 to each member and obtain

$$x^2 - 2x + 1 = y + 3 + 1$$
$$(x - 1)^2 = y + 4$$

This is an equation of a parabola with vertex at $(1, -4)$ and opening upward. A sketch is shown in Figure 2.7.1. The parabola intersects the $x$ axis at the points $(-1, 0)$ and $(3, 0)$. Observe from the figure that $y < 0$ if and only if $-1 < x < 3$. Thus the solution set of inequality (1) is $\{x \mid -1 < x < 3\}$.

Furthermore, from Figure 2.7.1 we see that $y \geq 0$ if and only if $x \leq -1$ or $x \geq 3$. Therefore, the inequality

$$x^2 - 2x - 3 \geq 0$$

has the solution set $\{x \mid x \leq -1\} \cup \{x \mid x \geq 3\}$.  •

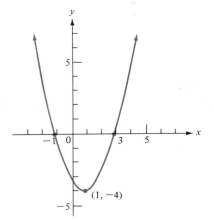

Figure 2.7.1

## Quadratic Inequalities

**EXAMPLE 1**
Find graphically the solution set of
$$3x^2 > 10x - 3$$

**SOLUTION**
The given inequality is equivalent to
$$3x^2 - 10x + 3 > 0 \tag{2}$$
We consider the equation
$$y = 3x^2 - 10x + 3$$
which is equivalent to
$$3x^2 - 10x = y - 3$$
$$x^2 - \frac{10}{3}x = \frac{1}{3}y - 1$$
$$x^2 - \frac{10}{3}x + \frac{25}{9} = \frac{1}{3}y - 1 + \frac{25}{9}$$
$$\left(x - \frac{5}{3}\right)^2 = \frac{1}{3}\left(y + \frac{16}{3}\right)$$

The graph of this equation is a parabola with vertex at $(\frac{5}{3}, -\frac{16}{3})$ and opening upward. To find the points of intersection of the parabola with the $x$ axis, we solve the equation
$$3x^2 - 10x + 3 = 0$$
$$(3x - 1)(x - 3) = 0$$
$$3x - 1 = 0 \qquad x - 3 = 0$$
$$x = \frac{1}{3} \qquad\qquad x = 3$$

Therefore, the parabola intersects the $x$ axis at $(\frac{1}{3}, 0)$ and $(3, 0)$. A sketch of the parabola is shown in Figure 2.7.2. From the figure it follows that $y > 0$ if and only if $x < \frac{1}{3}$ or $x > 3$. Therefore, the solution set of inequality (2) is $\{x \mid x < \frac{1}{3}\} \cup \{x \mid x > 3\}$.

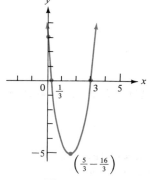

**Figure 2.7.2**

The solution set of a quadratic inequality in one variable can be found analytically. The following illustration shows the procedure.

**ILLUSTRATION 2.** To find the solution set of the inequality
$$x^2 + 2x > 15$$
we first obtain the equivalent inequality
$$x^2 + 2x - 15 > 0$$
$$(x - 3)(x + 5) > 0$$

This inequality will be satisfied only when both factors have the same sign, that is, when $x - 3 > 0$ and $x + 5 > 0$, or when $x - 3 < 0$ and $x + 5 < 0$. We consider two cases.

**CASE 1:** $x - 3 > 0$ and $x + 5 > 0$. That is,

$$x > 3 \quad \text{and} \quad x > -5$$

Both inequalities hold if $x > 3$, and hence the solution set for Case 1 is $\{x \mid x > 3\}$.

**CASE 2:** $x - 3 < 0$ and $x + 5 < 0$. That is,

$$x < 3 \quad \text{and} \quad x < -5$$

Both inequalities hold if $x < -5$, and so the solution set for Case 2 is $\{x \mid x < -5\}$.

The solution set of the given inequality is the union of the solution sets of Cases 1 and 2, which is

$$\{x \mid x < -5\} \cup \{x \mid x > 3\}$$

A sketch of the graph of this solution set is shown in Figure 2.7.3. •

**Figure 2.7.3**

EXAMPLE 2
Find the solution set of the inequality

$$3x^2 < 10x + 8$$

and draw a sketch of the graph of the solution set.

SOLUTION
The given inequality is equivalent to

$$3x^2 - 10x - 8 < 0$$
$$(x - 4)(3x + 2) < 0$$

The product of two factors is negative if and only if the two factors have opposite signs. Hence the inequality will be satisfied if and only if either $x - 4 < 0$ and $3x + 2 > 0$, or $x - 4 > 0$ and $3x + 2 < 0$. We consider two cases.

**CASE 1:** $x - 4 < 0$ and $3x + 2 > 0$. That is,

$$x < 4 \quad \text{and} \quad x > -\frac{2}{3}$$

Hence the solution set for Case 1 is $\{x \mid x > -\frac{2}{3}\} \cap \{x \mid x < 4\}$ or, equivalently, $\{x \mid -\frac{2}{3} < x < 4\}$.

**CASE 2:** $x - 4 > 0$ and $3x + 2 < 0$. That is,

$$x > 4 \quad \text{and} \quad x < -\frac{2}{3}$$

The solution set for Case 2 is $\{x \mid x > 4\} \cap \{x \mid x < -\frac{2}{3}\}$, which is $\emptyset$.

The solution set of the given inequality is the union of the solution sets of

## Quadratic Inequalities

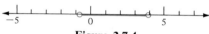

**Figure 2.7.4**

Cases 1 and 2, which is $\{x \mid -\frac{2}{3} < x < 4\} \cup \emptyset$ or, equivalently, $\{x \mid -\frac{2}{3} < x < 4\}$.

A sketch of the graph is shown in Figure 2.7.4.

**ILLUSTRATION 3:** The inequality of Example 6 in Section 1.6 is

$$\frac{3x}{x+2} < 5$$

An alternative method of finding the solution set is to multiply both members of the inequality by $(x + 2)^2$, which is positive when $x \neq -2$. Doing this, we have

$$\frac{3x}{x+2}(x+2)^2 < 5(x+2)^2$$

$$3x(x+2) < 5(x^2 + 4x + 4)$$
$$3x^2 + 6x < 5x^2 + 20x + 20$$
$$0 < 2x^2 + 14x + 20$$
$$x^2 + 7x + 10 > 0$$
$$(x+2)(x+5) > 0$$

We consider two cases.

**CASE 1:** $x + 2 > 0$ and $x + 5 > 0$. That is,

$$x > -2 \quad \text{and} \quad x > -5$$

The solution set for Case 1 is then $\{x \mid x > -2\}$.

**CASE 2:** $x + 2 < 0$ and $x + 5 < 0$. That is,

$$x < -2 \quad \text{and} \quad x < -5$$

The solution set for Case 2 is then $\{x \mid x < -5\}$.

The union of the solution sets of Cases 1 and 2 is

$$\{x \mid x < -5\} \cup \{x \mid x > -2\}$$

which agrees with the result in Example 6 of Section 1.6.  ●

**EXAMPLE 3**

Use the method of Illustration 3 to find the solution set of the inequality

$$\frac{x+1}{2x-1} > 3$$

**SOLUTION**

We multiply each member of the inequality by $(2x - 1)^2$ and we have

$$\frac{x+1}{2x-1}(2x-1)^2 > 3(2x-1)^2$$

$$(x+1)(2x-1) > 3(4x^2 - 4x + 1)$$
$$2x^2 + x - 1 > 12x^2 - 12x + 3$$
$$0 > 10x^2 - 13x + 4$$
$$(5x - 4)(2x - 1) < 0$$

**CASE 1:** $5x - 4 < 0$ and $2x - 1 > 0$. That is,

$$x < \frac{4}{5} \quad \text{and} \quad x > \frac{1}{2}$$

The solution set for Case 1 is $\{x \mid x > \frac{1}{2}\} \cap \{x \mid x < \frac{4}{5}\}$ or, equivalently, $\{x \mid \frac{1}{2} < x < \frac{4}{5}\}$.

**CASE 2:** $5x - 4 > 0$ and $2x - 1 < 0$. That is,

$$x > \frac{4}{5} \quad \text{and} \quad x < \frac{1}{2}$$

The solution set for Case 2 is $\emptyset$.

The solution set of the given inequality is the union of the solution sets of Cases 1 and 2 which is $\{x \mid \frac{1}{2} < x < \frac{4}{5}\}$.

### EXAMPLE 4

A firm manufactures and sells portable radios, and it can sell at a price of $75 each all the radios it produces. If $x$ radios are manufactured each day, then the number of dollars in the daily total cost of production is $x^2 + 25x + 96$. How many radios should be produced each day so that the firm is guaranteed a profit?

### SOLUTION

The number of dollars in the total revenue received each day from the sale of $x$ radios is $75x$. If $P$ dollars is the daily profit from the sale of $x$ radios, then because profit equals revenue minus cost, we have

$$P = 75x - (x^2 + 25x + 96)$$
$$= -x^2 + 50x - 96$$

For the firm to be guaranteed a profit, we must have $P > 0$; that is,

$$-x^2 + 50x - 96 > 0$$

or, equivalently,

$$x^2 - 50x + 96 < 0$$
$$(x - 2)(x - 48) < 0 \tag{3}$$

Inequality (3) will be satisfied in either of two cases.

**CASE 1:** $x - 2 < 0$ and $x - 48 > 0$; that is,

$$x < 2 \quad \text{and} \quad x > 48$$

Therefore, the solution set for Case 1 is $\{x \mid x < 2\} \cap \{x \mid x > 48\}$, which is $\emptyset$.

**CASE 2:** $x - 2 > 0$ and $x - 48 < 0$; that is,

$$x > 2 \quad \text{and} \quad x < 48$$

Hence the solution set for Case 2 is $\{x \mid x > 2\} \cap \{x \mid x < 48\}$ or, equivalently, $\{x \mid 2 < x < 48\}$.

The solution set of inequality (3) is the union of the solution sets of Cases 1 and 2, which is $\{x \mid 2 < x < 48\} \cup \emptyset$ or, equivalently, $\{x \mid 2 < x < 48\}$.

Thus we conclude that for the firm to be guaranteed a profit, the number of radios produced and sold each day must be greater than 2 and less than 48.

In Section 3.4 we return to the problem in Example 4 and learn how to determine the number of radios that should be produced and sold each day in order to have the greatest profit.

## EXERCISES 2.7

In Exercises 1 through 6, find graphically the solution set of the given inequality.

1. $x^2 - 4x + 3 < 0$
2. $x^2 - 6x + 5 > 0$
3. $x^2 + 2x - 8 > 0$
4. $x^2 + 8x + 15 < 0$
5. $2x^2 + x - 6 \leq 0$
6. $3x^2 - 4x - 4 \geq 0$

In Exercises 7 through 19, find the solution set of the given inequality. Draw a sketch of the graph of the solution set.

7. $x^2 > 9$
8. $x^2 < 4$
9. $(x + 3)(x - 4) \leq 0$
10. $(2x - 1)(3x - 5) > 0$
11. $x^2 - 4x + 3 > 0$
12. $2x^2 + x - 1 \leq 0$
13. $4 - 3x - x^2 \geq 0$
14. $4x^2 + 9x \geq 9$
15. $2x^2 < 6x - 3$
16. $x^2 + 3x + 1 < 0$
17. $x^2 > 2x + 1$
18. $\dfrac{5}{2x - 1} < x - 2$
19. $\dfrac{6}{3x - 4} > x + 1$

In Exercises 20 through 23, use the method of Illustration 3 to find the solution set of the given inequality.

20. Exercise 26 of Exercises 1.6
21. Exercise 27 of Exercises 1.6
22. $\dfrac{6 - 2x}{4 + x} > 5$
23. $\dfrac{2 - 7x}{5 - 4x} > 4$

24. A firm can sell at a price of $100 per unit all of a particular commodity it produces. If $x$ units are produced each day, the number of dollars in the total cost of each day's production is $x^2 + 20x + 700$. How many units should be produced each day so that the firm is guaranteed a profit?

25. A company that builds and sells desks can sell at a price of $200 per desk all the desks it produces. If $x$ desks are built and sold each week, then the number of dollars in the total cost of the week's production is $x^2 + 40x + 1500$. How many desks should be built each week in order that the manufacturer is guaranteed a profit?

## 2.8 Inequalities in Two Variables

Statements of the form

$$Ax + By + C > 0 \qquad Ax + By + C < 0$$
$$Ax + By + C \geq 0 \qquad Ax + By + C \leq 0$$

where $A$ and $B$ are not both zero, are inequalities of the first degree in two variables. By the graph of such an inequality, we mean the set of all points $(x, y)$ in a rectangular Cartesian coordinate system for which $(x, y)$ is an ordered pair satisfying the inequality.

Every line in a plane divides the plane into two regions, one on each side of the line, and each of these regions is called a *half plane*. The graph of an inequality of the form

$$Ax + By + C > 0 \quad \text{or} \quad Ax + By + C < 0$$

is a half plane. To show this, let $L$ be the line having the equation $Ax + By + C = 0$. If $B \neq 0$, then $L$ is nonvertical and its equation can be written as $y = mx + b$, where $m = -\dfrac{A}{B}$ and $b = -\dfrac{C}{B}$. If $(x, y)$ is any point in the plane, it follows from Theorem 1.2.15(i) that exactly one of the following statements holds: $y = mx + b$, $y > mx + b$, or $y < mx + b$. Now, $y > mx + b$ if and only if the point $(x, y)$ is above the point $(x, mx + b)$ on line $L$ (see Figure 2.8.1), and $y < mx + b$ if and only if the point $(x, y)$ is below the point $(x, mx + b)$ on $L$ (see Figure 2.8.2). Therefore, the line $L$ divides the plane into two regions. One region is the half plane above line $L$, which is the graph of the inequality $y > mx + b$, and the other region is the half plane below $L$, which is the graph of the inequality $y < mx + b$.

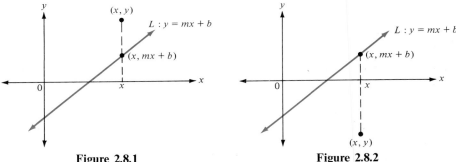

**Figure 2.8.1**     **Figure 2.8.2**

If $B = 0$, then $L$ is a vertical line and its equation can be written as $x = a$, where $a = -\dfrac{C}{A}$. If $(x, y)$ is any point in the plane, it follows from Theorem 1.2.15(i) that exactly one of the following statements is true: $x = a$, $x > a$, or $x < a$. The point $(x, y)$ is to the right of the point $(a, y)$ if and only if $x > a$; hence the graph of the inequality $x > a$ is the half plane lying to the right of the line $x = a$. Similarly, the graph of $x < a$ is the half plane lying to the left of the line $x = a$, because $x < a$ if and only if the point $(x, y)$ is to the left of the point $(a, y)$. We have proved the following theorem.

**2.8.1 THEOREM**   (i) The graph of $y > mx + b$ is the half plane lying above the line $y = mx + b$.
(ii) The graph of $y < mx + b$ is the half plane lying below the line $y = mx + b$.
(iii) The graph of $x > a$ is the half plane lying to the right of the line $x = a$.
(iv) The graph of $x < a$ is the half plane lying to the left of the line $x = a$.

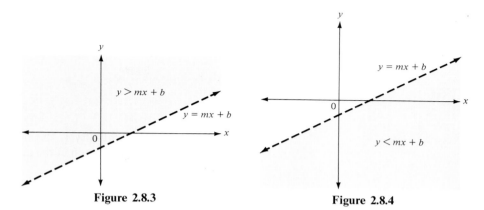

**Figure 2.8.3**  **Figure 2.8.4**

Figures 2.8.3 and 2.8.4 show sketches of the graphs of $y > mx + b$ and $y < mx + b$, respectively. Observe that the appropriate half plane is shaded. The graph of the line $y = mx + b$ is indicated by a dashed line to show that it is not part of the graph.

Sketches of the graphs of $x > a$ and $x < a$ are shown in Figures 2.8.5 and 2.8.6, respectively.

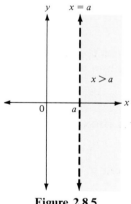
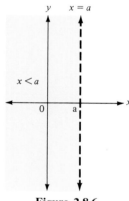

**Figure 2.8.5**  **Figure 2.8.6**

**ILLUSTRATION 1.** The inequality
$$3x + y - 6 > 0$$
is equivalent to
$$y > -3x + 6$$
The graph of this inequality is the half plane above the line having the equation $3x + y - 6 = 0$. A sketch of this graph is the shaded half plane shown in Figure 2.8.7. The graph of the line is indicated by a dashed line to show that it is not part of the graph. ●

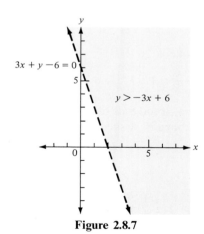

**Figure 2.8.7**

## EXAMPLE 1
Draw a sketch of the graph of the inequality

$$2x - 4y + 5 > 0$$

**SOLUTION**
The given inequality is equivalent to

$$-4y > -2x - 5$$

which is equivalent to

$$y < \tfrac{1}{2}x + \tfrac{5}{4}$$

The graph of this inequality is the half plane below the line having the equation $y = \tfrac{1}{2}x + \tfrac{5}{4}$. A sketch of the graph is the shaded half plane shown in Figure 2.8.8.

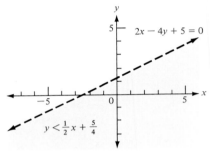

**Figure 2.8.8**

A *closed half plane* is a half plane together with the line bounding it, and it is the graph of an inequality of the form

$$Ax + By + C \geq 0 \quad \text{or} \quad Ax + By + C \leq 0$$

**ILLUSTRATION 2.** The inequality

$$4x + 5y - 20 \geq 0$$

is equivalent to

$$5y \geq -4x + 20$$
$$y \geq -\tfrac{4}{5}x + 4$$

Therefore, the graph of this inequality is the closed half plane consisting of the line $y = -\tfrac{4}{5}x + 4$ and the half plane above it. A sketch of the graph is shown in Figure 2.8.9. •

The inequalities

$$y > Ax^2 + Bx + C \qquad y < Ax^2 + Bx + C$$
$$y \geq Ax^2 + Bx + C \qquad y \leq Ax^2 + Bx + C$$

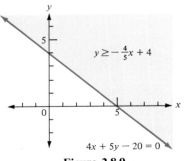

**Figure 2.8.9**

## Inequalities in Two Variables

are quadratic inequalities. We can draw a sketch of the graph of such inequalities by a method similar to that for a linear inequality. The next two examples show the procedure.

**EXAMPLE 2**
Draw a sketch of the graph of the inequality

$$y < 2x^2 + 3$$

**SOLUTION**
If we equate the two members of the inequality, we have the equation

$$y = 2x^2 + 3$$
$$y - 3 = 2x^2$$

A sketch of the graph of this parabola is shown in Figure 2.8.10 by a dashed curve because the parabola is not part of the graph of the inequality. The graph of the inequality is the region below the parabola, and it is shaded in the figure.

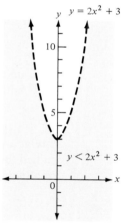

**Figure 2.8.10**

**EXAMPLE 3**
Draw a sketch of the graph of the inequality

$$y \geq 4x^2 - 24x + 34$$

**SOLUTION**
Equating the two members of the inequality, we obtain

$$y = 4x^2 - 24x + 34$$
$$y - 34 = 4(x^2 - 6x)$$
$$y - 34 + 36 = 4(x^2 - 6x + 9)$$
$$y + 2 = 4(x - 3)^2$$

The graph of this equation is a parabola, and the graph of the given inequality consists of the parabola and the region above the parabola. A sketch of the graph is shown in Figure 2.8.11.

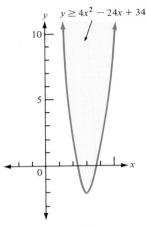

Figure 2.8.11

All the points $(x, y)$ whose coordinates satisfy the equation

$$\sqrt{(x - h)^2 + (y - k)^2} = r$$

are $r$ units from the point $C(h, k)$ and hence lie on the circle having its center at $C$ and a radius of $r$. The points whose coordinates satisfy the inequality

$$\sqrt{(x - h)^2 + (y - k)^2} < r \qquad (1)$$

are at a distance less than $r$ units from $C$, and these points are in the region called the *interior of the circle*. Similarly, the points whose coordinates satisfy the inequality

$$\sqrt{(x - h)^2 + (y - k)^2} > r \qquad (2)$$

are in the region called the *exterior of the circle* because these points are at a distance greater than $r$ units from $C$. By squaring both members of inequalities (1) and (2), we have the following theorem.

**2.8.2 THEOREM**   The graph of the inequality

$$(x - h)^2 + (y - k)^2 < r^2$$

is the interior of the circle having its center at $(h, k)$ and a radius of $r$, and the graph of the inequality

$$(x - h)^2 + (y - k)^2 > r^2$$

is the exterior of the circle.

## EXAMPLE 4

Draw a sketch of the graph of each of the following inequalities.

(a) $x^2 + y^2 - 8x + 6y + 21 > 0$
(b) $x^2 + y^2 + 4x - 10y + 20 < 0$

**SOLUTION**

(a) The inequality
$$x^2 + y^2 - 8x + 6y + 21 > 0$$
is equivalent to
$$x^2 - 8x + 16 + y^2 + 6y + 9 > -21 + 16 + 9$$
$$(x - 4)^2 + (y + 3)^2 > 4$$

The graph of this inequality is the exterior of the circle having its center at $(4, -3)$ and a radius of 2. A sketch of this graph is shown in Figure 2.8.12.

(b) The inequality
$$x^2 + y^2 + 4x - 10y + 20 < 0$$
is equivalent to
$$x^2 + 4x + 4 + y^2 - 10y + 25 < -20 + 4 + 25$$
$$(x + 2)^2 + (y - 5)^2 < 9$$

Figure 2.8.13 shows a sketch of the graph of this inequality. The graph is the interior of the circle having center at $(-2, 5)$ and radius 3.

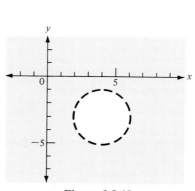

Figure 2.8.12

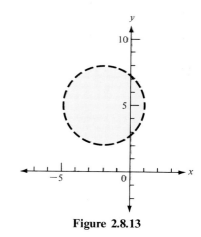

Figure 2.8.13

## EXERCISES 2.8

*In Exercises 1 through 16, draw a sketch of the graph of the given inequality.*

1. $y < 4x - 2$
2. $y \geq 2x - 3$
3. $2x - 7 \geq 0$
4. $y + 8 < 0$
5. $3x - 6y + 4 \leq 0$
6. $2y - 8x + 5 > 0$
7. $5x + 6 > 2y$
8. $9x + 3y \geq 7$
9. $y - 2x^2 < 0$

10. $2x^2 + 5y \leq 0$
11. $x^2 + 4y + 6x + 8 \geq 0$
12. $x^2 - 4y + 8 > 0$
13. $x^2 + y^2 < 9$
14. $x^2 + y^2 \geq 36$
15. $x^2 + y^2 - 6y + 4x \geq 0$
16. $3x^2 + 3y^2 + 8x - 6y < 1$

In Exercises 17 through 20, define the given region by an inequality.

17. The half plane below the line $3x + 2y - 12 = 0$.
18. The half plane above the line $5x - 4y - 20 = 0$.
19. The closed half plane bounded by $2x - 7y + 3 = 0$ and containing the point $(-3, 2)$.
20. The closed half plane bounded by $4x - y + 8 = 0$ and containing the point $(5, -3)$.

## REVIEW EXERCISES (CHAPTER 2)

In Exercises 1 through 14, find the solution set of the given equation.

1. $10x^2 + 7x - 6 = 0$
2. $5y^2 + 17y = 12$
3. $2x^2 = 4x + 5$
4. $x^2 + 2x - 1 = 0$
5. $3w^2 - 2w + 2 = 0$
6. $1 + 2t + 4t^2 = 0$
7. $(y - 4)(y + 2) = 4$
8. $(3x + 10)(x - 3) = 2x + 14$
9. $\dfrac{x}{x + 2} = \dfrac{x}{1 - 2x} - \dfrac{x}{x + 2}$
10. $\dfrac{x + 11}{x + 8} - \dfrac{3x - 2}{x - 2} = 0$
11. $\sqrt{3x + 7} + \sqrt{x + 6} = 3$
12. $\sqrt{x + 2} + \sqrt{x + 5} - \sqrt{8 - x} = 0$
13. $36x^4 - 13x^2 + 1 = 0$
14. $8z^3 + 27 = 0$

In Exercises 15 and 16, solve for $x$ in terms of the other symbols.

15. $6x^2 - 2xy - x + y - 1 = 0$
16. $rsx^2 + s^2x + rtx + st = 0$

In Exercises 17 through 28, draw a sketch of the graph of the given equation.

17. $y^2 - x + 1 = 0$
18. $y = x - 1$
19. $x^2 + 8y = 0$
20. $3x^2 - 16y = 0$
21. $x^2 + y^2 = 16$
22. $(3x + 4y - 15)(x - 2y + 8) = 0$
23. $(x - 4y)(x^2 - 4y) = 0$
24. $x^2 + 9y^2 = 9$
25. $x^2 + y^2 = 0$
26. $x^2 + y^2 = 1$
27. $x^2 - y^2 = 1$
28. $x^2 - y^2 = 0$

29. Show that the graph of the equation $x^2 + y^2 - 4x + 10y + 13 = 0$ is a circle and find its center and radius.

In Exercises 30 and 31, find graphically the solution set of the given inequality.

30. $x^2 - 8x + 12 \leq 0$
31. $2x^2 - x - 10 > 0$

*In Exercises 32 through 38, find the solution set of the given inequality. Draw a sketch of the graph of the solution set.*

32. $x^2 \geq 16$
33. $x^2 < 25$
34. $x^2 + 2x - 8 \leq 0$
35. $2x^2 + 7x \geq 15$
36. $x^2 > 7x - 4$
37. $2x - 3 \geq \dfrac{5}{x - 3}$
38. $\dfrac{3x - 1}{2x - 1} < \dfrac{3x + 4}{2}$

*In Exercises 39 through 42, draw a sketch of the graph of the given inequality.*

39. $3x - 4y \geq 12$
40. $5x + 3y < 15$
41. $x^2 + y^2 < 16$
42. $x^2 - 2y \geq 2$

43. Draw a sketch of the parabola having the equation $x^2 - 6y = 0$, and find the focus and an equation of the directrix.
44. Draw a sketch of the parabola having the equation $y = x^2 - 8x$. Find the vertex, the axis, the focus, and the directrix.
45. Use the definition of a parabola (2.5.1) to find an equation of the parabola whose focus is at $(0, -4)$ and whose directrix has the equation $y = 4$.
46. A small pipe takes 24 minutes longer to fill a tank than it takes a large pipe. The two pipes together can fill the tank in 9 minutes. How long does it take each pipe alone to fill the tank?
47. A train on its way east was delayed 1 hour when it was 560 kilometers west of New York City. By increasing its normal speed 10 kilometers per hour, the train arrived at New York on schedule. What is the train's normal speed?
48. In a long-distance race covering 42 kilometers, one runner finished 12 minutes before another runner. If the faster runner's speed was 1 kilometer per hour more than the slower runner's speed, what were their speeds?
49. On a river whose rate is 4 kilometers per hour, it takes a motorboat 1 hour longer to go 40 kilometers up the river than it takes to return the same distance. What is the rate of the boat in still water?
50. A carpenter can sell all the bookcases that are made at a price of $65 per bookcase. If $x$ bookcases are built and sold each week, then the number of dollars in the total cost of the week's production is $x^2 + 15x + 225$. How many bookcases should be constructed each week for the carpenter to be guaranteed a profit?

# 3
# Functions

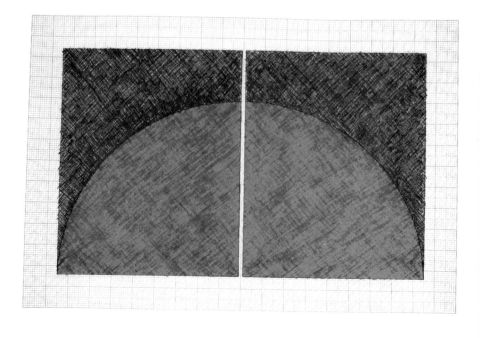

# 3.1 Relations and Functions, and Their Graphs

Sets of ordered pairs of real numbers were discussed in Section 1.8. We now refer to any such set as a "relation."

**3.1.1 DEFINITION** A set of ordered pairs of real numbers is called a *relation*.

**ILLUSTRATION 1.** The set $\{(1, 1), (2, 4), (3, 9)\}$ is a relation. This relation can be defined in words as the set of ordered pairs for which the first component is a positive integer less than 4, and the second component is the square of the first. The graph of this relation consists of the three points whose coordinates are given by the ordered pairs in the relation, and it is shown in Figure 3.1.1.

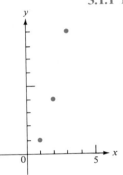

Figure 3.1.1

For the set of ordered pairs in a relation there is a correspondence between the set of first components and the set of second components of the ordered pairs. Because of this correspondence (or relationship) the term "relation" is appropriate for the set of ordered pairs.

**3.1.2 DEFINITION** If $S$ is a relation, then the *domain* of $S$ is the set of all first components of the ordered pairs in $S$, and the *range* of $S$ is the set of all second components of the ordered pairs in $S$.

The domain of the relation in Illustration 1 is the set $\{1, 2, 3\}$, and the range of this relation is $\{1, 4, 9\}$.

Because the solution set of an equation in two variables is a set of ordered pairs, the solution set is a relation and the equation is said to define the relation. The graph of such a relation is the same as the graph of the equation.

**ILLUSTRATION 2.** The relation defined by the equation $3x - 4y = 12$ is the solution set of the equation, and it can be written with symbols as $\{(x, y) | 3x - 4y = 12\}$. This relation is an infinite set. Both the domain and the range are the set $R$ of real numbers. Some of the ordered pairs in the relation are $(0, -3)$, $(4, 0)$, $(8, 3)$, $(\frac{4}{3}, -2)$, $(3, -\frac{3}{4})$, and so on. The graph of the relation is a line, and a sketch of it is shown in Figure 3.1.2.

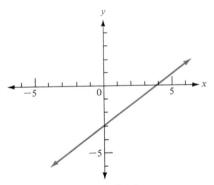

Figure 3.1.2

**ILLUSTRATION 3.** Consider the following three relations.

$$S = \{(x, y) | y^2 = x - 3\}$$
$$T = \{(x, y) | y = \sqrt{x - 3}\}$$
$$W = \{(x, y) | y = -\sqrt{x - 3}\}$$

The equation defining $S$ is equivalent to the two equations

$$y = \sqrt{x - 3} \quad \text{and} \quad y = -\sqrt{x - 3} \tag{1}$$

For these equations, we see that if $x - 3 < 0$, that is, if $x < 3$, a square root of a negative number is obtained and hence no real number $y$ exists. Therefore, we must restrict $x$ so that $x \geq 3$. Hence the domain of $S$ is $\{x \mid x \geq 3\}$. The range of $S$ is the set $R$ of real numbers. The graph of relation $S$ is the graph of the equation $y^2 = x - 3$. This graph is a parabola and a sketch of it is shown in Figure 3.1.3.

Relation $T$ has as its domain the set $\{x \mid x \geq 3\}$, and because $y$ is a nonnegative number, the range of $T$ is $\{y \mid y \geq 0\}$. The graph of $T$ is the upper half of the graph of $S$, and a sketch of it is shown in Figure 3.1.4.

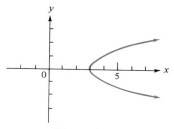

**Figure 3.1.3**

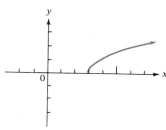

**Figure 3.1.4**

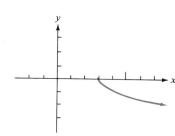

**Figure 3.1.5**

The domain of relation $W$ is also the set $\{x \mid x \geq 3\}$. Because $y$ is a nonpositive number, the range of $W$ is $\{y \mid y \leq 0\}$. The graph of $W$, a sketch of which is shown in Figure 3.1.5, is the lower half of the parabola of Figure 3.1.3.

Relation $S$ is the union of relations $T$ and $W$; that is,

$$\{(x, y) \mid y^2 = x - 3\} = \{(x, y) \mid y = \sqrt{x - 3}\} \cup \{(x, y) \mid y = -\sqrt{x - 3}\}$$

The domains of $S$, $T$, and $W$ are equal. However, for each number $x$, other than 3, in this common domain, there are two ordered pairs in $S$ having $x$ as the first component but only one ordered pair in each of the relations $T$ and $W$ for which $x$ is the first component. For instance, when $x$ is 4, then from equations (1) $y$ is either 1 or $-1$. Thus both the ordered pairs $(4, 1)$ and $(4, -1)$ are in relation $S$. However, $(4, 1)$ is the ony ordered pair in $T$ having 4 as its first component and $(4, -1)$ is the only ordered pair in $W$ having 4 as its first component. Similarly, both $(5, \sqrt{2})$ and $(5, -\sqrt{2})$ are ordered pairs in $S$, but the only ordered pair in $T$ having 5 as its first component is $(5, \sqrt{2})$ and the only one in $W$ is $(5, -\sqrt{2})$. •

**3.1.3 DEFINITION** A set of ordered pairs of real numbers $(x, y)$, in which no two ordered pairs have the same first component, is called a *function*. The set of all possible values of $x$ is called the *domain* of the function, and the set of all possible values of $y$ is called the *range* of the function.

By comparing Definitions 3.1.1 and 3.1.3, we see that a function is a particular kind of relation; it is a relation for which the second component is unique for a specific value of the first component. It follows from the discussion in Illustration 3 that each of the relations $T$ and $W$ is a function, but that relation $S$ is not a function.

In Definition 3.1.3 the restriction that no two ordered pairs can have the same first number implies that no vertical line can intersect the graph of a function at more than one point. Verify this statement for the graphs of the functions $T$ and $W$ of Illustration 3 by referring to Figures 3.1.4 and 3.1.5, respectively. Also refer to Figure 3.1.3, which shows the sketch of the graph of relation $S$ of Illustration 3. Relation $S$ is not a function, and note that any vertical line to the right of the line $x = 3$ intersects the graph in two points.

EXAMPLE 1
For each of the following relations, determine the domain and range, and state whether or not the relation is a function.

(a) $\{(-3, 1), (-1, 3), (1, 5), (3, 7)\}$
(b) $\{(1, 2), (2, 3), (3, 4), (3, 5), (5, 6)\}$
(c) $\{(-2, 2), (-1, 2), (0, 2), (1, 2)\}$
(d) $\{(1, 2), (1, 3), (1, 4), (1, 5)\}$

SOLUTION
(a) The domain, the set of all first components, is $\{-3, -1, 1, 3\}$. The range, the set of all second components, is $\{1, 3, 5, 7\}$. Because no two ordered pairs have the same first component, the relation is a function.
(b) The domain is $\{1, 2, 3, 5\}$ and the range is $\{2, 3, 4, 5, 6\}$. The relation is not a function because the two ordered pairs $(3, 4)$ and $(3, 5)$ have the same first component.
(c) The domain is $\{-2, -1, 0, 1\}$, and the range is $\{2\}$. The relation is a function because no two ordered pairs have the same first component.
(d) The domain is $\{1\}$, and the range is $\{2, 3, 4, 5\}$. The relation is not a function because all four of the ordered pairs have the same first component.

We consider $y$ to be a function of $x$ if there is some rule by which a unique value is assigned to $y$ by a corresponding value of $x$. Such relationships can be given by equations such as

$$y = 2x^2 - 4 \tag{2}$$

and

$$y = \sqrt{x^2 - 25} \tag{3}$$

**ILLUSTRATION 4.** Equation (2) defines a function. Let us call this function $f$. The equation gives the rule by which a unique value of $y$ can be determined whenever $x$ is given; that is, multiply the number $x$ by itself, then multiply that product by 2, and subtract 4. The function $f$ is the set of all ordered pairs $(x, y)$ such that $x$ and $y$ satisfy equation (2); that is,

$$f = \{(x, y) \mid y = 2x^2 - 4\}$$

A sketch of the graph of $f$ is shown in Figure 3.1.6. ●

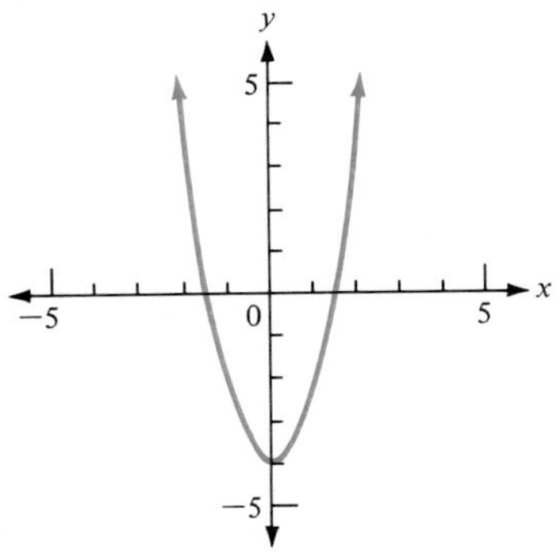

**Figure 3.1.6**

In Illustration 4, the numbers $x$ and $y$ are variables. Because for the function $f$, values are assigned to $x$ and because the value of $y$ is dependent upon the choice of $x$, we call $x$ the *independent variable* and $y$ the *dependent variable*. The domain of the function is the set of all values of the independent variable that yield real values for the dependent variable. The range of the function is the set of all these values of the dependent variable. For the function $f$ of Illustration 4, the domain is the set $R$ of real numbers. The smallest value that $y$ can assume is $-4$, which occurs when $x = 0$. The range of $f$ is then the set of all real numbers greater than or equal to $-4$.

**ILLUSTRATION 5.** Let $g$ be the function which is the set of all ordered pairs $(x, y)$ defined by equation (3); that is,

$$g = \{(x, y) \mid y = \sqrt{x^2 - 25}\}$$

Because the numbers in the range are confined to real numbers, $y$ is a function of $x$ only for $x \geq 5$ or $x \leq -5$ (or simply $|x| \geq 5$) because for any $x$ satisfying either of these inequalities a unique value of $y$ is determined. However, if $-5 < x < 5$, a square root of a negative number is obtained, and hence no real number $y$ exists. Therefore, we must restrict $x$, and so

$$g = \{(x, y) \mid y = \sqrt{x^2 - 25} \text{ and } |x| \geq 5\}$$

The domain of $g$ is $\{x \mid x \leq -5\} \cup \{x \mid x \geq 5\}$, and the range is $\{y \mid y \geq 0\}$. A sketch of the graph of $g$ is shown in Figure 3.1.7. •

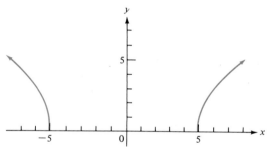

Figure 3.1.7

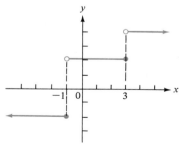

Figure 3.1.8

**ILLUSTRATION 6.** Let $h$ be the function which is the set of all ordered pairs $(x, y)$ such that

$$y = \begin{cases} -2, & \text{if } x \leq -1 \\ 2, & \text{if } -1 < x \leq 3 \\ 4, & \text{if } 3 < x \end{cases}$$

The domain of $h$ is the set $R$ of real numbers, while the range of $h$ consists of the three numbers, $-2$, $2$, and $4$. A sketch of the graph is shown in Figure 3.1.8. •

156  Functions  [Ch. 3

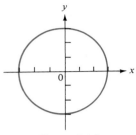

Figure 3.1.9

**ILLUSTRATION 7.** Consider the relation $\{(x, y) \mid x^2 + y^2 = 9\}$. This relation is not a function because for any $x$ greater than $-3$ and less than 3, there are two ordered pairs having that number as a first element. For example, both $(2, \sqrt{5})$ and $(2, -\sqrt{5})$ are ordered pairs in the relation. Furthermore, observe that the graph of the relation, shown in Figure 3.1.9, is a circle with center at the origin and radius 3, and a vertical line having the equation $x = a$, where $-3 < a < 3$, intersects the circle in two points. •

### EXAMPLE 2
Let $F$ be the function which is the set of all ordered pairs $(x, y)$ such that

$$y = \begin{cases} 2x - 1, & \text{if } x < 1 \\ x^2, & \text{if } 1 \leq x \end{cases}$$

Find the domain and range of $F$, and draw a sketch of the graph of $F$.

### SOLUTION
A sketch of the graph of $F$ is shown in Figure 3.1.10. The domain of $F$ is the set $R$ of real numbers, and the range of $F$ is also $R$.

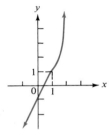

Figure 3.1.10

### EXAMPLE 3
Let $G$ be the function which is the set of all ordered pairs $(x, y)$ such that

$$y = \frac{x^2 - 4}{x - 2}$$

Find the domain and range of $G$, and draw a sketch of the graph of $G$.

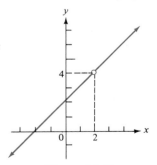

Figure 3.1.11

### SOLUTION
Because a value for $y$ is determined for each value of $x$ except $x = 2$, the domain of $G$ consists of all real numbers except 2. When $x = 2$, both the numerator and denominator are zero, and $0/0$ is undefined. Factoring the numerator into $(x - 2)(x + 2)$, we obtain

$$y = \frac{(x - 2)(x + 2)}{x - 2}$$

or $y = x + 2$, provided that $x \neq 2$. In other words, the function $G$ consists of all ordered pairs $(x, y)$ such that

$$y = x + 2 \quad \text{and} \quad x \neq 2$$

A sketch of the graph is shown in Figure 3.1.11. The range of $G$ is the set of all real numbers except 4. The graph consists of all points on the line $y = x + 2$ except the point $(2, 4)$.

## EXAMPLE 4

Let $H$ be the function which is the set of all ordered pairs $(x, y)$ such that

$$y = \begin{cases} x + 2, & \text{if } x \neq 2 \\ 1, & \text{if } x \leq 2 \end{cases}$$

Find the domain and range of $H$, and draw a sketch of the graph of $H$.

## SOLUTION

A sketch of the graph of $H$ is shown in Figure 3.1.12. The graph consists of the point $(2, 1)$ and all points on the line $y = x + 2$ except the point $(2, 4)$. Function $H$ is defined for all values of $x$, and therefore the domain of $H$ is the set $R$ of real numbers. The range of $H$ consists of all real numbers except 4.

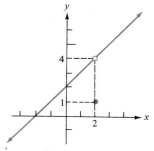

**Figure 3.1.12**

## EXAMPLE 5

Let $f$ be the function which is the set of all ordered pairs $(x, y)$ such that

$$y = \frac{(x^2 + x - 6)(x^2 - 16)}{(x^2 + 7x + 12)(x - 4)}$$

Find the domain and range of $f$, and draw a sketch of the graph of $f$.

## SOLUTION

Factoring the numerator and denominator, we obtain

$$y = \frac{(x + 3)(x - 2)(x + 4)(x - 4)}{(x + 4)(x + 3)(x - 4)}$$

We see that the denominator is zero for $x = -4, -3,$ and $4$; therefore $f$ is undefined for these three values of $x$. For values of $x \neq -4, -3,$ or $4$, we may divide numerator and denominator by the common factors and obtain

$$y = x - 2 \quad \text{if } x \neq -4, -3, \text{ or } 4$$

A sketch of the graph of $f$ is shown in Figure 3.1.13.

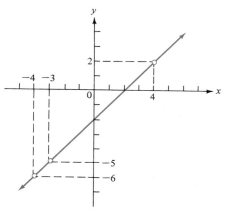

**Figure 3.1.13**

**158** Functions [Ch. 3]

The domain of $f$ is the set of all real numbers except $-4$, $-3$, and $4$; and the range of $f$ is the set of all real numbers except those values of $x - 2$ obtained by replacing $x$ by $-4$, $-3$, or $4$. That is, the range of $f$ is all real numbers except $-6$, $-5$, and $2$. The graph of this function is the straight line $y = x - 2$, with the points $(-4, -6)$, $(-3, -5)$, and $(4, 2)$ deleted.

### EXAMPLE 6
Let $g$ be the function which is the set of all ordered pairs $(x, y)$ such that
$$y = \begin{cases} x^2 - 1, & \text{if } x \neq 2 \\ 6, & \text{if } x = 2 \end{cases}$$
Find the domain and range of $g$, and draw a sketch of the graph of $g$.

### SOLUTION
A sketch of the graph of $g$ is shown in Figure 3.1.14. The graph consists of the point $(2, 6)$ and all points on the parabola $y = x^2 - 1$, except the point $(2, 3)$. Function $g$ is defined for all values of $x$, and so the domain of $g$ is the set $R$ of real numbers. The range of $g$ is $\{y \mid y \geq -1\}$.

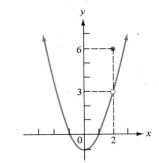

**Figure 3.1.14**

### EXAMPLE 7
Let $h$ be the function which is the set of all ordered pairs $(x, y)$ such that
$$y = \begin{cases} x + 5, & \text{if } x < -3 \\ \sqrt{9 - x^2}, & \text{if } -3 \leq x \leq 3 \\ 5 - x, & \text{if } 3 < x \end{cases}$$
Find the domain and range of $h$, and draw a sketch of the graph of $h$.

### SOLUTION
A sketch of the graph of $h$ is shown in Figure 3.1.15. The domain of $g$ is the set $R$ of real numbers, and the range is $\{y \mid y \leq 3\}$.

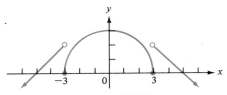

**Figure 3.1.15**

### EXAMPLE 8
Let
$$f = \{(x, y) \mid y = \sqrt{x(x - 2)}\}$$
Find the domain and range of $f$ and draw a sketch of the graph of $f$.

### SOLUTION
Because $\sqrt{x(x - 2)}$ is not a real number when $x(x - 2) < 0$, the domain of $f$ consists of the values of $x$ for which $x(x - 2) \geq 0$. This inequality will be satisfied when one of the following two cases holds: $x \geq 0$ and $x - 2 \geq 0$; or $x \leq 0$ and $x - 2 \leq 0$.

**CASE 1:** $x \geq 0$ and $x - 2 \geq 0$. That is,
$$x \geq 0 \quad \text{and} \quad x \geq 2$$

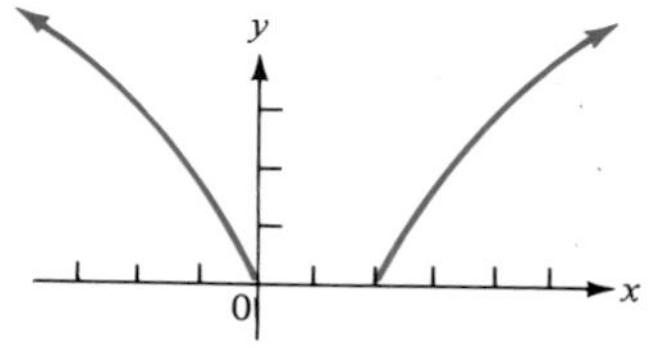

Both inequalities hold if $x \geq 2$, which is the set $\{x \mid x \geq 2\}$.

CASE 2: $x \leq 0$ and $x - 2 \leq 0$. That is,

$$x \leq 0 \quad \text{and} \quad x \leq 2$$

Both inequalities hold if $x \leq 0$, which is the set $\{x \mid x \leq 0\}$.
Combining the solutions for the two cases we obtain for the domain of $f$ the set $\{x \mid x \leq 0\} \cup \{x \mid x \geq 2\}$. The range of $f$ is $\{y \mid y \geq 0\}$. Figure 3.1.16 shows a sketch of the graph of $f$.

Figure 3.1.16

## EXERCISES 3.1

*In Exercises 1 through 10, find the domain and range of the given relation and determine if the given relation is a function.*

1. $\{(-3, 0), (-1, 2), (1, 4), (3, 6)\}$
2. $\{(-4, -5), (0, 2), (6, -1)\}$
3. $\{(0, 4), (2, 2), (4, 2), (6, 0)\}$
4. $\{(-2, 2), (-1, 2), (1, 2), (2, 2)\}$
5. $\{(-1, 1), (1, 3), (1, 4), (3, 5)\}$
6. $\{(-5, -4), (-1, -3), (-1, 3), (4, 3)\}$
7. $\{(-2, -1), (0, -1), (2, -1)\}$
8. $\{(1, 2), (1, 3), (1, 4), (1, 5)\}$
9. $\{(2, -2), (2, 0), (3, 3), (4, 1), (4, 5)\}$
10. $\{(3, -1), (6, -1), (9, 1), (12, 1)\}$

*In Exercises 11 through 18, draw a sketch of the graph of the given relation. Is the relation a function?*

11. $\{(x, y) \mid y = 2x + 5\}$
12. $\{(x, y) \mid y = -3x + 4\}$
13. $\{(x, y) \mid 3x - 6y = 2\}$
14. $\{(x, y) \mid 9x - 3y = 4\}$
15. $\{(x, y) \mid x^2 + 4y = 8\}$
16. $\{(x, y) \mid 2x^2 + y = 6\}$
17. $\{(x, y) \mid x + y^2 = 5\}$
18. $\{(x, y) \mid x - y^2 = 7\}$

*In Exercises 19 and 20, draw sketches of the graphs of relations R, S, and T. Determine if any of the relations are functions. Show that $R = S \cup T$.*

19. $R = \{(x, y) \mid y^2 = x + 9\}$, $S = \{(x, y) \mid y = \sqrt{x + 9}\}$, $T = \{(x, y) \mid y = -\sqrt{x + 9}\}$
20. $R = \{(x, y) \mid x^2 + y^2 = 16\}$, $S = \{(x, y) \mid y = \sqrt{16 - x^2}\}$, $T = \{(x, y) \mid y = -\sqrt{16 - x^2}\}$

*In Exercises 21 through 30, find the domain and range of the given function, and draw a sketch of the graph of the function.*

21. $f = \{(x, y) \mid y = 4x - 3\}$
22. $g = \{(x, y) \mid y = x^2 + 3\}$
23. $F = \{(x, y) \mid y = x^2 - 4\}$
24. $G = \{(x, y) \mid y = \sqrt{x - 2}\}$
25. $h = \{(x, y) \mid y = \sqrt{3x - 12}\}$
26. $H = \{(x, y) \mid y = \sqrt{x^2 - 36}\}$
27. $f = \{(x, y) \mid y = \sqrt{36 - x^2}\}$
28. $F = \{(x, y) \mid y = \sqrt{10 - x^2}\}$
29. $g = \{(x, y) \mid y = -5\}$
30. $f = \{(x, y) \mid y = \frac{8}{3}\}$

In Exercises 31 through 44, the function is the set of all ordered pairs $(x, y)$ satisfying the given equation. Find the domain and range, and draw a sketch of the graph of the function.

**31.** $f: y = \begin{cases} -3, & \text{if } x < 3 \\ 3, & \text{if } x \geq 2 \end{cases}$

**32.** $g: y = \begin{cases} -5, & \text{if } x < -3 \\ 0, & \text{if } -3 \leq x \leq 3 \\ 5, & \text{if } x > 3 \end{cases}$

**33.** $f: y = \dfrac{x^2 - 9}{x - 3}$

**34.** $f: y = \dfrac{(x - 2)(x^2 + 4x + 3)}{x^2 - x - 2}$

**35.** $g: y = \begin{cases} x + 3, & \text{if } x \neq 3 \\ 5, & \text{if } x = 3 \end{cases}$

**36.** $f: y = \dfrac{x^2 - 16}{x + 4}$

**37.** $h: y = \dfrac{(x - 1)(x^2 + x - 20)}{x^2 + 4x - 5}$

**38.** $g: y = \begin{cases} x - 4, & \text{if } x \neq -4 \\ 3, & \text{if } x = -4 \end{cases}$

**39.** $F: y = \begin{cases} x^2 - 4, & \text{if } x < 3 \\ 1, & \text{if } x \geq 3 \end{cases}$

**40.** $G: y = \begin{cases} x + 6, & \text{if } x \leq -2 \\ x^2, & \text{if } x > -2 \end{cases}$

**41.** $H: y = \begin{cases} x + 2, & \text{if } x \leq -4 \\ \sqrt{16 - x^2}, & \text{if } -4 < x < 4 \\ 2 - x, & \text{if } x \geq 4 \end{cases}$

**42.** $f: y = \sqrt{x^2 + 3x}$

**43.** $g: y = \sqrt{x^2 - 4x - 5}$

**44.** $h: y = \sqrt{2x^2 - 7x - 15}$

## 3.2 Function Notation and Types of Functions

If $f$ is the function having as its domain values of $x$ and as its range values of $y$, we use the symbol $f(x)$ (read "$f$ at $x$") to denote the particular value of $y$ that corresponds to the value of $x$. The number $f(x)$ is called the *function value* of $f$ corresponding to $x$. Therefore, the function of Illustration 4 in Section 3.1 can be defined by the equation $f(x) = 2x^2 - 4$. Because when $x = 1$, $2x^2 - 4 = -2$, we write $f(1) = -2$. Similarly, $f(-2) = 4$, $f(0) = -4$, and so on.

If the letter $g$ is used to denote a function, we would use the symbol $g(x)$ to denote the function value corresponding to $x$. Thus if $g(x) = 8 - 5x$, then $g(-4) = 28$, $g(0) = 8$, $g(4) = -12$, and so on.

When defining a function, the domain of the function must be either given or understood. For instance, if we are given

$$f(x) = 2x^2 + 3x - 5$$

it is implied that $x$ can be any real number. However, if we are given

$$f(x) = 2x^2 + 3x - 2 \qquad (0 \leq x \leq 5)$$

then the domain of $f$ consists of all real numbers between and including 0 and 5. If $g$ is defined by

$$g(x) = \frac{x+5}{x^2-4}$$

it is implied that $x^2 \neq 4$, because the rational expression is undefined when $x^2 = 4$. Therefore, the domain of $g$ is the set of all real numbers except 2 and $-2$. If $h$ is the function such that

$$h(x) = \sqrt{9-x^2}$$

it is implied that the domain of $h$ is $\{x \mid -3 \leq x \leq 3\}$ because $\sqrt{9-x^2}$ is undefined (that is, it is not a real number) for $x > 3$ or $x < -3$.

**EXAMPLE 1**
If $f(x) = 2x^2 + 3x - 5$, find each of the following.

(a) $f(0)$    (b) $f(3)$
(c) $f(-1)$    (d) $2f(h)$
(e) $f(2h)$    (f) $f(2x)$
(g) $f(x) + f(h)$    (h) $f(x+h)$

**SOLUTION**

(a) $f(0) = 2(0)^2 + 3(0) - 5$
$\phantom{f(0)} = -5$

(b) $f(3) = 2(3)^2 + 3(3) - 5$
$\phantom{f(3)} = 18 + 9 - 5$
$\phantom{f(3)} = 22$

(c) $f(-1) = 2(-1)^2 + 3(-1) - 5$
$\phantom{f(-1)} = 2 - 3 - 5$
$\phantom{f(-1)} = -6$

(d) $2f(h) = 2(2h^2 + 3h - 5)$
$\phantom{2f(h)} = 4h^2 + 6h - 10$

(e) $f(2h) = 2(2h)^2 + 3(2h) - 5$
$\phantom{f(2h)} = 8h^2 + 6h - 5$

(f) $f(2x) = 2(2x)^2 + 3(2x) - 5$
$\phantom{f(2x)} = 8x^2 + 6x - 5$

(g) $f(x) + f(h) = (2x^2 + 3x - 5) + (2h^2 + 3h - 5)$
$\phantom{f(x) + f(h)} = 2x^2 + 3x + (2h^2 + 3h - 10)$

(h) $f(x+h) = 2(x+h)^2 + 3(x+h) - 5$
$\phantom{f(x+h)} = 2x^2 + 4hx + 2h^2 + 3x + 3h - 5$
$\phantom{f(x+h)} = 2x^2 + (4hx + 3x) + (2h^2 + 3h - 5)$
$\phantom{f(x+h)} = 2x^2 + (4h + 3)x + (2h^2 + 3h - 5)$

Compare the computations in parts (g) and (h) of Example 1. In part (g) we compute $f(x) + f(h)$, which is the sum of the two function values $f(x)$ and $f(h)$. In part (h) we compute $f(x+h)$, which is the function value at the sum of $x$ and $h$.

**EXAMPLE 2**
If $g(x) = 3x^2 - 2x + 4$, find

$$\frac{g(x+h) - g(x)}{h}$$

where $h \neq 0$.

**SOLUTION**

$$\frac{g(x+h) - g(x)}{h} = \frac{3(x+h)^2 - 2(x+h) + 4 - (3x^2 - 2x + 4)}{h}$$

$$= \frac{3x^2 + 6hx + 3h^2 - 2x - 2h + 4 - 3x^2 + 2x - 4}{h}$$

$$= \frac{6hx - 2h + 3h^2}{h}$$

$$= 6x - 2 + 3h$$

**3.2.1 DEFINITION** (i) A function $f$ is said to be an *even* function if for every $x$ in the domain of $f$, $f(-x) = f(x)$.
(ii) A function $f$ is said to be an *odd* function if for every $x$ in the domain of $f$, $f(-x) = -f(x)$.

In both parts (i) and (ii) it is understood that $-x$ is in the domain of $f$ whenever $x$ is.

**ILLUSTRATION 1**

(a) If $f(x) = 3x^4 - 2x^2 + 7$, then
$$f(-x) = 3(-x)^4 - 2(-x)^2 + 7$$
$$= 3x^4 - 2x^2 + 7$$
$$= f(x)$$

Therefore, $f$ is an even function.

(b) If $g(x) = 3x^5 - 4x^3 - 9x$, then
$$g(-x) = 3(-x)^5 - 4(-x)^3 - 9(-x)$$
$$= -3x^5 + 4x^3 + 9x$$
$$= -(3x^5 - 4x^3 - 9x)$$
$$= -g(x)$$

Therefore, $g$ is an odd function.

(c) If $h(x) = 2x^4 + 7x^3 - x^2 + 9$, then
$$h(-x) = 2(-x)^4 + 7(-x)^3 - (-x)^2 + 9$$
$$= 2x^4 - 7x^3 - x^2 + 9$$

We see that the function $h$ is neither even nor odd. ●

From the definition of an even function and Theorem 2.4.3(ii), it follows that the graph of an even function is symmetric with respect to the $y$ axis. From Theorem 2.4.6 and the definition of an odd function, we see that the graph of an odd function is symmetric with respect to the origin.

**ILLUSTRATION 2**

(a) If $f(x) = x^4$, then $f(-x) = (-x)^4$, and so $f(-x) = f(x)$. Therefore, $f$ is an even function. The graph of $f$ is shown in Figure 3.2.1 and it is symmetric with respect to the $y$ axis.
(b) If $g(x) = x^3$, then $g(-x) = (-x)^3$, so $g(-x) = -g(x)$. Thus $g$ is an odd function. The graph of $g$ is shown in Figure 3.2.2, and it is symmetric with respect to the origin. ●

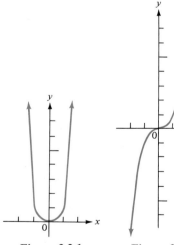

**Figure 3.2.1**     **Figure 3.2.2**

If the range of a function consists of only one number, then the function is called a *constant function*. Thus, if $f(x) = c$, where $c$ is a real number, then $f$

is a constant function and its graph is a straight line parallel to the $x$ axis at a directed distance of $c$ units from the $x$ axis.

**ILLUSTRATION 3**

(a) The function $f$ for which $f(x) = 5$ is a constant function, and a sketch of its graph is shown in Figure 3.2.3.
(b) The function $g$ for which $g(x) = -\tfrac{7}{2}$ is a constant function, and a sketch of its graph is shown in Figure 3.2.4. •

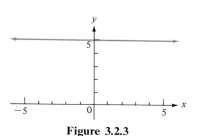

Figure 3.2.3

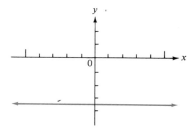

Figure 3.2.4

If $f$ is a function for which $f(x)$ is a polynomial of degree $n$, then $f$ is called a *polynomial function* of degree $n$. If the degree of the polynomial is 1, then the function is called a *linear function;* if the degree is 2, the function is called a *quadratic function;* and if the degree is 3, the function is called a *cubic function.*

**ILLUSTRATION 4**

(a) The function $f$ defined by
$$f(x) = 3x^4 - 2x^3 + 5x - 1$$
is a polynomial function of degree 4.
(b) The function $g$ defined by
$$g(x) = 6x - 3$$
is a linear function.
(c) The function $h$ defined by
$$h(x) = 5x^2 + 8x - 4$$
is a quadratic function.
(d) The function $F$ defined by
$$F(x) = 4x^3 - 7x^2 + 2$$
is a cubic function. •

The general linear function is defined by
$$f(x) = mx + b$$
where $m$ and $b$ are constants and $m \neq 0$. The graph of this function is a straight line having $m$ as its slope and $b$ as its $y$ intercept. The particular linear function defined by
$$f(x) = x$$
is called the *identity function*.

**ILLUSTRATION 5.** The identity function can be represented with ordered pair notation as
$$f = \{(x,y) \mid y = x\}$$
Because $f(x) = x$, then
$$f(-x) = -x$$
$$= -f(x)$$
Therefore, the identity function is an odd function and its graph is symmetric with respect to the origin. A sketch of the graph of the identity function is shown in Figure 3.2.5. ●

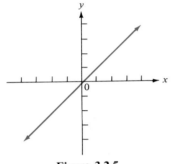

**Figure 3.2.5**

If a function can be expressed as the quotient of two polynomial functions, the function is called a *rational function*.

**ILLUSTRATION 6.** The function $f$ defined by
$$f(x) = \frac{x^4 + 2x^2 - 1}{x^2 - 9}$$
is a rational function. Because the denominator of the rational expression is zero when $x$ is 3 or $-3$, the domain of this function $f$ is the set of all real numbers except 3 and $-3$. ●

An *algebraic function* is a function formed by a finite number of algebraic operations on the identity function and the constant function. These algebraic operations include addition, subtraction, multiplication, division, raising to powers, and extracting roots.

**ILLUSTRATION 7.** The function $g$ defined by
$$g(x) = \frac{x \sqrt[3]{x^2 - 3}}{(x^2 + 5x - 2)^4}$$
is an algebraic function. ●

In addition to algebraic functions there are *transcendental functions*. Examples of transcendental functions are exponential and logarithmic functions; they are discussed in Chapter 4.

The function given in the next example is called the *absolute value function*.

**EXAMPLE 3**
Let $h$ be the function defined by
$$h(x) = |x|$$
Find the domain and range of $h$ and draw a sketch of its graph.

**SOLUTION**
From the definition of the absolute value of a real number, it follows that
$$h(x) = \begin{cases} x, & \text{if } x \geq 0 \\ -x, & \text{if } x < 0 \end{cases}$$
The domain of $h$ is the set $R$ of real numbers, and the range of $h$ is the set of nonnegative numbers. A sketch of the graph of $h$ is shown in Figure 3.2.6.

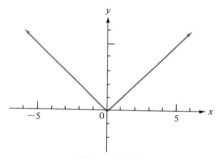

**Figure 3.2.6**

Observe in Example 3 that the graph of the absolute value function is symmetric with respect to the $y$ axis. This occurs because it is an even function, since $|-x| = |x|$.

In the following example, the symbol $[\![x]\!]$ is used, where $[\![x]\!]$ is defined as the greatest integer not greater than $x$. Hence, $[\![2]\!] = 2$, $[\![3.6]\!] = 3$, $[\![\tfrac{1}{2}]\!] = 0$, $[\![-3.6]\!] = -4$, $[\![-8]\!] = -8$, and so on. The function $F$, defined by $F(x) = [\![x]\!]$, is called the *greatest integer function*.

**EXAMPLE 4**
The function $F$ is the greatest integer function, defined by
$$F(x) = [\![x]\!]$$
$$= n \quad \text{if } n \leq x < n + 1$$
where $n$ is an integer. Draw a sketch of the graph of $F$, and state the domain and range of $F$.

**SOLUTION**
The function $F$ is the set $\{(x, y) \mid y = [\![x]\!]\}$. Therefore,

if $-5 \leq x < -4$, $\quad y = -5$
if $-4 \leq x < -3$, $\quad y = -4$
if $-3 \leq x < -2$, $\quad y = -3$
if $-2 \leq x < -1$, $\quad y = -2$
if $-1 \leq x < \phantom{-}0$, $\quad y = -1$

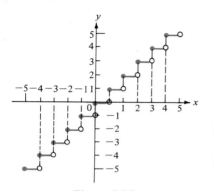

Figure 3.2.7

if $0 \leq x < 1$, $y = 0$
if $1 \leq x < 2$, $y = 1$
if $2 \leq x < 3$, $y = 2$
if $3 \leq x < 4$, $y = 3$
if $4 \leq x < 5$, $y = 4$

and so on.

A sketch of the graph of $F$ is shown in Figure 3.2.7. The solid dot on the left endpoint of each line segment is used to indicate that the endpoint is part of the graph, and the open dot on the right endpoint of each line segment is used to indicate that the endpoint is not part of the graph.

The domain of $F$ is the set $R$ of real numbers, and the range of $F$ is the set of all the integers.

## EXERCISES 3.2

In Exercises 1 through 14, find the given function value.

**1.** $f(3)$ if $f(x) = 6 - 2x$
**2.** $g(-2)$ if $g(x) = 4x + 7$
**3.** $g(-5)$ if $g(x) = 3x^2 + 10x + 5$
**4.** $f(4)$ if $f(x) = x^2 - 6x + 1$
**5.** $f(-4)$ if $f(x) = \sqrt{25 - x^2}$
**6.** $g(-1)$ if $g(x) = x^2 - 1$
**7.** $g(6)$ if $g(x) = \dfrac{3x + 7}{4x - 9}$
**8.** $f(-3)$ if $f(x) = \dfrac{5 + x}{4 - x}$
**9.** $h(-4)$ if $h(x) = |2x - 5|$
**10.** $F\left(\dfrac{2}{3}\right)$ if $F(x) = |3x - 2|$
**11.** $f(4)$ if $f(x) = \begin{cases} 2x - 3, & \text{if } x \leq 2 \\ x^2, & \text{if } x > 2 \end{cases}$
**12.** $f(-3)$ if $f(x) = \begin{cases} (x - 1)^2, & \text{if } x < 1 \\ 1 - x, & \text{if } x \geq 1 \end{cases}$
**13.** $f(-4)$ for the function of Exercise 11.
**14.** $f(3)$ for the function of Exercise 12.

In Exercises 15 through 22, find each of the following: (a) $2f(h)$; (b) $f(2h)$; (c) $f(2x)$; (d) $f(x + 2)$; (e) $f(x^2)$.

**15.** $f(x) = 4x - 5$
**16.** $f(x) = 8 + 3x$
**17.** $f(x) = 2x^2 + 1$
**18.** $f(x) = 9 - x^2$
**19.** $f(x) = 3 - 5x + 2x^2$
**20.** $f(x) = x^2 + x + 1$
**21.** $f(x) = \dfrac{2}{1 + x}$
**22.** $f(x) = \sqrt[3]{x^2}$

In Exercises 23 through 32, find $\dfrac{f(x + h) - f(x)}{h}$, where $h \neq 0$.

**23.** $f(x) = 6x + 7$
**24.** $f(x) = 3 - 2x$
**25.** $f(x) = 1 - x^2$
**26.** $f(x) = 4x^2$
**27.** $f(x) = 3x^2 + 4x + 1$
**28.** $f(x) = 1 - 5x + 4x^2$

29. $f(x) = x^3 + 3x - 1$
30. $f(x) = x^3 - 2x^2 + 4$
31. $f(x) = \dfrac{8-x}{3+x}$
32. $f(x) = \dfrac{2x+5}{3x-1}$

33. For each of the following functions, determine whether $f$ is even, odd, or neither.
(a) $f(x) = 3x^4 - 2x^2 + 5$
(b) $f(x) = 4x^3 - x$
(c) $f(t) = t^2 - 2t + 1$
(d) $f(y) = y^6 - 1$
(e) $f(x) = 3x^5 + 1$
(f) $f(x) = |x|$
(g) $f(x) = \dfrac{x^2+1}{x^3-x}$
(h) $f(r) = \dfrac{r-1}{r+1}$

34. There is one function that is both even and odd. What is it?

*In Exercises 35 through 50, draw a sketch of the graph of the function defined by the given equation.*

35. $f(x) = -3$
36. $g(x) = \frac{5}{2}$
37. $g(x) = 3|x|$
38. $f(x) = |x| - 2$
39. $F(x) = |x| + 3$
40. $G(x) = |x - 2|$
41. $f(x) = |3x + 9|$
42. $g(x) = |x| + |x - 1|$
43. $g(x) = |x + 2| - |x|$
44. $f(x) = x|x|$
45. $G(x) = [\![x]\!] - 1$
46. $F(x) = [\![x - 1]\!]$
47. $f(x) = 2 - [\![x]\!]$
48. $g(x) = [\![2 - x]\!]$
49. $F(x) = x + [\![x]\!]$
50. $G(x) = 2x - [\![x]\!]$

## 3.3 Operations on Functions

We now consider the operations of addition, subtraction, multiplication, and division on functions. The functions obtained from these operations—called the sum, the difference, the product, and the quotient of the original functions—are given in the following definition.

**3.3.1 DEFINITION** If $f$ and $g$ are functions, then

(i) The *sum* of $f$ and $g$, denoted by $f + g$, is the function defined by

$$(f + g)(x) = f(x) + g(x)$$

(ii) The *difference* of $f$ and $g$, denoted by $f - g$, is the function defined by

$$(f - g)(x) = f(x) - g(x)$$

(iii) The *product* of $f$ and $g$, denoted by $fg$, is the function defined by

$$(fg)(x) = f(x) \cdot g(x)$$

(iv) The *quotient* of $f$ by $g$, denoted by $f/g$, is the function defined by

$$(f/g)(x) = \dfrac{f(x)}{g(x)}$$

In each case the *domain* of the resulting function is the intersection of the domains of $f$ and $g$, with the exception that in part (iv) the values of $x$ for which $g(x) = 0$ are excluded.

Observe in part (i) of Definition 3.3.1 that, because $f$ and $g$ are functions and not numbers, the $+$ between $f$ and $g$ does not indicate the addition of real numbers. However, because $f(x)$ and $g(x)$ are real numbers, the $+$ between $f(x)$ and $g(x)$ does indicate the sum of real numbers.

**ILLUSTRATION 1.** Let $f(x) = 2x - 3$ and $g(x) = x^2 + 4x - 5$.

(a) The sum of $f$ and $g$ is the function defined by
$$(f + g)(x) = (2x - 3) + (x^2 + 4x - 5)$$
$$= x^2 + 6x - 8$$

(b) The difference of $f$ and $g$ is the function defined by
$$(f - g)(x) = (2x - 3) - (x^2 + 4x - 5)$$
$$= -x^2 - 2x + 2$$

(c) The product of $f$ and $g$ is the function defined by
$$(fg)(x) = (2x - 3)(x^2 + 4x - 5)$$
$$= 2x^3 + 5x^2 - 22x + 15$$

(d) The quotient of $f$ by $g$ is the function defined by
$$(f/g)(x) = \frac{2x - 3}{x^2 + 4x - 5}$$

(e) The quotient of $g$ by $f$ is the function defined by
$$(g/f)(x) = \frac{x^2 + 4x - 5}{2x - 3}$$

The domains of both $f$ and $g$ are the set $R$ of real numbers. Hence in parts (a), (b), and (c) the domain of the resulting function is $R$. In part (d) the denominator is $x^2 + 4x - 5 = (x + 5)(x - 1)$, and this product is zero when either $x = -5$ or $x = 1$. Thus the domain of $f/g$ is $\{x \mid x \in R$ and $x \neq -5, x \neq 1\}$. In part (e) the denominator is $2x - 3$, which is zero when $x = \frac{3}{2}$. Therefore, the domain of $g/f$ is $\{x \mid x \in R$ and $x \neq \frac{3}{2}\}$. ●

**EXAMPLE 1**

If $f$ and $g$ are the functions defined by
$$f(x) = \frac{1}{x - 3} \quad \text{and} \quad g(x) = \frac{x + 4}{x - 2}$$

**SOLUTION**

(a) $(f + g)(x) = \dfrac{1}{x - 3} + \dfrac{x + 4}{x - 2}$

$= \dfrac{x^2 + 2x - 14}{(x - 3)(x - 2)}$

(b) $(f - g)(x) = \dfrac{1}{x - 3} - \dfrac{x + 4}{x - 2}$

$= \dfrac{-x^2 + 10}{(x - 3)(x - 2)}$

find each of the following function values and in each case determine the domain of the resulting function.
(a) $(f + g)(x)$  (b) $(f - g)(x)$
(c) $(fg)(x)$    (d) $(f/g)(x)$
(e) $(g/f)(x)$

(c) $(fg)(x) = \dfrac{1}{x - 3} \cdot \dfrac{x + 4}{x - 2}$

$= \dfrac{x + 4}{(x - 3)(x - 2)}$

(e) $(g/f)(x) = \dfrac{x + 4}{x - 2} \div \dfrac{1}{x - 3}$

$= \dfrac{(x + 4)(x - 3)}{x - 2}$

(d) $(f/g)(x) = \dfrac{1}{x - 3} \div \dfrac{x + 4}{x - 2}$

$= \dfrac{x - 2}{(x - 3)(x + 4)}$

The domain of $f$ is $\{x \mid x \in R \text{ and } x \neq 3\}$, and the domain of $g$ is $\{x \mid x \in R \text{ and } x \neq 2\}$. Thus, in parts (a), (b), and (c), the domain of the resulting function is $\{x \mid x \in R \text{ and } x \neq 2, x \neq 3\}$. In part (d), because $g(x) = 0$ when $x = -4$, the domain of $f/g$ is $\{x \mid x \in R \text{ and } x \neq 2, x \neq 3, x \neq -4\}$. In part (e), because there is no value of $x$ for which $f(x) = 0$, the domain of $g/f$ is $\{x \mid x \in R \text{ and } x \neq 2, x \neq 3\}$.

To indicate the product of a function $f$ multiplied by itself, or $ff$, we write $f^2$.

**ILLUSTRATION 2.** If $f$ is defined by $f(x) = 4x - 3$, then $f^2$ is the function defined by

$$f^2(x) = (4x - 3)(4x - 3)$$
$$= 16x^2 - 24x + 9$$
●

In addition to combining two functions by the operations given in Definition 3.3.1, we shall consider the "composite function" of two given functions.

**3.3.2 DEFINITION** Given the two functions $f$ and $g$, the *composite function*, denoted by $f \circ g$, is defined by

$$(f \circ g)(x) = f(g(x))$$

and the domain of $f \circ g$ is the set of all numbers $x$ in the domain of $g$ such that $g(x)$ is in the domain of $f$.

**EXAMPLE 2**
Given that $f$ is defined by $f(x) = \sqrt{x}$, and $g$ is defined by $g(x) = x - 4$, find $F(x)$ if $F = f \circ g$, and find the domain of $F$.

**SOLUTION**

$$F(x) = (f \circ g)(x)$$
$$= f(g(x))$$
$$= f(x - 4)$$
$$= \sqrt{x - 4}$$

The domain of $g$ is the set $R$ of real numbers, and the domain of $f$ is the set of all nonnegative numbers. Therefore, the domain of $F$ is the set of real numbers for which $x - 4 \geq 0$; that is, $\{x \mid x \geq 4\}$.

**EXAMPLE 3**
Given that $f$ is defined by $f(x) = \sqrt{x}$ and $g$ is defined by $g(x) = x^2 - 1$, find

(a) $f \circ f$  (b) $g \circ g$
(c) $f \circ g$  (d) $g \circ f$

Also, find the domain of the composite function in each part.

**SOLUTION**
The domain of $f$ is $\{x \mid x \geq 0\}$, and the domain of $g$ is the set $R$ of real numbers.

(a) $(f \circ f)(x) = f(f(x))$
$\phantom{(f \circ f)(x)} = f(\sqrt{x})$
$\phantom{(f \circ f)(x)} = \sqrt{\sqrt{x}}$
$\phantom{(f \circ f)(x)} = \sqrt[4]{x}$

The domain of $f \circ f$ is $\{x \mid x \geq 0\}$.

(b) $(g \circ g)(x) = g(g(x))$
$\phantom{(g \circ g)(x)} = g(x^2 - 1)$
$\phantom{(g \circ g)(x)} = (x^2 - 1)^2 - 1$
$\phantom{(g \circ g)(x)} = x^4 - 2x^2$

The domain of $g \circ g$ is the set $R$ of real numbers.

(c) $(f \circ g)(x) = f(g(x))$
$\phantom{(f \circ g)(x)} = f(x^2 - 1)$
$\phantom{(f \circ g)(x)} = \sqrt{x^2 - 1}$

The domain of $f \circ g$ is $\{x \mid x \leq -1\} \cup \{x \mid x \geq 1\}$.

(d) $(g \circ f)(x) = g(f(x))$
$\phantom{(g \circ f)(x)} = g(\sqrt{x})$
$\phantom{(g \circ f)(x)} = (\sqrt{x})^2 - 1$
$\phantom{(g \circ f)(x)} = x - 1$

The domain of $g \circ f$ is $\{x \mid x \geq 0\}$.

Note in part (d) that even though $x - 1$ is defined for all values of $x$, the domain of $g \circ f$, by Definition 3.3.2, is the set of all numbers $x$ in the domain of $f$ such that $f(x)$ is in the domain of $g$.

**EXAMPLE 4**
Given that $f$ is defined by $f(x) = 3x + 2$, find a function $g$ such that $(f \circ g)(x) = x^2$.

**SOLUTION**
Because $(f \circ g)(x) = x^2$, then

$$f(g(x)) = x^2$$
$$3(g(x)) + 2 = x^2$$
$$3(g(x)) = x^2 - 2$$
$$g(x) = \frac{x^2 - 2}{3}$$

*EXERCISES 3.3*

*In Exercises 1 through 10, the functions $f$ and $g$ are defined. In each exercise, find*
(a) $(f + g)(x)$; (b) $(f - g)(x)$; (c) $(fg)(x)$; (d) $(f/g)(x)$; (e) $(g/f)(x)$; and (f) $f^2(x)$.
*In each case determine the domain of the resulting function.*

**1.** $f(x) = x - 2$; $g(x) = x + 7$
**2.** $f(x) = 3 - 2x$; $g(x) = 6 - 3x$
**3.** $f(x) = 4x + 2$; $g(x) = 4 - x^2$
**4.** $f(x) = x^2 + 3x - 5$; $g(x) = 2x + 1$

5. $f(x) = 2x^2 - x - 15$; $g(x) = 3x^2 - 8x - 3$
6. $f(x) = x^2 - 1$; $g(x) = 4x^2 + 7x + 3$
7. $f(x) = \dfrac{1}{x}$; $g(x) = \sqrt{x}$
8. $f(x) = \sqrt{x}$; $g(x) = -\dfrac{1}{x}$
9. $f(x) = \dfrac{1}{x}$; $g(x) = \dfrac{x^2 + x}{x - 1}$
10. $f(x) = \dfrac{x + 3}{x - 5}$; $g(x) = \dfrac{1}{x - 4}$

*In Exercises 11 through 20, the functions f and g are defined. In each exercise find
(a) $f \circ g$; (b) $g \circ f$; (c) $f \circ f$; (d) $g \circ g$. Also, find the domain of the composite function in each part.*

11. The functions of Exercise 1.
12. The functions of Exercise 2.
13. $f(x) = x - 1$; $g(x) = x^2$
14. $f(x) = \sqrt{x}$; $g(x) = x^2 + 1$
15. $f(x) = \sqrt{x - 2}$; $g(x) = x^2 - 2$
16. $f(x) = x^2 - 1$; $g(x) = \dfrac{1}{x}$
17. The functions of Exercise 7.
18. The functions of Exercise 8.
19. $f(x) = |x|$; $g(x) = |x + 2|$
20. $f(x) = \sqrt{x^2 - 1}$; $g(x) = \sqrt{x - 1}$

21. If $f(x) = 3x - 6$ and $g(x) = \dfrac{1}{3}x + 2$, show that $(f \circ g)(x) = x$ and $(g \circ f)(x) = x$. (Because of these relationships, $f$ and $g$ are "inverse functions." Inverse functions are discussed in Section 3.6.)

22. If $f(x) = 4 - 2x$, find a function $g$ such that $(f \circ g)(x) = x$ and $(g \circ f)(x) = x$.

23. If $f(x) = x^2$, find a function $g$ such that $(f \circ g)(x) = x^2 - 6x + 9$.

24. If $f(x) = \dfrac{1}{x}$, show that $(f \circ f)(x) = x$, if $x \neq 0$.

## 3.4 Quadratic Functions

The general quadratic function is defined by
$$f(x) = ax^2 + bx + c$$
where $a$, $b$, and $c$ are constants and $a \neq 0$. The function $f$ defined by this equation can be written as
$$f = \{(x, y) \mid y = ax^2 + bx + c\}$$
and the graph of $f$ is the same as the graph of the equation
$$y = ax^2 + bx + c$$

In Section 2.5 we learned that the graph of such an equation is a parabola whose axis is parallel to the $y$ axis. Also in Section 2.5, properties of the parabola were discussed, and these are used as an aid in drawing a sketch of the graph.

**ILLUSTRATION 1.** If the function $f$ is defined by
$$f(x) = -2x^2 + 8x - 5 \qquad (1)$$
the graph of $f$ is the same as the graph of the equation
$$y = -2x^2 + 8x - 5 \qquad (2)$$
This equation is equivalent to the equation
$$2(x^2 - 4x) = -y - 5$$
To complete the square of the binomial within the parentheses, we add $2(4)$ to each member and we have
$$2(x^2 - 4x + 4) = -y - 5 + 8$$
$$(x - 2)^2 = -\frac{1}{2}(y - 3)$$
This equation is of the form of the equation
$$(x - h)^2 = 4p(y - k)$$
where $(h, k)$ is $(2, 3)$ and $p = -\frac{1}{8}$. Therefore, the vertex of the parabola is at $(2, 3)$ and the axis is the line having the equation $x = 2$. Because $p < 0$, the parabola opens downward. We find a few more points on the parabola by substituting values of $x$ into equation (2). When $x = 0$, $y = -5$, and when $x = 1$, $y = 1$; thus the points $(0, -5)$ and $(1, 1)$ are on the parabola. Because a parabola is symmetric with respect to its axis, it follows that the points $(3, 1)$ and $(4, -5)$ are also on the parabola. A sketch of the parabola is shown in Figure 3.4.1. •

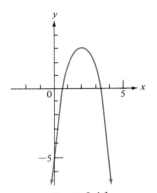

**Figure 3.4.1**

The *zeros* of a function $f$ are the values of $x$ for which $f(x) = 0$. For the function of Illustration 1 we can obtain the zeros by substituting $0$ for $f(x)$ in equation (1). Doing this, we have
$$-2x^2 + 8x - 5 = 0 \qquad (3)$$
$$2x^2 - 8x + 5 = 0$$

Solving this equation by the quadratic formula where $a$ is $2$, $b$ is $-8$, and $c$ is $5$, we have
$$x = \frac{-b \pm \sqrt{b^2 - 4ac}}{2a}$$
$$= \frac{8 \pm \sqrt{64 - 40}}{4}$$
$$= \frac{8 \pm \sqrt{24}}{4}$$

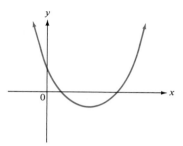

**Figure 3.4.2**

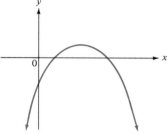

**Figure 3.4.3**

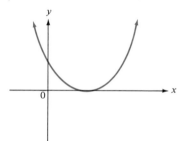

**Figure 3.4.4**

$$= \frac{8 \pm 2\sqrt{6}}{4}$$

$$= \frac{4 \pm \sqrt{6}}{2}$$

Hence the zeros of the function of Illustration 1 are

$$\frac{4 + \sqrt{6}}{2} \quad \text{and} \quad \frac{4 - \sqrt{6}}{2}$$

These two numbers are the roots of quadratic equation (3). These numbers are also the $x$ intercepts of the parabola that is the graph of the given function.

In general, if

$$f = \{(x,y) | y = ax^2 + bx + c\} \tag{4}$$

then the zeros of $f$ are the roots of the equation

$$ax^2 + bx + c = 0 \tag{5}$$

and the $x$ intercepts of the graph of $f$.

In Section 2.2 we learned that a quadratic equation in one variable can have two real roots, one real root (of multiplicity two) or no real roots (that is, two imaginary roots). If equation (5) has two real roots, the graph of the function $f$ defined in (4) intersects the $x$ axis at two distinct points. This situation is shown in Figure 3.4.2, where the parabola opens upward ($a > 0$), and in Figure 3.4.3, where the parabola opens downward ($a < 0$). If the equation has one real root, the graph is tangent to the $x$ axis, as shown in Figures 3.4.4 ($a > 0$) and 3.4.5 ($a < 0$). If the equation has no real roots, the graph does not intersect the $x$ axis as shown in Figures 3.4.6 ($a > 0$) and 3.4.7 ($a < 0$).

When the graph of a quadratic function opens upward, the function has a *minimum value* and this minimum function value occurs at the vertex of the

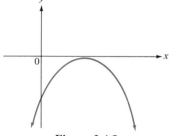

**Figure 3.4.5**

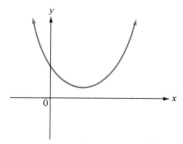

**Figure 3.4.6**

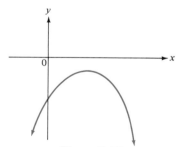

**Figure 3.4.7**

parabola. There is no maximum value for such a function. When the parabola opens downward, the function has a *maximum value* occurring at the vertex; it has no minimum value.

**ILLUSTRATION 2.** The function of Illustration 1 is

$$f = \{(x,y) \mid y = -2x^2 + 8x - 5\}$$

We showed that the graph of $f$ is a parabola opening downward and having its vertex at (2, 3). Therefore, this function $f$ has a maximum value of 3 and it occurs when $x = 2$ (that is, $f(2) = 3$). •

**EXAMPLE 1**
Find either a maximum value or a minimum value of the function $f$ if

$$f = \{(x,y) \mid y = 3x^2 + 3x + 2\}$$

**SOLUTION**
We write the equation of the given function in the form $(x - h)^2 = 4p(y - k)$. The equation

$$y = 3x^2 + 3x + 2$$

is equivalent to

$$3(x^2 + x) = y - 2$$

Completing the square of the binomial in parentheses in the left member, we have

$$3\left(x^2 + x + \frac{1}{4}\right) = y - 2 + \frac{3}{4}$$

$$\left(x + \frac{1}{2}\right)^2 = \frac{1}{3}\left(y - \frac{5}{4}\right)$$

The parabola opens upward and its vertex is at $(-\frac{1}{2}, \frac{5}{4})$. Therefore, $f(-\frac{1}{2})$ is the minimum value of $f$ and $f(-\frac{1}{2}) = \frac{5}{4}$.

**EXAMPLE 2**
Find either a maximum value or a minimum value of the function $g$ if

$$g = \{(x,y) \mid 3x^2 - 12x + 2y = -20\}$$

**SOLUTION**
The equation $3x^2 - 12x + 2y = -20$ is equivalent to

$$3x^2 - 12x = -2y - 20$$
$$3(x^2 - 4x) = -2y - 20$$
$$3(x^2 - 4x + 4) = -2y - 20 + 12$$
$$3(x - 2)^2 = -2y - 8$$
$$(x - 2)^2 = -\frac{2}{3}(y + 4)$$

The parabola opens downward and its vertex is at $(2, -4)$. Thus $g(2)$ is the maximum value of $g$, and $g(2) = -4$.

## Quadratic Functions

### EXAMPLE 3
The financial manager of a college magazine determines that 1000 copies of the magazine will be sold if the price is 50 cents and that the number of copies sold decreases by 10 for each 1 cent added to the price. What price will yield the largest gross income from sales?

### SOLUTION
The number of cents in the gross income depends upon the price per copy. Let $f(x)$ be the number of cents in the gross income when $x$ cents is the price per copy.

The amount by which $x$ exceeds 50 is $x - 50$. To determine the number of copies sold when $x$ cents is the price per copy, we must subtract from 1000 the product of 10 and this excess. Hence, when $x$ cents is the price per copy, the number of copies sold is $1000 - 10(x - 50)$.

We obtain an expression for the gross income by multiplying the number of copies sold by the price per copy. Hence,

$$f(x) = [1000 - 10(x - 50)]x$$
$$f(x) = (1500 - 10x)x$$
$$f(x) = 1500x - 10x^2$$

We wish to determine the maximum value of the function $f$ if

$$f = \{(x,y) \mid y = 1500x - 10x^2\}$$

The equation $y = 1500x - 10x^2$ is equivalent to

$$10(x^2 - 150x) = -y$$
$$10(x^2 - 150x + 5625) = -y + 56{,}250$$
$$(x - 75)^2 = -\frac{1}{10}(y - 56{,}250)$$

The graph of this equation is a parabola opening downward and having its vertex at $(75, 56250)$. Thus the function $f$ has a maximum value when $x = 75$; and $f(75) = 56{,}250$. Therefore, to yield the largest gross income from sales the price per copy should be 75 cents.

**CHECK:** If the price per copy is 75 cents, then the number of copies sold is $1000 - 10 \cdot 25 = 750$. Therefore, the number of cents in the gross income is $(750)(75) = 56{,}250$.

### EXAMPLE 4
In Example 4 of Section 2.7 we had the following problem: "A firm manufactures and sells portable radios, and it can sell at a price of $75 all the radios it produces. If $x$ radios are manufactured each day, then the number of dollars in the daily total cost of production is $x^2 +$

### SOLUTION
Let $P(x)$ dollars be the daily profit from the sale of $x$ radios. Then, as in Example 4 of Section 2.7, we have

$$P(x) = -x^2 + 50x - 96 \qquad (6)$$

The equation $y = -x^2 + 50x - 96$ is equivalent to

$$x^2 - 50x = -y - 96$$
$$x^2 - 50x + 625 = -y + 529$$
$$(x - 25)^2 = -(y - 529)$$

$25x + 96$." How many radios should be produced each day in order to have the greatest profit?

The graph of this equation is a parabola opening downward and having its vertex at $(25, 529)$. Therefore, the function $P$ has a maximum value of 529 when $x = 25$. Thus to have the greatest profit the number of radios produced and sold each day should be 25.

CHECK: From equation (6),

$$P(25) = -(25)^2 + 50(25) - 96$$
$$= -625 + 1250 - 96$$
$$= 529$$

## EXERCISES 3.4

In Exercises 1 through 10, draw a sketch of the graph of the given function and determine from the graph which of the following statements characterizes the roots of the corresponding quadratic equation: (a) two real roots; (b) one real root of multiplicity two; or (c) two imaginary roots.

1. $\{(x, y) | y = x^2 - 4x\}$
2. $\{(x, y) | y = x^2 - 3\}$
3. $\{(x, y) | y = -x^2 + 4\}$
4. $\{(x, y) | y = x^2 - 6x + 11\}$
5. $\{(x, y) | y = -4x^2 + 8x - 8\}$
6. $\{(x, y) | y = 2x^2 + 4x + 1\}$
7. $\{(x, y) | y = 9 - 6x + x^2\}$
8. $\{(x, y) | y = 1 - 4x - x^2\}$
9. $\{(x, y) | 8y = 4x^2 + 20x + 49\}$
10. $\{(x, y) | y = -4x^2 + 12x - 9\}$

In Exercises 11 through 14, find the zeros of the given function.

11. $f(x) = x^2 - 2x - 3$
12. $f(x) = 6x^2 - 7x - 5$
13. $f(x) = 2x^2 - 2x - 1$
14. $f(x) = x^2 - 3x + 1$

In Exercises 15 through 20, find either a maximum or minimum value of the given function.

15. $f = \{(x, y) | x^2 - 4x - 8y - 4 = 0\}$
16. $f = \{(x, y) | x^2 + 6x + 2y + 5 = 0\}$
17. $F = \{(x, y) | x^2 + 8x + 2y + 8 = 0\}$
18. $g = \{(x, y) | 3x^2 + 6x - y + 9 = 0\}$
19. $g = \{(x, y) | 3y = -9x^2 + 12x - 5\}$
20. $G = \{(x, y) | 8y = 4x^2 + 12x - 9\}$

21. Find two numbers whose sum is 10 and whose product is a maximum.
22. Find two numbers whose difference is 14 and whose product is a minimum.
23. An object is thrown straight upward from the ground with an initial velocity of 96 feet per second. If the height of the object is $s$ feet after $t$ seconds, and if air resistance is neglected, $s = 96t - 16t^2$. What is the maximum height reached by the object and how many seconds after it is thrown does it reach its maximum height?

24. A projectile is shot straight upward from a point 15 feet above the ground with an initial velocity of 176 feet per second. If the height of the projectile is $s$ feet after $t$ seconds and if air resistance is neglected, $s = 15 + 176t - 16t^2$. How long does it take the projectile to reach its maximum height and what is the maximum height?

25. If 200 yards of fencing is available to enclose a rectangular field, what should be the dimensions of the field having the largest possible area?

26. If, in Exercise 25, an existing straight stone wall is used as a fence for one side of the rectangular field, what should be the lengths of the other three sides?

27. A travel agency offers an organization an all-inclusive tour for $800 per person if not more than 100 people take the tour. However, the cost per person will be reduced $5 for each person in excess of 100. How many people should take the tour in order for the travel agency to receive the largest gross revenue, and what is this largest gross revenue?

28. A student club on a college campus charges annual membership dues of $10, less 5 cents for each member over 60. How many members would give the club the most revenue from annual dues?

29. A firm can sell at a price of $100 per unit all of a particular commodity it produces. If $x$ units are produced each day, the number of dollars in the total cost of each day's production is $x^2 + 20x + 700$. How many units should be produced each day in order for the firm to have the greatest daily total profit? What is the greatest daily total profit?

30. A company that builds and sells desks can sell at a price of $200 per desk all the desks it produces. If $x$ desks are built and sold each week, then the number of dollars in the total cost of the week's production is $x^2 + 40x + 1500$. How many desks should be built each week in order for the manufacturer to have the greatest weekly total profit? What is the greatest weekly total profit?

## 3.5 Graphs of Rational Functions

In Section 3.2 we defined a rational function as one that can be expressed as the quotient of two polynomial functions. Therefore, if $f$ and $g$ are polynomial functions, and $S$ is the function defined by

$$S(x) = \frac{f(x)}{g(x)} \tag{1}$$

then $S$ is a rational function. The domain of $S$ is the set of all real numbers except the zeros of $g$.

**ILLUSTRATION 1.** The following are rational functions.

$$T(x) = \frac{x+2}{x-3} \qquad S(x) = \frac{x^2-4}{x-2} \qquad F(x) = \frac{x^2}{x^2-9} \qquad G(x) = \frac{2x}{x^2+1}$$

The domain of $T$ is $\{x \mid x \neq 3\}$; the domain of $S$ is $\{x \mid x \neq 2\}$; the domain of $F$ is $\{x \mid x \neq 3, x \neq -3\}$; the domain of $G$ is the set $R$ of real numbers. ●

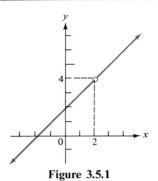

**Figure 3.5.1**

The function $S$ of Illustration 1 can be defined by equation (1) where $f(x) = x^2 - 4$ and $g(x) = x - 2$. Observe that 2 is a zero of both $f$ and $g$. In Example 3 of Section 3.1 we discussed the function $S$ and a sketch of its graph is shown in Figure 3.1.11, which is repeated here in Figure 3.5.1. This function $S$ is a special case of a rational function of the form $f(x)/g(x)$ for which a number $a$ is a zero of both $f$ and $g$.

We now consider function $T$ of Illustration 1. This function is a special case of a rational function of the form $f(x)/g(x)$, where $g(a) = 0$ and $f(a) \neq 0$.

$$T(x) = \frac{x+2}{x-3}$$

and when $x = 3$ the denominator is zero and the numerator is not zero. The domain of $T$ is the set of all real numbers except 3. Because $T(0) = -\frac{2}{3}$, the graph intersects the $y$ axis at $-\frac{2}{3}$, and because $T(-2) = 0$, the graph intersects the $x$ axis at $-2$.

We shall investigate the function values $T(x)$ when $x$ is close to 3 but not equal to 3. First let $x$ take on the values $4, \frac{7}{2}, \frac{10}{3}, \frac{13}{4}, \frac{31}{10}, \frac{301}{100}, \frac{3001}{1000}$, and so on. We are taking values of $x$ closer and closer to 3 but greater than 3; in other words, the variable $x$ is approaching 3 from the right. We illustrate this in Table 3.5.1. From the table we see intuitively that as $x$ gets closer and closer

Table 3.5.1

| $x$ | 4 | $\frac{7}{2}$ | $\frac{10}{3}$ | $\frac{13}{4}$ | $\frac{31}{10}$ | $\frac{301}{100}$ | $\frac{3001}{1000}$ |
|---|---|---|---|---|---|---|---|
| $T(x) = \dfrac{x+2}{x-3}$ | 6 | 11 | 16 | 21 | 51 | 501 | 5001 |

to 3 from the right, $T(x)$ increases without bound. In other words, we can make $T(x)$ greater than any preassigned positive number by taking $x$ close enough to 3 and $x$ greater than 3. To indicate that $T(x)$ increases without bound as $x$ approaches 3 from the right, we use the symbolism

$$T(x) \to +\infty \quad \text{as } x \to 3^+$$

The symbol "$+\infty$" (positive infinity) is not a real number; it is used to indicate the behavior of the function values $T(x)$ as $x$ gets closer and closer to 3. The "+" symbol as a superscript after the 3 indicates that $x$ is approaching 3 from the right.

Now let the variable $x$ approach 3 through values less than 3; that is, let $x$ take on the values $2, \frac{5}{2}, \frac{8}{3}, \frac{11}{4}, \frac{29}{10}, \frac{299}{100}, \frac{2999}{1000}$, and so on. Refer to Table 3.5.2. Notice that as $x$ gets closer and closer to 3 from the left, the values of $T(x)$ decrease without bound (the values of $T(x)$ are negative numbers whose absolute values increase without bound); that is, we can make $T(x)$ less than any preassigned negative number by taking $x$ close enough to 3 and $x$ less than 3. We use the following notation to indicate that $T(x)$ decreases without

### 3.5] Graphs of Rational Functions 179

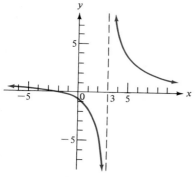

**Figure 3.5.2**

**Table 3.5.2**

| $x$ | 2 | $\frac{5}{2}$ | $\frac{8}{3}$ | $\frac{11}{4}$ | $\frac{29}{10}$ | $\frac{299}{100}$ | $\frac{2999}{1000}$ |
|---|---|---|---|---|---|---|---|
| $T(x) = \dfrac{x+2}{x-3}$ | $-4$ | $-9$ | $-14$ | $-19$ | $-49$ | $-499$ | $-4999$ |

bound as $x$ approaches 3 from the left:

$$T(x) \to -\infty \qquad \text{as } x \to 3^-$$

In Figure 3.5.2 we have a sketch of the graph of $T$ that shows the behavior of $T(x)$ near $x = 3$. As $x$ gets closer and closer to 3 from either the right or left, the absolute value of $T(x)$ gets larger and larger. Observe that the graph does not intersect the line $x = 3$, which is shown as a dashed line in the figure. The line $x = 3$ is called a "vertical asymptote" of the graph of $T$.

**3.5.1 DEFINITION** The line $x = a$ is said to be a *vertical asymptote* of the graph of the function $f$ if at least one of the following statements is true:

(i) $f(x) \to +\infty$ as $x \to a^+$
(ii) $f(x) \to -\infty$ as $x \to a^+$
(iii) $f(x) \to +\infty$ as $x \to a^-$
(iv) $f(x) \to -\infty$ as $x \to a^-$

In Figure 3.5.2 both statements (i) and (iv) of Definition 3.5.1 are true for the function $T$ and when $a$ is 3.

**ILLUSTRATION 2.** In Figure 3.5.3 we have a sketch of the graph of a function for which statements (i) and (iii) of Definition 3.5.1 are true, and in Figure 3.5.4 statements (ii) and (iii) are true. ●

The following theorem can be proved from Definition 3.5.1.

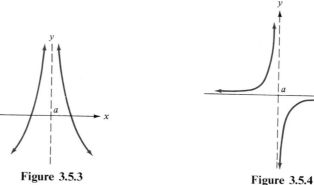

Figure 3.5.3          Figure 3.5.4

**THEOREM 3.5.2** The graph of a rational function of the form $f(x)/g(x)$ will have the line $x = a$ as a vertical asymptote if $g(a) = 0$ and $f(a) \neq 0$.

**ILLUSTRATION 3.** The function $F$ of Illustration 1 is defined by

$$F(x) = \frac{x^2}{(x-3)(x+3)}$$

Let $f(x) = x^2$ and $g(x) = (x-3)(x+3)$. Then $g(3) = 0$ and $f(3) \neq 0$; furthermore, $g(-3) = 0$ and $f(-3) \neq 0$. Therefore, from Theorem 3.5.2 it follows that the lines $x = 3$ and $x = -3$ are vertical asymptotes of the graph of $F$. In Example 1 we use this information as well as other facts to draw the sketch of the graph of $F$ shown in Figure 3.5.10. ●

In the following definition we use the notation "$f(x) \to b^+$" to mean that $f(x)$ approaches $b$ through values greater than $b$, and "$f(x) \to b^-$" means that $f(x)$ approaches $b$ through values less than $b$.

**3.5.3 DEFINITION** The line $y = b$ is said to be a *horizontal asymptote* of the graph of the function $f$ if at least one of the following statements is true:

(i) $f(x) \to b^+$ as $x \to +\infty$
(ii) $f(x) \to b^+$ as $x \to -\infty$
(iii) $f(x) \to b^-$ as $x \to +\infty$
(iv) $f(x) \to b^-$ as $x \to -\infty$

**ILLUSTRATION 4.** In Figure 3.5.5, we have a sketch of the graph of a function for which statement (iii) of Definition 3.5.3 is true, and in Figure 3.5.6, statement (ii) is true. Both statements (i) and (iv) are true for the graph of the function shown in Figure 3.5.7. The graph in Figure 3.5.7 also has the line $x = a$ as a vertical asymptote because $f(x) \to +\infty$ as $x \to a^+$ and $f(x) \to -\infty$ as $x \to a^-$. ●

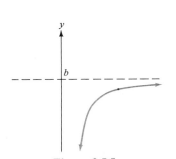

**Figure 3.5.5**

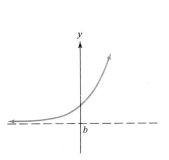

**Figure 3.5.6**

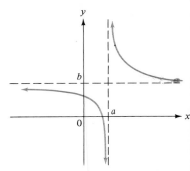

**Figure 3.5.7**

**3.5.4 THEOREM** The graph of a rotational function of the form $f(x)/g(x)$, where $f$ is a polynomial of degree $n$ and $g$ is a polynomial of degree $m$, has

(i) The $x$ axis as a horizontal asymptote if $n < m$.
(ii) The line $y = c$ $(c \neq 0)$ as a horizontal asymptote if $n = m$.
(iii) No horizontal asymptote if $n > m$

The proof of Theorem 3.5.5 is omitted, but the following illustration involving special cases should make the truth of the theorem plausible.

**ILLUSTRATION 5**

(a) The function $G$ of Illustration 1 is defined by

$$G(x) = \frac{2x}{x^2 + 1}$$

The degree of the numerator is less than the degree of the denominator. If we divide the numerator and denominator by $x^2$, we have

$$G(x) = \frac{\frac{2}{x}}{1 + \frac{1}{x^2}}$$

As $x \to +\infty$, $\quad \frac{2}{x} \to 0^+ \quad$ and $\quad \frac{1}{x^2} \to 0^+$

and

as $x \to -\infty$, $\quad \frac{2}{x} \to 0^- \quad$ and $\quad \frac{1}{x^2} \to 0^+$

Therefore,

$$G(x) \to 0^+ \quad \text{as } x \to +\infty$$

and

$$G(x) \to 0^- \quad \text{as } x \to -\infty$$

Hence, from Definition 3.5.3, the line $y = 0$ is a horizontal asymptote of the graph of $G$. This result agrees with part (i) of Theorem 3.5.4. Function $G$ is discussed in Example 2 and a sketch of its graph is shown in Figure 3.5.12.

(b) The function $F$ of Illustration 1 is defined by

$$F(x) = \frac{x^2}{x^2 - 9}$$

The degree of the numerator equals the degree of the denominator. If we divide the numerator and denominator by $x^2$, we have

$$F(x) = \frac{1}{1 - \dfrac{9}{x^2}}$$

As $x \to +\infty$ or $x \to -\infty$, $\dfrac{9}{x^2} \to 0^+$ and therefore,

$$F(x) \to 1^+ \qquad \text{as } x \to +\infty \quad \text{or} \quad x \to -\infty$$

From Definition 3.5.3 it follows that the line $y = 1$ is a horizontal asymptote. This result agrees with part (ii) of Theorem 3.5.4.

(c) Consider the function $H$ defined by

$$H(x) = \frac{x^2 + 1}{x - 4}$$

The degree of the numerator is greater than the degree of the denominator. If we divide the numerator and denominator by $x^2$, we obtain

$$H(x) = \frac{1 + \dfrac{1}{x^2}}{\dfrac{1}{x} - \dfrac{4}{x^2}} \qquad (2)$$

As $|x|$ increases without bound,

$$\frac{1}{x^2} \to 0, \qquad \frac{1}{x} \to 0, \qquad \text{and} \qquad \frac{4}{x^2} \to 0$$

Therefore, as $x \to +\infty$ or $x \to -\infty$, the numerator of the fraction in (2) approaches 1 and the denominator approaches 0. Consequently, $|H(x)|$ increases without bound as $|x|$ increases without bound, and so the graph of $H$ does not have a horizontal asymptote. This result agrees with part (iii) of Theorem 3.5.4. A sketch of the graph of $H$ is shown in Figure 3.5.8. ●

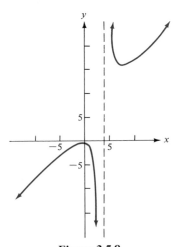

**Figure 3.5.8**

Another aid in sketching the graph of a rational function is to determine if there are any regions where there are no points on the graph. The following illustration shows the procedure involved.

**ILLUSTRATION 6.** We wish to determine if there are any regions excluded by the graph of the function $F$ of Illustration 1.

$$F(x) = \frac{x^2}{x^2 - 9} \qquad (3)$$

First we find the numbers that make either the numerator or the denominator zero. These numbers are $-3$, $0$, and $3$. We then consider the $x$-intervals excluding these numbers: $x < -3$; $-3 < x < 0$; $0 < x < 3$; and $x > 3$. In Table 3.5.3, we ascertain the sign ("+" or "−") of $F(x)$ for each of these intervals.

## Table 3.5.3

|  | Sign of $x^2$ | Sign of $x^2 - 9$ | Sign of $F(x)$ |
|---|---|---|---|
| $x < -3$ | + | + | + |
| $-3 < x < 0$ | + | − | − |
| $0 < x < 3$ | + | − | − |
| $x > 3$ | + | + | + |

From Table 3.5.3 we learn that $F(x) > 0$ when $x < -3$, so $F(x)$ cannot be negative for these values of $x$. Because there can be no graph in the third quadrant when $x < -3$, we crosshatch this region. See Figure 3.5.9. In this figure we have also crosshatched the following regions: the portion of the second quadrant when $-3 < x < 0$ because $F(x)$ cannot be positive there; the portion of the first quadrant when $0 < x < 3$ because $F(x)$ cannot be positive there; and the portion of the fourth quadrant when $x > 3$ because $F(x)$ cannot be negative there.

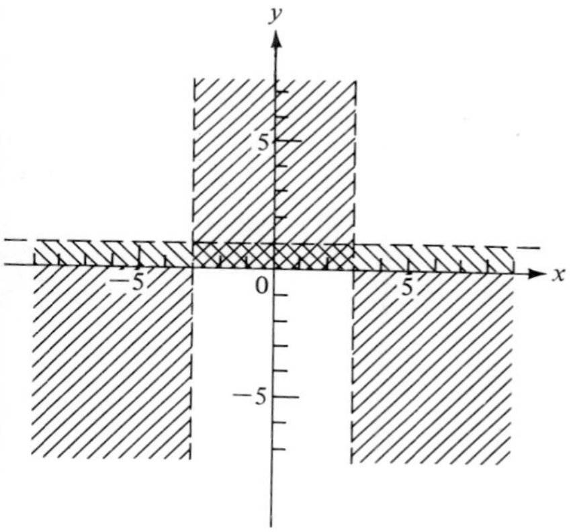

In equation (3) that defines $F(x)$, we replace $F(x)$ by $y$ and solve the equation for $x$. Doing this, we have

$$y = \frac{x^2}{x^2 - 9}$$
$$x^2 y - 9y = x^2$$
$$x^2 y - x^2 = 9y$$
$$x^2(y - 1) = 9y$$
$$x^2 = \frac{9y}{y - 1}$$
$$x = \pm \sqrt{\frac{9y}{y - 1}} \tag{4}$$

**Figure 3.5.9**

When the fraction under the radical sign in equation (4) is negative, $x$ will be imaginary. Therefore, any value of $y$ for which the numerator and denominator of this fraction have opposite signs will give us a region that is excluded by the graph. The numerator changes sign when $y = 0$ and the denominator changes sign when $y = 1$. We consider the $y$-regions excluding these numbers: $y < 0$; $0 < y < 1$; and $y > 1$. We make use of Table 3.5.4 to learn when

## Table 3.5.4

|  | Sign of $9y$ | Sign of $y - 1$ | Sign of $\dfrac{9y}{y - 1}$ |
|---|---|---|---|
| $y < 0$ | − | − | + |
| $0 < y < 1$ | + | − | − |
| $y > 1$ | + | + | + |

the fraction is positive and when it is negative. When $y < 0$ or $y > 1$, the fraction under the radical sign is positive, and therefore $x$ is real; hence, for these values of $y$ there will be points on the graph. When $0 < y < 1$, the fraction under the radical sign is negative; thus $x$ is imaginary and therefore the corresponding excluded region is crosshatched in Figure 3.5.9. •

EXAMPLE 1
Draw a sketch of the graph of the function $F$ defined by

$$F(x) = \frac{x^2}{x^2 - 9}$$

SOLUTION
The domain of $F$ is the set of all real numbers except 3 and $-3$. Because $F(0) = 0$, the graph has an intercept at the origin. In Illustration 3 we obtained the vertical asymptotes $x = 3$ and $x = -3$, and in Illustration 5(b) we obtained the horizontal asymptote $y = 1$. Furthermore, in Illustration 6, we determined the crosshatched regions shown in Figure 3.5.9. Because $F$ is an even function, its graph is symmetric with respect to the $y$ axis. We find a few points on the graph. These points are obtained from Table 3.5.5. We plot these points and use the asymptotes as guides to draw the portion of the graph in the first and fourth quadrants. Using the symmetry property, we complete the graph in the second and third quadrants. The sketch is shown in Figure 3.5.10.

Table 3.5.5

| $x$ | 0 | 1 | 2 | 4 | 5 | 6 |
|---|---|---|---|---|---|---|
| $F(x) = \frac{x^2}{x^2 - 9}$ | 0 | $-\frac{1}{8}$ | $-\frac{4}{5}$ | $\frac{16}{7}$ | $\frac{25}{16}$ | $\frac{4}{3}$ |

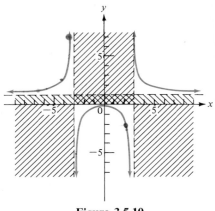

Figure 3.5.10

## EXAMPLE 2

Draw a sketch of the graph of the function $G$ defined by

$$G(x) = \frac{2x}{x^2 + 1}$$

## SOLUTION

The domain of $G$ is the set $R$ of real numbers. Because $G(0) = 0$, the graph has an intercept at the origin. There are no vertical asymptotes because the denominator is never zero. In Illustration 5(a) we ascertained that the $x$ axis is a horizontal asymptote.

We now determine if there are excluded regions. The denominator is never zero, but the numerator is zero when $x = 0$. Thus we consider the $x$ intervals $x < 0$ and $x > 0$. The denominator is always positive; so when $x < 0$, $G(x) < 0$, and when $x > 0$, $G(x) > 0$. Because $G(x)$ cannot be positive when $x < 0$, there are no points of the graph in the second quadrant; this region is crosshatched in Figure 3.5.11. Because $G(x)$ cannot be negative

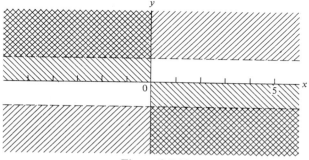

**Figure 3.5.11**

when $x > 0$, there are no points of the graph in the fourth quadrant. This region is also crosshatched in Figure 3.5.11. In the equation defining $G(x)$, we replace $G(x)$ by $y$ and solve for $x$. We have

$$yx^2 + y = 2x$$
$$yx^2 - 2x + y = 0$$
$$x = \frac{2 \pm \sqrt{4 - 4y^2}}{2y}$$
$$x = \frac{1 \pm \sqrt{1 - y^2}}{y} \qquad (5)$$

From equation (5) it follows that $x$ will be imaginary when $1 - y^2 < 0$. We solve this inequality for $y$.

$$1 - y^2 < 0$$
$$-y^2 < -1$$
$$y^2 > 1$$

The solution set of this inequality is $\{y \mid y > 1\} \cup \{y \mid y < -1\}$. In Figure 3.5.11 we crosshatch the regions for $y > 1$ and $y < -1$. The graph then lies in the first and third quadrants between the lines $y = 1$ and $y = -1$. Because

$G$ is an odd function, the graph of $G$ is symmetric with respect to the origin. We find a few points on the graph where $x$ is positive and these are obtained from Table 3.5.6.

Table 3.5.6

| $x$ | 1 | 2 | 3 | 4 | 6 |
|---|---|---|---|---|---|
| $G(x) = \dfrac{2x}{x^2+1}$ | 1 | $\frac{4}{5}$ | $\frac{3}{5}$ | $\frac{8}{17}$ | $\frac{12}{37}$ |

We plot these points and use the fact that the $x$ axis is an asymptote to draw the portion of the graph in the first quadrant. From the symmetry property, we complete the graph in the third quadrant. Figure 3.5.12 shows the required sketch.

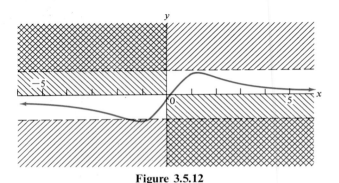

Figure 3.5.12

## EXERCISES 3.5

*In Exercises 1 through 20, a rational function $f$ is given. (a) Determine the domain of $f$. (b) Find the asymptotes of the graph of $f$ if there are any. (c) Determine if there are any regions excluded by the graph of $f$. (d) Draw a sketch of the graph of $f$.*

1. $f(x) = \dfrac{1}{x}$
2. $f(x) = \dfrac{1}{x-3}$
3. $f(x) = \dfrac{3}{x^2}$
4. $f(x) = \dfrac{4}{(x-1)^2}$
5. $f(x) = \dfrac{x+1}{x-4}$
6. $f(x) = \dfrac{x-4}{x+1}$
7. $f(x) = \dfrac{x^2}{x-1}$
8. $f(x) = -\dfrac{2}{x^3}$
9. $f(x) = \dfrac{x^2}{x^2-4}$
10. $f(x) = \dfrac{x+1}{x^2+2}$
11. $f(x) = \dfrac{4x}{x^2+4}$
12. $f(x) = \dfrac{4}{x^2-4}$
13. $f(x) = \dfrac{x}{x^2-4}$
14. $f(x) = \dfrac{x^2-1}{x^2+1}$
15. $f(x) = \dfrac{x^2+1}{x^2-1}$
16. $f(x) = \dfrac{2x}{x^2-1}$
17. $f(x) = \dfrac{x^2}{x^2-3x+2}$
18. $f(x) = \dfrac{x^2+1}{x-1}$
19. $f(x) = \dfrac{8x^2}{4x^2+8x-5}$
20. $f(x) = \dfrac{9x^2}{9x^2+3x-20}$

## 3.6 The Inverse of a Function

If $S$ is a given relation, another relation can be obtained by interchanging the components of each ordered pair in $S$. This new relation is called the "inverse" of $S$. The inverse of $S$ is denoted by $S^{-1}$ (read "$S$ inverse"). Note that in using $-1$ to denote the inverse of a relation, it should not be confused with the exponent $-1$.

**3.6.1 DEFINITION** If $S$ is a relation, then the *inverse* of $S$ is the relation defined by

$$S^{-1} = \{(x, y) | (y, x) \in S\}$$

From Definition 3.6.1 it follows that the domain of $S^{-1}$ is the range of $S$ and the range of $S^{-1}$ is the domain of $S$.

**ILLUSTRATION 1.** If $S = \{(1, 3), (2, 5), (3, 7), (4, 1), (6, 5)\}$, then $S^{-1} = \{(3, 1), (5, 2), (7, 3), (1, 4), (5, 6)\}$. Figure 3.6.1 shows the graphs of both $S$ and $S^{-1}$ on the same coordinate system. The graph of $S^{-1}$ is in color.

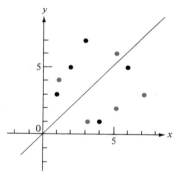

Figure 3.6.1

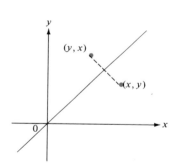

Figure 3.6.2

Because the points $(x, y)$ and $(y, x)$ are symmetric with respect to the line $y = x$ (see Figure 3.6.2) the graphs of a relation and its inverse are symmetric with respect to this line. Observe in Figure 3.6.1 that such is the case for the relations $S$ and $S^{-1}$.

**ILLUSTRATION 2.** If $T = \{(x, y) | y = 4 - 3x\}$, then

$$T^{-1} = \{(x, y) | (y, x) \in T\}$$
$$= \{(x, y) | x = 4 - 3y\}$$

If the equation $x = 4 - 3y$ is solved for $y$, we obtain the equivalent equation $y = \tfrac{1}{3}(4 - x)$. Thus $T^{-1}$ can also be given by

$$T^{-1} = \left\{(x, y) \bigg| y = \frac{1}{3}(4 - x)\right\}$$

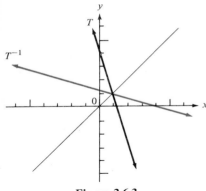

Figure 3.6.3

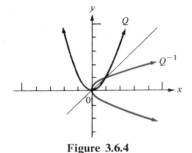

Figure 3.6.4

The graphs of $T$ and $T^{-1}$ are shown in Figure 3.6.3; the graph of $T^{-1}$ is in color. Observe the symmetry of the graphs with respect to the line $y = x$. ●

Notice in Illustration 2 that the equation defining $T^{-1}$ is $x = 4 - 3y$, which is obtained from the equation defining $T$ by interchanging $x$ and $y$.

**ILLUSTRATION 3.** If $Q = \{(x, y) \mid y = x^2\}$, then

$$Q^{-1} = \{(x, y) \mid (y, x) \in Q\}$$
$$= \{(x, y) \mid x = y^2\}$$

Figure 3.6.4 shows the graphs of $Q$ and $Q^{-1}$, and the graph of $Q^{-1}$ is in color. Again, note the symmetry of the graphs with respect to the line $y = x$. ●

The relations $S$, $T$, and $Q$ of Illustrations 1, 2, and 3, respectively, are all functions. However, $S^{-1}$ is not a function because the ordered pairs (5, 2) and (5, 6) have the same first element. The relation $T^{-1}$ is a function because from the equation $y = \frac{1}{3}(4 - x)$, a given value of $x$ determines a distinct value of $y$. The relation $Q^{-1}$ is not a function because for any positive value of $x$, the equations $y = \pm \sqrt{x}$ determine two different values for $y$. We see then that the inverse of a function is not necessarily a function.

A question arises: Under what conditions is the inverse of a function also a function? We know that if $f$ is a function, then each element in the domain of $f$ corresponds to only one element in the range. Because the domain of $f^{-1}$ is the range of $f$ and the range of $f^{-1}$ is the domain of $f$, it follows that in order for $f^{-1}$ to be a function, each element in the range of $f$ must correspond to only one element in the domain. Hence we have the following theorem, which follows from this statement.

**3.6.2 THEOREM** The function $f$ has a function as its inverse if and only if whenever $x_1$ and $x_2$ are in the domain of $f$

$$f(x_1) = f(x_2) \quad \text{implies} \quad x_1 = x_2 \qquad (1)$$

or, equivalently,

$$x_1 \neq x_2 \quad \text{implies} \quad f(x_1) \neq f(x_2) \qquad (2)$$

Recall that no vertical line can intersect the graph of a function in more than one point. If a horizontal line intersects the graph of a function in more than one point, then statements (1) and (2) of Theorem 3.6.2 do not hold, and therefore the function cannot have a function as its inverse. Note in Figure 3.6.4 that all horizontal lines above the $x$ axis intersect the graph of $Q$ in two points and the function $Q$ does not have a function as its inverse. However, in Figure 3.6.3 observe that no horizontal line intersects the graph of $T$ in more than one point and function $T$ has a function as its inverse.

## Illustration 4

(a) Relation $S$ of Illustration 1 is a function and

$$S = \{(1, 3), (2, 5), (3, 7), (4, 1), (6, 5)\}$$

Because $S(2) = 5$ and $S(6) = 5$, it follows that $S(2) = S(6)$. However, $2 \neq 6$. Hence statement (1) of Theorem 3.6.2 does not hold, and therefore $S$ does not have a function as its inverse.

(b) Relation $T$ of Illustration 2 is a function and

$$T = \{(x, y) \mid y = 4 - 3x\}$$

In order to apply Theorem 3.6.2 to show that $T$ has a function as its inverse, we show that statement (1) of the theorem is valid. Suppose that $x_1$ and $x_2$ are two numbers in the domain of $T$. Then

$$T(x_1) = 4 - 3x_1 \quad \text{and} \quad T(x_2) = 4 - 3x_2$$

Assume that $T(x_1) = T(x_2)$. Then

$$4 - 3x_1 = 4 - 3x_2$$

Adding $-4$ to each member of this equation, we obtain

$$-3x_1 = -3x_2$$

Now dividing each member by $-3$, we have

$$x_1 = x_2$$

Therefore, the assumption that $T(x_1) = T(x_2)$ implies that $x_1 = x_2$. Thus statement (1) is valid and $T$ has a function as its inverse.

(c) Relation $Q$ of Illustration 3 is a function and

$$Q = \{(x, y) \mid y = x^2\}$$

Because $Q(3) = 9$ and $Q(-3) = 9$, it follows that $Q(3) = Q(-3)$. But $3 \neq -3$, and therefore statement (1) of Theorem 3.6.2 does not hold. Hence $Q$ does not have a function as its inverse. ●

**EXAMPLE 1**

Given

$$f(x) = 3x + 5$$

(a) Prove that the inverse of $f$ is a function.
(b) Find $f^{-1}(x)$.
(c) Find $f^{-1}(-4)$.

**SOLUTION**

(a) We wish to show that statement (1) of Theorem 3.6.2 holds; that is, we wish to prove that

$$f(x_1) = f(x_2) \quad \text{implies} \quad x_1 = x_2$$

Assume that $f(x_1) = f(x_2)$. Then

$$3x_1 + 5 = 3x_2 + 5$$
$$3x_1 = 3x_2$$
$$x_1 = x_2$$

Therefore, statement (1) holds and so $f$ has a function as its inverse.

(b) Because
$$f = \{(x,y) \mid y = 3x + 5\}$$
then
$$\begin{aligned} f^{-1} &= \{(x,y) \mid (y,x) \in f\} \\ &= \{(x,y) \mid x = 3y + 5\} \\ &= \left\{(x,y) \mid y = \frac{x-5}{3}\right\} \end{aligned}$$
Therefore,
$$f^{-1}(x) = \frac{x-5}{3}$$

(c)
$$f^{-1}(-4) = \frac{-4-5}{3}$$
$$= -3$$

**EXAMPLE 2**
If $f$ is the function defined by $f(x) = x^3$, prove that the inverse of $f$ is a function.

**SOLUTION**
We show that statement (1) of Theorem 3.6.2 is valid. The domain of $f$ is the set $R$ of real numbers. Let $x_1$ and $x_2$ be any two real numbers. Then $f(x_1) = x_1^3$ and $f(x_2) = x_2^3$. Assume that $f(x_1) = f(x_2)$. Then
$$x_1^3 = x_2^3$$
$$x_1^3 - x_2^3 = 0$$
Factoring the left member as the difference of two cubes, we have the equivalent equation
$$(x_1 - x_2)(x_1^2 + x_1 x_2 + x_2^2) = 0 \tag{3}$$
The factor $x_1^2 + x_1 x_2 + x_2^2$ is zero only if $x_1 = x_2 = 0$. Therefore, if $x_1$ and $x_2$ are not both zero, equation (3) implies that
$$x_1 - x_2 = 0$$
$$x_1 = x_2$$
Hence $f(x_1) = f(x_2)$ implies that $x_1 = x_2$. Therefore, by statement (1) of Theorem 3.6.2, the inverse of $f$ is a function.

**EXAMPLE 3**
The inverse of the function $f$ of Example 2 is the function denoted by $f^{-1}$. Find $f^{-1}(x)$.

**SOLUTION**
$$f = \{(x,y) \mid y = x^3\}$$
Hence
$$\begin{aligned} f^{-1} &= \{(x,y) \mid (y,x) \in f\} \\ &= \{(x,y) \mid x = y^3\} \end{aligned}$$

Because there is only one real number that is a cube root of $x$, it follows that
$$\{(x,y)\,|\,x=y^3\} = \{(x,y)\,|\,y=\sqrt[3]{x}\}$$
Therefore,
$$f^{-1} = \{(x,y)\,|\,y=\sqrt[3]{x}\}$$
and thus $f^{-1}(x) = \sqrt[3]{x}$.

**EXAMPLE 4**
On the same coordinate system draw sketches of the graphs of the functions $f$ and $f^{-1}$ of Examples 2 and 3, respectively.

**SOLUTION**
$$f = \{(x,y)\,|\,y=x^3\} \quad \text{and} \quad f^{-1} = \{(x,y)\,|\,y=\sqrt[3]{x}\}$$

The graph of $f$ is the same as the graph of the equation $y=x^3$, and the graph of $f^{-1}$ is the same as the graph of the equation $y=\sqrt[3]{x}$. Sketches of the graphs are shown in Figure 3.6.5. Note that the graphs are symmetric with respect to the line $y=x$. Also observe that no horizontal line intersects the graph of $f$ in more than one point.

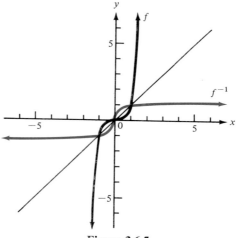

**Figure 3.6.5**

If $f$ is the function defined by the equation
$$y = f(x) \tag{4}$$
and if $f^{-1}$ is a function, then
$$f^{-1}(y) = x \tag{5}$$
If we substitute the value of $y$ from equation (4) into equation (5), we have
$$f^{-1}(f(x)) = x \tag{6}$$

Substituting the value of $x$ from equation (5) into equation (4), we obtain

$$y = f(f^{-1}(y))$$

or, equivalently, if we replace $y$ by $x$,

$$f(f^{-1}(x)) = x \qquad (7)$$

**EXAMPLE 5**
Verify equations (6) and (7) for the function of Example 1.

**SOLUTION**
In Example 1, we have

$$f(x) = 3x + 5 \quad \text{and} \quad f^{-1}(x) = \frac{x-5}{3}$$

Therefore,

$$\begin{aligned} f^{-1}(f(x)) &= f^{-1}(3x+5) \\ &= \frac{(3x+5)-5}{3} \\ &= \frac{3x}{3} \\ &= x \end{aligned}$$

and

$$\begin{aligned} f(f^{-1}(x)) &= f\left(\frac{x-5}{3}\right) \\ &= 3\left(\frac{x-5}{3}\right) + 5 \\ &= (x-5) + 5 \\ &= x \end{aligned}$$

We introduce the concepts of "increasing" and "decreasing" functions because we can prove a theorem which states that any such function has a function as its inverse.

**3.6.3 DEFINITION**   Let $f$ be a function. Then

(i) $f$ is said to be an *increasing function* if

$$x_1 < x_2 \quad \text{implies} \quad f(x_1) < f(x_2)$$

where $x_1$ and $x_2$ are any numbers in the domain of $f$.

(ii) $f$ is said to be a *decreasing function* if

$$x_1 < x_2 \quad \text{implies} \quad f(x_1) > f(x_2)$$

where $x_1$ and $x_2$ are any numbers in the domain of $f$.

The concept of an increasing function is illustrated by the sketch of the graph of such a function in Figure 3.6.6. In this figure, we have a sketch of the graph of a function $f$ and two points, $(x_1, f(x_1))$ and $(x_2, f(x_2))$, on the graph. We see that $x_1 < x_2$ and $f(x_1) < f(x_2)$. In Figure 3.6.7 we have a sketch of the graph of a decreasing function $g$. The two points $(x_1, g(x_1))$ and $(x_2, g(x_2))$ indicate that $x_1 < x_2$ and $g(x_1) > g(x_2)$.

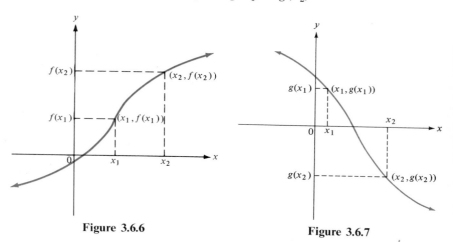

**Figure 3.6.6**      **Figure 3.6.7**

**ILLUSTRATION 5.** The function $f$ of Example 2 is defined by $f(x) = x^3$. Because

$$x_1 < x_2 \quad \text{implies} \quad x_1^3 < x_2^3$$

it follows from Definition 3.6.3(i) that $f$ is an increasing function. A sketch of the graph of $f$ appears in Figure 3.6.5. •

**ILLUSTRATION 6.** The function $T$ of Illustration 2 is defined by

$$T(x) = 4 - 3x$$

From a property of inequalities it follow that

$$x_1 < x_2 \quad \text{implies} \quad -3x_1 > -3x_2 \tag{8}$$

If we add 4 to each member of the second inequality in statement (8), we have

$$x_1 < x_2 \quad \text{implies} \quad 4 - 3x_1 > 4 - 3x_2$$

But $4 - 3x_1 = T(x_1)$ and $4 - 3x_2 = T(x_2)$, and hence

$$x_1 < x_2 \quad \text{implies} \quad T(x_1) > T(x_2)$$

Therefore, by Definition 3.6.3(ii), $T$ is a decreasing function. Refer to Figure 3.6.3, which shows a sketch of the graph of $T$. •

**3.6.4 THEOREM** (i) If $f$ is an increasing function, then $f^{-1}$ is a function.
(ii) If $f$ is a decreasing function, then $f^{-1}$ is a function.

**Proof.** Suppose that we have an increasing function. Let $x_1$ and $x_2$ be two numbers in the domain of $f$ such that $x_1 \neq x_2$ and choose $x_1$ as the smaller of the two numbers. Then $x_1 < x_2$. Because $f$ is an increasing function, it follows from Definition 3.6.3(i) that $f(x_1) < f(x_2)$. Therefore, $f(x_1) \neq f(x_2)$. Thus

$$x_1 \neq x_2 \quad \text{implies} \quad f(x_1) \neq f(x_2)$$

But this statement is the same as statement (2) of Theorem 3.6.2, and hence $f$ has a function as its inverse. This proves part (i). The proof of part (ii) is similar and is omitted.

A special case of part (i) of Theorem 3.6.4 is given by the function $f$ of Example 2. In Illustration 5 we showed that this function is an increasing function, and in Example 2 we showed that $f^{-1}$ is a function.

A special case of part (ii) of Theorem 3.6.4 is given by the function $T$ of llustration 2. In Illustration 6 we showed that $T$ is a decreasing function, and we have seen that $T^{-1}$ is a function?

**EXAMPLE 6**
Let $F$ be the function defined by

$$F = \{(x, y) | y = x^2 \text{ and } x \geq 0\}$$

(a) Find $F^{-1}$.
(b) Draw sketches of the graphs of $F$ and $F^{-1}$ on the same coordinate system.
(c) Prove that $F^{-1}$ is a function by showing that $F$ is an increasing function.

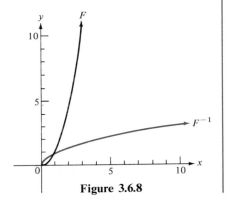

Figure 3.6.8

**SOLUTION**
(a) 
$$F^{-1} = \{(x, y) | (y, x) \in F\}$$
$$= \{(x, y) | x = y^2 \text{ and } y \geq 0\}$$

The equation $x = y^2$ is equivalent to the two equations $y = \sqrt{x}$ and $y = -\sqrt{x}$. However, in the definition of $F^{-1}$ we have the restriction that $y \geq 0$. Therefore, the set $\{(x, y) | x = y^2 \text{ and } y \geq 0\}$ is equivalent to the set $\{(x, y) | y = \sqrt{x}\}$, so

$$F^{-1} = \{(x, y) | y = \sqrt{x}\}$$

(b) Sketches of the graphs of $F$ and $F^{-1}$ on the same coordinate system are shown in Figure 3.6.8.
(c) It is apparent from the sketch of the graph of $F$ in Figure 3.6.8 that $F$ is an increasing function. However, to prove this fact analytically, we show that Definition 3.6.3(i) holds.

Let $x_1$ and $x_2$ be two numbers in the domain of $F$ so that $x_1 < x_2$. Because the domain of $F$ is the set of nonnegative numbers, it follows that $x_1$ is nonnegative and $x_2$ is positive. Therefore,

$$x_1 < x_2 \quad \text{implies} \quad x_1^2 < x_2^2$$

But $x_1^2 = F(x_1)$ and $x_2^2 = F(x_2)$. Therefore,

$$x_1 < x_2 \quad \text{implies} \quad F(x_1) < F(x_2)$$

Hence, by Definition 3.6.3(i), $F$ is an increasing function.

## EXERCISES 3.6

In Exercises 1 through 12, a relation S is given. Find $S^{-1}$ and draw sketches of the graphs of S and $S^{-1}$ on the same coordinate system. Determine if $S^{-1}$ is a function.

1. $S = \{(-4, 3), (-2, -3), (0, -1), (2, 1), (4, 0)\}$
2. $S = \{(-3, -2), (-1, 2), (1, -1), (3, 2), (4, 3)\}$
3. $S = \{(-1, 5), (0, -2), (1, -1), (1, 2), (2, 4)\}$
4. $S = \{(x, y) \mid y = 4 - 2x\}$
5. $S = \{(x, y) \mid y = 4 - 2x^2\}$
6. $S = \{(x, y) \mid x = y^2 + 3\}$
7. $S = \{(x, y) \mid x^2 + y^2 = 25\}$
8. $S = \{(x, y) \mid 4x^2 + 9y^2 = 36\}$
9. $S = \{(x, y) \mid y = \sqrt{9 - x^2}\}$
10. $S = \{(x, y) \mid x = \sqrt{4 - y^2}\}$
11. $S = \{(x, y) \mid xy = 6\}$
12. $S = \{(x, y) \mid y = |x - 3|\}$

In Exercises 13 through 24, a function f is given. Find the inverse of f and determine if it is a function. Draw sketches of the graphs of f and its inverse on the same coordinate system.

13. $f = \{(-5, 7), (-2, 4), (1, 1), (4, -2)\}$
14. $f = \{(-2, -4), (0, -2), (2, 0), (4, 2), (6, 3)\}$
15. $f = \{(x, y) \mid 5x + 2y = 10\}$
16. $f = \{(x, y) \mid 3x - 4y = 12\}$
17. $f = \{(x, y) \mid y = |x + 1|\}$
18. $f = \{(x, y) \mid y = [\![x]\!]\}$
19. $f = \{(x, y) \mid y = -\sqrt{x + 4}\}$
20. $f = \{(x, y) \mid y = \sqrt{4 - x}\}$
21. $f = \{(x, y) \mid x^2 - 6x - 6y + 15 = 0\}$
22. $f = \{(x, y) \mid x^2 + 4x + 4y - 8 = 0\}$
23. $f = \{(x, y) \mid y = (x - 2)^3\}$
24. $f = \{(x, y) \mid y = x^{2/3}\}$

In Exercises 25 through 28, $f(x)$ is given. (a) Use the method of Example 1 to prove that the inverse of f is a function. (b) Find $f^{-1}(x)$. (c) Find $f^{-1}(a)$ for the given value of a. (d) Verify equations (6) and (7) for f and $f^{-1}$.

25. $f(x) = 4x - 3;\ a = -5$
26. $f(x) = 2x + 3;\ a = \frac{2}{3}$
27. $f(x) = x^3 + 2;\ a = -6$
28. $f(x) = (x + 2)^3;\ a = 1$

In Exercises 29 through 34, a function f is given. (a) Find the range of f. (b) Determine $f^{-1}$ and state the domain of $f^{-1}$. (c) Is $f^{-1}$ a function?

29. $f = \{(x, y) \mid y = x^2 - 5,\ x \geq 0\}$
30. $f = \{(x, y) \mid y = 2 - x^2,\ x \leq 0\}$
31. $f = \{(x, y) \mid x^2 - 4x - 3y - 8 = 0,\ x \leq 0\}$
32. $f = \{(x, y) \mid x^2 + 6x + 4y + 5 = 0,\ x \geq -3\}$
33. $f = \{(x, y) \mid y = \sqrt{16 - x^2}, |x| \leq 4\}$
34. $f = \{(x, y) \mid y = \sqrt{x^2 - 9}, |x| \geq 3\}$

*In Exercises 35 through 38, for the given function F do the following.* (a) *Find* $F^{-1}$. (b) *Draw sketches of the graphs of F and* $F^{-1}$ *on the same coordinate system.* (c) *Prove that* $F^{-1}$ *is a function by showing that F is either an increasing or a decreasing function.*

**35.** $F = \{(x,y) \mid y = (x-2)^2 \; x \geq 2\}$
**37.** $F = \{(x,y) \mid y = (2-x)^3\}$
**36.** $F = \{(x,y) \mid y = (x+3)^2, \; x \leq -3\}$
**38.** $F = \{(x,y) \mid y = (x+1)^3\}$

## 3.7 Variation

Problems involving the dependence of one variable upon another variable occur in both the physical and social sciences. The formulas used in such problems often determine functions. For instance, if $I$ dollars is the simple interest for one year earned by a principal of $P$ dollars at the rate of 6 per cent per year, then

$$I = 0.06P \tag{1}$$

For a given nonnegative value of $P$ there corresponds a unique value of $I$, and so the value of $I$ depends upon the value of $P$. If $f$ is the function defined by $f(P) = 0.06P$ and the domain of $f$ is the set of nonnegative real numbers, then equation (1) can be written as

$$I = f(P) \tag{2}$$

In equation (2) the symbol $P$, which represents a number in the domain of $f$, is called the *independent variable*, and the symbol $I$, which represents a number in the range of $f$, is called the *dependent variable*.

Equation (1) is an example of "direct variation," and $I$ is said to "vary directly" as $P$.

**3.7.1 DEFINITION**  A variable $y$ is said to *vary directly* as a variable $x$ if

$$y = kx \tag{3}$$

where $k$ is a nonzero constant. More generally, a variable $y$ is said to *vary directly* as the $n$th power of $x$ ($n > 0$) if

$$y = kx^n \tag{4}$$

The constant $k$ in equations (3) and (4) is called the *constant of variation* (or the *constant of proportionality*). Sometimes the terminology associated with equation (3) is "$y$ is *directly proportional* to $x$" and the terminology associated with equation (4) is "$y$ is *directly proportional* to the $n$th power of $x$."

**ILLUSTRATION 1.** If an object falls $s$ feet from rest in $t$ seconds and air resistance is neglected, then

$$s = 16t^2$$

Hence $s$ varies directly as the square of $t$ and the constant of variation is 16. The independent variable is $t$ and the dependent variable is $s$. ●

**EXAMPLE 1**

The variable $u$ varies directly as the variable $v$, and $u = 3$ when $v = 12$.

(a) Find a formula involving $u$ and $v$.
(b) Find the value of $u$ when $v = 7$.

**SOLUTION**

(a) Because $u$ varies directly as $v$, we have the equation

$$u = kv \qquad (5)$$

Equation (5) is satisfied when $u$ is 3 and $v$ is 12; thus we have

$$3 = k \cdot 12$$
$$k = \frac{1}{4}$$

Substituting this value of $k$ into equation (5), we obtain the formula

$$u = \frac{1}{4}v \qquad (6)$$

(b) Let $\bar{u}$ be the value of $u$ when $v = 7$. Substituting $\bar{u}$ for $u$ and 7 for $v$ in equation (6), we obtain

$$\bar{u} = \frac{7}{4}$$

**EXAMPLE 2**

The period (the time for one complete oscillation) of a pendulum varies directly as the square root of the length of the pendulum. If a pendulum of length 8 feet has a period of 3 seconds, what is the period of a pendulum of length 2 feet?

**SOLUTION**

Let $T$ seconds be the period of a pendulum of length $L$ feet. Then

$$T = k\sqrt{L} \qquad (7)$$

When $L = 8$, $T = 3$, and substituting these values into equation (7), we have

$$3 = k\sqrt{8}$$
$$k = \frac{3}{2\sqrt{2}} \qquad (8)$$

Replacing $k$ in equation (7) by its value from equation (8), we have

$$T = \frac{3}{2\sqrt{2}}\sqrt{L} \qquad (9)$$

Let $\bar{T}$ seconds be the period of a pendulum of length 2 feet. Then from equation (9), we have

$$\bar{T} = \frac{3}{2\sqrt{2}}\sqrt{2}$$
$$= \frac{3}{2}$$

Therefore, the period of a pendulum of length 2 feet is $\frac{3}{2}$ seconds.

**3.7.2 DEFINITION**   A variable $y$ is said to *vary inversely* as a variable $x$ if

$$y = \frac{k}{x}$$

where $k$ is a nonzero constant. More generally, a variable $y$ is said to *vary inversely* as the $n$th power of $x$ ($n > 0$) if

$$y = \frac{k}{x^n}$$

The terminology "$y$ is *inversely proportional* to $x$" and "$y$ is *inversely proportional* to the $n$th power of $x$" may be associated with the equations of Definition 3.7.2.

**ILLUSTRATION 2.** If $t$ hours is the time required for an automobile to travel a distance of 60 miles at the rate of $r$ miles per hour, then

$$t = \frac{60}{r}$$

Hence $t$ varies inversely as $r$. The constant of variation is 60, the independent variable is $r$, and the dependent variable is $t$.   •

**EXAMPLE 3**
The weight of a body varies inversely as the square of its distance from the center of the earth. If a body weighs 200 pounds on the earth's surface, how much does it weigh at a distance of 400 miles above the earth's surface? Assume that the radius of the earth is 4000 miles.

**SOLUTION**
Let $w$ pounds be the weight of the body when its distance from the center of the earth is $x$ miles. Then

$$w = \frac{k}{x^2} \qquad (10)$$

When $x = 4000$, $w = 200$. Substituting these values into equation (10), we have

$$200 = \frac{k}{4000^2}$$

$$k = 3{,}200{,}000{,}000$$

Substituting this value of $k$ into equation (10), we have

$$w = \frac{3{,}200{,}000{,}000}{x^2} \qquad (11)$$

Let $\bar{w}$ pounds be the weight of the body when $x = 4400$ (that is, when the body is 400 miles above the earth's surface). Substituting these values for $w$ and $x$ in equation (11), we have

$$\bar{w} = \frac{3{,}200{,}000{,}000}{4400^2}$$

$$= \frac{20{,}000}{121}$$

$$= 165.3$$

Therefore, the body weighs 165.3 pounds at a distance of 400 miles above the earth's surface.

**3.7.3 DEFINITION** A variable $z$ is said to *vary jointly* as variables $x$ and $y$ if

$$z = kxy$$

where $k$ is a nonzero constant. More generally, a variable $z$ is said to *vary jointly* as the $n$th power of $x$ and the $m$th power of $y$ ($n > 0$ and $m > 0$) if

$$z = kx^n y^m$$

In Definition 3.7.3 each of the variables $x$ and $y$ is an independent variable and $z$ is the dependent variable.

**ILLUSTRATION 3.** The formula for computing $V$, the number of cubic units in the volume of a right-circular cone of radius $r$ units and height $h$ units, is

$$V = \frac{1}{3}\pi r^2 h$$

Therefore, the volume of the cone varies jointly as the square of the radius and the height. The constant of variation is $\frac{1}{3}\pi$. The independent variables are $r$ and $h$, and the dependent variable is $V$. ●

**EXAMPLE 4**
For a work crew the cost of labor varies jointly as the number of workers and the number of days of work. If a crew of twelve workers in 5 days earns a total payroll of $2700, how many workers are employed if the payroll for 10 days work is $3600?

**SOLUTION**
Let $C$ dollars be the cost of labor if $x$ workers are employed for $y$ days. Then

$$C = kxy \qquad (12)$$

When $x = 12$, and $y = 5$, then $C = 2700$. Substituting these values into equation (12), we have

$$2700 = k(12)(5)$$
$$k = 45$$

Substituting 45 for $k$ in equation (12), we obtain

$$C = 45xy \qquad (13)$$

Let $\bar{x}$ be the number of workers employed if $C = 3600$ and $y = 10$. Then from equation (13), we have

$$3600 = 45(\bar{x})(10)$$
$$\bar{x} = 8$$

Thus there are eight workers if the payroll for 10 days' work is $3600.

In Example 4, the constant $k$ is the number of dollars earned by a worker for 1 day's work.

Sometimes, a combination of joint variation and inverse variation occurs, as shown in the next illustration.

ILLUSTRATION 4. If a variable $w$ varies jointly as the square of $x$ and the cube of $y$ and inversely as the fourth power of $z$, we have the equation

$$w = k\left(\frac{x^2 y^3}{z^4}\right)$$

where $k$ is a nonzero constant. There are three independent variables, and they are $x$, $y$, and $z$. The dependent variable is $w$. •

EXAMPLE 5
The safe load of a rectangular beam varies jointly as the width and the square of the depth, and inversely as the length of the beam between its supports. If a beam 12 centimeters wide, 20 centimeters deep, and 6 meters long has a safe load of 1000 kilograms, find the safe load of a beam 10 centimeters wide, 25 centimeters deep, and 5 meters long.

SOLUTION
Let $S$ kilograms be the safe load of a beam having a width of $w$ centimeters, a depth of $d$ centimeters, and a length of $l$ meters. Then

$$S = k\left(\frac{wd^2}{l}\right) \tag{14}$$

When $w = 12$, $d = 20$, and $l = 6$, then $S = 1000$. Substituting these values into equation (14), we have

$$1000 = k\left(\frac{12 \cdot 20^2}{6}\right)$$

$$k = \frac{5}{4}$$

Substituting $\frac{5}{4}$ for $k$ in equation (14), we obtain

$$S = \frac{5}{4}\left(\frac{wd^2}{l}\right) \tag{15}$$

Let $\bar{S}$ be the value of $S$ when $w = 10$, $d = 25$, and $l = 5$. Then substituting these values into equation (15), we have

$$\bar{S} = \frac{5}{4}\left(\frac{10 \cdot 25^2}{5}\right)$$

$$= 1562.5$$

Thus the safe load for a beam 10 centimeters wide, 25 centimeters deep, and 5 meters long is 1562.5 kilograms.

*EXERCISES 3.7*

1. If $y$ varies directly as $x$, and $y = 6$ when $x = 2$, find (a) a formula involving $y$ and $x$; and (b) the value of $y$ when $x = 3$.
2. If $s$ varies directly as $t$, and $s = 10$ when $t = 15$, find (a) a formula involving $s$ and $t$; and (b) the value of $s$ when $t = 7$.
3. If $v$ varies directly as the square of $u$, and $v = 5$ when $u = 3$, find (a) a formula involving $v$ and $u$; and (b) the value of $v$ when $u = 6$.
4. If $y$ varies directly as the cube of $x$, and $y = 6$ when $x = 10$, find (a) a formula involving $y$ and $x$; and (b) the value of $y$ when $x = 2$.

5. If $p$ varies inversely as the square root of $q$, and $p = 3$ when $q = 8$, find (a) a formula involving $p$ and $q$; and (b) the value of $p$ when $q = 18$.
6. If $m$ varies inversely as the square of $n$, and $m = 3$ when $n = 4$; find (a) a formula involving $m$ and $n$; and (b) the value of $m$ when $n = 16$.
7. If $z$ varies directly as the cube of $x$, and inversely as the square of $y$, and $z = 8$ when $x = 4$ and $y = 6$, find (a) a formula involving $x$, $y$, and $z$; and (b) the value of $z$ when $x = 2$ and $y = 2$.
8. If $t$ varies directly as the square of $s$, and inversely as $r$, and $t = 8$ when $s = 4$ and $r = 3$, find (a) a formula involving $r$, $s$, and $t$; and (b) the value of $t$ when $s = 9$ and $r = 6$.
9. The variable $r$ varies jointly as $s$ and the cube of $t$ and inversely as the square of $u$. If $r = 24$ when $s = 15$, $t = 4$, and $u = 10$, find (a) a formula involving $r$, $s$, $t$, and $u$; and (b) the value of $t$ when $r = 15$, $s = 8$, and $u = 6$.
10. The variable $w$ varies jointly as the fourth power of $x$ and the square of $y$ and inversely as $z$. If $w = 10$ when $x = 3$, $y = 2$, and $z = 36$, find (a) a formula involving $w$, $x$, $y$, and $z$; and (b) the value of $y$ when $w = 4$, $x = 2$, and $z = 40$.
11. The intensity of light from a given source varies inversely as the square of the distance from it. If the intensity is 225 candlepower at a distance of 5 meters from the source, find the intensity at a point 12 meters from the source.
12. For a vibrating string, the number of vibrations per second varies directly as the square root of the number of kilograms in the tension on the string. If a particular string vibrates 864 times per second under a tension of 24 kilograms, find the number of vibrations per second under a tension of 6 kilograms.
13. If $s$ feet is the distance a body falls from rest in $t$ seconds, then $s$ varies directly as the square of $t$. If a body falls 64 feet in 2 seconds, find how long it takes a body to fall 100 feet.
14. If $V$ cubic meters is the volume of a gas having a pressure of $P$ kilograms per square centimeter, then Boyle's law states that at a constant temperature, $V$ varies inversely as $P$. If a particular amount of gas occupies 100 cubic meters at a pressure of 24 kilograms per square centimeter, find (a) the volume of the gas when the pressure is 16 kilograms per square centimeter; and (b) the amount of pressure needed to compress the gas to 30 cubic meters.
15. If, in an electric circuit, $i$ amperes is the current, $R$ ohms is the resistance, and $E$ volts is the electromotive force, then $i$ varies directly as $E$ and inversely as $R$. If, in a particular circuit, a current of 30 amperes flows through a resistance of 5 ohms with an electromotive force of 30 volts, find the current that an electromotive force of 100 volts sends through a resistance of 12 ohms.
16. If $V$ kilometers per hour is the velocity of the wind and if $P$ kilograms is the force of the wind on a plane surface at right angles to the wind, then if $A$ square meters is the area of the plane surface, $P$ varies jointly as $A$ and the square of $V$. If the force on a plane surface of area 9 square meters is 48 kilograms when the wind's velocity is 40 kilometers per hour, find the force on a plane surface of area 24 square meters when the wind's velocity is 20 kilometers per hour.
17. If $V$ cubic meters is the volume of a gas having a pressure of $P$ kilograms per square centimeter at an absolute temperature of $T$ degrees, then $V$ varies directly as $T$ and inversely as $P$. If at a temperature of 180 degrees a particular amount of gas occupies 100 cubic meters at a pressure of 18 kilograms per square centimeter, find the volume of the gas at a temperature of 240 degrees when the pressure is 12 kilograms per square centimeter.
18. If $R$ ohms is the electrical resistance of a cable of length $L$ meters and diameter $d$ centimeters, then $R$ varies directly as $L$ and inversely as the square of $d$. If a cable 500 meters long and $\frac{1}{2}$ centimeter in diameter has a resistance of 0.1 ohm, what is the resistance of a cable 4000 meters long and 1 centimeter in diameter?
19. If $z$ varies directly as the square of $x$ and inversely as $y$, what is the effect on $z$ if $x$ is doubled and $y$ is tripled?
20. If $s$ varies directly as the square of $t$ and inversely

as the square root of $r$, what is the effect on $s$ if both $t$ and $r$ are doubled?

21. The safe load of a horizontal beam of a given length varies jointly as the width and the square of the depth. If a beam of width 4 inches and depth 8 inches is tipped over to make 4 inches the depth, what is the per cent loss of load?

22. Solve Exercise 21 if the beam has a width of 3 inches and a depth of 15 inches and is tipped over to make 3 inches the depth.

23. The illumination of an object from a lamp varies directly as the intensity of the light and inversely as the square of the distance of the object from the lamp. A student reading a book that is $2\frac{1}{2}$ feet from a lamp decides to increase the illumination by doubling the intensity of the light and bringing the book 6 inches closer. What is the per cent increase in illumination?

24. The stiffness of a rectangular beam varies jointly as its width and depth and inversely as the square of the length. If each of the three dimensions is increased by 20 per cent, what is the change in the stiffness?

## REVIEW EXERCISES (CHAPTER 3)

In Exercises 1 through 6, draw a sketch of the graph of the given relation. Is the relation a function?

1. $\{(x,y) \mid 5x - y = 3\}$
2. $\{(x,y) \mid 2x + y = 8\}$
3. $\{(x,y) \mid y^2 = x - 4\}$
4. $\{(x,y) \mid y = x^2 + 4x\}$
5. $\{(x,y) \mid xy = 10\}$
6. $\{(x,y) \mid x^2 + y^2 = 16\}$

In Exercises 7 through 14, find the domain and range of the given function and draw a sketch of the graph of the function.

7. $f(x) = x^2 - 4$
8. $g(x) = \sqrt{x - 4}$
9. $h(x) = \sqrt{x^2 - 4}$
10. $F(x) = \sqrt{4 - x^2}$
11. $f(x) = \dfrac{x^2 - 16}{x + 4}$
12. $g(x) = \begin{cases} 3 - x, & \text{if } x < 0 \\ 3 + 2x, & \text{if } x \geq 0 \end{cases}$
13. $F(x) = \begin{cases} x^2, & \text{if } x < -1 \\ (x + 2)^2, & \text{if } x \geq -1 \end{cases}$
14. $G(x) = \dfrac{(x + 2)(x^2 - 4x + 3)}{x^2 + x - 2}$

In Exercises 15 through 18, find the given function value.

15. $h(-2)$ for the function of Exercise 9.
16. $F(0)$ for the function of Exercise 10.
17. $f(2)$ for the function of Exercise 11.
18. $g(-1)$ for the function of Exercise 12.

In Exercises 19 through 23, $f(x) = 3x^2 - 2x + 1$. Find the indicated quantity.

19. (a) $f(2)$; (b) $f(-4)$
20. (a) $f(3t)$; (b) $3f(t)$
21. (a) $f(4x)$; (b) $4f(x)$
22. (a) $f(x^2)$; (b) $[f(x)]^2$
23. (a) $f(x - 2)$; (b) $f(x) - f(2)$
24. If $f(x) = x^2 - 4$, find $\dfrac{f(x + h) - f(x)}{h}$, where $h \neq 0$.

In Exercises 25 through 27, the functions $f$ and $g$ are defined. In each exercise, find (a) $(f + g)(x)$; (b) $(f - g)(x)$; (c) $(fg)(x)$; (d) $(f/g)(x)$; (e) $(g/f)(x)$; (f) $f^2(x)$. In each case determine the domain of the resulting function.

**25.** $f(x) = x - 5$; $g(x) = x^2 - 1$
**26.** $f(x) = \sqrt{x}$; $g(x) = x^2 + 1$
**27.** $f(x) = \sqrt{x}$; $g(x) = \dfrac{1}{x^2}$

In Exercises 28 through 30, for the functions $f$ and $g$, find (a) $(f \circ g)(x)$; (b) $(g \circ f)(x)$; (c) $(f \circ f)(x)$; (d) $(g \circ g)(x)$. In each case determine the domain of the resulting function.

**28.** The functions of Exercise 25.
**29.** The functions of Exercise 26.
**30.** The functions of Exercise 27.

In Exercises 31 through 33, draw a sketch of the graph of the given function and determine from the graph which of the following statements characterizes the roots of the corresponding quadratic equation: (a) two real roots; (b) one real root of multiplicity two; or (c) two imaginary roots.

**31.** $\{(x, y) \mid y = x^2 - 6x\}$
**32.** $\{(x, y) \mid y = 2x^2 + 8x + 11\}$
**33.** $\{(x, y) \mid y = -x^2 + 8x - 16\}$

In Exercises 34 through 36, find either a maximum or minimum value of the given function.

**34.** $f = \{(x, y) \mid x^2 - 6x + 6y - 15 = 0\}$
**35.** $g = \{(x, y) \mid x^2 + 2x - 4y - 11 = 0\}$
**36.** $h = \{(x, y) \mid y = 3x^2 + 6x + 7\}$

In Exercises 37 through 40, a rational function is given. (a) Determine the domain of $f$. (b) Find the asymptotes of the graph of $f$ if there are any. (c) Determine if there are any regions excluded by the graph of $f$. (d) Draw a sketch of the graph of $f$.

**37.** $f(x) = \dfrac{x^2}{x - 3}$
**38.** $f(x) = \dfrac{x - 1}{x^2 + 1}$
**39.** $f(x) = \dfrac{x^2}{25 - x^2}$
**40.** $f(x) = \dfrac{x^2 + 4}{x^2 - 4}$

In Exercises 41 through 46, a function is given. Find the inverse of $f$ and determine if it is a function. Draw sketches of the graphs of $f$ and its inverse on the same coordinate axes.

**41.** $f = \{(x, y) \mid y = 3x\}$
**42.** $f = \{(x, y) \mid y = 3|x|\}$
**43.** $f = \{(x, y) \mid y = \sqrt{x - 4}\}$
**44.** $f = \{(x, y) \mid y = -\sqrt{x^2 - 4}\}$
**45.** $f = \{(x, y) \mid y = 9 - x^2, x \geq 0\}$
**46.** $f = \{(x, y) \mid y = (x - 3)^2, x \geq 3\}$

47. Find two numbers whose sum is 18 and whose product is a maximum.
48. A travel agency offers an organization an all-inclusive tour for $1000 per person if not more than 200 people take the tour. However, the cost per person will be reduced $10 for each person in excess of 200. How many people should take the tour in order for the travel agency to receive the largest gross revenue, and what is this largest gross revenue?
49. A carpenter can sell all the bookcases that are made at a price of $64 per bookcase. If $x$ bookcases are built and sold each week, then the number of dollars in the total cost of the week's production is $x^2 + 15x + 225$. How many bookcases should be constructed each week in order for the carpenter to have the greatest weekly total profit? What is the greatest weekly total profit?
50. If $z$ varies directly as the square root of $x$ and inversely as the square of $y$, and $z = 30$, when $x = 9$ and $y = 2$, find (a) a formula involving $x$, $y$, and $z$; (b) the value of $x$ when $z = 50$ and $y = 4$.
51. If in an electric circuit, $i$ amperes is the current and $R$ ohms is the resistance, then when the electromotive force is constant, $i$ varies inversely as $R$. If in a particular circuit, the current is 12 amperes when the resistance is 2 ohms, find (a) the current when the resistance is 0.3 ohm; (b) the resistance when the current is 96 amperes.
52. For a horizontal beam of fixed length, if $S$ kilograms is the safe load, $w$ centimeters is the width, and $d$ centimeters is the depth, $S$ varies jointly as $w$ and the square of $d$. If such a beam of width 6 centimeters and depth 24 centimeters has a safe load of 1500 kilograms, find the safe load of a beam 9 centimeters wide and 16 centimeters deep.

# 4
# Exponential and Logarithmic Functions

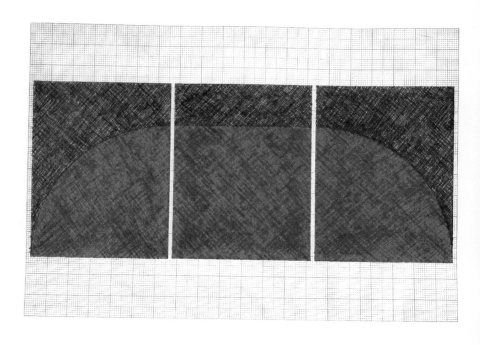

# 4.1 Properties of Exponents

The $n$th power of $a$, denoted by $a^n$, where $a$ is a real number and $n$ is a positive integer has been defined as follows:

$$a^n = a \cdot a \cdot a \cdot \cdots \cdot a \qquad (n \text{ factors of } a) \qquad (1)$$

The fundamental laws of exponents are stated in the following theorem.

**4.1.1 THEOREM** If $a$ and $b$ are real numbers, and $m$ and $n$ are positive integers, then

(i) $\quad a^n \cdot a^m = a^{n+m}$

(ii) $\quad (a^n)^m = a^{nm}$

(iii) $\quad (ab)^n = a^n b^n$

(iv) $\quad \left(\dfrac{a}{b}\right)^n = \dfrac{a^n}{b^n} \quad (b \neq 0)$

(v) $\quad \dfrac{a^n}{a^m} = \begin{cases} a^{n-m}, & \text{if } n > m \\ \dfrac{1}{a^{m-n}}, & \text{if } n < m \\ 1, & \text{if } n = m \end{cases} \quad (a \neq 0)$

The proofs of parts (i) through (v) of Theorem 4.1.1 require mathematical induction which is discussed in Chapter 8. However, informal arguments can be given in place of a formal proof. In the following illustration, an informal argument is given for part (i).

**ILLUSTRATION 1.** By the definition of a positive integer exponent

$$a^n = a \cdot a \cdot a \cdot \cdots \cdot a \qquad (n \text{ factors of } a)$$
$$a^m = a \cdot a \cdot a \cdot \cdots \cdot a \qquad (m \text{ factors of } a)$$

Therefore, the product of $a^n$ and $a^m$ gives us $n + m$ factors of $a$; that is,

$$a^n \cdot a^m = a \cdot a \cdot a \cdot \cdots \cdot a \qquad (n + m \text{ factors of } a)$$
$$= a^{n+m} \qquad \bullet$$

**ILLUSTRATION 2.** Following are special cases of parts (i) through (v) of Theorem 4.1.1.

$$x^3 \cdot x^4 = x^7$$
$$(x^3)^4 = x^{12}$$
$$(xy)^4 = x^4 y^4$$
$$\left(\frac{x}{y}\right)^3 = \frac{x^3}{y^3}$$
$$\frac{x^5}{x^2} = x^3$$
$$\frac{x^3}{x^9} = \frac{1}{x^6}$$
$$\frac{x^4}{x^4} = 1$$

By applying the associative law for multiplication, the formulas of Theorem 4.1.1 can be extended. For instance,

$$x^4 \cdot x^8 \cdot x \cdot x^6 = x^{4+8+1+6}$$
$$= x^{19}$$

and

$$(a^4 b^5 c d^2)^3 = (a^4)^3 (b^5)^3 c^3 (d^2)^3$$
$$= a^{12} b^{15} c^3 d^6$$

**EXAMPLE 1**
Find each of the following products.

(a) $(-5r^3 s^4 t^2)(-6s^5 t^4)(-r^2 s)$
(b) $(nx^n)(nx^n)$, $n$ is a positive integer
(c) $(2x^3 y^2 z)^3 (-x^2 y^3 z^4)^4$
(d) $(-a^2 c)^3 (-ab^4 c^2)^4 (3b^5 c^3)^2$

**SOLUTION**
(a) $(-5r^3 s^4 t^2)(-6s^5 t^4)(-r^2 s) = [(-5)(-6)(-1)] r^{3+2} s^{4+5+1} t^{2+4}$
$$= -30 r^5 s^{10} t^6$$

(b) $(nx^n)(nx^n) = n^{1+1} x^{n+n}$
$$= n^2 x^{2n}$$

(c) $(2x^3 y^2 z)^3 (-x^2 y^3 z^4)^4 = [2^3 (x^3)^3 (y^2)^3 z^3][(-1)^4 (x^2)^4 (y^3)^4 (z^4)^4]$
$$= [8 x^9 y^6 z^3][1 x^8 y^{12} z^{16}]$$
$$= 8 x^{17} y^{18} z^{19}$$

(d) $(-a^2 c)^3 (-ab^4 c^2)^4 (3b^5 c^3)^2$
$$= [(-1)^3 (a^2)^3 c^3][(-1)^4 a^4 (b^4)^4 (c^2)^4][3^2 (b^5)^2 (c^3)^2]$$
$$= [(-1) a^6 c^3][1 a^4 b^{16} c^8][9 b^{10} c^6]$$
$$= -9 a^{10} b^{26} c^{17}$$

**ILLUSTRATION 3.** If $x \neq 0$, $y \neq 0$, and $z \neq 0$, then

$$\frac{56 x^5 y^3 z^4}{8 x^3 y^2 z} = \left(\frac{56}{8}\right) x^{5-3} y^{3-2} z^{4-1}$$
$$= 7 x^2 y z^3$$

Note that $(8x^3 y^2 z)(7 x^2 y z^3) = 56 x^5 y^3 z^4$, which verifies the computation. ●

## EXAMPLE 2

Find each of the following quotients, where none of the variables is zero. In part (b) $n$ is a positive integer.

(a) $\dfrac{(-2x^2y^3)^2(-4x^5y^2)}{-8y^3(x^2)^3}$

(b) $\dfrac{a(a^n b^{n+1})^2}{b(ab^2)^n}$

### SOLUTION

(a) $\dfrac{(-2x^2y^3)^2(-4x^5y^2)}{-8y^3(x^2)^3} = \dfrac{(-2)^2(x^2)^2(y^3)^2(-4x^5y^2)}{-8y^3 x^6}$

$= \dfrac{4x^4 y^6(-4x^5 y^2)}{-8y^3 x^6}$

$= \dfrac{-16 x^9 y^8}{-8 y^3 x^6}$

$= \left(\dfrac{-16}{-8}\right) x^{9-6} y^{8-3}$

$= 2x^3 y^5$

(b) $\dfrac{a(a^n b^{n+1})^2}{b(ab^2)^n} = \dfrac{a(a^{2n} b^{2n+2})}{b(a^n b^{2n})}$

$= \dfrac{a^{2n+1} b^{2n+2}}{a^n b^{2n+1}}$

$= a^{(2n+1)-n} b^{(2n+2)-(2n+1)}$

$= a^{n+1} b^{2n+2-2n-1}$

$= a^{n+1} b$

We are now concerned with integer exponents other than positive integers. The definition given by formula (1) has meaning only when the exponent $n$ is a positive integer. Therefore, when the exponent is zero or a negative integer, a different definition must be given. We want these definitions to be such that the same formulas (Theorem 4.1.1(i) through (v)) that apply for positive integer exponents also hold for zero and negative integer exponents. In particular, if (i) of Theorem 4.1.1 is to hold for a zero exponent, then if $a \neq 0$,

$$a^0 \cdot a^n = a^{0+n}$$

that is,

$$a^0 \cdot a^n = a^n$$

and dividing on both sides of the equation by $a^n$, we get

$$a^0 = 1$$

Therefore, we must define $a^0$ as 1.

Now, suppose that $n$ is a positive integer, and therefore $-n$ is a negative integer. If (i) of Theorem 4.1.1 is to hold for a negative integer exponent, then

$$a^n \cdot a^{-n} = a^0$$

that is,
$$a^n \cdot a^{-n} = 1$$
and dividing on both sides of the equation by $a^n$, we get
$$a^{-n} = \frac{1}{a^n}$$
Thus, we must define $a^{-n} = 1/a^n$. We have the following formal definition.

**4.1.2 DEFINITION** If $a \in R$, $a \neq 0$, and $-n$ is a negative integer, then

$$a^0 = 1$$

$$a^{-n} = \frac{1}{a^n}$$

It can be shown that formulas (i) through (v) of Theorem 4.1.1 hold for zero and negative integer exponents; thus, they hold for all integer exponents.

**EXAMPLE 3**
Find the numerical value in lowest terms.

(a) $(2^{-3} \cdot 3^{-2})^{-1}$
(b) $(2^{-3} + 3^{-2})^{-1}$

**SOLUTION**

(a) $(2^{-3} \cdot 3^{-2})^{-1} = 2^{-3(-1)} \cdot 3^{-2(-1)}$
$= 2^3 \cdot 3^2$
$= 72$

(b) $(2^{-3} + 3^{-2})^{-1} = \dfrac{1}{(2^{-3} + 3^{-2})^1}$

$= \dfrac{1}{\dfrac{1}{2^3} + \dfrac{1}{3^2}}$

$= \dfrac{(2^3 \cdot 3^2)(1)}{(2^3 \cdot 3^2)\dfrac{1}{2^3} + (2^3 \cdot 3^2)\dfrac{1}{3^2}}$

$= \dfrac{8 \cdot 9}{9 + 8}$

$= \dfrac{72}{17}$

**EXAMPLE 4**

Write each algebraic expression as a simple fraction with positive exponents only.

(a) $\left(\dfrac{x^{-3}y^4z^{-5}}{x^6y^{-2}z^{-4}}\right)^{-2}$

(b) $\dfrac{ab^{-1} - a^{-1}b}{a^{-3} - b^{-3}}$

**SOLUTION**

(a) $\left(\dfrac{x^{-3}y^4z^{-5}}{x^6y^{-2}z^{-4}}\right)^{-2} = \dfrac{x^{-3(-2)}y^{4(-2)}z^{-5(-2)}}{x^{6(-2)}y^{-2(-2)}z^{-4(-2)}}$

$= \dfrac{x^6}{x^{-12}} \cdot \dfrac{y^{-8}}{y^4} \cdot \dfrac{z^{10}}{z^8}$

$= x^{6-(-12)}y^{-8-4}z^{10-8}$

$= x^{18}y^{-12}z^2$

$= x^{18} \cdot \dfrac{1}{y^{12}} \cdot z^2$

$= \dfrac{x^{18}z^2}{y^{12}}$

(b) $\dfrac{ab^{-1} - a^{-1}b}{a^{-3} - b^{-3}} = \dfrac{a \cdot \dfrac{1}{b} - \dfrac{1}{a} \cdot b}{\dfrac{1}{a^3} - \dfrac{1}{b^3}}$

$= \dfrac{(a^3b^3)\dfrac{a}{b} - (a^3b^3)\dfrac{b}{a}}{(a^3b^3)\dfrac{1}{a^3} - (a^3b^3)\dfrac{1}{b^3}}$

$= \dfrac{a^4b^2 - a^2b^4}{b^3 - a^3}$

$= \dfrac{a^2b^2(a^2 - b^2)}{(-1)(a^3 - b^3)}$

$= \dfrac{a^2b^2(a+b)(a-b)}{(-1)(a-b)(a^2 + ab + b^2)}$

$= -\dfrac{a^2b^2(a+b)}{a^2 + ab + b^2}$

We now wish to define a rational exponent of the form $\dfrac{1}{n}$, where $n$ is a positive integer. Formula (ii) of Theorem 4.1.1 states that $(a^n)^m = a^{nm}$. If this formula is to hold when the exponent is $\dfrac{1}{n}$, then we must have

$$(a^{1/n})^n = a^{n/n}$$
$$= a$$

If this equality is to hold, then from Definition 1.3.2, $a^{1/n}$ must be defined to be an $n$th root of $a$. We therefore have the following definition.

**4.1.3 DEFINITION**  If $a \in R$ and $n$ is a positive integer, then

$$a^{1/n} = \sqrt[n]{a}$$

**ILLUSTRATION 4**

(a) $25^{1/2} = \sqrt{25}$     (b) $(-8)^{1/3} = \sqrt[3]{-8}$     (c) $\left(\dfrac{1}{81}\right)^{1/4} = \sqrt[4]{\dfrac{1}{81}}$

  $\quad\quad\quad\ = 5$         $\quad\quad\quad\quad\ \ = -2$         $\quad\quad\quad\quad\quad\ \ = \dfrac{1}{3}$ •

Consider now how we should define $a^{m/n}$, where $m$ and $n$ are positive integers and $a$ is a real number such that $\sqrt[n]{a}$ is a real number. We place the restriction that $m$ and $n$ are relatively prime (that is, $m$ and $n$ contain no common positive integer factors other than 1). Hence we are considering expressions such as

$$9^{3/2} \quad 8^{2/3} \quad (-27)^{4/3} \quad 7^{3/4}$$

If formula (ii) of Theorem 4.1.1 is to hold for rational exponents, as well as for integer exponents, then $a^{m/n}$ must be defined in such a way that

$$a^{m/n} = (a^{1/n})^m$$

Therefore, we have the following definition.

**4.1.4 DEFINITION**  If $a \in R$, and $m$ and $n$ are positive integers that are relatively prime, then if $\sqrt[n]{a}$ is a real number

$$a^{m/n} = (\sqrt[n]{a})^m$$

or, equivalently,

$$a^{m/n} = (a^{1/n})^m$$

**ILLUSTRATION 5.** We use Definition 4.1.4 to find the value of the expression.

(a) $9^{3/2} = (\sqrt{9})^3$         (b) $8^{2/3} = (\sqrt[3]{8})^2$
$\quad\quad\ \ = 3^3$                $\quad\quad\quad\ = 2^2$
$\quad\quad\ \ = 27$                 $\quad\quad\quad\ = 4$

(c) $(-27)^{4/3} = (\sqrt[3]{-27})^4$     (d) $-27^{4/3} = -(\sqrt[3]{27})^4$
$\quad\quad\quad\quad = (-3)^4$           $\quad\quad\quad\quad = -(3)^4$
$\quad\quad\quad\quad = 81$               $\quad\quad\quad\quad = -81$ •

It can be shown that the commutative law holds for rational exponents and, therefore,

$$(a^m)^{1/n} = (a^{1/n})^m$$

from which it follows that

$$\sqrt[n]{a^m} = (\sqrt[n]{a})^m \qquad (2)$$

The next theorem follows immediately from Definition 4.1.4 and equality (2).

**4.1.5 THEOREM** If $a \in R$ and $m$ and $n$ are positive integers that are relatively prime, then if $\sqrt[n]{a}$ is a real number

$$\boxed{a^{m/n} = \sqrt[n]{a^m}}$$

or, equivalently,

$$\boxed{a^{m/n} = (a^m)^{1/n}}$$

Definition 4.1.4 and Theorem 4.1.5 give two alternatives for computing $a^{m/n}$. Compare the computation in the following illustration with that of Illustration 5 and you will see that the computation in Illustration 5 (where Definition 4.1.4 is used) is simpler than that in Illustration 6 (where Theorem 4.1.5 is used).

**ILLUSTRATION 6.** We use Theorem 4.1.5 to find the value of the expressions in Illustration 5.

(a) $9^{3/2} = \sqrt{9^3}$
$\phantom{9^{3/2}} = \sqrt{729}$
$\phantom{9^{3/2}} = 27$

(b) $8^{2/3} = \sqrt[3]{8^2}$
$\phantom{8^{2/3}} = \sqrt[3]{64}$
$\phantom{8^{2/3}} = 4$

(c) $(-27)^{4/3} = \sqrt[3]{(-27)^4}$
$\phantom{(-27)^{4/3}} = \sqrt[3]{531{,}441}$
$\phantom{(-27)^{4/3}} = 81$

(d) $-27^{4/3} = -\sqrt[3]{(27)^4}$
$\phantom{-27^{4/3}} = -\sqrt[3]{531{,}441}$
$\phantom{-27^{4/3}} = -81$ •

The laws of exponents (Theorem 4.1.1(i) through (v)) are satisfied by positive rational exponents with one exception: (ii) $(a^p)^q = a^{pq}$ when $a < 0$, $p$ is a positive even integer, and $q$ is the reciprocal of a positive even integer. For instance, consider the expression $[(-9)^2]^{1/4}$. By first computing $(-9)^2$, we have

$$[(-9)^2]^{1/4} = 81^{1/4}$$
$$= 3$$

However, if (ii) is applied first, we have
$$(-9)^{2(1/4)} = (-9)^{1/2}$$
But $(-9)^{1/2}$ is not a real number. Therefore,
$$[(-9)^2]^{1/4} \neq (-9)^{2(1/4)}$$
If we have $[(-9)^2]^{1/2}$, then by first computing $(-9)^2$, we have
$$[(-9)^2]^{1/2} = 81^{1/2}$$
$$= 9$$
If we first apply (ii), we get
$$(-9)^{2(1/2)} = (-9)^1$$
$$= -9$$
Hence
$$[(-9)^2]^{1/2} \neq (-9)^{2(1/2)}$$
In order not to have this ambiguity, we make the following definition

**4.1.6 DEFINITION** If $a \in R$, and $m$ and $n$ are positive even integers,
$$(a^m)^{1/n} = |a|^{m/n}$$

A particular case of Definition 4.1.6 occurs when $m = n$. We have then
$$(a^n)^{1/n} = |a| \quad \text{(if } n \text{ is a positive even integer)}$$
or, equivalently,
$$\sqrt[n]{a^n} = |a| \quad \text{(if } n \text{ is even)}$$
If $n$ is 2, we have
$$\sqrt{a^2} = |a| \tag{3}$$

**ILLUSTRATION 7.** From Definition 4.1.6 we have
$$[(-9)^2]^{1/4} = |-9|^{2/4}$$
$$= 9^{1/2}$$
$$= \sqrt{9}$$
$$= 3$$
From equality (3) we have
$$\sqrt{(-9)^2} = |-9|$$
$$= 9$$

We now wish to define a negative rational exponent. Suppose that $m$ and $n$ are positive integers that are relatively prime. Then

$$-\frac{m}{n}$$

represents a negative rational number. Another representation for the same rational number is

$$\frac{-m}{n}$$

Therefore, in order for (ii) of Theorem 4.1.1 to hold for negative rational exponents, we must have

$$a^{-m/n} = (a^{1/n})^{-m}$$

By the definition of a negative integer exponent, it follows that if $a \neq 0$, then

$$(a^{1/n})^{-m} = \frac{1}{(a^{1/n})^m}$$

Therefore, we give the following definition.

**4.1.7 DEFINITION**  If $a \in R$, $a \neq 0$, and $m$ and $n$ are positive integers, then if $\sqrt[n]{a}$ is a real number,

$$a^{-m/n} = \frac{1}{a^{m/n}}$$

**ILLUSTRATION 8.** We compute $8^{-2/3}$ by three different methods. In part (a) we first use Definition 4.1.7, and in parts (b) and (c) we apply laws of exponents to negative rational exponents.

(a) $8^{-2/3} = \dfrac{1}{8^{2/3}}$  (b) $8^{-2/3} = (8^{-1/3})^2$  (c) $8^{-2/3} = (8^{-2})^{1/3}$

$\phantom{(a) 8^{-2/3}} = \dfrac{1}{(\sqrt[3]{8})^2}$  $\phantom{(b) 8^{-2/3}} = \left(\dfrac{1}{8^{1/3}}\right)^2$  $\phantom{(c) 8^{-2/3}} = \left(\dfrac{1}{8^2}\right)^{1/3}$

$\phantom{(a) 8^{-2/3}} = \dfrac{1}{2^2}$  $\phantom{(b) 8^{-2/3}} = \left(\dfrac{1}{2}\right)^2$  $\phantom{(c) 8^{-2/3}} = \left(\dfrac{1}{64}\right)^{1/3}$

$\phantom{(a) 8^{-2/3}} = \dfrac{1}{4}$  $\phantom{(b) 8^{-2/3}} = \dfrac{1}{4}$  $\phantom{(c) 8^{-2/3}} = \dfrac{1}{4}$  ●

Rational exponents (positive, negative, and zero) satisfy the laws of exponents (Theorem 4.1.1(i) through (v)) with the understanding that $(a^m)^{1/n} = |a|^{m/n}$ when $m$ and $n$ are even integers.

## EXAMPLE 5

Simplify so that each variable appears only once and all the exponents are positive, where $r$, $s$, and $t$ are positive real numbers.

$$\left(\frac{r^{-3/4}s^{-2}t^{2/3}}{r^3 s^{1/2} t^{-1/6}}\right)^{-1/5}$$

**SOLUTION**

$$\left(\frac{r^{-3/4}s^{-2}t^{2/3}}{r^3 s^{1/2} t^{-1/6}}\right)^{-1/5} = \left(\frac{r^{-3/4}}{r^3} \cdot \frac{s^{-2}}{s^{1/2}} \cdot \frac{t^{2/3}}{t^{-1/6}}\right)^{-1/5}$$

$$= (r^{(-3/4)-3} s^{-2-(1/2)} t^{(2/3)-(-1/6)})^{-1/5}$$

$$= (r^{-15/4} s^{-5/2} t^{5/6})^{-1/5}$$

$$= r^{(-15/4)(-1/5)} s^{(-5/2)(-1/5)} t^{(5/6)(-1/5)}$$

$$= r^{3/4} s^{1/2} t^{-1/6}$$

$$= r^{3/4} s^{1/2} \cdot \frac{1}{t^{1/6}}$$

$$= \frac{r^{3/4} s^{1/2}}{t^{1/6}}$$

## EXAMPLE 6

Find the product and express the result with positive exponents, where $a > 0$ and $b > 0$.

$$(a^{1/2} - b^{1/2})(a^{-1/2} + b^{-1/2})$$

**SOLUTION**

$$(a^{1/2} - b^{1/2})(a^{-1/2} + b^{-1/2}) = a^0 + a^{1/2} b^{-1/2} - a^{-1/2} b^{1/2} - b^0$$

$$= 1 + a^{1/2} \cdot \frac{1}{b^{1/2}} - \frac{1}{a^{1/2}} \cdot b^{1/2} - 1$$

$$= \frac{a^{1/2}}{b^{1/2}} - \frac{b^{1/2}}{a^{1/2}}$$

$$= \frac{a^{1/2} \cdot a^{1/2}}{b^{1/2} \cdot a^{1/2}} - \frac{b^{1/2} \cdot b^{1/2}}{a^{1/2} \cdot b^{1/2}}$$

$$= \frac{a - b}{a^{1/2} b^{1/2}}$$

## EXAMPLE 7

Simplify the expression, where each variable can be any real number.

(a) $(u^2 v^4)^{1/4}$
(b) $[(-3x)^2 (y - 3)^2]^{1/2}$

**SOLUTION**

(a) $(u^2 v^4)^{1/4} = (u^2)^{1/4} (v^4)^{1/4}$
$= |u|^{2/4} |v|^{4/4}$
$= |u|^{1/2} |v|$

(b) $[(-3x)^2 (y - 3)^2]^{1/2} = [(-3)^2]^{1/2} [x^2]^{1/2} [(y - 3)^2]^{1/2}$
$= |-3| |x| |y - 3|$
$= 3|x| |y - 3|$

## EXERCISES 4.1

*In Exercises 1 through 22, find the numerical value in lowest terms.*

1. $(-5)^{-3}$
2. $\dfrac{2}{4^{-3}}$
3. $(-6)^{-2}$
4. $9 \cdot 3^0$
5. $36^{1/2}$
6. $(-8)^{-1/3}$
7. $27^{2/3}$
8. $\left(\dfrac{2}{5}\right)^{-4}$

9. $\left(-\dfrac{1}{8}\right)^{-2/3}$
10. $\left(-\dfrac{1}{125}\right)^{-4/3}$
11. $-0.16^{3/2}$
12. $-0.0016^{-3/4}$
13. $2^{-3} \cdot 7^{-1}$
14. $\dfrac{(-3)^{-4}}{(-4)^{-3}}$
15. $2^{-4} - 2^4$
16. $(4^{-1} \cdot 2^{-3})^{-1}$
17. $(4^{-1} + 2^{-3})^{-1}$
18. $\dfrac{2^{-3} + 3^{-2}}{2^{-4} + 3^{-1}}$
19. $\dfrac{6^{-1} + 2^{-3}}{5^0 + 4^{-2}}$
20. $\dfrac{(3 \cdot 10^{-1})(6 \cdot 10^{-2})}{2 \cdot 10^{-4}}$
21. $[(-9)^2]^{1/2}$
22. $[(-36)^2]^{1/4}$

*In Exercises 23 through 46, write the expression so that each variable occurs only once and the exponents are positive. Assume that all the variables are positive.*

23. $x^{-4}x^3$
24. $\dfrac{a^4}{a^{-6}}$
25. $(t^{-2})^{-3}$
26. $(-3a^3)^{-2}$
27. $a^{2/3}a^{-1/4}$
28. $\dfrac{y^{1/6}}{y^{2/3}}$
29. $x^{-3/4}x^{5/6}x^{-1/3}$
30. $7x^{-6}yz^{-7}$
31. $(x^2y^{-1})^{-2}$
32. $\dfrac{8^{-1}s^{-3}t^0}{(2st)^{-5}}$
33. $\left(\dfrac{x^{-3/4}}{x^{5/4}}\right)^{-1/8}$
34. $\dfrac{4a^2b^{-1/3}}{2a^{-1/2}b^3}$
35. $\left(\dfrac{3x^{-3}y^{-2}z^{-1}}{2xy^2z^3}\right)^{-1}$
36. $\dfrac{(-x^{-3/5})^{-2/3}}{(x^{-3/4})^{1/6}}$
37. $a^{-1} - b^{-1}$
38. $(a - b)^{-1}$
39. $\dfrac{2x^{-1}}{y^{-1} - 2x^{-1}}$
40. $\dfrac{a^{-1} + b^{-1}}{a^{-2} - b^{-2}}$
41. $\dfrac{a^{-1} - b^{-1}}{a^2 - b^2}$
42. $\dfrac{(x+y)^{-1}}{x^{-1} + y^{-1}}$
43. $\dfrac{x^{-2} + y^{-2}}{x^{-2} - y^{-2}}$
44. $\dfrac{2x^{-1} + 3xy^{-2}}{4x^{-2} - 9x^2y^{-4}}$
45. $\left(\dfrac{x^{7/2}y^{4/3}z^{-9}}{x^0y^{-1}z^{-2}}\right)^{-1/7}$
46. $\left(\dfrac{u^{-2/3}v^{-4/3}w^{-4}}{u^{-1/3}v^{2/3}w^{-7/3}}\right)^{-3}$

*In Exercises 47 through 54, find the given product and express the result with positive exponents. Assume that all the variables are positive.*

47. $y^{-1/4}(y^{5/2} + y^{3/8})$
48. $t^{-2/3}(t^{2/3} - t^{-1/3})$
49. $(x^{1/2} + y^{1/2})^2$
50. $(a^{1/2} - b^{1/2})^2$
51. $(a^{-1/2} + b^{-1/2})^2$
52. $(x^{1/4} - x^{1/2})(x^{-1/4} + x^{-1/2})$
53. $(a^{1/3} + b^{1/3})^3$
54. $(a^{1/3} - b^{1/3})^3$

*In Exercises 55 through 62, write the given expression as either a monomial or a simple fraction in lowest terms with positive exponents only; n is a positive integer. Assume that all the variables are positive.*

55. $\dfrac{x^n x^{n+1}}{x^{2n-1}}$
56. $\dfrac{a^{2n-1}a^{n+1}}{a^{3n}}$
57. $(x^4)^{n/4}(x^{4n})^{4/n}$
58. $\dfrac{y^{3-3n}y^{n+2}}{y^{3-2n}}$
59. $\left(\dfrac{a^{2n}}{a^{n+1}}\right)^{-2}$
60. $\left(\dfrac{x^n}{x^{n-1}}\right)^{-1}$
61. $\dfrac{(10^{1-n})^n}{(1000^n)^{n+1}} \cdot \dfrac{10{,}000^{n^2+2}}{100^{3-n}}$
62. $\dfrac{9^{2n-1}}{(3^{n+1})^n} \cdot \dfrac{(81^{n-1})^{n+1}}{(27^{n+2})^{n-1}}$

In Exercises 63 through 68, simplify the given expression. Each variable can be any real number.

63. $(16x^4y^8)^{1/4}$
64. $(4a^4b^{10})^{1/2}$
65. $[(-5)^4(x-5)^4]^{1/2}$
66. $[(-3)^6x^2(x^2+9)^2]^{1/2}$
67. $[(-2)^8(x-2)^8(2-y)^4]^{1/4}$
68. $[(t+1)^4(t-1)^8]^{1/4}$

69. (a) Simplify the expression:
$$(x^2+6x+9)^{1/2} - (x^2-6x+9)^{1/2}$$
(b) For what values of $x$ is the expression in part (a) equivalent to 6?

70. (a) Simplify the expression:
$$(x^2-8x+16)^{1/2} + (x^2+8x+16)^{1/2}$$
(b) For what values of $x$ is the expression in part (a) equivalent to $2x$?

## 4.2 Properties of Radicals

By using the fact that $\sqrt[n]{a} = a^{1/n}$, the following theorem can be proved using laws of exponents.

**4.2.1 THEOREM** If $a, b \in R$, then

(i) $$\sqrt[n]{a} \cdot \sqrt[n]{b} = \sqrt[n]{ab}$$

and

(ii) $$\frac{\sqrt[n]{a}}{\sqrt[n]{b}} = \sqrt[n]{\frac{a}{b}} \quad (b \neq 0)$$

where $a \geq 0$ and $b \geq 0$ if $n$ is even.

In Theorem 4.2.1 we have the condition that $a \geq 0$ and $b \geq 0$ when $n$ is even. These restrictions require that $\sqrt[n]{a}$ and $\sqrt[n]{b}$ be real numbers in order for equalities (i) and (ii) to hold. In Illustration 7 of Section 5.1 we show why the equalities do not hold for $a < 0$ and $b < 0$ when $n$ is even.

**ILLUSTRATION 1.** From Theorem 4.2.1(i), we have

(a) $\sqrt{4} \cdot \sqrt{25} = \sqrt{4 \cdot 25}$
$= \sqrt{100}$
$= 10$

(b) $\sqrt[3]{-9} \cdot \sqrt[3]{-3} = \sqrt[3]{(-9)(-3)}$
$= \sqrt[3]{27}$
$= 3$

From Theorem 4.2.1(ii), it follows that

(c) $\dfrac{\sqrt[4]{96}}{\sqrt[4]{6}} = \sqrt[4]{\dfrac{96}{6}}$
$= \sqrt[4]{16}$
$= 2$

(d) $\dfrac{\sqrt[3]{-54}}{\sqrt[3]{2}} = \sqrt[3]{\dfrac{-54}{2}}$
$= \sqrt[3]{-27}$
$= -3$ ●

Formula (i) of Theorem 4.2.1 can be written as

$$\sqrt[n]{ab} = \sqrt[n]{a} \cdot \sqrt[n]{b} \qquad (a \geq 0 \text{ and } b \geq 0 \text{ if } n \text{ is even}) \qquad (1)$$

The next illustration shows how equality (1) can be used to simplify a radical whose radicand contains as a factor a power having an exponent greater than or equal to the index of the radical.

**ILLUSTRATION 2.** To simplify the radical $\sqrt{540}$, we first write 540 as a product of prime factors. Because $540 = 10 \cdot 54 = 2 \cdot 5 \cdot 6 \cdot 9 = 2 \cdot 5 \cdot 2 \cdot 3 \cdot 3^2$, we have

$$\sqrt{540} = \sqrt{2^2 \cdot 3^3 \cdot 5}$$
$$= \sqrt{(2^2 \cdot 3^2)(3 \cdot 5)}$$
$$= \sqrt{(2 \cdot 3)^2} \sqrt{3 \cdot 5}$$
$$= (2 \cdot 3) \sqrt{15}$$
$$= 6\sqrt{15} \qquad \bullet$$

**EXAMPLE 1**
Simplify the radical where $x > 0$, $y > 0$, and $z > 0$.

(a) $\sqrt[4]{405 x^5 y^8 z^3}$
(b) $\sqrt[3]{-16 x^5 y^3 z^6}$

**SOLUTION**

(a) $\sqrt[4]{405 x^5 y^8 z^3} = \sqrt[4]{3^4 \cdot 5 x^5 y^8 z^3}$
$$= \sqrt[4]{(3^4 x^4 y^8)(5 x z^3)}$$
$$= \sqrt[4]{(3xy^2)^4} \sqrt[4]{5xz^3}$$
$$= 3xy^2 \sqrt[4]{5xz^3}$$

(b) $\sqrt[3]{-16 x^5 y^3 z^6} = \sqrt[3]{(-1)2^4 x^5 y^3 z^6}$
$$= \sqrt[3]{[(-1)^3 2^3 x^3 y^3 z^6](2x^2)}$$
$$= \sqrt[3]{(-1 \cdot 2xyz^2)^3} \sqrt[3]{2x^2}$$
$$= -2xyz^2 \sqrt[3]{2x^2}$$

When Theorem 4.2.1(i) is applied to find the product of radicals having the same order, simplification of the product can be facilitated by expressing any constant as a product of prime factors before performing the multiplication.

**EXAMPLE 2**
Find the product and simplify the result.

$$\sqrt[3]{126 r^2 s^2 t} \cdot \sqrt[3]{36 rs^2 t^2}$$

**SOLUTION**

$$\sqrt[3]{126 r^2 s^2 t} \cdot \sqrt[3]{36 rs^2 t^2} = \sqrt[3]{2 \cdot 3^2 \cdot 7 r^2 s^2 t} \sqrt[3]{2^2 \cdot 3^2 rs^2 t^2}$$
$$= \sqrt[3]{2^3 \cdot 3^4 \cdot 7 r^3 s^4 t^3}$$
$$= \sqrt[3]{(2^3 \cdot 3^3 r^3 s^3 t^3)(3 \cdot 7s)}$$
$$= \sqrt[3]{(2 \cdot 3rst)^3} \sqrt[3]{3 \cdot 7s}$$
$$= 6rst \sqrt[3]{21s}$$

Because the distributive law

$$a(b + c) = ab + ac \qquad (2)$$

## Properties of Radicals

is valid for any real numbers $a$, $b$, and $c$, we can apply it to the product of a monomial and a binomial that contains radicals.

**ILLUSTRATION 3**

$$\sqrt{5}(\sqrt{7} + \sqrt{3}) = \sqrt{5} \cdot \sqrt{7} + \sqrt{5} \cdot \sqrt{3}$$
$$= \sqrt{35} + \sqrt{15} \qquad \bullet$$

From equality (2), we have

$$ab + ac = a(b + c) \tag{3}$$

Equality (3) can be used to factor an expression having a common radical factor.

**ILLUSTRATION 4**

$$\sqrt[3]{ab} + \sqrt[3]{4b} = \sqrt[3]{a}\sqrt[3]{b} + \sqrt[3]{4}\sqrt[3]{b}$$
$$= \sqrt[3]{b}(\sqrt[3]{a} + \sqrt[3]{4}) \qquad \bullet$$

If we apply the commutative law for multiplication to both members of equality (3), we obtain

$$ba + ca = (b + c)a \tag{4}$$

In the next illustration equality (4) is used to simplify a sum of the form of the left member of equality (4), where $a$ is a radical and $b$ and $c$ are rational numbers.

**ILLUSTRATION 5**

$$5\sqrt[3]{6} + 9\sqrt[3]{6} = (5 + 9)\sqrt[3]{6}$$
$$= 14\sqrt[3]{6} \qquad \bullet$$

Note that each term in Illustration 5 contains exactly the same radical factor; that is, for each radical the radicands are equal and the orders are equal. Sometimes it is possible to simplify a sum of terms involving different radicals by replacing them with equivalent terms having the same radical factor. This procedure is shown in the following example.

**EXAMPLE 3**

Simplify the sum

$$\sqrt[3]{48} + \sqrt[3]{162} + \sqrt[3]{384}$$

**SOLUTION**

$$\sqrt[3]{48} + \sqrt[3]{162} + \sqrt[3]{384} = \sqrt[3]{2^4 \cdot 3} + \sqrt[3]{2 \cdot 3^4} + \sqrt[3]{2^7 \cdot 3}$$
$$= \sqrt[3]{2^3} \cdot \sqrt[3]{2 \cdot 3} + \sqrt[3]{3^3} \cdot \sqrt[3]{2 \cdot 3} + \sqrt[3]{(2^2)^3} \cdot \sqrt[3]{2 \cdot 3}$$
$$= 2\sqrt[3]{6} + 3\sqrt[3]{6} + 4\sqrt[3]{6}$$
$$= 9\sqrt[3]{6}$$

It is easier to compute with expressions containing radicals if the radicals are simplified before performing the operations.

**EXAMPLE 4**

Find the product

$(\sqrt{200} + \sqrt{108})(\sqrt{18} - \sqrt{147})$

**SOLUTION**

$(\sqrt{200} + \sqrt{108})(\sqrt{18} - \sqrt{147})$
$= (\sqrt{2^3 \cdot 5^2} + \sqrt{2^2 \cdot 3^3})(\sqrt{2 \cdot 3^2} - \sqrt{3 \cdot 7^2})$
$= (\sqrt{(2 \cdot 5)^2}\sqrt{2} + \sqrt{(2 \cdot 3)^2}\sqrt{3})(\sqrt{3^2}\sqrt{2} - \sqrt{7^2}\sqrt{3})$
$= (10\sqrt{2} + 6\sqrt{3})(3\sqrt{2} - 7\sqrt{3})$
$= 30\sqrt{2^2} - 70\sqrt{6} + 18\sqrt{6} - 42\sqrt{3^2}$
$= 60 - 52\sqrt{6} - 126$
$= -66 - 52\sqrt{6}$

Formula (ii) of Theorem 4.2.1 can be written as

$$\sqrt[n]{\frac{a}{b}} = \frac{\sqrt[n]{a}}{\sqrt[n]{b}} \qquad (a \geq 0 \text{ and } b > 0 \text{ if } n \text{ is even}) \qquad (5)$$

If the radicand of a radical is a fraction having a monomial in the denominator, equality (5) can be used to replace the radical by an equivalent expression for which the radicand contains no fraction. This process is called *rationalizing the denominator,* and for a radical of order $n$ the procedure consists of first building the fraction to one in which the denominator is an $n$th power of a monomial. The next illustration and example show the computation involved.

**ILLUSTRATION 6.** To rationalize the denominator of the radical

$$\sqrt{\frac{3}{5}}$$

we wish to build the fraction in the radicand to one in which the denominator is the square of an integer. Hence we first multiply the numerator and denominator by 5 and then we apply equality (5). Doing this, we have

$$\sqrt{\frac{3}{5}} = \sqrt{\frac{3 \cdot 5}{5 \cdot 5}}$$
$$= \frac{\sqrt{3 \cdot 5}}{\sqrt{5^2}}$$
$$= \frac{\sqrt{15}}{5}$$

The result can also be written as $\frac{1}{5}\sqrt{15}$.

## EXAMPLE 5
Rationalize the denominator.

(a) $\sqrt[3]{\dfrac{2x}{3y^2}}$, $y \neq 0$

(b) $\dfrac{4}{\sqrt{7xy}}$, $x > 0$ and $y > 0$

(c) $\dfrac{75cd^2}{\sqrt[3]{225c^4d^2}}$, $c \neq 0$ and $d \neq 0$

### SOLUTION

(a) $\sqrt[3]{\dfrac{2x}{3y^2}} = \sqrt[3]{\dfrac{2x}{3y^2} \cdot \dfrac{3^2 y}{3^2 y}}$

$= \sqrt[3]{\dfrac{2 \cdot 3^2 xy}{3^3 y^3}}$

$= \dfrac{\sqrt[3]{2 \cdot 3^2 xy}}{\sqrt[3]{(3y)^3}}$

$= \dfrac{\sqrt[3]{18xy}}{3y}$

(c) $\dfrac{75cd^2}{\sqrt[3]{225c^4d^2}} = \dfrac{3 \cdot 5^2 cd^2}{\sqrt[3]{3^2 \cdot 5^2 c^4 d^2}} \cdot \dfrac{\sqrt[3]{3 \cdot 5c^2 d}}{\sqrt[3]{3 \cdot 5c^2 d}}$

$= \dfrac{3 \cdot 5^2 cd^2 \sqrt[3]{15c^2 d}}{\sqrt[3]{3^3 \cdot 5^3 c^6 d^3}}$

$= \dfrac{3 \cdot 5^2 cd^2 \sqrt[3]{15c^2 d}}{\sqrt[3]{(3 \cdot 5c^2 d)^3}}$

$= \dfrac{3 \cdot 5^2 cd^2 \sqrt[3]{15c^2 d}}{3 \cdot 5c^2 d}$

$= \dfrac{5d \sqrt[3]{15c^2 d}(3 \cdot 5cd)}{c(3 \cdot 5cd)}$

$= \dfrac{5d \sqrt[3]{15c^2 d}}{c}$

(b) $\dfrac{4}{\sqrt{7xy}} = \dfrac{4\sqrt{7xy}}{\sqrt{7xy}\sqrt{7xy}}$

$= \dfrac{4\sqrt{7xy}}{\sqrt{(7xy)^2}}$

$= \dfrac{4\sqrt{7xy}}{7xy}$

The quotient of a polynomial involving radicals divided by a monomial involving a radical can be obtained in one of two ways, as shown in the next illustration.

**ILLUSTRATION 7.** We find the quotient $(8\sqrt{35} + 4\sqrt{15}) \div 4\sqrt{5}$ by two methods.

(a) We divide each term of the dividend by the divisor.

$$\dfrac{8\sqrt{35} + 4\sqrt{15}}{4\sqrt{5}} = \dfrac{8\sqrt{5 \cdot 7}}{4\sqrt{5}} + \dfrac{4\sqrt{3 \cdot 5}}{4\sqrt{5}}$$

$$= 2\sqrt{\dfrac{5 \cdot 7}{5}} + \sqrt{\dfrac{3 \cdot 5}{5}}$$

$$= 2\sqrt{7} + \sqrt{3}$$

(b) We multiply the dividend and the divisor by a radical that gives an equivalent fraction having no radical in the denominator.

$$\frac{8\sqrt{35} + 4\sqrt{15}}{4\sqrt{5}} = \frac{(8\sqrt{5\cdot 7} + 4\sqrt{3\cdot 5})\sqrt{5}}{4\sqrt{5}\cdot\sqrt{5}}$$

$$= \frac{8\sqrt{5^2\cdot 7} + 4\sqrt{3\cdot 5^2}}{4\sqrt{5^2}}$$

$$= \frac{8\sqrt{5^2}\sqrt{7} + 4\sqrt{5^2}\sqrt{3}}{20}$$

$$= \frac{40\sqrt{7} + 20\sqrt{3}}{20}$$

$$= 2\sqrt{7} + \sqrt{3} \qquad \bullet$$

Recall the product

$$(a + b)(a - b) = a^2 - b^2$$

Each of the two factors is called the *conjugate* of the other factor. The concept of the conjugate is used to rationalize the denominator of a fraction when the denominator is a binominal containing a radical of index two in either or both of the terms. For instance, to rationalize the denominator of the fraction

$$\frac{5}{\sqrt{7} + \sqrt{3}}$$

we multiply the numerator and denominator by $\sqrt{7} - \sqrt{3}$, which is the conjugate of $\sqrt{7} + \sqrt{3}$, and we have

$$\frac{5}{\sqrt{7} + \sqrt{3}} = \frac{5(\sqrt{7} - \sqrt{3})}{(\sqrt{7} + \sqrt{3})(\sqrt{7} - \sqrt{3})}$$

$$= \frac{5\sqrt{7} - 5\sqrt{3}}{\sqrt{7^2} - \sqrt{3^2}}$$

$$= \frac{5\sqrt{7} - 5\sqrt{3}}{7 - 3}$$

$$= \frac{5\sqrt{7} - 5\sqrt{3}}{4}$$

**EXAMPLE 6**
Rationalize the denominator.

(a) $\dfrac{\sqrt{5} - \sqrt{2}}{\sqrt{5} + \sqrt{2}}$

(b) $\dfrac{\sqrt{a} + \sqrt{b}}{\sqrt{a} - 3\sqrt{b}}$

($a > 0$, $b > 0$ and $a \neq 9b$)

**SOLUTION**

(a) $\dfrac{\sqrt{5} - \sqrt{2}}{\sqrt{5} + \sqrt{2}} = \dfrac{(\sqrt{5} - \sqrt{2})(\sqrt{5} - \sqrt{2})}{(\sqrt{5} + \sqrt{2})(\sqrt{5} - \sqrt{2})}$

$= \dfrac{\sqrt{5^2} - \sqrt{10} - \sqrt{10} + \sqrt{2^2}}{\sqrt{5^2} - \sqrt{2^2}}$

$= \dfrac{5 - 2\sqrt{10} + 2}{5 - 2}$

$= \dfrac{7 - 2\sqrt{10}}{3}$

(b) $\dfrac{\sqrt{a} + \sqrt{b}}{\sqrt{a} - 3\sqrt{b}} = \dfrac{(\sqrt{a} + \sqrt{b})(\sqrt{a} + 3\sqrt{b})}{(\sqrt{a} - 3\sqrt{b})(\sqrt{a} + 3\sqrt{b})}$

$= \dfrac{\sqrt{a^2} + 3\sqrt{ab} + \sqrt{ab} + 3\sqrt{b^2}}{\sqrt{a^2} - 9\sqrt{b^2}}$

$= \dfrac{a + 4\sqrt{ab} + 3b}{a - 9b}$

## EXERCISES 4.2

*In Exercises 1 through 10, find the indicated root.*

1. $\sqrt{81}$
2. $\sqrt[3]{-64}$
3. $\sqrt[3]{-0.001}$
4. $\sqrt[4]{\frac{16}{625}}$
5. $\sqrt{\frac{16}{625}}$
6. $-\sqrt[5]{-32}$
7. $\sqrt[3]{\frac{216}{125}}$
8. $\sqrt[5]{\frac{243}{100,000}}$
9. $\sqrt{(-5)^2}$
10. $\sqrt{(-3)^4}$

*In Exercises 11 through 20, simplify the given radical. All the variables represent positive numbers.*

11. $\sqrt{48}$
12. $\sqrt[3]{-81}$
13. $\sqrt[3]{54}$
14. $\sqrt{16x^{16}}$
15. $\sqrt[3]{8c^8}$
16. $\sqrt[3]{-27x^{10}y^8}$
17. $\sqrt[5]{-96x^{25}y^{12}}$
18. $\sqrt[4]{16x^{16}y^4z^9}$
19. $\sqrt{b^2}$
20. $\sqrt[4]{x^4}$

*In Exercises 21 through 28, find the product and simplify the result. All the variables represent positive numbers.*

21. $\sqrt{10}\,\sqrt{30}$
22. $\sqrt{18}\,\sqrt{12}$
23. $(2\sqrt[3]{9})(4\sqrt[3]{-6})$
24. $\sqrt[4]{24x^3}\,\sqrt[4]{270x^2}$
25. $\sqrt[3]{-6s^2t^4}\,\sqrt[3]{9s^5t^2}$
26. $\sqrt[4]{\frac{10}{3}a^3b^5}\,\sqrt[4]{24a^2b^3}$
27. $\sqrt[3]{9xy^2}\,\sqrt[3]{6x^2y^4}\,\sqrt[3]{60x^5y}$
28. $\sqrt{2uv}\,\sqrt{3v}\,\sqrt{6uw}\,\sqrt{12vw}$

*In Exercises 29 through 38, write the given sum or difference as a single term. All radicands and variables represent positive numbers.*

29. $3\sqrt{7} + 4\sqrt{7}$
30. $4\sqrt{125} - 3\sqrt{45}$
31. $5\sqrt[3]{81} + 3\sqrt[3]{192}$
32. $6\sqrt[3]{54} - 2\sqrt[3]{128} - 3\sqrt[3]{16}$
33. $\sqrt{72a^3} + \sqrt{8a^3} - \sqrt{18a^3}$
34. $\frac{1}{3}x\sqrt{x^3y} - \frac{1}{2}\sqrt{xy^3} - \frac{1}{6}xy\sqrt{4xy}$
35. $s\sqrt{\frac{t}{s}} - \sqrt{\frac{s^3t}{9}}$
36. $5\sqrt[5]{\frac{3}{16}} - 2\sqrt[5]{192} - 8\sqrt[5]{\frac{2}{81}}$
37. $b^3\sqrt{\frac{1}{4c^3}} + \frac{b^2}{c^2}\sqrt{b^2c} + b\sqrt{\frac{4b^4}{c^3}} - \frac{b}{2c}\sqrt{\frac{9b^4}{c}}$
38. $\sqrt{1 - \frac{s}{t}} + 2\sqrt{\frac{t^2 - st}{t^2}}$

*In Exercises 39 through 48, find the product and express each radical in simplest form. All the radicands and all the variables represent positive numbers.*

39. $3\sqrt{2}(\sqrt{6} - 2\sqrt{3})$
40. $\sqrt{10}(5\sqrt{2} - 4\sqrt{5} - 3\sqrt{10})$
41. $(\sqrt{2} + 2\sqrt{6})(\sqrt{2} - \sqrt{6})$
42. $(5 - \sqrt{7})(5 + \sqrt{7})$
43. $(2\sqrt{3} + 3\sqrt{2})^2$
44. $(4\sqrt{5} - 5\sqrt{2})(3\sqrt{2} - 2\sqrt{5})$
45. $(\sqrt[3]{4} - \sqrt[3]{3})(\sqrt[3]{16} + \sqrt[3]{9})$
46. $(\sqrt{x+y} + \sqrt{x-y})^2$
47. $(3\sqrt{a+b} + 2\sqrt{a-b})(3\sqrt{a+b} - 2\sqrt{a-b})$
48. $(3\sqrt{3x+2} - 2\sqrt{3x-2})^2$

*In Exercises 49 through 62, rationalize the denominator. All the variables represent positive numbers.*

49. $\sqrt{\frac{5}{7}}$
50. $\sqrt{\frac{11}{3}}$
51. $\frac{2\sqrt{3}}{\sqrt{5}}$
52. $\frac{-5\sqrt{2}}{4\sqrt{3}}$
53. $\sqrt[3]{\frac{2}{25}}$
54. $\frac{\sqrt[3]{16}}{\sqrt[3]{9x}}$
55. $\frac{-3s}{\sqrt{2t}}$
56. $\sqrt[3]{\frac{x}{3y^2}}$
57. $\sqrt[3]{-\frac{1}{36y^2}}$
58. $\sqrt{\frac{8x^6y^3}{15z^7}}$
59. $\sqrt[4]{\frac{64a^6}{27b^2}}$
60. $\frac{\sqrt[4]{c^3d^2}}{3\sqrt[4]{cd}}$
61. $\frac{21}{\sqrt[3]{-98x}}$
62. $\frac{270a^2bc^3}{\sqrt[4]{288a^5b^6c^2}}$

*In Exercises 63 through 70, find the quotient and express the result in simplest form. All the radicands and variables represent positive numbers.*

63. $\frac{14\sqrt{14} - 21\sqrt{21}}{\sqrt{42}}$
64. $\frac{3a\sqrt{5} + 9\sqrt{3a^2}}{a\sqrt{3}}$
65. $\frac{4}{3 - \sqrt{2}}$
66. $\frac{\sqrt{3}}{\sqrt{6} + \sqrt{3}}$
67. $\frac{\sqrt{5} - 3}{2 - \sqrt{5}}$
68. $\frac{4\sqrt{2} - 3\sqrt{3}}{5\sqrt{3} - 3\sqrt{2}}$
69. $\frac{2\sqrt{2} - 3\sqrt{7}}{3\sqrt{2} + 2\sqrt{7}}$
70. $\frac{\sqrt{t}}{\sqrt{s} - \sqrt{t}}$

## 4.3 Exponential Functions

We have defined the power of a positive number when the exponent is a rational number. In particular, $2^x$ has been defined for any rational value of $x$. For instance,

$$2^5 = 2 \cdot 2 \cdot 2 \cdot 2 \cdot 2 \qquad 2^0 = 1 \qquad 2^{-3} = \frac{1}{2^3} \qquad 2^{2/3} = \sqrt[3]{2^2}$$
$$= 32 \qquad\qquad\qquad\qquad = \frac{1}{8} \qquad\qquad = \sqrt[3]{4}$$

But we have not defined $2^x$ when $x$ is an irrational number. Actually, the definition of an irrational exponent requires a knowledge of more advanced mathematics than is covered in this book. However, we can give an intuitive indication that irrational powers of positive numbers can exist by showing how we can interpret the meaning of $2^{\sqrt{3}}$. To do this, we make use of the following theorem, which we state without proof.

**4.3.1 THEOREM** If $r$ and $s$ are rational numbers, then

(i) if $b > 1$: $r < s$ implies $b^r < b^s$
(ii) if $0 < b < 1$: $r < s$ implies $b^r > b^s$

A decimal approximation for $\sqrt{3}$ can be obtained accurate to any number of decimal places desired. Using four decimal places, we have $\sqrt{3} \approx 1.7321$. Because $1 < 1.7 < 2$, then from Theorem 4.3.1 part (i) it follows that

$$2^1 < 2^{1.7} < 2^2$$

Because $1.7 < 1.73 < 1.8$, then

$$2^{1.7} < 2^{1.73} < 2^{1.8}$$

Because $1.73 < 1.732 < 1.74$, then

$$2^{1.73} < 2^{1.732} < 2^{1.74}$$

Because $1.732 < 1.7321 < 1.733$, then

$$2^{1.732} < 2^{1.7321} < 2^{1.733}$$

and so on. In each inequality we have a power of 2 for which the exponent is a decimal approximation of the value of $\sqrt{3}$, and in each successive inequality, the exponent contains one more decimal place than the exponent in the previous inequality. By following this procedure indefinitely, the difference between the left member of the inequality and the right member of the inequality can be made as small as we please. Hence our intuitition leads us to assume that there is a value of $2^{\sqrt{3}}$ that satisfies each successive inequality as we continue the procedure indefinitely. A similar discussion can be given for any irrational power of any positive number. Furthermore, Theorem 4.3.1 is valid if $r$ and $s$ are any real numbers.

We now can define an "exponential function."

**4.3.2 DEFINITION**  If $b > 0$ and $b \neq 1$, then the *exponential function with base b* is the function $f$ defined by

$$f(x) = b^x \qquad (1)$$

The domain of $f$ is the set $R$ of real numbers and the range of $f$ is the set of positive numbers.

Observe that if $b = 1$, equation (1) becomes $f(x) = 1^x$. But if $x$ is any real number, then $1^x = 1$, and thus we have a constant function. For this reason we impose the condition that $b \neq 1$ in Definition 4.3.2.

In the following two illustrations we consider the graphs of the exponential functions with bases 2 and $\frac{1}{2}$, respectively.

**ILLUSTRATION 1.** The exponential function with base 2 is the function $F$ such that

$$F(x) = 2^x$$

or, equivalently,

$$F = \{(x, y) \mid y = 2^x\}$$

Some of the ordered pairs in $F$ are given in Table 4.3.1.

**Table 4.3.1**

| $x$ | $-3$ | $-2$ | $-1$ | 0 | 1 | 2 | 3 |
|---|---|---|---|---|---|---|---|
| $y$ | $\frac{1}{8}$ | $\frac{1}{4}$ | $\frac{1}{2}$ | 1 | 2 | 4 | 8 |

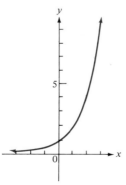

**Figure 4.3.1**

A sketch of the graph is shown in Figure 4.3.1; it is drawn by plotting the points whose coordinates are the ordered pairs given in Table 4.3.1 and connecting these points with a smooth curve. The graph indicates that the function is an increasing function.

Observe that

$$F(x) \to 0^+ \qquad \text{as } x \to -\infty$$

that is, $F(x)$ approaches zero as $x$ decreases without bound. Therefore, by Definition 3.5.3(ii) it follows that the $x$ axis is a horizontal asymptote of the graph of $F$. Furthermore, notice that

$$F(x) \to +\infty \qquad \text{as } x \to +\infty$$

that is, $F(x)$ increases without bound as $x$ increases without bound.  ●

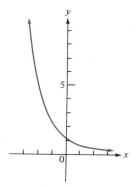

Figure 4.3.2

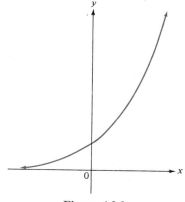

Figure 4.3.3

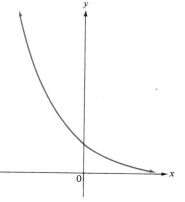

Figure 4.3.4

**ILLUSTRATION 2.** The exponential function with base $\frac{1}{2}$ is the function $G$ such that

$$G(x) = \left(\frac{1}{2}\right)^x$$

or, equivalently,

$$G = \left\{(x, y) \mid y = \left(\frac{1}{2}\right)^x\right\}$$

Some of the ordered pairs in $G$ are given in Table 4.3.2.

Table 4.3.2

| $x$ | $-3$ | $-2$ | $-1$ | $0$ | $1$ | $2$ | $3$ |
|---|---|---|---|---|---|---|---|
| $y$ | $8$ | $4$ | $2$ | $1$ | $\frac{1}{2}$ | $\frac{1}{4}$ | $\frac{1}{8}$ |

By plotting the points whose coordinates are the ordered pairs given in Table 4.3.2 and connecting these points with a smooth curve, we obtain the sketch of the graph of $G$ shown in Figure 4.3.2. The function is a decreasing function as indicated by the graph.

Because

$$G(x) \to 0^+ \quad \text{as } x \to +\infty$$

then from Definition 3.5.3(i), the $x$ axis is a horizontal asymptote of the graph of $G$. Also,

$$G(x) \to +\infty \quad \text{as } x \to -\infty \qquad \bullet$$

Figure 4.3.3 shows a sketch of the graph of the function $f$ for which $f(x) = b^x$ and $b > 1$. The exponential function with base $b$, for which $b > 1$, is an increasing function. This fact follows from Theorem 4.3.1, part (i), with $r$ and $s$ as real numbers, and the definition of an increasing function (3.6.3, part (i)).

In Figure 4.3.4 we have a sketch of the graph of the exponential function with base $b$, when $0 < b < 1$. This function is a decreasing function, which follows from Theorem 4.3.1 part (ii), with $r$ and $s$ as real numbers, and the definition of a decreasing function (3.6.3, part (ii)).

The laws of exponents that are valid for rational exponents also hold if the exponents are any real numbers. These laws are summarized in the following theorem.

**4.3.3 THEOREM** If $a$ and $b$ are any positive numbers, and $x$ and $y$ are any real numbers, then

(i) $\quad a^x a^y = a^{x+y}$

(ii) $\quad \dfrac{a^x}{a^y} = a^{x-y}$

(iii) $\quad (a^x)^y = a^{xy}$

(iv) $\quad (ab)^x = a^x b^x$

(v) $\quad \left(\dfrac{a}{b}\right)^x = \dfrac{a^x}{b^x}$

The proofs of properties (i) through (v) for real number exponents are beyond the scope of this book and therefore they are omitted.

**EXAMPLE 1** Simplify each of the following by applying laws of exponents.

(a) $2^{\sqrt{3}} \cdot 2^{\sqrt{12}}$  (b) $(7^{\sqrt{5}})^{\sqrt{20}}$

**SOLUTION**

(a) $2^{\sqrt{3}} \cdot 2^{\sqrt{12}} = 2^{\sqrt{3}} \cdot 2^{2\sqrt{3}}$
$= 2^{\sqrt{3}+2\sqrt{3}}$
$= 2^{3\sqrt{3}}$

(b) $(7^{\sqrt{5}})^{\sqrt{20}} = 7^{\sqrt{5} \cdot \sqrt{20}}$
$= 7^{\sqrt{100}}$
$= 7^{10}$

From Definition 4.3.2 we see that the base $b$ of an exponential function can be any positive number other than 1. For each value of $b$ we have a different exponential function. There is a particular value of $b$ that is very important in mathematics. It is an irrational number, denoted by $e$, and it arises in applications of mathematics in many fields. The value of $e$ to seven decimal places is 2.7182818. Thus we write

$$e \approx 2.7182818$$

A discussion of the origin of the number $e$ and how it is computed belongs to a calculus text. Approximations of some powers of $e$ are given in Table 3 in the Appendix.

The exponential function with base $e$ is often denoted by $\exp$. Hence

$$\exp = \{(x, y) \mid y = e^x\}$$

and

$$\exp(x) = e^x$$

**EXAMPLE 2** Draw a sketch of the graph of the exponential function with base $e$.

**SOLUTION** Some of the ordered pairs in $\exp$ are given in Table 4.3.3. The approximations of the powers of $e$ are found in Table 3 in the Appendix.

Table 4.3.3

| $x$ | 0 | 0.5 | 1 | 1.5 | 2 | 2.5 | −0.5 | −1 | −2 |
|---|---|---|---|---|---|---|---|---|---|
| $y$ | 1 | 1.6 | 2.7 | 4.5 | 7.4 | 12.2 | 0.6 | 0.4 | 0.1 |

The points whose coordinates are the ordered pairs given in Table 4.3.3 are plotted, and these points are connected with a smooth curve to give the sketch of the graph of exp shown in Figure 4.3.5.

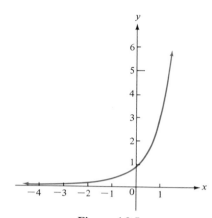

**Figure 4.3.5**

### EXAMPLE 3

In a certain culture, if $A$ is the number of bacteria present at $t$ minutes, then

$$A = ke^{0.06t} \qquad (2)$$

where $k$ is a constant. If there are 1000 bacteria present initially, how many bacteria will be present after 1 hour?

### SOLUTION

Because there are 1000 bacteria present initially, we know that $A = 1000$ when $t = 0$. Substituting these values into equation (2), we obtain

$$1000 = ke^{(0.06)(0)}$$
$$1000 = ke^0$$

Because $e^0 = 1$, we have $k = 1000$. Replacing $k$ by 1000 in equation (2), we have

$$A = 1000e^{0.06t} \qquad (3)$$

Let $\bar{A}$ be the number of bacteria present after 1 hour. Then when $t = 60$, $A = \bar{A}$. Substituting these values into equation (3), we obtain

$$\bar{A} = 1000e^{(0.06)(60)}$$
$$\bar{A} = 1000e^{3.6}$$

From Table 3 in the Appendix we obtain an approximation of $e^{3.6}$ to three decimal places: $e^{3.6} \approx 36.598$. Therefore,

$$\bar{A} \approx 1000(36.598)$$
$$\bar{A} \approx 36,598$$

Hence there are 36,598 bacteria present in the culture after 1 hour.

## EXERCISES 4.3

*In Exercises 1 through 10, draw a sketch of the graph of the given exponential function.*

1. $f = \{(x, y) \mid y = 3^x\}$
2. $g = \{(x, y) \mid y = 4^x\}$
3. $F = \{(x, y) \mid y = 3^{-x}\}$
4. $G = \{(x, y) \mid y = 4^{-x}\}$
5. $g = \{(x, y) \mid y = (\frac{1}{5})^x\}$
6. $f = \{(x, y) \mid y = 10^x\}$
7. $f = \{(x, y) \mid y = 2^{x+1}\}$
8. $g = \{(x, y) \mid y = 3^{x-1}\}$
9. $G = \{(x, y) \mid y = e^{2x}\}$
10. $F = \{(x, y) \mid y = e^{-x}\}$

*In Exercises 11 through 16, simplify the given expression by applying laws of exponents.*

11. $3^{\sqrt{2}} \cdot 3^{\sqrt{50}}$
12. $2^{\sqrt{12}} \cdot 2^{\sqrt{27}}$
13. $(5^{\sqrt{15}})^{\sqrt{6}}$
14. $(10^{\sqrt{10}})^{\sqrt{5}}$
15. $\dfrac{4^{\sqrt{32}}}{2^{\sqrt{18}}}$
16. $\dfrac{3^{\sqrt{45}}}{9^{\sqrt{20}}}$

17. Suppose that $A$ is the number of bacteria present in a certain culture at $t$ minutes and
$$A = ke^{0.05t}$$
where $k$ is a constant. If 5000 bacteria are present after 10 minutes have elapsed, how many bacteria were present initially?

18. If $p$ pounds per square foot is the atmospheric pressure at a height $h$ feet above sea level, then
$$P = ke^{-0.00003h}$$
where $k$ is a constant. If the atmospheric pressure at sea level is 2116 pounds per square foot, find the atmospheric pressure outside of an airplane that is 10,000 feet high.

19. If $A$ grams of a radioactive substance are present after $t$ seconds, then
$$A = ke^{-0.3t}$$
where $k$ is a constant. If 100 grams of the substance are present initially, how much is present after 5 seconds?

20. A tank contains a mixture of brine and water and initially the mixture contains 70 pounds of dissolved salt. Fresh water is running into the tank and the mixture, kept uniform by stirring, is running out at the same rate that the water is running in. After $t$ minutes the mixture contains $x$ pounds of salt and
$$x = ke^{-0.03t}$$
where $k$ is a constant. How many pounds of salt are in the tank at the end of 1 hour?

## 4.4 Logarithmic Functions

In Example 3 of Section 4.3 we had the equation
$$A = 1000e^{0.06t}$$
where $A$ is the number of bacteria present in a certain culture at $t$ minutes. Suppose that we wish to find in how many minutes there will be 50,000 bacteria present. If $T$ is the number of minutes to be determined, then we have the equation

$$50{,}000 = 1000e^{0.06T}$$

In this equation the unknown $T$ appears in an exponent. At the present we cannot solve such an equation, but the concept of a logarithm will give us the means to do so. We will develop this concept and then return to the problem in Example 6.

When $b > 1$, the exponential function with base $b$ is an increasing function, and when $0 < b < 1$, it is a decreasing function. Therefore, from Theorem 3.6.4, it follows that the exponential function with base $b$ has a function as its inverse, which is called the "logarithmic function with base $b$."

**4.4.1 DEFINITION** The *logarithmic function with base b* is the inverse of the exponential function with base $b$.

We use the notation "$\log_b$" to denote the logarithmic function with base $b$. Hence, if

$$F = \{(x, y) \mid y = b^x, b > 0, b \neq 1\}$$

then by Definition 4.4.1, $\log_b$ is the inverse of function $F$, and hence

$$\log_b = \{(x, y) \mid (y, x) \in F\}$$

or, equivalently,

$$\log_b = \{(x, y) \mid x = b^y, b > 0, b \neq 1\} \tag{1}$$

The function values of the function $\log_b$ are denoted by $\log_b(x)$, or more simply $\log_b x$ (read "logarithm with base $b$ of $x$"). Hence

$$\log_b = \{(x, y) \mid y = \log_b x, b > 0, b \neq 1\} \tag{2}$$

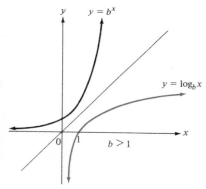

**Figure 4.4.1**

The domain of the exponential function with base $b$ is the set $R$ of real numbers, and its range is the set of positive numbers. Therefore, the domain of $\log_b$ is the set of positive numbers and the range is the set $R$. Figure 4.4.1 shows in color a sketch of the graph of $\log_b$, where $b > 1$. It is the graph that is symmetric, with respect to the line $y = x$, to the graph of the exponential function with base $b$ ($b > 1$), a sketch of which is shown in Figure 4.4.1. A sketch of the graph of $\log_b$, where $0 < b < 1$, is shown in color in Figure 4.4.2. Also shown in Figure 4.4.2 is a sketch of the graph of the exponential function with base $b$ ($0 < b < 1$), and we observe that the two graphs are symmetric with respect to the line $y = x$.

From the sketches of the graphs of $\log_b$ in Figures 4.4.1 and 4.4.2 we note the following properties of the logarithmic function with base $b$.

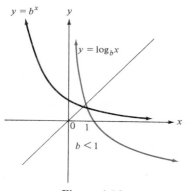

**Figure 4.4.2**

1. If $b > 1$, $\log_b$ is an increasing function. If $0 < b < 1$, $\log_b$ is a decreasing function.

2. If $b > 1$, $\log_b x$ is positive if $x > 1$, and $\log_b x$ is negative if $0 < x < 1$. If $0 < b < 1$, $\log_b x$ is negative if $x > 1$, and $\log_b x$ is positive if $0 < x < 1$. Furthermore, $\log_b x$ is not defined if $x$ is negative.
3. The only zero of the function $\log_b$ is 1; that is, $\log_b x = 0$ if and only if $x = 1$.
4. $\log_b x = 1$ if and only if $x = b$.
5. If $b > 1$, $\log_b x \to -\infty$ as $x \to 0^+$; and if $0 < b < 1$, $\log_b x \to +\infty$ as $x \to 0^+$.

By comparing the sets in equations (1) and (2) that define $\log_b$, it follows that

$$x = b^y \quad \text{is equivalent to} \quad y = \log_b x$$

**ILLUSTRATION 1**

$$3^2 = 9 \quad \text{is equivalent to} \quad \log_3 9 = 2$$
$$2^3 = 8 \quad \text{is equivalent to} \quad \log_2 8 = 3$$
$$\left(\frac{1}{16}\right)^{1/2} = \frac{1}{4} \quad \text{is equivalent to} \quad \log_{1/16} \frac{1}{4} = \frac{1}{2}$$
$$5^{-2} = \frac{1}{25} \quad \text{is equivalent to} \quad \log_5 \frac{1}{25} = -2$$

**ILLUSTRATION 2**

$$\log_{10} 10{,}000 = 4 \quad \text{is equivalent to} \quad 10^4 = 10{,}000$$
$$\log_8 2 = \frac{1}{3} \quad \text{is equivalent to} \quad 8^{1/3} = 2$$
$$\log_6 1 = 0 \quad \text{is equivalent to} \quad 6^0 = 1$$
$$\log_9 \frac{1}{3} = -\frac{1}{2} \quad \text{is equivalent to} \quad 9^{-1/2} = \frac{1}{3}$$

**EXAMPLE 1**
Find the value of each of the following logarithms.

(a) $\log_7 49$   (b) $\log_5 \sqrt{5}$

(c) $\log_6 \frac{1}{6}$   (d) $\log_3 81$

(e) $\log_{10} 0.001$

**SOLUTION**
In each part we let $y$ represent the given logarithm and obtain an equivalent equation in exponential form. We then solve for $y$ by making use of the fact that if $a > 0$ and $a \neq 1$, $a^y = a^n$ implies $y = n$.
(a) Let $\log_7 49 = y$. This equation is equivalent to $7^y = 49$. Because $49 = 7^2$, we have

$$7^y = 7^2$$

Therefore, $y = 2$; that is, $\log_7 49 = 2$.

(b) Let $\log_5 \sqrt{5} = y$. Therefore, $5^y = \sqrt{5}$ or, equivalently,
$$5^y = 5^{1/2}$$
Hence $y = \frac{1}{2}$; that is, $\log_5 \sqrt{5} = \frac{1}{2}$.
(c) Let $\log_6 \frac{1}{6} = y$. Thus $6^y = \frac{1}{6}$ or, equivalently,
$$6^y = 6^{-1}$$
Therefore, $y = -1$; that is, $\log_6 \frac{1}{6} = -1$.
(d) Let $\log_3 81 = y$. Thus $3^y = 81$ or, equivalently,
$$3^y = 3^4$$
Hence $y = 4$; that is, $\log_3 81 = 4$.
(e) Let $\log_{10} 0.001 = y$. Then $10^y = 0.001$. Because $10^{-3} = 0.001$, we have
$$10^y = 10^{-3}$$
Therefore, $y = -3$; that is, $\log_{10} 0.001 = -3$.

EXAMPLE 2
Solve each of the following equations for $x$.
(a) $\log_6 x = 2$
(b) $\log_{27} x = \dfrac{2}{3}$
(c) $\log_{1/2} x = -4$

SOLUTION
(a) The equation $\log_6 x = 2$ is equivalent to
$$6^2 = x$$
Therefore, $x = 36$.
(b) The equation $\log_{27} x = \frac{2}{3}$ is equivalent to
$$27^{2/3} = x$$
Hence
$$\begin{aligned} x &= (\sqrt[3]{27})^2 \\ &= 3^2 \\ &= 9 \end{aligned}$$
(c) The equation $\log_{1/2} x = -4$ is equivalent to
$$\left(\frac{1}{2}\right)^{-4} = x$$
Therefore,
$$\begin{aligned} x &= \frac{1}{\left(\frac{1}{2}\right)^4} \\ &= \frac{1}{\frac{1}{16}} \\ &= 16 \end{aligned}$$

**EXAMPLE 3**
Solve each of the following equations for $b$.

(a) $\log_b 4 = \dfrac{1}{3}$

(b) $\log_b 81 = -2$

(c) $\log_b 125 = \dfrac{3}{2}$

**SOLUTION**

(a) The equation $\log_b 4 = \frac{1}{3}$ is equivalent to
$$b^{1/3} = 4$$
Hence
$$(b^{1/3})^3 = 4^3$$
$$b = 64$$

(b) The equation $\log_b 81 = -2$ is equivalent to
$$b^{-2} = 81$$
Hence
$$(b^{-2})^{-1/2} = 81^{-1/2}$$
$$b = \frac{1}{81^{1/2}}$$
$$= \frac{1}{9}$$

(c) The equation $\log_b 125 = \frac{3}{2}$ is equivalent to
$$b^{3/2} = 125$$
Therefore,
$$(b^{3/2})^{2/3} = 125^{2/3}$$
$$b = (\sqrt[3]{125})^2$$
$$= 5^2$$
$$= 25$$

Because the equation
$$b^y = x \tag{3}$$
is equivalent to the equation
$$y = \log_b x \tag{4}$$
we can substitute the value of $y$ from equation (4) into equation (3) and obtain
$$b^{\log_b x} = x \tag{5}$$
where $b > 0$, $b \neq 1$, and $x > 0$.

From equation (5) we note that *a logarithm is an exponent;* that is, $\log_b x$ is the exponent of the power to which we must raise $b$ to obtain $x$.

**ILLUSTRATION 3.** From equation (5) it follows that
$$3^{\log_3 7} = 7$$
and
$$10^{\log_{10} 5} = 5$$

If we substitute the value of $x$ from equation (3) into equation (4), we have
$$\log_b b^y = y \tag{6}$$
where $b > 0$, $b \neq 1$, and $y$ is any real number.

**ILLUSTRATION 4.** From equation (6) it follows that
$$\log_{10} 10^{-4} = -4$$
and
$$\log_e e^3 = 3$$

**EXAMPLE 4**
Simplify each of the following expressions.

(a) $\log_3 (\log_4 64)$
(b) $\log_2 (\log_5 625)$

**SOLUTION**
(a) Because $4^3 = 64$, $\log_4 64 = 3$, and so
$$\log_3 (\log_4 64) = \log_3 3$$
$$= 1$$

(b) Because $5^4 = 625$, $\log_5 625 = 4$, and so
$$\log_2 (\log_5 625) = \log_2 4$$
$$= 2$$

**EXAMPLE 5**
Draw a sketch of the graph of the logarithmic function with base 3.

**SOLUTION**
$$\log_3 = \{(x, y) \mid y = \log_3 x\}$$

Table 4.4.1 gives some of the ordered pairs in $\log_3$. These ordered pairs are obtained from the equation $x = 3^y$, which is equivalent to the equation $y = \log_3 x$.

Table 4.4.1

| $x$ | 1 | 3 | 9 | 27 | $\frac{1}{3}$ | $\frac{1}{9}$ | $\frac{1}{27}$ |
|---|---|---|---|---|---|---|---|
| $y$ | 0 | 1 | 2 | 3 | $-1$ | $-2$ | $-3$ |

By plotting the points whose coordinates are the ordered pairs given in Table 4.4.1 and connecting these points with a smooth curve, we obtain the sketch of the graph of $\log_3$ shown in Figure 4.4.3.

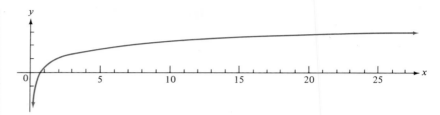

**Figure 4.4.3**

The logarithmic function with base $e$ is called the *natural logarithmic function*. It can be denoted by $\log_e$, but a more customary notation is $\ln$. Therefore,

$$\ln = \{(x,y) \mid x = e^y\} \tag{7}$$

The function values of $\ln$ are denoted by $\ln x$ (read "natural logarithm of $x$"). Hence equation (7) is equivalent to

$$\ln = \{(x,y) \mid y = \ln x\}$$

The natural logarithmic function is very important in the calculus and its applications. A sketch of its graph is shown in Figure 4.4.4. In the figure the graph of $\ln$ is shown in color and a sketch of the graph of the exponential function with base $e$ is also shown. The two graphs are seen to be symmetric with respect to the line $y = x$.

Table 4 in the Appendix gives approximations to four decimal places of the natural logarithm of numbers between 0.1 and 190.

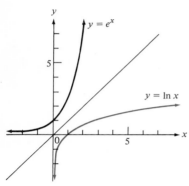

**Figure 4.4.4**

**EXAMPLE 6**

In Example 3 of Section 4.3 we obtained the equation

$$A = 1000 e^{0.06t}$$

where $A$ is the number of bacteria present in a certain culture at $t$ minutes when there are 1000 bacteria present initially. Determine how many minutes elapse until there are 50,000 bacteria present in the culture.

**SOLUTION**

Let $T$ represent the number of minutes that elapse until there are 50,000 bacteria present. Then in the given equation, we substitute 50,000 for $A$ and $T$ for $t$, and we have

$$50{,}000 = 1000 e^{0.06T}$$
$$50 = e^{0.06T} \tag{8}$$

Because the equation $x = e^y$ is equivalent to the equation $y = \ln x$, it follows that equation (8) is equivalent to

$$0.06T = \ln 50$$

From Table 4 in the Appendix, we have $\ln 50 = 3.9120$; therefore,

$$0.06T = 3.9120$$
$$T = \frac{3.9120}{0.06}$$
$$T = 65.20$$

Thus 1 hour, 5 minutes, and 12 seconds elapse until there are 50,000 bacteria present.

## EXERCISES 4.4

*In Exercises 1 through 10, express the relationship in the given equation by using logarithmic notation.*

1. $3^4 = 81$
2. $2^5 = 32$
3. $5^3 = 125$
4. $7^2 = 49$
5. $10^{-3} = 0.001$
6. $5^{-2} = \frac{1}{25}$
7. $8^{2/3} = 4$
8. $81^{3/4} = 27$
9. $625^{-3/4} = \frac{1}{125}$
10. $10^0 = 1$

*In Exercises 11 through 20, express the relationship in the given equation by using exponential notation.*

11. $\log_8 64 = 2$
12. $\log_{10} 10{,}000 = 4$
13. $\log_3 81 = 4$
14. $\log_5 125 = 3$
15. $\log_1 1 = 0$
16. $\log_8 2 = \frac{1}{3}$
17. $\log_{1/3} 9 = -2$
18. $\log_{1/2} 64 = -6$
19. $\log_9 \frac{1}{3} = -\frac{1}{2}$
20. $\log_{16} \frac{1}{8} = -\frac{3}{4}$

*In Exercises 21 through 32, find the value of the given logarithm.*

21. $\log_{10} 100$
22. $\log_4 64$
23. $\log_{27} 9$
24. $\log_6 \sqrt{6}$
25. $\log_2 \frac{1}{8}$
26. $\log_3 \frac{1}{81}$
27. $\log_8 \frac{1}{2}$
28. $\log_7 7$
29. $\log_{27} \frac{1}{81}$
30. $\log_{1/4} \frac{1}{32}$
31. $\log_e \sqrt[3]{e}$
32. $\log_{10} 1$

*In Exercises 33 through 42, solve the given equation for either b or x.*

33. $\log_7 x = 3$
34. $\log_4 x = 3$
35. $\log_{1/3} x = -4$
36. $\log_{1/4} x = -3$
37. $\log_b 144 = 2$
38. $\log_b 0.01 = -2$
39. $\log_b 6 = \frac{1}{3}$
40. $\log_2 x = \frac{3}{2}$
41. $\log_{1/4} x = \frac{7}{2}$
42. $\log_b 27 = -3$

*In Exercises 43 through 50, simplify the given expression.*

43. $\log_6 (\log_5 5)$
44. $\log_5 (\log_2 32)$
45. $\log_2 (\log_9 81)$
46. $\log_2 (\log_3 81)$
47. $\log_2 (\log_2 256)$
48. $\log_3 (\log_3 3)$
49. $\log_b (\log_b b), b > 0$
50. $\log_b (\log_a a^b), a > 0$ and $b > 0$

*In Exercises 51 through 55, draw a sketch of the graph of the given function.*

51. $\{(x, y) \mid y = \log_{10} x\}$
52. $\{(x, y) \mid y = \log_2 x\}$
53. $\{(x, y) \mid y = \log_3 x^2\}$
54. $\{(x, y) \mid y = \ln (x + 1)\}$
55. $\{(x, y) \mid y = \ln (x - 1)\}$

## 4.5 Properties of Logarithmic Functions

We now state and prove three theorems that give properties of logarithms that follow from corresponding properties of exponents. After the statement of each theorem, an illustration is given to show the property of exponents involved. In the proofs we make use of the fact that

$$x = b^y \quad \text{is equivalent to} \quad y = \log_b x$$

We refer to the equation $x = b^y$ as the *exponential form* of the equation $y = \log_b x$, and we refer to the equation $y = \log_b x$ as the *logarithmic form* of the equation $x = b^y$.

**4.5.1 THEOREM**  If $b > 0$, $b \neq 1$, and $u$ and $v$ are positive numbers, then

$$\log_b uv = \log_b u + \log_b v \qquad (1)$$

**ILLUSTRATION 1.** Suppose in the statement of Theorem 4.5.1, $b$ is 2, $u$ is 4, and $v$ is 8. Then

$$\begin{aligned}\log_b uv &= \log_2 4 \cdot 8 \\ &= \log_2 2^2 \cdot 2^3 \\ &= \log_2 2^{2+3} \\ &= \log_2 2^5 \\ &= 5 \quad \text{(because } \log_b b^y = y\text{)}\end{aligned}$$

$$\begin{aligned}\log_b u + \log_b v &= \log_2 4 + \log_2 8 \\ &= \log_2 2^2 + \log_2 2^3 \\ &= 2 + 3 \\ &= 5\end{aligned}$$

Therefore, when $b$ is 2, $u$ is 4, and $v$ is 8, equation (1) is valid. ●

*Proof of Theorem 4.5.1.* Let

$$r = \log_b u \quad \text{and} \quad s = \log_b v \qquad (2)$$

The exponential forms of equations (2) are, respectively,

$$u = b^r \quad \text{and} \quad v = b^s$$

Therefore,

$$uv = b^r \cdot b^s$$

Applying a law of exponents, we have

$$uv = b^{r+s}$$

The logarithmic form of this equation is

$$\log_b uv = r + s \qquad (3)$$

Substituting the values of $r$ and $s$ from equations (2) into equation (3), we obtain

$$\log_b uv = \log_b u + \log_b v$$

**ILLUSTRATION 2.** If we are given $\log_{10} 2 = 0.3010$ and $\log_{10} 3 = 0.4771$, we can apply Theorem 4.5.1 to find $\log_{10} 6$.

$$\begin{aligned}\log_{10} 6 &= \log_{10}(2 \cdot 3) \\ &= \log_{10} 2 + \log_{10} 3 \\ &= 0.3010 + 0.4771 \\ &= 0.7781\end{aligned}$$

●

Because $\log_{10} 2$, $\log_{10} 3$, and $\log_{10} 6$ are irrational numbers, the values given for them in Illustration 2 are only decimal approximations. Hence the symbol $\approx$ (approximately equals) is more appropriate than the symbol $=$

(equals). However, in computations such as Illustration 2, the conventional practice is to use the equals symbol.

**4.5.2 THEOREM** If $b > 0$, $b \neq 1$, and $u$ and $v$ are positive numbers, then

$$\log_b \frac{u}{v} = \log_b u - \log_b v \qquad (4)$$

**ILLUSTRATION 3.** Suppose that in the statement of Theorem 4.5.2, $b$ is 2, $u$ is 128, and $v$ is 16. Then

$$\log_b \frac{u}{v} = \log_2 \frac{128}{16} \qquad \log_b u - \log_b v = \log_2 128 - \log_2 16$$
$$= \log_2 \frac{2^7}{2^4} \qquad\qquad\qquad\qquad = \log_2 2^7 - \log_2 2^4$$
$$= \log_2 2^{7-4} \qquad\qquad\qquad\qquad = 7 - 4$$
$$= \log_2 2^3 \qquad\qquad\qquad\qquad\quad = 3$$
$$= 3$$

Hence, when $b$ is 2, $u$ is 128, and $v$ is 16, equation (4) holds. •

***Proof of Theorem 4.5.2.*** As in the proof of Theorem 4.5.1, we let

$$r = \log_b u \quad \text{and} \quad s = \log_b v$$

The exponential forms of these equations are, respectively,

$$u = b^r \quad \text{and} \quad v = b^s$$

Hence

$$\frac{u}{v} = \frac{b^r}{b^s}$$

Applying a law of exponents, we have

$$\frac{u}{v} = b^{r-s}$$

The logarithmic form of this equation is

$$\log_b \frac{u}{v} = r - s$$

Substituting $\log_b u$ for $r$ and $\log_b v$ for $s$, we have

$$\log_b \frac{u}{v} = \log_b u - \log_b v$$

**ILLUSTRATION 4.** From Theorem 4.5.2 it follows that

$$\log_{10} \frac{3}{2} = \log_{10} 3 - \log_{10} 2$$

Substituting the values of $\log_{10} 3$ and $\log_{10} 2$ given in Illustration 2, we obtain

$$\log_{10} \frac{3}{2} = 0.4771 - 0.3010$$

$$= 0.1761 \qquad \bullet$$

**4.5.3 THEOREM** If $b > 0$, $b \neq 1$, $n$ is any real number, and $u$ is a positive number, then

$$\boxed{\log_b u^n = n \log_b u} \tag{5}$$

**ILLUSTRATION 5.** Suppose in the statement of Theorem 4.5.3, $b$ is 2, $n$ is 3, and $u$ is 4. Then

$$\begin{aligned}
\log_b u^n &= \log_2 4^3 & n \log_b u &= 3 \log_2 4 \\
&= \log_2 (2^2)^3 & &= 3 \log_2 2^2 \\
&= \log_2 2^{2 \cdot 3} & &= 3 \cdot 2 \\
&= \log_2 2^6 & &= 6 \\
&= 6
\end{aligned}$$

Thus, when $b$ is 2, $n$ is 3, and $u$ is 4, equation (5) is valid. $\bullet$

*Proof of Theorem 4.5.3.* Let

$$r = \log_b u \quad \text{or, equivalently,} \quad u = b^r$$

Then

$$u^n = (b^r)^n$$

Applying a law of exponents, we obtain

$$u^n = b^{nr}$$

The logarithmic form of this equation is

$$\log_b u^n = nr$$

Substituting $\log_b u$ for $r$, we have

$$\log_b u^n = n \log_b u$$

**ILLUSTRATION 6.** Because $\log_{10} 2 = 0.3010$, it follows from Theorem 4.5.3 that

$$\log_{10} 32 = \log_{10} 2^5 \qquad \text{and} \qquad \log_{10} \sqrt[3]{2} = \log_{10} 2^{1/3}$$
$$= 5 \log_{10} 2 \qquad\qquad\qquad\qquad = \tfrac{1}{3} \log_{10} 2$$
$$= 5(0.3010) \qquad\qquad\qquad\qquad = \tfrac{1}{3}(0.3010)$$
$$= 1.5050 \qquad\qquad\qquad\qquad\quad = 0.1003 \qquad \bullet$$

**EXAMPLE 1**
Express each of the following in terms of logarithms of $x$, $y$, and $z$, where the variables represent positive numbers.

(a) $\log_b x^2 y^3 z^4$
(b) $\log_b \dfrac{x}{yz^2}$
(c) $\log_b \sqrt[5]{\dfrac{xy^2}{z^3}}$

**SOLUTION**
(a) Using Theorem 4.5.1, we have

$$\log_b x^2 y^3 z^4 = \log_b x^2 + \log_b y^3 + \log_b z^4$$

Applying Theorem 4.5.3 to each of the logarithms in the right member, we obtain

$$\log_b x^2 y^3 z^4 = 2 \log_b x + 3 \log_b y + 4 \log_b z$$

(b) From Theorem 4.5.2 it follows that

$$\log_b \frac{x}{yz^2} = \log_b x - \log_b yz^2$$

Applying Theorem 4.5.1 to the second logarithm in the right member, we have

$$\log_b \frac{x}{yz^2} = \log_b x - (\log_b y + \log_b z^2)$$
$$= \log_b x - \log_b y - 2 \log_b z$$

(c) From Theorem 4.5.3 it follows that

$$\log_b \sqrt[5]{\frac{xy^2}{z^3}} = \frac{1}{5} \log_b \frac{xy^2}{z^3}$$

Applying Theorem 4.5.2 to the right member, we obtain

$$\log_b \sqrt[5]{\frac{xy^2}{z^3}} = \frac{1}{5} (\log_b xy^2 - \log_b z^3)$$
$$= \frac{1}{5} (\log_b x + \log_b y^2 - \log_b z^3)$$
$$= \frac{1}{5} (\log_b x + 2 \log_b y - 3 \log_b z)$$
$$= \frac{1}{5} \log_b x + \frac{2}{5} \log_b y - \frac{3}{5} \log_b z$$

## EXAMPLE 2
Write each of the following expressions as a single logarithm with a coefficient of 1.

(a) $\log_b x + 2 \log_b y - 3 \log_b z$

(b) $\dfrac{1}{3}(\log_b 4 - \log_b 3 + 2 \log_b x - \log_b y)$

## SOLUTION

(a) $\log_b x + 2 \log_b y - 3 \log_b z = (\log_b x + \log_b y^2) - \log_b z^3$
$= \log_b xy^2 - \log_b z^3$
$= \log_b \dfrac{xy^2}{z^3}$

(b) $\dfrac{1}{3}(\log_b 4 - \log_b 3 + 2 \log_b x - \log_b y)$
$= \dfrac{1}{3}[(\log_b 4 + \log_b x^2) - (\log_b 3 + \log_b y)]$
$= \dfrac{1}{3}[\log_b 4x^2 - \log_b 3y]$
$= \dfrac{1}{3} \log_b \dfrac{4x^2}{3y}$
$= \log_b \sqrt[3]{\dfrac{4x^2}{3y}}$

## EXAMPLE 3
Given $\log_{10} 2 = 0.3010$, $\log_{10} 3 = 0.4771$, and $\log_{10} 7 = 0.8451$, use the properties of logarithms from Theorems 4.5.1, 4.5.2, and 4.5.3 to find the value of each of the following logarithms.

(a) $\log_{10} 5$  (b) $\log_{10} 28$
(c) $\log_{10} 2100$  (d) $\log_{10} \sqrt[3]{4.2}$

## SOLUTION
In addition to the given logarithms we can easily determine the logarithm with base 10 of any integer power of 10; for instance, $\log_{10} 10 = 1$, $\log_{10} 10^2 = 2$, $\log_{10} 10^3 = 3$, $\log_{10} 10^{-1} = -1$, and so on.

(a) $\log_{10} 5 = \log_{10} \dfrac{10}{2}$
$= \log_{10} 10 - \log_{10} 2$
$= 1 - 0.3010$
$= 0.6990$

(b) $\log_{10} 28 = \log_{10} 2^2 \cdot 7$
$= \log_{10} 2^2 + \log_{10} 7$
$= 2 \log_{10} 2 + \log_{10} 7$
$= 2(0.3010) + 0.8451$
$= 0.6020 + 0.8451$
$= 1.4471$

(c) $\log_{10} 2100 = \log_{10} 3 \cdot 7 \cdot 10^2$
$= \log_{10} 3 + \log_{10} 7 + \log_{10} 10^2$
$= 0.4771 + 0.8451 + 2$
$= 3.3222$

(d) $\log_{10} \sqrt[3]{4.2} = \log_{10} \left( \dfrac{2 \cdot 3 \cdot 7}{10} \right)^{1/3}$
$= \dfrac{1}{3}(\log_{10} 2 + \log_{10} 3 + \log_{10} 7 - \log_{10} 10)$
$= \dfrac{1}{3}(0.3010 + 0.4771 + 0.8451 - 1)$
$= \dfrac{1}{3}(0.6232)$
$= 0.2077$

## Properties of Logarithmic Functions

**EXAMPLE 4**

Use the values of $\log_{10} 2$ and $\log_{10} 7$ given in Example 3 to find the value of each of the following.

(a) $\log_{10} \dfrac{7}{2}$   (b) $\dfrac{\log_{10} 7}{\log_{10} 2}$

**SOLUTION**

(a) $\log_{10} \dfrac{7}{2} = \log_{10} 7 - \log_{10} 2$

$= 0.8451 - 0.3010$

$= 0.5441$

(b) $\dfrac{\log_{10} 7}{\log_{10} 2} = \dfrac{0.8451}{0.3010}$

$\approx 2.808$

Compare the computations in parts (a) and (b) of Example 4. In part (a) we have the logarithm of a quotient, which, upon applying Theorem 4.5.2 is the difference of two logarithms. In part (b) we have the quotient of the logarithms of two numbers. The computation is performed by dividing 0.8451 by 0.3010.

In the next example we have equations involving logarithms.

**EXAMPLE 5**

Find the solution set of each of the following equations.
(a) $\log_{10} (x + 3) = 2$
(b) $\log_2 (x + 4) - \log_2 (x - 3) = 3$
(c) $\log_3 x + \log_3 (2x - 3) = 4 - \log_3 3$

**SOLUTION**

(a) $\log_{10} (x + 3) = 2$

The exponential form of this equation is

$$x + 3 = 10^2$$

Therefore,

$$x = 100 - 3$$
$$x = 97$$

Thus the solution set is $\{97\}$.

(b) $\log_2 (x + 4) - \log_2 (x - 3) = 3$

Applying Theorem 4.5.2 to the left member, we have

$$\log_2 \dfrac{x + 4}{x - 3} = 3$$

Writing this equation in the equivalent exponential form, we have

$$\dfrac{x + 4}{x - 3} = 2^3$$

Therefore,

$$x + 4 = 8(x - 3)$$
$$x + 4 = 8x - 24$$
$$-7x = -28$$
$$x = 4$$

Thus the solution set is $\{4\}$.

(c) The given equation is equivalent to

$$\log_3 3 + \log_3 x + \log_3 (2x - 3) = 4$$

Applying Theorem 4.5.1 to the left member, we obtain

$$\log_3 3x(2x - 3) = 4$$

Writing the equation in the equivalent exponential form, we have

$$6x^2 - 9x = 3^4$$
$$6x^2 - 9x - 81 = 0$$
$$2x^2 - 3x - 27 = 0$$
$$(2x - 9)(x + 3) = 0$$
$$2x - 9 = 0 \qquad x + 3 = 0$$
$$x = \frac{9}{2} \qquad x = -3$$

When $x = -3$, neither $\log_3 x$ nor $\log_3 (2x - 3)$ exist; hence we reject the root $-3$. Therefore, the solution set is $\{\frac{9}{2}\}$.

## EXERCISES 4.5

In Exercises 1 through 18, express the given logarithm in terms of logarithms of x, y, and z, where the variables represent positive numbers.

1. $\log_b (5xy)$
2. $\log_b (3xyz)$
3. $\log_b \left(\frac{y}{z}\right)$
4. $\log_b \left(\frac{xy}{z}\right)$
5. $\log_b \left(\frac{x}{yz}\right)$
6. $\log_b (x^4 y^2)$
7. $\log_b (xy^5)$
8. $\log_b z^{1/3}$
9. $\log_b \sqrt{xy}$
10. $\log_b \sqrt[3]{yz^2}$
11. $\log_b (x^{1/3} z^3)$
12. $\log_b (x^2 y^3 z)$
13. $\log_b \left(\frac{xy^{1/2}}{z^4}\right)$
14. $\log_b \left(\frac{y^2}{x^5 z^{1/4}}\right)$
15. $\log_b \sqrt[3]{\frac{x^2}{yz^2}}$
16. $\log_b \sqrt[5]{\frac{x^3 y^4}{z^2}}$
17. $\log_b (\sqrt[3]{x^2} \sqrt{yz})$
18. $\log_b (\sqrt[4]{xy^3} \sqrt{z})$

In Exercises 19 through 24, write the given expression as a single logarithm with a coefficient of 1.

19. $4 \log_{10} x + \frac{1}{2} \log_{10} y$
20. $5 \log_{10} x + \frac{1}{2} \log_{10} y - \frac{1}{3} \log_{10} z$
21. $\frac{3}{4} \log_b x - 6 \log_b y - \frac{4}{5} \log_b z$
22. $\frac{2}{3} \log_b x - 4 \log_b y + \log_b z$
23. $\ln \pi + \ln h + 2 \ln r - \ln 3$
24. $\log_{10} 2 + \log_{10} \pi + \frac{1}{2} \log_{10} t - \frac{1}{2} \log_{10} g$

In Exercises 25 through 38, find the value of the given quantity if $\log_{10} 2 = 0.3010$, $\log_{10} 3 = 0.4771$, and $\log_{10} 7 = 0.8451$.

25. $\log_{10} 14$
26. $\log_{10} 18$
27. $\log_{10} 15$
28. $\log_{10} 42$
29. $\log_{10} 63$
30. $\log_{10} 120$
31. $\log_{10} 140$
32. $\log_{10} 0.21$
33. $\log_{10} \sqrt[3]{10.5}$
34. $\log_{10} \sqrt[3]{126}$
35. $\log_{10} \left(\frac{\sqrt[5]{49}}{36^2}\right)$
36. $\log_{10} \left(\frac{14}{\sqrt[3]{84}}\right)$
37. $\frac{\log_{10} 2}{\log_{10} 3}$
38. $\frac{\log_{10} 7}{\log_{10} 2}$

*In Exercises 39 through 44, find the solution set of the given equation.*

**39.** $\log_{10} x + 3\log_{10} 2 = 3$
**40.** $\log_{10} x + \log_{10}(x + 15) = 2$
**41.** $\log_3(x + 6) - \log_3(x - 2) = 2$
**42.** $\log_2(11 - x) = \log_2(x + 1) + 3$
**43.** $\log_2(x + 1) + \log_2(3x - 5) = \log_2(5x - 3) + 2$
**44.** $\log_3(2x - 1) - \log_3(5x + 2) = \log_3(x - 2) - 2$

## 4.6 Common Logarithms

One application of logarithms is to facilitate certain numerical computations. The most convenient logarithms to use for this purpose are those with base 10.

The function values of the logarithmic function with base 10 are called *common logarithms*.

**4.6.1 DEFINITION**

The common logarithm of a positive number $x$ can be written as $\log_{10} x$. However, when writing common logarithms it is customary to omit the subscript 10. Thus, when we write $\log x$, it is understood to represent the same number as $\log_{10} x$, and the function *log* denotes the logarithmic function with base 10. Hence

$$\log x = y \quad \text{is equivalent to} \quad 10^y = x$$

**ILLUSTRATION 1**

$$\begin{aligned}
\log 10 &= 1 &\quad \text{because } 10^1 &= 10 \\
\log 100 &= 2 &\quad \text{because } 10^2 &= 100 \\
\log 1000 &= 3 &\quad \text{because } 10^3 &= 1000 \\
\log 10{,}000 &= 4 &\quad \text{because } 10^4 &= 10{,}000
\end{aligned}$$

and so on. Furthermore,

$$\begin{aligned}
\log 1 &= 0 &\quad \text{because } 10^0 &= 1 \\
\log 0.1 &= -1 &\quad \text{because } 10^{-1} &= 0.1 \\
\log 0.01 &= -2 &\quad \text{because } 10^{-2} &= 0.01 \\
\log 0.001 &= -3 &\quad \text{because } 10^{-3} &= 0.001 \\
\log 0.0001 &= -4 &\quad \text{because } 10^{-4} &= 0.0001
\end{aligned}$$

and so on. ●

Common logarithms are useful because any positive number $x$ can be written in the form

$$x = a \cdot 10^c \quad \text{where } 1 \leq a < 10 \text{ and } c \text{ is an integer} \tag{1}$$

When a number is expressed in this form it is said to be written in *scientific notation*.

**Illustration 2.** Each of the following numbers is written in scientific notation.

$$582 = (5.82)10^2$$
$$97{,}136 = (9.7136)10^4$$
$$485{,}000 = (4.85)10^5$$
$$0.627 = (6.27)10^{-1}$$
$$0.003916 = (3.916)10^{-3}$$
$$2.04 = (2.04)10^0$$

●

To write a number in scientific notation, the first factor is obtained by placing a decimal point after the first left-hand nonzero digit. The second factor is a power of 10, and the exponent is obtained by counting the number of digits that must be passed over to move from the new position of the decimal point to the original position of the decimal point; if the movement of the decimal point from the new position to the original position is to the right, then the exponent is positive; if the movement is to the left, then the exponent is negative. You should verify this rule by applying it to the numbers in Illustration 2.

If a number is written in scientific notation, it can be written in standard form by moving the decimal point in the first factor the number of places indicated by the exponent of the power of 10; the decimal point is moved to the right if the exponent is positive and it is moved to the left if the exponent is negative. This rule is applied in the next illustration.

**Illustration 3**

$$(3.659)10^4 = 36{,}590$$
$$(8.007)10^2 = 800.7$$
$$(3.92)10^{-3} = 0.00392$$
$$(4.018)10^{-1} = 0.4018$$

●

From equation (1) it follows that if $x$ is any positive number

$$\log x = \log(a \cdot 10^c)$$

where $1 \leq a < 10$ and $c$ is an integer. Applying Theorem 4.5.1 to the right member of this equation, we have

$$\log x = \log a + \log 10^c$$

or, equivalently,

$$\log x = \log a + c \qquad \text{where } 1 \leq a < 10 \text{ and } c \text{ is an integer} \qquad (2)$$

From equation (2) it follows that the common logarithm of any positive number $x$ can be written as the sum of $c$, an integer and $\log a$, the common logarithm of a number between 1 and 10. The integer $c$ is called the *charac-*

*teristic* of the logarithm, and the number $\log a$ ($1 \leq a < 10$) is called the *mantissa* of the logarithm.

The characteristic of the common logarithm of a number is the same as the exponent of 10 when the number is written in scientific notation; hence the characteristic is determined by the position of the decimal point in the number.

### ILLUSTRATION 4

(a) The characteristic of $\log 281$ is 2 because $281 = (2.81)10^2$.
(b) The characteristic of $\log 0.00281$ is $-3$ because $0.00281 = (2.81)10^{-3}$.
(c) The characteristic of $\log 2.81$ is 0 because $2.81 = (2.81)10^0$. •

Because log is an increasing function and

$$1 \leq a < 10$$

it follows that

$$\log 1 \leq \log a < \log 10 \qquad (3)$$

But $\log 1 = 0$ and $\log 10 = 1$. Hence inequality (3) is equivalent to

$$0 \leq \log a < 1$$

Therefore, the mantissa of a logarithm is a nonnegative number less than 1.

Table 2 in the Appendix gives approximations, to four decimal places, of common logarithms of numbers between 1.00 and 9.99. The entries in this table are computed by methods using advanced mathematics. A portion of Table 2 is given in Table 4.6.1.

Table 4.6.1

| $N$ | 0 | 1 | 2 | 3 | 4 | 5 | 6 | 7 | 8 | 9 |
|---|---|---|---|---|---|---|---|---|---|---|
| 5.8 | .7634 | .7642 | .7649 | .7657 | .7664 | .7672 | .7679 | .7686 | .7694 | .7701 |
| 5.9 | .7709 | .7716 | .7723 | .7731 | .7738 | .7745 | .7752 | .7760 | .7767 | .7774 |
| 6.0 | .7782 | .7789 | .7796 | .7803 | .7810 | .7818 | .7825 | .7832 | .7839 | .7846 |
| 6.1 | .7853 | .7860 | .7868 | .7875 | .7882 | .7889 | .7896 | .7903 | .7910 | .7917 |
| 6.2 | .7924 | .7931 | .7938 | .7945 | .7952 | .7959 | .7966 | .7973 | .7980 | .7987 |
| 6.3 | .7993 | .8000 | .8007 | .8014 | .8021 | .8028 | .8035 | .8041 | .8048 | .8055 |
| 6.4 | .8062 | .8069 | .8075 | .8082 | .8089 | .8096 | .8102 | .8109 | .8116 | .8122 |
| 6.5 | .8129 | .8136 | .8142 | .8149 | .8156 | .8162 | .8169 | .8176 | .8182 | .8189 |
| 6.6 | .8195 | .8202 | .8209 | .8215 | .8222 | .8228 | .8235 | .8241 | .8248 | .8254 |
| 6.7 | .8261 | .8267 | .8274 | .8280 | .8287 | .8293 | .8299 | .8306 | .8312 | .8319 |
| 6.8 | .8325 | .8331 | .8338 | .8344 | .8351 | .8357 | .8363 | .8370 | .8376 | .8382 |
| 6.9 | .8388 | .8395 | .8401 | .8407 | .8414 | .8420 | .8426 | .8432 | .8439 | .8445 |
| 7.0 | .8451 | .8457 | .8463 | .8470 | .8476 | .8482 | .8488 | .8494 | .8500 | .8506 |
| 7.1 | .8513 | .8519 | .8525 | .8531 | .8537 | .8543 | .8549 | .8555 | .8561 | .8567 |

When using the table to find log $N$ ($1 \leq N < 10$), the first two digits of the numeral for $N$ are located in the first column (headed $N$) and the third digit is located in the top row (containing $N$). Then log $N$ is the number appearing in the same row as the first two digits of $N$ and in the same column as the third digit of $N$.

**ILLUSTRATION 5.** We use Table 4.6.1 to find a value for log 6.23. It is the number appearing in the row containing 6.2 and in the column with heading 3. Hence

$$\log 6.23 = 0.7945$$

In a similar way we find the following logarithms.

$$\log 5.98 = 0.7767$$
$$\log 6.60 = 0.8195$$
$$\log 6.00 = 0.7782$$

and so on. •

**EXAMPLE 1**

Find values, to four decimal places, of each of the following logarithms.

(a) log 62.3
(b) log 623
(c) log 62,300

**SOLUTION**

We first write each of the numbers in scientific notation and then apply Theorem 4.5.4. The mantissa of the logarithm is found in the table.

(a) $\log 62.3 = \log (6.23 \cdot 10^1)$
$= \log 6.23 + \log 10^1$
$= 0.7945 + 1$
$= 1.7945$

(b) $\log 623 = \log (6.23 \cdot 10^2)$
$= \log 6.23 + \log 10^2$
$= 0.7945 + 2$
$= 2.7945$

(c) $\log 62{,}300 = \log (6.23 \cdot 10^4)$
$= \log 6.23 + \log 10^4$
$= 0.7945 + 4$
$= 4.7945$

Recall that if $b > 1$, and $0 < x < 1$, then $\log_b x$ is negative (refer to Figure 4.4.1). Hence $\log x < 0$ if $0 < x < 1$. However, even though log $x$ is negative, it can still be written in the form given in equation (2), that is, as the sum of an integer $c$ (the characteristic) and the logarithm of a number between 1 and 10 (the mantissa, which is a nonnegative number less than 1). For instance,

$$\log 0.00623 = \log (6.23 \cdot 10^{-3})$$
$$= \log 6.23 + \log 10^{-3}$$
$$= 0.7945 + (-3)$$

We do not add the numbers 0.7945 and ($-3$) in the third step of computing the value for log 0.00623 because doing this gives log $0.00623 = -2.2055$ (note that $-2.2055 = -2 - 0.2055$, which is not in the form of the sum of an integer and a nonnegative mantissa). However, there is a convenient way of writing a negative logarithm so that the nonnegative mantissa is apparent. If we write $7 - 10$ in place of $-3$, we have

$$\log 0.00623 = 0.7945 + (7 - 10)$$
$$= 7.7945 - 10$$

Observe that we could just as well write

$$\log 0.00623 = 0.7945 + (1 - 4)$$
$$= 1.7945 - 4$$

and

$$\log 0.00623 = 0.7945 + (6 - 9)$$
$$= 6.7945 - 9$$

However, the more conventional notation is $7.7945 - 10$. In a similar way,

$$\log 0.623 = \log (6.23 \cdot 10^{-1})$$
$$= \log 6.23 + \log 10^{-1}$$
$$= 0.7945 + (9 - 10)$$
$$= 9.7945 - 10$$

$$\log 0.000623 = \log (6.23 \cdot 10^{-4})$$
$$= \log 6.23 + \log 10^{-4}$$
$$= 0.7945 + (6 - 10)$$
$$= 6.7945 - 10$$

The next illustration shows how $x$ can be found if $\log x$ is given.

### ILLUSTRATION 6

(a) We wish to find $x$ if $\log x = 1.7731$. The sequence of digits in $x$ is determined from the mantissa 0.7731. Refer to the *body* of Table 4.6.1 to find that the mantissa 0.7731 is associated with 5.93; that is,

$$\log 5.93 = 0.7731$$

Because the characteristic of $\log x$ is 1,

$$\log x = \log (5.93 \cdot 10^1)$$
$$= \log 59.3$$

Because $\log M = \log N$ implies $M = N$, it follows that $x = 59.3$.

(b) If $\log x = 8.7731 - 10$, then because $\log 5.93 = 0.7731$ and the characteristic of $\log x$ is $-2$, we have

$$\log x = \log (5.93 \cdot 10^{-2})$$
$$= \log 0.0593$$

Therefore, $x = 0.0593$

**4.6.2 DEFINITION** The *antilogarithm* of a number $y$ (written *antilog y*) is the number $x$ such that $\log x = y$; that is,

$$\text{antilog } y = x \quad \text{if and only if} \quad \log x = y$$

In Illustration 6 we found that if $\log x = 1.7731$, then $x = 59.3$. Therefore, from Definition 4.6.2, it follows that

$$\text{antilog } 1.7731 = 59.3$$

Also, because $\log x = 8.7731 - 10$ implies that $x = 0.0593$, then

$$\text{antilog } (8.7731 - 10) = 0.0593$$

EXAMPLE 2
Find each of the following antilogarithms.

(a) antilog 3.3892
(b) antilog 0.9741
(c) antilog (5.6946 − 10)

SOLUTION
We use Table 2 in the Appendix.

(a) We locate the mantissa 0.3892 in the body of the table and we find that it is associated with the number 2.45. Hence

$$\text{antilog } 0.3892 = 2.45$$

Therefore,

$$\text{antilog } 3.3892 = 2.45 \cdot 10^3$$
$$= 2450$$

(b) From the table we find that the mantissa 0.9741 is associated with the number 9.42. Hence

$$\text{antilog } 0.9741 = 9.42$$

(c) Because the mantissa 0.6946 is associated with the number 4.95,

$$\text{antilog } 0.6946 = 4.95$$

Hence

$$\text{antilog } (5.6946 - 10) = 4.95 \cdot 10^{-5}$$
$$= 0.0000495$$

In numerical computations we are often involved with approximations whose accuracy is indicated by the "significant digits" in the numerals representing the approximations. By the *significant digits* of a numeral we mean all the digits beginning with the first nonzero digit on the left and ending with the last digit on the right, unless otherwise stated. For instance, the numeral 325 has three significant digits, the numeral 0.005271 has four significant digits, and the numeral 864.00 has five significant digits (the reason for writing the two zeros after the decimal point is to indicate that the digits are significant).

Scientific notation affords a convenient way of indicating the significant digits in a numeral. If we write 83,200 as $(8.32)10^4$, then there are three significant digits. However, if we write $(8.320)10^4$, then there are four significant digits. Similarly, if we write $(8.3200)10^4$, there are five significant digits.

From Table 2 we can determine the mantissa of the logarithm of any number having three significant digits. The mantissas in the table have been "rounded off to 4 significant digits." A numeral is said to be *rounded off to k significant digits* if it is replaced by the number, having $k$ significant digits, to which it is closest. For instance, the numeral 0.52368 is rounded off to four significant digits as 0.5237, while the numeral 0.78142 is rounded off to four significant digits as 0.7814. To round off to four significant digits a five-digit numeral whose fifth digit is 5, we adopt the following convention: if the fifth digit is 5 and the fourth digit is even, we round off to the fourth digit; if the fifth digit is 5 and the fourth digit is odd, we increase the fourth digit by one. Hence we round off 0.26185 to 0.2618, and 0.39235 is rounded off to 0.3924. A similar convention is used to round off to any number of significant digits.

An approximation of the logarithm of a number represented by a numeral having four significant digits can be found from Table 2 by using the method of *linear interpolation*. We demonstrate the method by the following illustration.

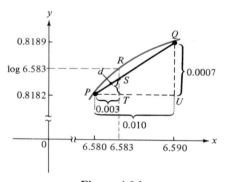

**Figure 4.6.1**

**ILLUSTRATION 7.** We wish to find log 6.583 to four decimal places by using Table 2. Because log is an increasing function,

$$\log 6.580 < \log 6.583 < \log 6.590$$

From Table 2 we find $\log 6.580 = 0.8182$ and $\log 6.590 = 0.8189$. Figure 4.6.1 shows a portion of the graph of log from the point $P$ (6.580, 0.8182) to the point $Q$(6.590, 0.8189) where the units on the axes are magnified and the portion of the graph is distorted in order to demonstrate the procedure. In a more accurate figure, the line segment from $P$ to $Q$ would be much closer to the graph than it is in Figure 4.6.1.

The point $R$ on the graph has an abscissa ($x$ coordinate) of 6.583, and the ordinate of $R$ is the exact value of log 6.583. An approximation of the ordinate of $R$ is the ordinate of the point $S$ on the line segment from $P$ to $Q$ that has an abscissa of 6.583. The ordinate of $S$ is represented by $0.8182 + d$, where $d$ units is the length of the line segment $TS$ shown in Figure 4.6.1. To compute $d$, we use a property of similar triangles which states that the lengths of corresponding sides are proportional. Because triangle $PTS$ is similar to triangle $PUQ$, we have

$$\frac{d}{0.0007} = \frac{0.003}{0.010}$$

Therefore,
$$d = \frac{3}{10}(0.0007)$$
$$= 0.0002$$

Thus
$$\log 6.583 = 0.8182 + 0.0002$$
$$= 0.8184$$

Henceforth, when performing the calculations involved in linear interpolation, we shall not draw a portion of the graph and show the similar triangles as we did in Illustration 7. Instead, we arrange the computation as shown in the following example.

EXAMPLE 3

Find
$$\log 30.46$$

SOLUTION

We use Table 2 of the Appendix to find the mantissas of log 30.40 and log 30.50.

$$0.10 \left\{ 0.06 \left\{ \begin{array}{l} \log 30.40 = 1.4829 \\ \log 30.46 = ? \\ \log 30.50 = 1.4843 \end{array} \right\} d \right\} 0.0014$$

Then
$$\frac{d}{0.0014} = \frac{0.06}{0.10}$$

Therefore,
$$d = \frac{6}{10}(0.0014)$$
$$= 0.0008$$

Hence
$$\log 30.46 = 1.4829 + 0.0008$$
$$= 1.4837$$

In the next example we use linear interpolation to find an antilogarithm.

EXAMPLE 4

Find
$$\text{antilog } 0.7351$$

SOLUTION

Let antilog $0.7351 = x$. Then $\log x = 0.7351$. The mantissa 0.7351 does not appear in the body of Table 2 of the Appendix. However, the mantissas 0.7348 and 0.7356 do appear and they are associated with the numbers 5.43 and 5.44, respectively.

$$0.010 \left\{ n \left\{ \begin{array}{l} \log 5.430 = 0.7348 \\ \log x = 0.7351 \\ \log 5.440 = 0.7356 \end{array} \right\} 0.0003 \right\} 0.0008$$

Thus
$$\frac{n}{0.010} = \frac{0.0003}{0.0008}$$

Therefore,
$$n = \frac{3}{8}(0.010)$$
$$= 0.004$$

Hence
$$x = 5.430 + 0.004$$
$$= 5.434$$

One of the reasons for considering common logarithms is their use in simplifying numerical computations. Because of the present extensive use of various types of automatic calculators, the importance of this application of logarithms is not as great as it was in the past. However, there are certain operations, such as determining approximations to roots of numbers and calculating the value of exponentials like $2^{1.73}$ (see Example 5(b)), that you should be able to perform by using logarithms.

**EXAMPLE 5**

Compute each of the following by logarithms.

(a) $(5.38)^6$    (b) $2^{1.73}$

**SOLUTION**

In each part we apply Theorem 4.5.3.

(a) $\log (5.38)^6 = 6 \log 5.38$
$= 6(0.7308)$
$= 4.3848$

Hence
$(5.38)^6 = (2.43)10^4$

(b) $\log 2^{1.73} = (1.73) \log 2$
$= (1.73)(0.3010)$
$= 0.5207$

Hence
$2^{1.73} = 3.32$

In part (a) of Example 5 the answer is written in scientific notation to indicate that there are three significant digits.

**EXAMPLE 6**

Compute each of the following by logarithms.

(a) $\sqrt[4]{72.65}$    (b) $\sqrt[3]{0.0349}$

**SOLUTION**

In each part we apply Theorem 4.5.3.

(a) $\log \sqrt[4]{72.65} = \log (72.65)^{1/4}$
$= \frac{1}{4} \log 72.65$
$= \frac{1}{4}(1.8612)$
$= 0.4653$

Therefore,
$\sqrt[4]{72.65} = 2.919$

(b) $\log \sqrt[3]{0.0349} = \log (0.0349)^{1/3}$
$= \frac{1}{3} \log 0.0349$
$= \frac{1}{3}(28.5428 - 30)$
$= 9.5143 - 10$

Therefore,
$\sqrt[3]{0.0349} = 0.327$

In part (b) of Example 6, log 0.0349 is written as $28.5428 - 30$ instead of as $8.5428 - 10$, so that the negative part of the characteristic is exactly divisible by 3, and we avoid decimals (remember, by definition, the characteristic must be an integer).

**EXAMPLE 7**
Find the value of
$$2^{\sqrt{3}}$$
to two decimal places.

**SOLUTION**
Let
$$x = 2^{\sqrt{3}}$$
Then
$$\log x = \sqrt{3} \log 2$$
$$= 1.732(0.3010)$$
$$= 0.5213$$
Therefore,
$$x = \text{antilog } 0.5213$$
$$= 3.322$$

Thus, to two decimal places, $2^{\sqrt{3}} = 3.32$.

Logarithms are often useful in the computation involved in solving investment problems pertaining to "compound interest." Interest is called *compound interest* if, during the term of a loan or investment, the interest earned each period is added to the principal and then earns interest itself. The rate of interest is usually given as an annual rate, but often the interest is computed and then added to the principal more frequently than once a year. If the interest is compounded $m$ times per year, then the annual rate must be divided by $m$ to determine the interest for each period. For instance, if $200 is deposited in a savings account that pays 8 per cent interest compounded quarterly, then the number of dollars in the account at the end of the first 3-month period will be

$$200 + 200\left(\frac{0.08}{4}\right) = 200(1 + 0.02)$$
$$= 200(1.02)$$

The number of dollars in the account at the end of the second 3-month period will be

$$200(1.02) + 200(1.02)(0.02) = 200(1.02)(1 + 0.02)$$
$$= 200(1.02)^2$$

and so forth. More generally, if $P$ dollars is invested at an interest rate of $100i$ per cent, compounded $m$ times per year and if $A_n$ is the number of dollars in the amount of the investment at the end of $n$ interest periods, then

$$A_n = P\left(1 + \frac{i}{m}\right)^n \tag{4}$$

## EXAMPLE 8

Suppose that $1000 is invested in a trust fund on the first New Year's Day after a child is born. If the interest is 8 per cent compounded semiannually, how much will be in the trust fund on New Year's Day when the child is 21 years old?

## SOLUTION

We use formula (4) where $P = 1000$, $i = 0.8$, $m = 2$, and $n = 42$. We have

$$A_{42} = 1000\left(1 + \frac{0.08}{2}\right)^{42}$$
$$= 1000(1.04)^{42}$$

Then

$$\log A_{42} = \log 1000 + \log (1.04)^{42}$$
$$= 3 + 42 \log 1.04$$
$$= 3 + 42(0.0170)$$
$$= 3.7140$$

Hence

$$A_{42} = 5176$$

Therefore, to the nearest hundred dollars there will be $5200 in the trust fund on New Year's Day when the child is 21 years old.

In Example 8 a four-place table of logarithms does not give an accuracy of four significant digits. The reason for this situation is that log 1.04 is given only to three significant digits and the third digit has been rounded off. Furthermore, log 1.04 is multiplied by 42, which affects the accuracy of the third digit. In fact, the correct value of $A_{42}$ to four significant digits is 5193, which has only two digits that agree with the result obtained by using logarithms.

## EXERCISES 4.6

In Exercises 1 through 10, write the given number in scientific notation.

1. 52.60
2. 43851
3. 0.0061
4. 0.276
5. 172,000 (three significant digits)
6. 172,000 (four significant digits)
7. 0.03960
8. 0.00006405
9. 0.0000080022
10. 0.0001030

In Exercises 11 through 20, find the given logarithm.

11. log 364
12. log 4.27
13. log 51.8
14. log 395
15. log 0.27
16. log 0.0041
17. log 0.0913
18. log 62,400
19. log 348,000
20. log 0.256

In Exercises 21 through 30, find the given antilogarithm.

21. antilog 2.4014
22. antilog 1.6590
23. antilog 0.9258
24. antilog 3.7767
25. antilog 8.4742 − 10
26. antilog 9.8014 − 10
27. antilog 6.2900 − 10
28. antilog 7.9004 − 10
29. antilog 5.7642
30. antilog 6.8500

In Exercises 31 through 40, find the given logarithm.

**31.** log 2.754
**32.** log 68.37
**33.** log 4589
**34.** log 0.1621
**35.** log 0.009262
**36.** log 0.08596
**37.** log 0.00003333
**38.** log 534.5
**39.** log 779,800
**40.** log 20,040

In Exercises 41 through 50, find the given antilogarithm.

**41.** antilog 0.5471
**42.** antilog 2.8772
**43.** antilog 3.9690
**44.** antilog 4.3228
**45.** antilog 9.7089 − 10
**46.** antilog 8.9670 − 10
**47.** antilog 7.1601 − 10
**48.** antilog 5.5608 − 10
**49.** antilog 6.8690
**50.** antilog 1.0770

In Exercises 51 through 60, use logarithms to perform the computation.

**51.** $(2.36)^5$
**52.** $(78.1)^4$
**53.** $(0.06100)^3$
**54.** $(1.099)^5$
**55.** $\sqrt[5]{76.98}$
**56.** $\sqrt[7]{2.677}$
**57.** $\sqrt[3]{0.003379}$
**58.** $\sqrt[4]{0.09700}$
**59.** $\sqrt{0.05100}$
**60.** $(0.4873)^{2/3}$

In Exercises 61 through 64, find the value to two decimal places.

**61.** $3^{\sqrt{2}}$
**62.** $4^{\sqrt{5}}$
**63.** $\sqrt{5}^{\sqrt{3}}$
**64.** $\sqrt{3}^{\pi}$

**65.** If $T$ seconds is the time for one complete oscillation of a simple pendulum of length $l$ feet, then

$$T = 2\pi \sqrt{\frac{l}{g}}$$

where $g = 32.16$. Find the time for a complete oscillation of a pendulum 2.18 feet long.

**66.** Use the formula of Exercise 65 to compute the value of $g$ at a point on the surface of the earth where a pendulum 3.002 feet long requires 1.923 seconds for a complete oscillation.

**67.** Use the formula of Exercise 65 to find the length of a simple pendulum whose time for one complete oscillation is 2 seconds.

**68.** On his twenty-fifth birthday, a man inherited $5000. If he invested this amount at 8 per cent, compounded annually, how much would he receive when he retires at the age of 65?

**69.** A man borrowed $10,000 at 9 per cent with the understanding that interest was to be paid monthly. However, the borrower did not make the monthly interest payments and so the principal with interest at 9 per cent compounded monthly was due at the end of the year. What was the amount due at the end of the year?

## 4.7 Exponential Equations

An *exponential equation* is one in which a variable occurs in an exponent. Sometimes an exponential equation can be solved by considering the equivalent equation obtained by equating the common logarithms of the two members and then solving the resulting equation.

**ILLUSTRATION 1.** In order to solve the equation

$$3^x = 16 \tag{1}$$

we equate the common logarithms of the two members, and we have

$$\log 3^x = \log 16$$
$$x \log 3 = \log 16$$
$$x = \frac{\log 16}{\log 3} \qquad (2)$$

We use Table 2 of the Appendix to find log 16 and log 3, and we obtain

$$x = \frac{1.2041}{0.4771}$$

This quotient can be computed by logarithms. Because in Table 2 we can find only mantissas of logarithms of numbers represented by four-digit numerals, we round off the numerator to four significant digits. The computation by logarithms is as follows.

$$\log 1.204 = 10.0806 - 10$$
$$\log 0.4771 = \underline{9.6786 - 10}\,(-)$$
$$\log x = 0.4020$$
$$x = 2.524$$

Therefore, the solution set of equation (1) is $\{2.524\}$. Note that 2.524 is an approximation to three decimal places of the value of $x$. The exact value of $x$ is given by equation (2), and the solution set can be expressed in logarithmic notation as $\left\{\dfrac{\log 16}{\log 3}\right\}$. •

**EXAMPLE 1**
Find the solution set of the equation
$$5^{3x-1} = 0.08$$

**SOLUTION**
Equating the common logarithms of the two members of the given equation, we have

$$\log 5^{3x-1} = \log 0.08$$
$$(3x - 1) \log 5 = \log 0.08$$
$$3x \log 5 - \log 5 = \log 0.08$$
$$3x \log 5 = \log 0.08 + \log 5$$
$$x = \frac{\log 0.08 + \log 5}{3 \log 5}$$

From Table 2 of the Appendix we find $\log 0.08 = 8.9031 - 10$ and $\log 5 = 0.6990$. Therefore,

$$x = \frac{8.9031 - 10 + 0.6990}{3(0.6990)}$$
$$= \frac{-0.3979}{2.0970}$$
$$= -0.1897$$

Thus the solution set is $\{-0.1897\}$.

The logarithm of a number to any base can be found by solving an exponential equation. The next example shows the method.

**EXAMPLE 2**
Find the value of $\log_4 19$.

**SOLUTION**
Let
$$y = \log_4 19$$
Writing this equation in exponential form, we have
$$4^y = 19$$
Therefore,
$$\log 4^y = \log 19$$
$$y \log 4 = \log 19$$
$$y = \frac{\log 19}{\log 4}$$
Hence
$$y = \frac{1.2788}{0.6021}$$
$$= 2.124$$
Thus to four significant digits, $\log_4 19 = 2.124$.

The procedure applied in Example 2 can be used to obtain a formula relating $\log_a x$ and $\log_b x$, that is, logarithms with different bases of a given number. Let
$$y = \log_a x$$
Writing this equation in exponential form, we have
$$a^y = x$$
Equating the logarithms, with base $b$, of each member of this equation, we have
$$\log_b a^y = \log_b x$$
$$y \log_b a = \log_b x$$
$$y = \frac{\log_b x}{\log_b a}$$
Replacing $y$ by $\log_a x$, we have
$$\log_a x = \frac{\log_b x}{\log_b a} \tag{3}$$

If we are given a table of logarithms with base $b$, we can form a table of logarithms with base $a$ by using equation (3); simply divide each entry in the given table by $\log_b a$.

If in equation (3), $a = e$ and $b = 10$, we have

$$\log_e x = \frac{\log_{10} x}{\log_{10} e} \qquad (4)$$

Using the notation ln instead of $\log_e$ and log instead of $\log_{10}$, equation (4) can be written as

$$\ln x = \frac{\log x}{\log e} \qquad (5)$$

The value of $e$ to four significant digits is 2.718 and $\log 2.718 = 0.4343$. Therefore, if in equation (5) $\log e$ is replaced by 0.4343 (and we retain the symbol $=$ instead of using $\approx$), we have

$$\ln x = \frac{\log x}{0.4343} \qquad (6)$$

Because $\frac{1}{0.4343} = 2.303$, equation (6) is equivalent to

$$\ln x = 2.303 \log x \qquad (7)$$

Equation (7) gives a formula for computing the natural logarithm of a number if a table of common logarithms is available.

### ILLUSTRATION 2

(a) Using formula (7), we have

$$\ln 12.53 = (2.303) \log 12.53$$
$$= (2.303)(1.0980)$$
$$= 2.529$$

(b) The value of ln 12.53 can also be found by using the method of Example 2. Let

$$y = \ln 12.53$$

Writing this equation in exponential form, we have

$$e^y = 12.53$$

Equating the common logarithms of each member of this equation, we have

$$\log e^y = \log 12.53$$
$$y \log e = \log 12.53$$
$$y = \frac{\log 12.53}{\log e}$$

Therefore,
$$y = \frac{1.0980}{0.4343}$$

Hence
$$\ln 12.53 = 2.529$$

If both members of equation (6) are multiplied by 0.4343, we obtain

$$\log x = 0.4343 \ln x \qquad (8)$$

Equation (8) can be used to find the common logarithm of a number if the natural logarithm is known.

The following example shows an application of exponential equations to the field of biology.

**EXAMPLE 3**

In Example 3 of Section 4.3 we obtained the equation

$$A = 1000e^{0.06t} \qquad (9)$$

where $A$ is the number of bacteria present in a certain culture at $t$ minutes, when there are 1000 bacteria present initially. Determine how many minutes elapse until there are 1350 bacteria present in the culture.

**SOLUTION**

Let $T$ represent the number of minutes that elapse until there are 1350 bacteria present. Then, in equation (9), we substitute 1350 for $A$ and $T$ for $t$, and we have

$$1350 = 1000e^{0.06T}$$

or, equivalently,

$$1.35 = e^{0.06T} \qquad (10)$$

Because 1.35 does not appear in Table 4 of natural logarithms, we equate the common logarithms of each member of equation (10), and we have

$$\log 1.35 = \log e^{0.06T}$$
$$= 0.06T \log e$$

Solving this equation for $T$, we obtain

$$T = \frac{\log 1.35}{0.06 \log e}$$

Because $\log 1.35 = 0.1303$ and $\log e = 0.4343$, we have

$$T = \frac{0.1303}{0.06(0.4343)}$$
$$= 5.000$$

Therefore, 5 minutes elapse until there are 1350 bacteria present.

## Example 4

If $100 is deposited into a savings account that pays 6 per cent interest compounded semiannually, and no withdrawals or additional deposits are made, how long will it take until there is $150 on deposit?

## Solution

We use formula (4) of Section 4.6, which is

$$A_n = P\left(1 + \frac{i}{m}\right)^n$$

where $A_n$ dollars is the amount at the end of $n$ interest periods of an investment of $P$ dollars at an interest rate of $100i$ per cent compounded $m$ times per year. In this problem, $A_n = 150$, $P = 100$, $i = 0.06$, and $m = 2$. We wish to find $n$. We have then

$$150 = 100\left(1 + \frac{0.06}{2}\right)^n$$

$$1.5 = (1.03)^n$$

Thus

$$\log 1.5 = \log (1.03)^n$$
$$\log 1.5 = n \log 1.03$$
$$n = \frac{\log 1.5}{\log 1.03}$$

Because $\log 1.5 = 0.1761$ and $\log 1.03 = 0.0128$, we have

$$n = \frac{0.1761}{0.0128}$$
$$= 13.76$$

Because $n$ is the number of interest periods, and interest is compounded semiannually, the number of years is $\frac{1}{2}(13.76) = 6.88$. Therefore, it will take seven years until there is $150 on deposit.

## EXERCISES 4.7

*In Exercises 1 through 12, find the solution set of the given equation.*

1. $4^x = 7$
2. $3^x = 25$
3. $5^{2x} = 4$
4. $100^x = 65$
5. $3^{2+x} = 5^x$
6. $10^{3x-2} = 37$
7. $3^{x+1} = 4^{x-1}$
8. $3^{2x+1} = 5^{3x-1}$
9. $(1.02)^x = 1.892$
10. $(1.04)^x = 0.932$
11. $e^{3x} = 21$
12. $10^{3x} = 57$

*In Exercises 13 through 20, find the value of the given logarithm to four significant digits.*

13. $\log_3 12$
14. $\log_5 200$
15. $\log_2 18$
16. $\log_6 54$
17. $\ln 155$
18. $\ln 28$
19. $\log_{100} 75$
20. $\log_{20} 100$

21. How long will it take $100 to triple itself if it is drawing interest at 6 per cent compounded semiannually?

22. The half-life of radium is 1690 years; that is, half of a given amount of radium will decay in 1690 years. Suppose that $A$ milligrams is the amount of radium present $t$ years from now and

$$A = A_0 e^{kt}$$

where $A_0$ milligrams is the amount of radium present now and $k$ is a constant. Find $k$.

23. It is determined statistically that the population of a certain city $t$ years from now will be $P$, where

$$P = 40{,}000 e^{kt}$$

and $k$ is a constant. If the population increases from 40,000 to 60,000 in 40 years, when will the population be 80,000?

24. If $A$ grams of a radioactive substance are present after $t$ seconds, then

$$A = 100 e^{-0.3t}$$

How long will it be until only 50 grams of the substance are present?

25. One kind of United States government bond is sold at $74 to be redeemed 12 years later at a maturity value of $100. What is the rate of interest earned on the investment under the assumption that interest is compounded annually?

26. In a certain speculative investment a piece of real estate was purchased three years ago for $2000 and sold today for $10,000. What is the rate of interest, compounded monthly, that has been obtained?

## REVIEW EXERCISES (CHAPTER 4)

In Exercises 1 through 4, find the numerical value in simplest form of the given expression.

1. $(-64)^{2/3}$

2. $\left(\dfrac{1}{16}\right)^{-3/4}$

3. $(1^0 + 2^0 + 3^0 + 4^0)^{-5/2}$

4. $\dfrac{2^{-1} + 3^{-2}}{2^{-2}}$

In Exercises 5 through 8, simplify the given expression. Express fractions in lowest terms with positive exponents only. All variables represent positive numbers.

5. $\dfrac{x^{-2} y^3}{x^4 y^{-5}}$

6. $\left(\dfrac{a^{4/3} b^{-3}}{a^{-1} b^{1/9}}\right)^{-3}$

7. $\dfrac{x^{-1} - y^{-1}}{y^{-1}}$

8. $(s + t)^{-1}(s^{-2} - t^{-2})$

In Exercises 9 and 10, simplify the given expression. Each variable can be any real number.

9. $(4x^2 y^4 z^6)^{1/2}$

10. $[(a^4 + 1)^2 (b - 1)^2]^{1/2}$

In Exercises 11 through 18, simplify the given expression. All variables represent positive numbers.

11. $2\sqrt{75} - 3\sqrt{12}$

12. $\sqrt[3]{\dfrac{3x}{4}} + \sqrt[3]{\dfrac{2x}{9}} + \sqrt[3]{\dfrac{x}{36}}$

13. $(\sqrt[3]{4} + \sqrt[3]{2})(\sqrt[3]{4} - \sqrt[3]{2})$

14. $(2\sqrt{5} - 3\sqrt{2})(5\sqrt{5} + 2\sqrt{2})$

15. $(\sqrt{3x} - 3\sqrt{xy})(2\sqrt{3x} + 4\sqrt{xy})$

16. $\dfrac{3 + \sqrt{6}}{2 + \sqrt{6}}$

17. $\dfrac{x - y}{\sqrt{x} - \sqrt{y}}$

18. $\dfrac{4}{\sqrt{w + 2} - \sqrt{w}}$

*In Exercises 19 and 20, draw a sketch of the graph of the given exponential function.*

19. $f = \{(x, y) \mid y = 6^x\}$

20. $g = \{(x, y) \mid y = 5^{-x}\}$

*In Exercises 21 through 26, solve the equation for y, x, or b.*

21. $\log_5 x = 4$
22. $\log_b 4 = \tfrac{1}{3}$
23. $\log_8 16 = y$
24. $\log_{27} 81 = y$
25. $\log_b 256 = \tfrac{4}{3}$
26. $\log_9 x = \tfrac{3}{2}$

*In Exercises 27 and 28, express the given logarithm in terms of logarithms of x, y, and z, where the variables represent positive numbers.*

27. $\log_b \dfrac{x \sqrt[3]{y}}{z^4}$

28. $\log_b \sqrt[5]{\dfrac{x^4 y^2}{z^3}}$

*In Exercises 29 and 30, write the given expression as a single logarithm with a coefficient of 1.*

29. $\log_b 4 + \log_b \pi + 2\log_b r + \log_b h - \log_b 3$

30. $\tfrac{2}{3}\log_b y - 4\log_b x - \tfrac{1}{3}\log_b z$

*In Exercises 31 through 34, find the solution set of the given equation.*

31. $\log_4 (2x + 3) - 2\log_4 x = 2$
32. $\log_3 (2x - 3) + \log_3 (x + 3) = 4$
33. $5^x = 6$
34. $3^{x-2} = 8$

*In Exercises 35 and 36, use logarithms to perform the computation.*

35. $\sqrt[3]{0.09632}$

36. $(2.743)^{0.8}$

*In Exercises 37 and 38, find the value to two decimal places.*

37. $2^\pi$

38. $\pi^{\sqrt{2}}$

*In Exercises 39 and 40, find the value of the given logarithm to four significant digits.*

39. $\log_8 7$

40. $\log_7 10$

**41.** If $V$ cubic units is the volume of a right circular cone of altitude $h$ units and base radius $r$ units, then

$$V = \tfrac{1}{3}\pi r^2 h$$

Use logarithms to find to four significant digits the value of $V$ if $h = 8.271$ and $r = 2.345$.

**42.** If $s = 8.60e^{-2.53t}$, find $t$ when $s = 0.570$.

**43.** How long will it take $100 to double itself if it is drawing interest at 6 per cent compounded annually?

**44.** An amount of $500 is deposited into a savings account and earns interest for 7 years at a rate of 6 per cent compounded quarterly. If there are no withdrawals or additional deposits, how much is in the account after 7 years?

**45.** If $A$ milligrams of radium are present after $t$ years, then

$$A = ke^{-0.0004t}$$

where $k$ is a constant. If 60 milligrams of radium are present now, how much radium will be present 100 years from now?

# 5
# Polynomial Functions, the Theory of Polynomial Equations, and Complex Numbers

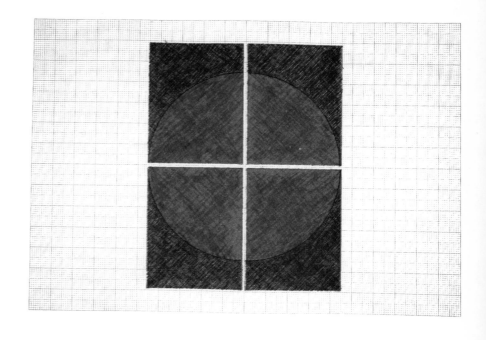

## 5.1 The Set of Complex Numbers

In Section 1.3 we gave a brief introduction to complex numbers. The set $C$ of complex numbers was introduced as a set which has the set $R$ of real numbers as a subset, and which contains numbers whose squares are negative numbers. Set $C$ was defined (Definition 1.3.4) as the set of all numbers of the form $a + bi$ where $a, b \in R$ and $i^2 = -1$. Using set notation, we have

$$C = \{a + bi \,|\, a, b \in R, i^2 = -1\}$$

If we let $z = a + bi$, then $a$ is called the real part of $z$ and $b$ is called the imaginary part of $z$. If the imaginary part of a complex number is zero, the number is real; that is, $a + 0i$ is the real number $a$. Therefore, $R$ is a subset of $C$. Another subset of $C$ is the set $I$ of imaginary numbers defined by

$$I = \{a + bi \,|\, a, b \in R, i^2 = -1, b \neq 0\}$$

The number $0 + bi$ can be written as $bi$ and it is called a pure imaginary number. The set $K$ of pure imaginary numbers is a subset of $I$ and it is defined by

$$K = \{bi \,|\, b \in R, i^2 = -1\}$$

**ILLUSTRATION 1**

(a) The complex number $-6 + 0i$ is a real number, and it can be written as $-6$.
(b) The complex number $-4 + 7i$ is an imaginary number.
(c) The complex number $0 + 2i$ is a pure imaginary number, and it can be written as $2i$.
(d) The complex number $0 + 0i$ is a real number, and it can be written as $0$.

●

**5.1.1 DEFINITION**  Two complex numbers $a + bi$ and $c + di$ are said to be *equal* if and only if $a = c$ and $b = d$.

**ILLUSTRATION 2**

(a) If
$$a + bi = -10 + 7i$$
then, by Definition 5.1.1, $a = -10$ and $b = 7$.
(b) If
$$-7x + 2yi = 35 + 6i$$

then, by Definition 5.1.1

$$-7x = 35 \qquad 2y = 6$$
$$x = -5 \qquad y = 3$$

●

We wish to define addition and multiplication of complex numbers so that the axioms for these operations on the set of real numbers are valid. To arrive at such definitions, we consider two complex numbers, $a + bi$ and $c + di$, as if they were polynomials in $i$ and then simplify the result by letting $i^2 = -1$. Thus

$$(a + bi) + (c + di) = a + c + bi + di$$
$$= (a + c) + (b + d)i$$

and

$$(a + bi)(c + di) = ac + adi + bci + bdi^2$$
$$= ac + (ad + bc)i + bd(-1)$$
$$= (ac - bd) + (ad + bc)i$$

We have then the following definition.

**5.1.2 DEFINITION** If $a + bi$ and $c + di$ are complex numbers, then

$$(a + bi) + (c + di) = (a + c) + (b + d)i$$

and

$$(a + bi)(c + di) = (ac - bd) + (ad + bc)i$$

By using Definition 5.1.2, it can be shown that the set $C$ is closed under the operations of addition and multiplication. In the next example we prove this for the operation of addition and leave as an exercise the proof for the operation of multiplication (see Exercise 51 in Exercises 5.1).

**EXAMPLE 1**
Prove that the set $C$ is closed under the operation of addition.

**SOLUTION**
We must prove that the sum of any two complex numbers is a complex number. Let $a + bi$ and $c + di$ be complex numbers. Then, by Definition 5.1.2,

$$(a + bi) + (c + di) = (a + c) + (b + d)i$$

The numbers $a, b, c, d \in R$, and the set $R$ is closed under addition. Therefore, $a + c$ and $b + d$ are real numbers. Hence the number $(a + c) + (b + d)i$ is a complex number.

It also can be proved that addition and multiplication on $C$ are commutative and associative, and that multiplication is distributive over addition.

**EXAMPLE 2**
Find the sum and product of the complex numbers $5 - 4i$ and $-2 + 6i$.

**SOLUTION**

$$(5 - 4i) + (-2 + 6i) = 5 - 2 - 4i + 6i$$
$$= 3 + 2i$$

$$(5 - 4i)(-2 + 6i) = -10 + 30i + 8i - 24i^2$$
$$= -10 + 38i - 24(-1)$$
$$= -10 + 38i + 24$$
$$= 14 + 38i$$

It is advised that you compute with complex numbers, as in the discussion preceding Definition 5.1.2, rather than memorize the definitions.

The additive identity element in the set of complex numbers is 0, which can be written as $0 + 0i$. The additive inverse of the complex number $a + bi$ is $-a - bi$ because

$$(a + bi) + (-a - bi) = [a + (-a)] + [b + (-b)]i$$
$$= 0 + 0i$$

Therefore,

$$-(a + bi) = -a - bi$$

As with real numbers, subtraction of complex numbers is defined in terms of addition; that is,

$$(a + bi) - (c + di) = (a + bi) + [-(c + di)]$$
$$= (a + bi) + (-c - di)$$
$$= (a - c) + (b - d)i$$

**EXAMPLE 3**
Find the difference of the two complex numbers of Example 2.

**SOLUTION**

$$(5 - 4i) - (-2 + 6i) = (5 - 4i) + (2 - 6i)$$
$$= 5 + 2 - 4i - 6i$$
$$= 7 - 10i$$

Because

$$(a + bi)(1 + 0i) = a \cdot 1 + a \cdot 0i + 1 \cdot bi + bi \cdot 0i$$
$$= a + 0i + bi + 0i^2$$
$$= a + bi$$

it follows that the multiplicative identity in $C$ is $1 + 0i$, which is the real number 1.

We consider an expression of the form

$$\frac{s + ti}{u + vi} \qquad (u \text{ and } v \text{ are not both } 0) \qquad (1)$$

to indicate the quotient of $(s + ti) \div (u + vi)$ and assume that such an expression satisfies the same properties as a corresponding rational expression. To write expression (1) in the form $a + bi$, we multiply the numerator and denominator by $u - vi$, which is the *conjugate* of $u + vi$.

**ILLUSTRATION 3**

$$\begin{aligned}
\frac{4 - 7i}{5 + 2i} &= \frac{(4 - 7i)(5 - 2i)}{(5 + 2i)(5 - 2i)} \\
&= \frac{20 - 8i - 35i + 14i^2}{25 - 10i + 10i - 4i^2} \\
&= \frac{20 - 43i + 14(-1)}{25 - 4(-1)} \\
&= \frac{6 - 43i}{29} \\
&= \frac{6}{29} - \frac{43}{29}i
\end{aligned}$$

●

The computation in Illustration 3 is similar to the method used to rationalize the denominator of a fraction when the denominator is a binomial containing a radical. More generally, when $u$ and $v$ are not both 0,

$$\begin{aligned}
\frac{s + ti}{u + vi} &= \frac{(s + ti)(u - vi)}{(u + vi)(u - vi)} \\
&= \frac{su - svi + tui - tvi^2}{u^2 - uvi + uvi - v^2i^2} \\
&= \frac{su + (tu - sv)i - tv(-1)}{u^2 - v^2(-1)} \\
&= \frac{(su + tv) + (tu - sv)i}{u^2 + v^2} \\
&= \left(\frac{su + tv}{u^2 + v^2}\right) + \left(\frac{tu - sv}{u^2 + v^2}\right)i
\end{aligned}$$

**EXAMPLE 4**
Find the quotient of the two complex numbers of Example 2.

**SOLUTION**

$$\begin{aligned}
\frac{5 - 4i}{-2 + 6i} &= \frac{(5 - 4i)(-2 - 6i)}{(-2 + 6i)(-2 - 6i)} \\
&= \frac{-10 - 30i + 8i + 24i^2}{4 + 12i - 12i - 36i^2} \\
&= \frac{-10 - 22i + 24(-1)}{4 - 36(-1)} \\
&= \frac{-34 - 22i}{40} \\
&= -\frac{17}{20} - \frac{11}{20}i
\end{aligned}$$

**ILLUSTRATION 4.** If $a$ and $b$ are not both zero, then

$$(a + bi)\left[\left(\frac{a}{a^2 + b^2}\right) - \left(\frac{b}{a^2 + b^2}\right)i\right] = \frac{a + bi}{1} \cdot \frac{a - bi}{a^2 + b^2}$$

$$= \frac{(a + bi)(a - bi)}{a^2 + b^2}$$

$$= \frac{a^2 - abi + abi - b^2i^2}{a^2 + b^2}$$

$$= \frac{a^2 - b^2(-1)}{a^2 + b^2}$$

$$= \frac{a^2 + b^2}{a^2 + b^2}$$

$$= 1 \qquad \bullet$$

From Illustration 4, we see that

$$(a + bi)\left[\left(\frac{a}{a^2 + b^2}\right) - \left(\frac{b}{a^2 + b^2}\right)i\right] = 1 \quad (a \text{ and } b \text{ are not both } 0) \quad (2)$$

From this equality it follows that every complex number, except $0 + 0i$, has a multiplicative inverse, and if we denote the multiplicative inverse of $a + bi$ by $(a + bi)^{-1}$, we have

$$(a + bi)^{-1} = \left(\frac{a}{a^2 + b^2}\right) + \left(\frac{-b}{a^2 + b^2}\right)i \qquad (a \text{ and } b \text{ are not both } 0)$$

**ILLUSTRATION 5.** The multiplicative inverse (reciprocal) of $4 - 3i$ is $(4 - 3i)^{-1}$ or, equivalently, $\frac{1}{4 - 3i}$. To find the standard form of this complex number, we multiply the numerator and denominator by $4 + 3i$, which is the conjugate of $4 - 3i$. We have then

$$\frac{1}{4 - 3i} = \frac{1 \cdot (4 + 3i)}{(4 - 3i)(4 + 3i)}$$

$$= \frac{4 + 3i}{16 - 9i^2}$$

$$= \frac{4 + 3i}{16 - 9(-1)}$$

$$= \frac{4}{25} + \frac{3}{25}i$$

Thus the multiplicative inverse of $4 - 3i$ is $\frac{4}{25} + \frac{3}{25}i$. $\qquad \bullet$

## The Set of Complex Numbers

In summary, we have the following facts about the set $C$.

1. The set $C$ is closed under the operations of addition and multiplication.
2. Addition and multiplication on $C$ are commutative and associative; multiplication is distributive over addition.
3. There is an identity element for addition and an identity element for multiplication.
4. Every element in $C$ has an additive inverse and every element in $C$, except $0 + 0i$, has a multiplicative inverse.

Because of these facts, it follows that the set $C$ is a field under the operations of addition and multiplication. Consequently, the laws of exponents apply to positive integer powers of $i$.

**ILLUSTRATION 6**

$$
\begin{array}{llll}
i^3 = i^2 i & i^4 = i^2 i^2 & i^5 = i^4 i & i^6 = i^4 i^2 \\
\phantom{i^3} = (-1)i & \phantom{i^4} = (-1)(-1) & \phantom{i^5} = (1)i & \phantom{i^6} = (1)(-1) \\
\phantom{i^3} = -i & \phantom{i^4} = 1 & \phantom{i^5} = i & \phantom{i^6} = -1
\end{array}
$$

In Illustration 6 we see that we obtain the results $i$, $-i$, $1$, and $-1$. By noting that $i^4 = 1$, we can find any positive integer power of $i$, and it will be one of these four numbers obtained in Illustration 6. ●

**EXAMPLE 5**
Find each of the following.
(a) $i^9$   (b) $i^{23}$   (c) $i^{-3}$

**SOLUTION**

(a) $i^9 = i^8 i$
$\phantom{i^9} = (i^4)^2 i$
$\phantom{i^9} = (1)^2 i$
$\phantom{i^9} = i$

(b) $i^{23} = i^{20} i^2 i$
$\phantom{i^{23}} = (i^4)^5 (-1)i$
$\phantom{i^{23}} = (1)^5 (-1)i$
$\phantom{i^{23}} = -i$

(c) $i^{-3} = \dfrac{1}{i^3}$
$\phantom{i^{-3}} = \dfrac{1}{-i}$
$\phantom{i^{-3}} = \dfrac{1 \cdot i}{-i^2}$
$\phantom{i^{-3}} = \dfrac{i}{-(-1)}$
$\phantom{i^{-3}} = \dfrac{i}{1}$
$\phantom{i^{-3}} = i$

In Section 1.3 we defined (Definition 1.3.6) the principal square root of a negative number as follows: If $p \geq 0$, then

$$\sqrt{-p} = i\sqrt{p}$$

**ILLUSTRATION 7**

$$\sqrt{-4}\sqrt{-25} = (i\sqrt{4})(i\sqrt{25})$$
$$= (2i)(5i)$$
$$= 10i^2$$
$$= -10 \qquad \bullet$$

Note in Illustration 7 that, before multiplying, we expressed $\sqrt{-4}$ and $\sqrt{-25}$ as $i\sqrt{4}$ and $i\sqrt{25}$, respectively. We did not apply Theorem 4.2.1, which states that if $a, b \in R$, then

$$\sqrt[n]{a}\sqrt[n]{b} = \sqrt[n]{ab} \qquad (3)$$

because the theorem further states that $a \geq 0$ and $b \geq 0$ when $n$ is even. As a matter of fact, equality (3) is not valid if $a < 0$ and $b < 0$ when $n$ is even. If, in the first step of the solution of Illustration 7, we use equality (3), we obtain

$$\sqrt{(-4)(-25)} = \sqrt{100}$$
$$= 10$$

which is an incorrect result for the product $\sqrt{-4}\sqrt{-25}$. To avoid making such an error, you should replace the symbol $\sqrt{-p}$ (when $p > 0$) by $i\sqrt{p}$ before performing any multiplication or division.

**EXAMPLE 6**
Perform the indicated operations and express the result in the form $a + bi$.

(a) $\sqrt{-5}(\sqrt{15} - \sqrt{-5})$
(b) $(2 - \sqrt{-9}) \div (2 + \sqrt{-9})$

**SOLUTION**

(a) $\sqrt{-5}(\sqrt{15} - \sqrt{-5}) = i\sqrt{5}(\sqrt{3 \cdot 5} - i\sqrt{5})$
$$= i\sqrt{3 \cdot 5^2} - i^2\sqrt{5^2}$$
$$= i\sqrt{5^2}\sqrt{3} - (-1)\sqrt{5^2}$$
$$= 5 + 5i\sqrt{3}$$

(b) $\dfrac{2 - \sqrt{-9}}{2 + \sqrt{-9}} = \dfrac{(2 - 3i)(2 - 3i)}{(2 + 3i)(2 - 3i)}$
$$= \dfrac{4 - 6i - 6i + 9i^2}{4 - 9i^2}$$
$$= \dfrac{4 - 12i + 9(-1)}{4 - 9(-1)}$$
$$= \dfrac{-5 - 12i}{13}$$
$$= -\dfrac{5}{13} - \dfrac{12}{13}i$$

The next theorem gives some properties of the conjugates of complex numbers. We use the notation $\bar{z}$ to denote the conjugate of the complex number $z$. That is,

$$\text{if } z = a + bi \qquad \text{then} \qquad \bar{z} = a - bi$$

**5.1.3 THEOREM**  If $z_1, z_2, z^n \in C$, then
  (i) $\overline{z_1} + \overline{z_2} = \overline{z_1 + z_2}$
  (ii) $\overline{z_1} \cdot \overline{z_2} = \overline{z_1 \cdot z_2}$
  (iii) $\overline{z^n} = \overline{z}^n$, for all positive integers $n$

The proof of part (i) is given below and the proof of part (ii) is left as an exercise (see Exercise 52 in Exercises 5.1). The proof of part (iii) requires mathematical induction which is discussed in Section 8.2 (see Exercise 19 in Exercises 8.2).

**PROOF OF PART (I):** Let $z_1 = a + bi$ and $z_2 = c + di$. Then $\overline{z_1} = a - bi$ and $\overline{z_2} = c - di$.

$$\overline{z_1} + \overline{z_2} = (a - bi) + (c - di)$$
$$= (a + c) - (b + d)i \qquad (4)$$

However,

$$z_1 + z_2 = (a + bi) + (c + di)$$
$$= (a + c) + (b + d)i$$

Thus

$$\overline{z_1 + z_2} = (a + c) - (b + d)i \qquad (5)$$

From equations (4) and (5) and Definition 5.1.1 (equality of complex numbers), it follows that

$$\overline{z_1} + \overline{z_2} = \overline{z_1 + z_2}$$

**ILLUSTRATION 8.** We verify Theorem 5.1.3 for particular complex numbers.

(a) $\overline{(3 + 5i)} + \overline{(2 - 4i)} = (3 - 5i) + (2 + 4i)$
$\phantom{\overline{(3 + 5i)} + \overline{(2 - 4i)}} = 5 - i$

and

$\overline{(3 + 5i) + (2 - 4i)} = \overline{(5 + i)}$
$\phantom{\overline{(3 + 5i) + (2 - 4i)}} = 5 - i$

Thus part (i) of Theorem 5.1.3 is verified if $z_1 = 3 + 5i$ and $z_2 = 2 - 4i$.

(b) $\overline{(3 + 5i)} \cdot \overline{(2 - 4i)} = (3 - 5i) \cdot (2 + 4i)$
$\phantom{\overline{(3 + 5i)} \cdot \overline{(2 - 4i)}} = 6 - 10i + 12i - 20i^2$
$\phantom{\overline{(3 + 5i)} \cdot \overline{(2 - 4i)}} = 26 + 2i$

and

$\overline{(3 + 5i) \cdot (2 - 4i)} = \overline{(6 + 10i - 12i - 20i^2)}$
$\phantom{\overline{(3 + 5i) \cdot (2 - 4i)}} = \overline{26 - 2i}$
$\phantom{\overline{(3 + 5i) \cdot (2 - 4i)}} = 26 + 2i$

(c) $\overline{(2+5i)^3} = \overline{[2^3 + 3\cdot 2^2(5i) + 3\cdot 2(5i)^2 + (5i)^3]}$
$= \overline{[8 + 60i + 150i^2 + 125i^3]}$
$= \overline{[8 + 60i - 150 - 125i]}$
$= \overline{-142 - 65i}$
$= -142 + 65i$

and

$\overline{(2+5i)}^3 = (2-5i)^3$
$= 2^3 - 3\cdot 2^2(5i) + 3\cdot 2(5i)^2 - (5i)^3$
$= 8 - 60i + 150i^2 - 125i^3$
$= 8 - 60i - 150 + 125i$
$= -142 + 65i$

Therefore, part (iii) of Theorem 5.1.3 is verified if $z = 2 + 5i$ and $n = 3$.
●

Parts (i) and (ii) of Theorem 5.1.3 can be extended to more than two complex numbers. For instance, if $z_1, z_2, z_3 \in C$, then

$$\overline{z_1} + \overline{z_2} + \overline{z_3} = \overline{z_1 + z_2 + z_3} \tag{6}$$

and

$$\overline{z_1}\cdot\overline{z_2}\cdot\overline{z_3} = \overline{z_1\cdot z_2\cdot z_3} \tag{7}$$

Formula (6) can be proved by applying part (i) of Theorem 5.1.3 as follows.

$\overline{z_1} + \overline{z_2} + \overline{z_3} = (\overline{z_1} + \overline{z_2}) + \overline{z_3}$
$= \overline{(z_1 + z_2)} + \overline{z_3}$
$= \overline{(z_1 + z_2) + z_3}$
$= \overline{z_1 + z_2 + z_3}$

Formula (7) can be proved in a similar way.

More generally, we have the following formulas, where $z_1, z_2, \ldots, z_n \in C$.

$$\overline{z_1} + \overline{z_2} + \cdots + \overline{z_n} = \overline{z_1 + z_2 + \cdots + z_n} \tag{8}$$

and

$$\overline{z_1}\cdot\overline{z_2}\cdot\cdots\cdot\overline{z_n} = \overline{z_1\cdot z_2\cdot\cdots\cdot z_n} \tag{9}$$

Formulas (8) and (9) can be proved by mathematical induction, and the proofs are omitted.

## EXERCISES 5.1

In Exercises 1 through 36, perform the indicated operations and express the result in the form $a + bi$.

1. $(5 + 2i) + (7 + i)$
2. $(4 - 3i) + (-6 + 8i)$
3. $(-9 - 4i) + (3 + 4i)$
4. $7i + (5 - i)$
5. $(3 - 4i) - (2 - 7i)$
6. $(4 - 3\sqrt{-16}) + (-1 - \sqrt{-4})$
7. $(5 + 2\sqrt{-9}) + (3 + 4\sqrt{-25})$
8. $(5 + 8i) - (9 - 6i)$
9. $(-3 - \sqrt{-20}) - (6 - \sqrt{-45})$
10. $(4 - \sqrt{-18}) - (2 - \sqrt{-2})$
11. $\sqrt{-9}\sqrt{-25}$
12. $\sqrt{-4}\sqrt{-16}$
13. $\sqrt{-2}\sqrt{-8}$
14. $\sqrt{-5}\sqrt{-75}$
15. $\sqrt{-12}\sqrt{-16}\sqrt{-27}$
16. $\sqrt{-27}\sqrt{-54}\sqrt{-162}$
17. $\sqrt{-8}(3\sqrt{-9} - \sqrt{-8})$
18. $\sqrt{-18}(\sqrt{-2} - 9\sqrt{-18})$
19. $(2 - 7i)(2 + 7i)$
20. $(4 - 3i)(-1 + 2i)$
21. $(3 + 2\sqrt{-3})(-2 + 3\sqrt{-3})$
22. $(2 - \sqrt{-2})(2 - 3\sqrt{-2})$
23. $(-3 - 3\sqrt{-3})^2$
24. $(\sqrt{-3} - \sqrt{-2})^2$
25. $-5 \div i$
26. $7 \div 3i$
27. $1 \div (2i - 3)$
28. $-4 \div (6 + i)$
29. $(3 + 2i) \div (2 - i)$
30. $(2 - 5i) \div 3i$
31. $(3 + 2i) \div 4i$
32. $(2 - 6i) \div (2i - 3)$
33. $1 \div (3 + \sqrt{-2})$
34. $(\sqrt{-5} - 3)(2\sqrt{-5} + 4)$
35. $(3 + 2i)^{-2}$
36. $(4 - 2i)^{-2}$

In Exercises 37 through 46, simplify the given expression.

37. $i^{11}$
38. $i^{22}$
39. $i^{33}$
40. $i^{43}$
41. $i^{-5}$
42. $i^{-6}$
43. $i^{-15}$
44. $(i^4 + i^3 - i^2 + 1)^2$
45. $(2i + 3i^2 + 4i^3 - i^6)^3$
46. $(i - 1)^2 - (-i - 1)^2 + i^4$

In Exercises 47 through 50, find the value of the given expression for the indicated value of $x$.

47. $(x^2 - 2x + 3); x = 1 - i\sqrt{2}$
48. $(x^2 - 2x + 4); x = 1 - \sqrt{-3}$
49. $(4x^2 + 4x + 3); x = \frac{1}{2}(-1 + \sqrt{-2})$
50. $(3x^2 - 2x + 2); x = \frac{1}{3}(1 - \sqrt{-5})$

51. Prove that the set $C$ is closed under the operation of multiplication.
52. Prove Theorem 5.1.3(ii).

## 5.2 Geometric Representation of Complex Numbers

The set of complex numbers can be represented geometrically by points in a rectangular coordinate system. The geometric representation of the complex number $a + bi$ or, equivalently, $(a, b)$ is the point $P(a, b)$ in a rectangular coordinate system. Refer to Figure 5.2.1.

**276** Polynomial Functions, the Theory of Polynomial Equations, and Complex Numbers [Ch. 5

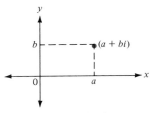

Figure 5.2.1

**EXAMPLE 1**
Show the geometric representation of each of the following complex numbers as a point in a rectangular coordinate system: $3 + 5i$; $-3 + 5i$; $-3 - 5i$; $3 - 5i$; $i$; $-2i$; $4$; and $-6$.

**SOLUTION**
The points are shown in Figure 5.2.2.

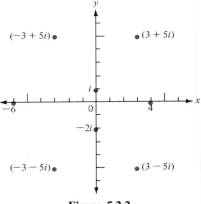

Figure 5.2.2

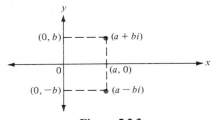

Figure 5.2.3

Observe that the geometric representations of a complex number $a + bi$ and its conjugate $a - bi$ are points that are symmetric with respect to the $x$ axis. See Figure 5.2.3.

When a rectangular coordinate system is used to represent the set of complex numbers, the horizontal axis is called the *axis of real numbers* (or the *real axis*), and the vertical axis is called the *axis of pure imaginary numbers* (or the *imaginary axis*); furthermore, the entire plane, whose points are placed in one-to-one correspondence with the complex numbers, is called the *complex plane*. Observe that any real number is a complex number which is represented by a point on the real axis and any pure imaginary number is represented by a point on the imaginary axis.

The sum of two complex numbers has an interesting geometric representation. By definition

$$(a + bi) + (c + di) = (a + c) + (b + d)i$$

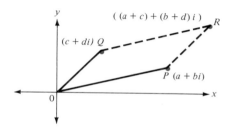

Figure 5.2.4

It can easily be shown that the point representing $(a + c) + (b + d)i$ is the end point of the diagonal of the parallelogram having as sides the line segments from the origin to the points $(a + bi)$ and $(c + di)$. See Figure 5.2.4, where $P$ is the point representing $(a + bi)$, $Q$ is the point representing $(c + di)$, and $R$ is the point representing $(a + c) + (b + d)i$. If $m_1$ is the slope of the line through $O$ and $P$, then

$$m_1 = \frac{b - 0}{a - 0}$$
$$= \frac{b}{a}$$

If $m_2$ is the slope of the line through $O$ and $Q$, then

$$m_2 = \frac{d - 0}{c - 0}$$
$$= \frac{d}{c}$$

If $m_3$ is the slope of the line through $P$ and $R$, then

$$m_3 = \frac{(b + d) - b}{(a + c) - a}$$
$$= \frac{d}{c}$$

If $m_4$ is the slope of the line through $Q$ and $R$, then

$$m_4 = \frac{(b + d) - d}{(a + c) - c}$$
$$= \frac{b}{a}$$

Because $m_1 = m_4$, it follows that the line through $O$ and $P$ is parallel to the line through $Q$ and $R$. Because $m_2 = m_3$, it follows that the line through $O$ and $Q$ is parallel to the line through $P$ and $R$. Because the quadrilateral $OPRQ$ has opposite sides parallel, the quadrilateral is a parallelogram.

## EXAMPLE 2

Perform geometrically the following addition.

$(3 + 4i) + (-5 + 6i)$

**SOLUTION**

Refer to Figure 5.2.5. The points representing $(3 + 4i)$ and $(-5 + 6i)$ are labeled $P$ and $Q$, respectively. We draw line segments $OP$ and $OQ$, and then draw line segments $PR$, and $QR$ parallel to $OQ$ and $OP$, respectively. The point of intersection, $R$, is seen to be at $(-2 + 10i)$. Thus

$$(3 + 4i) + (-5 + 6i) = -2 + 10i$$

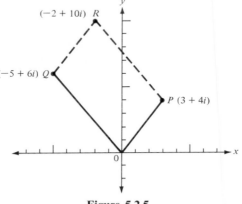

**Figure 5.2.5**

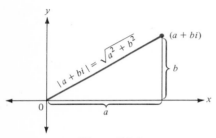

**Figure 5.2.6**

If $a + bi$ is an arbitrary complex number, then the distance in the complex plane from the origin to the graph of $a + bi$ is $\sqrt{a^2 + b^2}$. Refer to Figure 5.2.6. This number is called the "absolute value" or "modulus" of the number $a + bi$.

**5.2.1 DEFINITION**  The *absolute value,* or *modulus,* of the complex number $a + bi$, denoted by $|a + bi|$, is given by

$$|a + bi| = \sqrt{a^2 + b^2} \qquad (1)$$

If in equation (1), $b = 0$, we have

$$|a| = \sqrt{a^2}$$

which is consistent with Definition 1.7.1 (the absolute value of a real number) and equation (3) in Section 4.1. Furthermore, the absolute value of a real number $a$ is the distance on the real axis from the origin to the graph of $a$; this geometric interpretation is consistent with the geometric interpretation of $|a + bi|$.

## Geometric Representation of Complex Numbers

**EXAMPLE 3**

Write each of the following expressions without using absolute value bars.
(a) $|6 - i|$  (b) $|-3 + 2i|$
(c) $|-4 - 3i|$  (d) $|5i|$

**SOLUTION**

(a) $|6 - i| = \sqrt{6^2 + (-1)^2}$
$= \sqrt{37}$

(c) $|-4 - 3i| = \sqrt{(-4)^2 + (-3)^2}$
$= \sqrt{25}$
$= 5$

(b) $|-3 + 2i| = \sqrt{(-3)^2 + 2^2}$
$= \sqrt{13}$

(d) $|5i| = \sqrt{0^2 + 5^2}$
$= \sqrt{25}$
$= 5$

Recall that the set $R$ of real numbers is an ordered field. Even though the set $C$ of complex numbers is a field, it is not an ordered field. The reason for this is that it is impossible to define an order relation (such as $<$) for the set $C$ that has the properties of the order relation for the set $R$. Therefore, we do not refer to one complex number, with nonzero imaginary part, being greater than, or less than, another complex number.

We can use the order relation with the absolute values of complex numbers because they are real numbers. If $|z_1| < |z_2|$, then the point in the complex plane representing $z_1$ is closer to the origin than the point representing $z_2$. Furthermore, if $|z_1| = |z_2|$, the points representing $z_1$ and $z_2$ are at the same distance from the origin of the complex plane.

## EXERCISES 5.2

In Exercises 1 through 8, show the geometric representation of the given complex number as a point in the complex plane.

1. $7 - 8i$
2. $2 - 6i$
3. $-4 + 9i$
4. $-5 - 3i$
5. $-3i$
6. $-1 + 4i$
7. $-6 - i$
8. $-6i$

In Exercises 9 through 18, perform geometrically the addition.

9. $(2 + 3i) + (4 + 2i)$
10. $(3 + 7i) + (4 + i)$
11. $(-4 + 5i) + (1 + 3i)$
12. $(-6 + 6i) + (-2 - 5i)$
13. $(6 - 4i) + (-3 + 2i)$
14. $(-7 + 5i) + (4 - 3i)$
15. $(4 + 3i) + 5i$
16. $(-4 + 4i) + 2$
17. $(-7 - 3i) + (8 - 6i)$
18. $(-2 - 8i) + (-3 + 5i)$

In Exercises 19 through 26, write the given expression without absolute value bars.

19. $|5 + 2i|$
20. $|8 - 3i|$
21. $|-1 + 2i|$
22. $|-2 - 5i|$
23. $|2i|$
24. $|7 + i|$
25. $|-6 - 8i|$
26. $|-6i|$

In Exercises 27 through 31, prove the given equality if $z, z_1, z_2 \in C$.

27. $|-z| = |z|$
28. $|\bar{z}| = |z|$
29. $|z| = \sqrt{z \cdot \bar{z}}$
30. $|z_1 \cdot z_2| = |z_1| \cdot |z_2|$
31. $\left|\dfrac{z_1}{z_2}\right| = \dfrac{|z_1|}{|z_2|}$, if $z_2 \neq 0$

## 5.3 Equations Having Complex Roots

Equations having complex roots appeared in Chapter 2. We now give a review of some of the types of equations considered there, and then we extend our discussion to include other equations having complex roots.

If $k$ is a positive real number, then

$$(i\sqrt{k})^2 = i^2(\sqrt{k})^2 \qquad \text{and} \qquad (-i\sqrt{k})^2 = i^2(-\sqrt{k})^2$$
$$= (-1)k \qquad\qquad\qquad\qquad\quad = (-1)k$$
$$= -k \qquad\qquad\qquad\qquad\quad\;\; = -k$$

Hence solutions of the equation

$$x^2 = -k \qquad (1)$$

where $k$ is a positive real number, are $i\sqrt{k}$ and $-i\sqrt{k}$. We can show that these numbers are the only solutions because if $z$ is any complex number in the solution set of equation (1), then

$$z^2 + k = 0$$

or, equivalently,

$$(z - i\sqrt{k})(z + i\sqrt{k}) = 0 \qquad (2)$$

Theorem 2.1.1 states that if $r, s \in C$, then

$$rs = 0 \quad \text{if and only if} \quad r = 0 \text{ or } s = 0$$

Thus, from equation (2), it follows that either

$$z - i\sqrt{k} = 0 \quad \text{or} \quad z + i\sqrt{k} = 0$$

or, equivalently,

$$z = i\sqrt{k} \quad \text{or} \quad z = -i\sqrt{k}$$

Hence the solution set of equation (1) is $\{i\sqrt{k}, -i\sqrt{k}\}$.

**ILLUSTRATION 1.** The equation

$$x^2 + 5 = 0$$

is equivalent to

$$(x - i\sqrt{5})(x + i\sqrt{5}) = 0$$

This equation gives a true statement if and only if

$$x - i\sqrt{5} = 0 \quad \text{or} \quad x + i\sqrt{5} = 0$$

or, equivalently,

$$x = i\sqrt{5} \quad \text{or} \quad x = -i\sqrt{5}$$

Thus the solution set of the given equation is $\{i\sqrt{5}, -i\sqrt{5}\}$

Every quadratic equation with real coefficients has two complex numbers as solutions. The roots of the quadratic equation $ax^2 + bx + c = 0$ are given by the quadratic formula and they are

$$\frac{-b + \sqrt{b^2 - 4ac}}{2a} \quad \text{and} \quad \frac{-b - \sqrt{b^2 - 4ac}}{2a}$$

If $b^2 - 4ac < 0$, the two roots are imaginary numbers; observe that these numbers are conjugates of each other.

**ILLUSTRATION 2.** To find the roots of the equation $2x^2 - 5x + 4 = 0$, we apply the quadratic formula where $a$ is 2, $b$ is $-5$, and $c$ is 4. We obtain

$$\begin{aligned} x &= \frac{-b \pm \sqrt{b^2 - 4ac}}{2a} \\ &= \frac{-(-5) \pm \sqrt{(-5)^2 - 4 \cdot 2 \cdot 4}}{2 \cdot 2} \\ &= \frac{5 \pm \sqrt{-7}}{4} \\ &= \frac{5 \pm i\sqrt{7}}{4} \end{aligned}$$

The roots are, therefore, $\frac{5}{4} + \frac{\sqrt{7}}{4}i$ and $\frac{5}{4} - \frac{\sqrt{7}}{4}i$. •

We have seen that every polynomial equation of the second degree in one variable has two solutions, and every polynomial equation of the first degree in one variable has one solution. These equations are illustrations of the theorem (Theorem 5.6.3) that guarantees that a polynomial equation of degree $n$ in one variable has exactly $n$ solutions, some of which may be repeated. Related to this theorem is the fact that if $n$ is a positive integer, every nonzero real number has exactly $n$ distinct $n$th roots, some or all of which may be imaginary.

**ILLUSTRATION 3.** To find the four fourth roots of 81, we let $x$ represent a fourth root of 81 and form the equation

$$x^4 = 81$$

$$x^4 - 81 = 0$$

We factor the left member of this equation as the difference of two squares and obtain

$$(x^2 - 9)(x^2 + 9) = 0 \tag{3}$$

From Theorem 2.1.1, equation (3) gives a true statement if and only if $x^2 - 9 = 0$ or $x^2 + 9 = 0$. Hence we set each of the factors in equation (3) equal to zero and solve the equations.

$$x^2 - 9 = 0 \qquad x^2 + 9 = 0$$
$$x^2 = 9 \qquad x^2 = -9$$
$$x = \pm 3 \qquad x = \pm 3i$$

Therefore, the four fourth roots of 81 are $-3, 3, -3i$, and $3i$. We can verify these results by finding the fourth power of each of these numbers.

$$(-3)^4 = 81 \qquad 3^4 = 81 \qquad \begin{aligned}(-3i)^4 &= 3^4 i^4 \\ &= 81(i^2)(i^2) \\ &= 81(-1)(-1) \\ &= 81\end{aligned} \qquad \begin{aligned}(3i)^4 &= 3^4 i^4 \\ &= 81\end{aligned}$$

In the following example we have an equation in which the left member is the product of four factors and the right member is zero. We use an extension of Theorem 2.1.1 to a product of more than two factors.

**EXAMPLE 1**
Find the six sixth roots of 64.

**SOLUTION**
Let $x$ represent a sixth root of 64. Then

$$x^6 = 64 \qquad (4)$$
$$x^6 - 64 = 0$$

We factor the left member by first considering it as the difference of two squares $(x^3)^2 - 8^2$, and we have

$$(x^3 - 8)(x^3 + 8) = 0 \qquad (5)$$

We now factor the first factor as the difference of two cubes and the second factor as the sum of two cubes. Recall that

$$a^3 - b^3 = (a - b)(a^2 + ab + b^2)$$
$$a^3 + b^3 = (a + b)(a^2 - ab + b^2)$$

Therefore, equation (5) is equivalent to

$$(x - 2)(x^2 + 2x + 4)(x + 2)(x^2 - 2x + 4) = 0$$

We set each factor equal to zero and solve the resulting equations.

$$x - 2 = 0 \qquad x + 2 = 0$$
$$x = 2 \qquad x = -2$$

$$x^2 + 2x + 4 = 0$$
$$x = \frac{-2 \pm \sqrt{2^2 - 4(1)(4)}}{2(1)}$$
$$= \frac{-2 \pm \sqrt{4 - 16}}{2}$$

$$= \frac{-2 \pm \sqrt{-12}}{2}$$
$$= \frac{-2 \pm i\sqrt{4 \cdot 3}}{2}$$
$$= \frac{-2 \pm 2i\sqrt{3}}{2}$$
$$= -1 \pm i\sqrt{3}$$

$x^2 - 2x + 4 = 0$
$$x = \frac{-(-2) \pm \sqrt{(-2)^2 - 4(1)(4)}}{2(1)}$$
$$= \frac{2 \pm \sqrt{4 - 16}}{2}$$
$$= \frac{2 \pm \sqrt{-12}}{2}$$
$$= \frac{2 \pm i\sqrt{4 \cdot 3}}{2}$$
$$= \frac{2 \pm 2i\sqrt{3}}{2}$$
$$= 1 \pm i\sqrt{3}$$

Hence the six cube roots of 64 are the numbers in the solution set of equation (4) which is $\{-2, 2, -1 + i\sqrt{3}, -1 - i\sqrt{3}, 1 + i\sqrt{3}, 1 - i\sqrt{3}\}$.

In Section 5.6 we prove a theorem (Theorem 5.6.4) which states that if $f(x)$ is a polynomial of degree $n$ with real coefficients and if $z$ is a complex root of the equation $f(x) = 0$, then the conjugate $\bar{z}$ is also a root of the equation. We apply this theorem in the following illustration and example.

**ILLUSTRATION 4.** Suppose we wish to find an equation with real coefficients for which the complex number $3 - 2i$ is a root. Then $3 + 2i$ (the conjugate of $3 - 2i$) must also be a root of the required equation. We know that the quadratic equation

$$(x - r)(x - s) = 0$$

has the solution set $\{r, s\}$. Hence an equation having the given roots is

$$[x - (3 - 2i)][x - (3 + 2i)] = 0$$
$$[(x - 3) + 2i][(x - 3) - 2i] = 0$$
$$(x - 3)^2 - (2i)^2 = 0$$
$$(x^2 - 6x + 9) - (-4) = 0$$
$$x^2 - 6x + 13 = 0$$

●

## EXAMPLE 2

Find an equation with real coefficients for which the numbers 3 and $2 + i\sqrt{5}$ are roots.

**SOLUTION**

Because $2 + i\sqrt{5}$ is to be a root of the required equation, its conjugate, $2 - i\sqrt{5}$ must also be a root. Thus an equation having the given roots is

$$(x - 3)[x - (2 + i\sqrt{5})][x - (2 - i\sqrt{5})] = 0$$
$$(x - 3)[(x - 2)^2 - (i\sqrt{5})^2] = 0$$
$$(x - 3)[(x^2 - 4x + 4) - (-5)] = 0$$
$$(x - 3)(x^2 - 4x + 9) = 0$$
$$x^3 - 7x^2 + 21x - 27 = 0$$

So far all of the equations we have discussed have had real coefficients. We now consider some equations having complex coefficients, and such equations have solutions in the set of complex numbers.

## EXAMPLE 3

Find the solution set of the equation

$$2ix = 6 - ix$$

**SOLUTION**

The given equation is equivalent to

$$2ix + ix = 6$$

Thus

$$3ix = 6$$
$$x = \frac{6}{3i}$$
$$x = \frac{2i}{i^2}$$
$$x = -2i$$

Thus the solution set is $\{-2i\}$.

The quadratic formula was derived to solve a quadratic equation having real coefficients. Because only properties of a field were used in the derivation, and because the set of complex numbers is a field, it follows that the quadratic formula can be used to obtain the solutions of a quadratic equation having complex coefficients.

## EXAMPLE 4

Find the solution set of the equation

$$3ix^2 + 2x - 2i = 0$$

**SOLUTION**

Applying the quadratic formula where $a$ is $3i$, $b$ is 2, and $c$ is $-2i$, we obtain

$$x = \frac{-b \pm \sqrt{b^2 - 4ac}}{2a}$$
$$= \frac{-2 \pm \sqrt{4 - 4(3i)(-2i)}}{6i}$$
$$= \frac{-2 \pm \sqrt{4 + 24i^2}}{6i}$$

$$= \frac{-2 \pm \sqrt{-20}}{6i}$$

$$= \frac{-2 \pm 2i\sqrt{5}}{6i}$$

$$= -\frac{1}{3i} \pm \frac{\sqrt{5}}{3}$$

$$= -\frac{i}{3i^2} \pm \frac{\sqrt{5}}{3}$$

$$= \frac{1}{3}i \pm \frac{\sqrt{5}}{3}$$

Therefore, the solution set is $\left\{\frac{\sqrt{5}}{3} + \frac{1}{3}i, -\frac{\sqrt{5}}{3} + \frac{1}{3}i\right\}$.

## EXERCISES 5.3

*In Exercises 1 through 10, find the solution set of the given equation.*

1. $x^2 + 9 = 0$
2. $4x^2 + 25 = 0$
3. $3x^2 + 2 = 0$
4. $2x^2 + 7 = 0$
5. $x^2 + 2x + 6 = 0$
6. $x^2 - 4x - 7 = 0$
7. $4x^2 - 4x + 5 = 0$
8. $9x^2 + 12x + 13 = 0$
9. $3x^2 + 2x + 1 = 0$
10. $5x^2 - 6x + 2 = 0$

*In Exercises 11 through 20, find an equation with real coefficients for which the given numbers (or number) are roots.*

11. $5 + i$
12. $2 - i\sqrt{3}$
13. $\frac{\sqrt{3} - 4i}{5}$
14. $\frac{-1 + i}{3}$
15. $0, \frac{1}{2} + \frac{2}{3}i$
16. $1, \frac{1}{4} - \frac{1}{3}i$
17. $i, i\sqrt{3}$
18. $i\sqrt{2}, 1 - i$
19. $\sqrt{5}, -\sqrt{5}, \frac{2 - i\sqrt{5}}{3}$
20. $0, -4, -4i$

*In Exercises 21 through 34, find the solution set of the given equation. Write the solutions in standard form.*

21. $ix = 8 - 3ix$
22. $3ix - 20 = 5 - 2ix$
23. $ix - i = 4x$
24. $2ix - 3 = x$
25. $3(x - 1) = i(x + 1)$
26. $4(x - 1 + i) = i(x + 5)$
27. $x - 2 = i + 3ix$
28. $7x + 6i = 5 - 2ix$
29. $2ix^2 + 3x + 5i = 0$
30. $7ix^2 - 5x + 2i = 0$
31. $x^2 - 3ix + 4 = 0$
32. $3x^2 + 8ix + 3 = 0$
33. $5ix^2 + 4x - 3i = 0$
34. $4ix^2 + 8x + i = 0$

35. Find the three cube roots of 1.
36. Find the three cube roots of $-1$.
37. Find the three cube roots of $-27$.
38. Find the three cube roots of 64.
39. Find the four fourth roots of 625.
40. Find the four fourth roots of 16.
41. Find the four fourth roots of $-4$.
42. Find the six sixth roots of 1.

## 5.4 The Remainder Theorem and the Factor Theorem

A polynomial in a variable $x$ with real coefficients was defined in Section 1.3 as a sum of terms which are either a constant or the product of a constant and a positive integer power of $x$. Therefore, such a polynomial of degree $n$ can be written in the form

$$a_0 x^n + a_1 x^{n-1} + a_2 x^{n-2} + \cdots + a_{n-1} x + a_n$$

where $a_0 \neq 0$, $n$ is a nonnegative integer, and $a_0, a_1, \ldots, a_n$ are real numbers. In Section 3.2 we stated that a function $f$ is called a polynomial function if $f(x)$ is a polynomial of degree $n$. We now extend our definition of a polynomial function to allow the coefficients to be complex numbers.

**5.4.1 DEFINITION** The function $P$ defined by the equation

$$P(x) = a_0 x^n + a_1 x^{n-1} + a_2 x^{n-2} + \cdots + a_{n-1} + a_n \qquad (1)$$

and written

$$P = \{(x, P(x)) \mid P(x) = a_0 x^n + a_1 x^{n-1} + a_2 x^{n-2} + \cdots + a_{n-1} x + a_n\}$$

where $a_0 \neq 0$, $n$ is a nonnegative integer, and $a_0, a_1, \ldots, a_n$ are complex numbers, is called a *polynomial function of the nth degree with complex coefficients*. If $a_0, a_1, \ldots, a_n$ are real numbers, then $P$ is called a *polynomial function of the nth degree with real coefficients*.

The function value $P(x)$ denotes a polynomial and the equation $P(x) = 0$ denotes a *polynomial equation*.

**ILLUSTRATION 1.** Consider the polynomial $P(x) = 2x^3 - 5x^2 + 6x - 3$, and divide $P(x)$ by $(x - 2)$.

$$\begin{array}{r}
2x^2 - x + 4 \phantom{000} \\
x - 2 \overline{\smash{\big)}\, 2x^3 - 5x^2 + 6x - 3} \\
\underline{2x^3 - 4x^2 \phantom{0000000000}} \\
-x^2 + 6x \phantom{000} \\
\underline{-x^2 + 2x \phantom{000}} \\
4x - 3 \\
\underline{4x - 8} \\
5
\end{array}$$

Hence the quotient is $2x^2 - x + 4$, and the remainder is 5. Therefore, we can write

$$2x^3 - 5x^2 + 6x - 3 = (x - 2)(2x^2 - x + 4) + 5$$

●

Illustration 1 gives a special case of the following theorem, which we state without proof.

**5.4.2 THEOREM** If $P(x)$ is a polynomial with complex coefficients, and $r \in C$, then, when $P(x)$ is divided by $(x - r)$, we obtain as the quotient a unique polynomial $Q(x)$ with complex coefficients, and as the remainder a complex number $R$, such that for all values of $x$

$$P(x) = (x - r)Q(x) + R$$

**ILLUSTRATION 2.** For the polynomial $P(x)$ in Illustration 1,

$$P(2) = 2(2)^3 - 5(2)^2 + 6(2) - 3$$
$$= 5$$

Observe that this value $P(2)$ is the same number as the remainder obtained in Illustration 1 when $P(x)$ is divided by $(x - 2)$. •

In Illustration 2 we have a special case of the following theorem, known as the "remainder theorem."

**5.4.3 THEOREM** *The Remainder Theorem.* If $P(x)$ is a polynomial with complex coefficients, and $r \in C$, then if $P(x)$ is divided by $(x - r)$, the remainder is $P(r)$.

**Proof.** From Theorem 5.4.2 it follows that when $P(x)$ is divided by $(x - r)$, we obtain a polynomial $Q(x)$ as the quotient and a complex number $R$ as the remainder such that for all values of $x$

$$P(x) = (x - r)Q(x) + R \tag{2}$$

Because equation (2) is an identity (it is true for all values of $x$), it is satisfied when $x = r$. Thus

$$P(r) = (r - r)Q(r) + R$$
$$P(r) = R$$

and the theorem is proved.

**EXAMPLE 1**
Verify the remainder theorem if $P(x) = x^4 + x^3 - 5$ and $r = -2$.

**SOLUTION**
We wish to show that when $P(x)$ is divided by $[x - (-2)]$ or, equivalently, $(x + 2)$, the remainder is $P(-2)$. We first find $P(-2)$ by substituting into the expression for $P(x)$.

$$P(-2) = (-2)^4 + (-2)^3 - 5$$
$$= 16 - 8 - 5$$
$$= 3$$

We now divide $P(x)$ by $(x + 2)$.

$$
\begin{array}{r}
x^3 - \phantom{0}x^2 + 2x - 4\phantom{)} \\
x + 2 \overline{\smash{)}x^4 + \phantom{0}x^3 + 0x^2 + 0x - 5} \\
\underline{x^4 + 2x^3\phantom{+ 0x^2 + 0x - 5}} \\
-x^3 + 0x^2\phantom{+ 0x - 5} \\
\underline{-x^3 - 2x^2\phantom{+ 0x - 5}} \\
2x^2 + 0x\phantom{- 5} \\
\underline{2x^2 + 4x\phantom{- 5}} \\
-4x - 5\phantom{)} \\
\underline{-4x - 8\phantom{)}} \\
3\phantom{)}
\end{array}
$$

The remainder is 3, which is $P(-2)$.

The next theorem is the "factor theorem"; it is a consequence of the remainder theorem.

**5.4.4 THEOREM**   *The Factor Theorem.* If $P(x)$ is a polynomial with complex coefficients and $r \in C$, then $P(x)$ has $(x - r)$ as a factor if and only if $P(r) = 0$.

*Proof.* Because the statement of the theorem has an "if and only if" qualification, there are two parts to be proved. Part 1: $(x - r)$ is a factor of $P(x)$ if $P(r) = 0$; Part 2: $(x - r)$ is a factor of $P(x)$ only if $P(r) = 0$.

**PROOF OF PART 1:** From Theorems 5.4.2 and 5.4.3 it follows that for the polynomial $P(x)$ and the complex number $r$ there exists a unique polynomial $Q(x)$ such that

$$P(x) = (x - r)Q(x) + P(r)$$

If $P(r) = 0$, then

$$P(x) = (x - r)Q(x)$$

Thus $(x - r)$ is a factor of $P(x)$.

**PROOF OF PART 2:** We wish to prove that if $(x - r)$ is a factor of $P(x)$, then $P(r) = 0$. If $(x - r)$ is a factor of $P(x)$ then, when $P(x)$ is divided by $(x - r)$, the remainder must be zero. Thus, from the remainder theorem, it follows that $P(r) = 0$.

## EXAMPLE 2
Show that $(x - 4)$ is a factor of
$$2x^3 - 6x^2 - 5x - 12$$

**SOLUTION**
If $P(x) = 2x^3 - 6x^2 - 5x - 12$, then
$$\begin{aligned}P(4) &= 2(4)^3 - 6(4)^2 - 5(4) - 12 \\ &= 2(64) - 6(16) - 20 - 12 \\ &= 128 - 96 - 32 \\ &= 0\end{aligned}$$
Therefore, by the factor theorem it follows that $(x - 4)$ is a factor of $P(x)$.

## EXAMPLE 3
Determine if $(x - 3)$ is a factor of
$$2x^4 - 3x^3 + 6x^2 - 7x - 3$$

**SOLUTION**
Let $P(x) = 2x^4 - 3x^3 + 6x^2 - 7x - 3$. Then
$$\begin{aligned}P(3) &= 2(3)^4 - 3(3)^3 + 6(3)^2 - 7(3) - 3 \\ &= 2(81) - 3(27) + 6(9) - 21 - 3 \\ &= 162 - 81 + 54 - 21 - 3 \\ &= 111\end{aligned}$$
Because $P(3) \neq 0$, we conclude from the factor theorem that $(x - 3)$ is not a factor of $P(x)$.

## EXAMPLE 4
Find a value of $k$ so that $(x + 3)$ is a factor of $(3x^3 + kx^2 - 7x + 6)$.

**SOLUTION**
Let $P(x) = 3x^3 + kx^2 - 7x + 6$. By the factor theorem, $(x + 3)$ is a factor of $P(x)$ if $P(-3) = 0$. Equating $P(-3)$ to zero, we have
$$\begin{aligned}3(-3)^3 + k(-3)^2 - 7(-3) + 6 &= 0 \\ -81 + 9k + 21 + 6 &= 0 \\ 9k &= 54 \\ k &= 6\end{aligned}$$
Thus $(x + 3)$ is a factor of $P(x)$ if $k = 6$.

## EXERCISES 5.4

*In Exercises 1 through 8, use long division to find the remainder when the given polynomial is divided by the given linear expression. Then find the remainder by the remainder theorem.*

1. $(3x^2 - 4x + 5) \div (x - 3)$
2. $(4x^2 + 7x - 5) \div (x - 1)$
3. $(3x^4 + 7x^3 + x^2 + x + 9) \div (x + 1)$
4. $(x^3 - 4x^2 + 5) \div (x + 3)$
5. $(x^3 + 9) \div (x + 2)$
6. $(x^4 - 8) \div (x - 2)$
7. $(3x^5 - 7x^4 - 5x^3 - 4x^2 + 1) \div (x - 3)$
8. $(8x^5 + 7x^2 - 3) \div (x - \frac{1}{2})$

*In Exercises 9 through 16, use the factor theorem to answer the question.*

9. Is $(x - 3)$ a factor of $(2x^3 - 6x^2 - 5x + 15)$?
10. Is $(x + 3)$ a factor of $(3x^3 - x^2 - 22x - 24)$?
11. Is $(x + 2)$ a factor of $(x^4 + 2x^3 - 12x^2 - 11x + 6)$?
12. Is $(x - 2)$ a factor of $(x^7 - 128)$?

13. Is $(x + 3)$ a factor of $(x^5 + 243)$?
14. Is $(x - a)$ a factor of $(x^8 + a^8)$?
15. Is $(x - 2i)$ a factor of $(x^6 + 64)$?
16. Is $(x + i)$ a factor of $(x^7 - i)$?

17. Find a value of $k$ so that $(x + 2)$ is a factor of $(3x^3 + 5x^2 + kx - 10)$.
18. Find a value of $k$ so that $(x - 5)$ is a factor of $(kx^3 - 17x^2 - 4kx + 5)$.
19. Find a value of $k$ so that $(x - i)$ is a factor of $(2kx^4 + 7x^3 + kx^2 + 7x - 1)$.
20. Find a value of $k$ so that $(x + 3i)$ is a factor of $(kx^4 - x^3 + 28x^2 - 3kx + 18)$.
21. Find values of $k$ so that $(x - 4)$ is a factor of $(x^3 - k^2x^2 - 8kx - 16)$.
22. Find values of $k$ so that $(x + 1)$ is a factor of $(5x^3 + k^2x^2 + 2kx - 3)$.
23. For what integer values of $n$ is $(x - y)$ a factor of $(x^n - y^n)$?
24. For what integer values of $n$ is $(x + y)$ a factor of $(x^n - y^n)$?
25. For what integer values of $n$ is $(x + y)$ a factor of $(x^n + y^n)$?
26. For what integer values of $n$ is $(x - iy)$ a factor of $(x^n - y^n)$?
27. For what integer values of $n$ is $(x - iy)$ a factor of $(x^n + y^n)$?

## 5.5 Synthetic Division

Applications of the remainder theorem and the factor theorem involve the division of a polynomial by linear expressions of the form $(x - r)$. To simplify the computation of such divisions, we use a procedure called *synthetic division*, which we now explain.

In Illustration 1 of Section 5.4 we used long division to divide $(2x^3 - 5x^2 + 6x - 3)$ by $(x - 2)$. The computation is as follows.

$$\begin{array}{r} 2x^2 - x + 4 \phantom{000} \\ x - 2 \overline{\smash{)}2x^3 - 5x^2 + 6x - 3} \\ \underline{2x^3 - 4x^2 \phantom{000000000}} \\ -x^2 + 6x \phantom{000} \\ \underline{-x^2 + 2x \phantom{000}} \\ 4x - 3 \\ \underline{4x - 8} \\ 5 \end{array}$$

The writing can be shortened by omitting the powers of $x$ and recording only the coefficients. By doing this, the computation takes the following form.

$$\begin{array}{r} 2 \phantom{0} -1 \phantom{0} 4 \phantom{000} \\ 1 - 2 \overline{\smash{)}2 \phantom{0} -5 \phantom{0} 6 \phantom{0} -3} \\ \underline{2 \phantom{0} -4 \phantom{000000}} \\ -1 \phantom{0} 6 \phantom{000} \\ \underline{-1 \phantom{0} 2 \phantom{000}} \\ 4 \phantom{0} -3 \\ \underline{4 \phantom{0} -8} \\ 5 \end{array}$$

In the divisor, $x - 2$, the coefficient of $x$ is 1. Thus the coefficient of the first term in each remainder is the same as that of the succeeding term of the quotient. Furthermore, the first term of the next partial product is the same as the coefficient of the first term in each remainder. Hence we can omit the terms of the quotient, as well as the first terms of the partial products. With these terms omitted, we have

$$\begin{array}{r|rrrr} 1 - 2 & 2 & -5 & 6 & -3 \\ & & -4 & & \\ \hline & & -1 & 6 & \\ & & & 2 & \\ \hline & & & 4 & -3 \\ & & & & -8 \\ \hline & & & & 5 \end{array}$$

In synthetic division the divisor is a polynomial of the form $x - r$, and so the first coefficient in the divisor is always 1; thus we delete the coefficient 1. We can also move the numbers up so that they are arranged in three lines; doing this, we have

$$\begin{array}{r|rrrr} -2 & 2 & -5 & 6 & -3 \\ & & -4 & 2 & -8 \\ \hline & & -1 & 4 & 5 \end{array}$$

We now write 2, the first coefficient in the dividend, in the first position in the bottom row, and we have

$$\begin{array}{r|rrrr} -2 & 2 & -5 & 6 & -3 \\ & & -4 & 2 & -8 \\ \hline & 2 & -1 & 4 & 5 \end{array}$$

We notice that the first three numbers in the bottom row are the coefficients 2, $-1$, and 4 of the quotient; the last number in the bottom row is 5, and 5 is the remainder. The numbers in the second row are obtained by multiplying the number in the bottom row of the preceding column by $-2$, and the numbers in the bottom row are found by subtracting the numbers in the second row from those of the top row. If the multiplier, $-2$, is replaced by 2, the numbers in the second row can then be added to those of the top row to obtain the numbers in the bottom row. We make this change and the work appears as follows.

$$\begin{array}{r|rrrr} 2 & 2 & -5 & 6 & -3 \\ & & 4 & -2 & 8 \\ \hline & 2 & -1 & 4 & 5 \end{array}$$

This arrangement of the computation is the synthetic division of the polynomial $(2x^3 - 5x^2 + 6x - 3)$ by $(x - 2)$, with the quotient $(2x^2 - x + 4)$ and the remainder 5.

In general, the following steps give the procedure for synthetic division of a polynomial $P(x)$ by $(x - r)$.

1. Write $P(x)$ in the form $(a_0 x^n + a_1 x^{n-1} + a_2 x^{n-2} + \cdots + a_{n-1} x + a_n)$ and insert a zero coefficient for any missing term.
2. Write the coefficients of $P(x)$ in order in a horizontal row.
3. Bring down the first coefficient $a_0$ of $P(x)$ to the bottom row.
4. Multiply $a_0$ by $r$, and write the product in the second row below the coefficient $a_1$; then add the product to $a_1$ and write the sum in the bottom row.
5. Multiply this sum by $r$ and write the product in the second row below the coefficient $a_2$; add the product to $a_2$ and write the sum in the bottom row.
6. Continue the process of steps 4 and 5 as long as possible.
7. The last number in the bottom row is the remainder and the preceding numbers are the coefficients of the successive terms of the quotient.

**ILLUSTRATION 1.** We use synthetic division to find the quotient and remainder when $(x^4 - 7x^2 + 2x - 6)$ is divided by $(x + 3)$. Because we are dividing by $(x + 3)$ or, equivalently, $[x - (-3)]$, $r$ is $-3$. The coefficients of $P(x)$ are $1, 0, -7, 2, -6$ (we insert a zero coefficient for the missing term involving $x^3$). The computation has the following form.

$$\begin{array}{r|rrrrr} -3 & 1 & 0 & -7 & 2 & -6 \\ & & -3 & 9 & -6 & 12 \\ \hline & 1 & -3 & 2 & -4 & 6 \end{array}$$

Thus the quotient is $(x^3 - 3x^2 + 2x - 4)$, and the remainder is 6. •

**EXAMPLE 1**
Use synthetic division to find the quotient and remainder when
$$x^5 + 2x^4 - 5x^3 - 6x^2 - x + 5$$
is divided by $(x - 2)$.

**SOLUTION**

$$\begin{array}{r|rrrrrr} 2 & 1 & 2 & -5 & -6 & -1 & 5 \\ & & 2 & 8 & 6 & 0 & -2 \\ \hline & 1 & 4 & 3 & 0 & -1 & 3 \end{array}$$

The quotient is $x^4 + 4x^3 + 3x^2 - 1$, and the remainder is 3.

**EXAMPLE 2**
Use synthetic division to find the quotient and remainder when $(2x^4 - 3x^3 + x - 5)$ is divided by $(x - i)$.

**SOLUTION**

$$\begin{array}{r|rrrrr} i & 2 & -3 & 0 & 1 & -5 \\ & & 2i & -2-3i & 3-2i & 2+4i \\ \hline & 2 & -3+2i & -2-3i & 4-2i & -3+4i \end{array}$$

The quotient is $[2x^3 + (-3 + 2i)x^2 + (-2 - 3i)x + (4 - 2i)]$, and the remainder is $(-3 + 4i)$.

The remainder theorem states that for a given polynomial $P(x)$, the value of $P(r)$ is the remainder when $P(x)$ is divided by $(x - r)$. Because synthetic

## Synthetic Division

division provides a fast way of obtaining this remainder, it is usually easier to compute $P(r)$ by synthetic division than by direct substitution.

**EXAMPLE 3**
If $P(x) = 2x^5 + 4x^4 - 10x^3 - 20x - 10$, find $P(0)$, $P(-1)$, $P(3)$, and $P(-4)$.

**SOLUTION**
We obtain $P(0)$ and $P(-1)$ by direct substitution.
$$P(0) = -10 \qquad P(-1) = -2 + 4 + 10 + 20 - 10 = 22$$

We obtain $P(3)$ and $P(-4)$ by synthetic division.

$$\begin{array}{r|rrrrrr} 3 & 2 & 4 & -10 & 0 & -20 & -10 \\ & & 6 & 30 & 60 & 180 & 480 \\ \hline & 2 & 10 & 20 & 60 & 160 & 470 \end{array}$$

$$\begin{array}{r|rrrrrr} -4 & 2 & 4 & -10 & 0 & -20 & -10 \\ & & -8 & 16 & -24 & 96 & -304 \\ \hline & 2 & -4 & 6 & -24 & 76 & -314 \end{array}$$

Thus $P(3) = 470$ and $P(-4) = -314$.

**EXAMPLE 4**
Use synthetic division to determine if the linear expression is a factor of the given $P(x)$.

(a) $(x - 2)$; $P(x) = 4x^3 - 7x^2 + x - 2$
(b) $(x + 3)$; $P(x) = 2x^4 + 5x^3 + 11x + 6$

**SOLUTION**
(a) We use synthetic division to find $P(2)$.

$$\begin{array}{r|rrrr} 2 & 4 & -7 & 1 & -2 \\ & & 8 & 2 & 6 \\ \hline & 4 & 1 & 3 & 4 \end{array}$$

Therefore, $P(2) = 4$. Because $P(2) \neq 0$, we conclude from the factor theorem that $(x - 2)$ is not a factor of $P(x)$.

(b) We compute $P(-3)$ by synthetic division.

$$\begin{array}{r|rrrrr} -3 & 2 & 5 & 0 & 11 & 6 \\ & & -6 & 3 & -9 & -6 \\ \hline & 2 & -1 & 3 & 2 & 0 \end{array}$$

Because $P(-3) = 0$, it follows from the factor theorem that $(x + 3)$ is a factor of $P(x)$.

**EXAMPLE 5**
Show by synthetic division that $(x + 2i)$ is a factor of $x^3 - x^2 + 4x - 4$.

**SOLUTION**
Let $P(x) = x^3 - x^2 + 4x - 4$. We use synthetic division to find $P(-2i)$.

$$\begin{array}{r|rrrr} -2i & 1 & -1 & 4 & -4 \\ & & -2i & -4+2i & 4 \\ \hline & 1 & -1-2i & 2i & 0 \end{array}$$

Hence $P(-2i) = 0$. Therefore, by the factor theorem it follows that $(x + 2i)$ is a factor of $P(x)$.

## EXERCISES 5.5

*In Exercises 1 through 16, use synthetic division to find the quotient and remainder.*

1. $(2x^3 - x^2 + 3x + 12) \div (x - 4)$
2. $(y^3 + 4y^2 + 3y - 6) \div (y - 2)$
3. $(2x^4 + 5x^3 - 2x - 1) \div (x + 4)$
4. $(x^3 + 4x^2 - 7) \div (x + 3)$
5. $(3z^5 + z^4 - 4z^2 + 7) \div (z - 2)$
6. $(4x^6 + 21x^5 - 26x^3 + 27x) \div (x + 5)$
7. $(x^5 - 5x^4 + 15x^2) \div (x - 4)$
8. $(8x^3 - 6x^2 + 5x - 3) \div (x - \frac{1}{4})$
9. $(6x^3 - x^2 + 2x + 2) \div (x + \frac{1}{3})$
10. $(2x^4 - x^3 - 2x^2 - 4) \div (x + \frac{3}{2})$
11. $(x^7 - 1) \div (x - 1)$
12. $(x^7 + 1) \div (x + 1)$
13. $(x^4 + 3ix^3 - 2x^2 + ix - 3) \div (x + i)$
14. $(x^4 - ix^2 + 1) \div (x - 1 + i)$
15. $(x^3 + 2x^2 + 2) \div (x + 1 - i)$
16. $(x^4 - 2x^3 + 5x - 2) \div (x - 2i)$

*In Exercises 17 through 26, use synthetic division.*

17. If $P(x) = 4x^3 - 5x^2 - 4$, find $P(2)$ and $P(-3)$.
18. If $P(x) = 3x^3 + 4x^2 - 9$, find $P(-2)$ and $P(1)$.
19. If $P(x) = x^4 + 3x^3 - 5x^2 + 9$, find $P(-4)$ and $P(3)$.
20. If $P(x) = 2x^4 - 7x^3 - 15x + 1$, find $P(4)$ and $P(-2)$.
21. If $P(x) = 6x^3 - x^2 - 7x + 2$, find $P(-\frac{1}{3})$ and $P(\frac{3}{2})$.
22. If $P(x) = x^3 + 2x + 4$, find $P(-1.3)$ and $P(2.1)$.
23. If $P(x) = 2x^3 + 6x + 3$, find $P(2i)$ and $P(-i)$.
24. If $P(x) = x^4 + 8x^2 + 2x$, find $P(-3i)$ and $P(2i)$.
25. If $P(x) = 3x^3 - 3x^2 + 10x - 5$, find $P(1 - 2i)$ and $P(1 + 2i)$.
26. If $P(x) = x^3 - 4x^2 - 2x + 5$, find $P(3 + i)$ and $P(3 - i)$.

*In Exercises 27 through 34, show by synthetic division that the linear expression is a factor of $P(x)$.*

27. $(x + 3)$; $P(x) = 4x^3 + 9x^2 - 8x + 3$
28. $(x + 5)$; $P(x) = 2x^3 + 9x^2 - 3x + 10$
29. $(2x - 1)$; $P(x) = 6x^3 - 7x^2 + 4x - 1$ (*Hint:* $2x - 1 = 2(x - \frac{1}{2})$.)
30. $(3x + 2)$; $P(x) = 12x^3 + 5x^2 - 11x - 6$ (*Hint:* $3x + 2 = 3(x + \frac{2}{3})$.)
31. $(x + 3i)$; $P(x) = x^4 - 3x^3 + 11x^2 - 27x + 18$
32. $(x - 4i)$; $P(x) = x^4 - x^3 + 15x^2 - 16x - 16$
33. $(3x - 2 - i)$; $P(x) = 9x^3 - 21x^2 + 17x - 5$
34. $(4x + 1 + i)$; $P(x) = 8x^3 - 12x^2 - 7x - 2$

*In Exercises 35 through 38, use synthetic division to determine if the linear expression is a factor of $P(x)$.*

35. $(x - 4)$; $P(x) = 2x^4 - 7x^3 - 14x + 8$
36. $(x - 3)$; $P(x) = x^4 - 6x^2 - 5x - 12$
37. $(2x + 3)$; $P(x) = 4x^3 - 4x^2 - 11x + 6$
38. $(3x - 1)$; $P(x) = 9x^3 + 3x^2 - 5x - 1$

## 5.6 Complex Zeros of Polynomial Functions

The zeros of a polynomial function $P$ are the roots of the corresponding polynomial equation $P(x) = 0$; that is, $r$ is a zero of $P$ if $P(r) = 0$.

**ILLUSTRATION 1.** In Example 5, Section 5.5 we showed that if $P(x) = x^3 - x^2 + 4x - 4$, then $P(-2i) = 0$. Therefore, $-2i$ is a zero of the polynomial function $\{(x, P(x)) | P(x) = x^3 - x^2 + 4x - 4\}$. •

It is difficult to find the zeros of a polynomial function, except in special cases. For instance, the zeros of a quadratic function can be obtained by solving the corresponding polynomial equation by the quadratic formula (as shown in Section 3.4); however, for a polynomial function of degree three or four, the general method for obtaining the zeros is quite complicated. Furthermore, for the zeros of a polynomial function of degree greater than four, there is no general formula in terms of a finite number of operations on the coefficients.

Even though there is difficulty in determining zeros of polynomial functions, there is a theorem, called the "fundamental theorem of algebra," which guarantees that every polynomial function of nonzero degree has at least one complex zero.

**5.6.1 THEOREM** *Fundamental Theorem of Algebra.* Every polynomial function of degree greater than zero, with complex coefficients, has at least one complex zero.

The proof of this theorem is omitted because it requires concepts beyond the level of this book. It should be noted that if the complex zero of a polynomial is $a + bi$, where $b = 0$, then the number is referred to as a *real zero*.

Theorem 5.6.1 and the factor theorem are used to prove the next theorem.

**5.6.2 THEOREM** If $P(x)$ is the polynomial with complex coefficients defined by

$$P(x) = a_0 x^n + a_1 x^{n-1} + a_2 x^{n-2} + \cdots + a_{n-1} x + a_n$$

where $n \geq 1$, then

$$P(x) = a_0(x - r_1)(x - r_2) \cdots (x - r_n) \qquad a_0 \neq 0 \qquad (1)$$

where each $r_i$ ($i = 1, 2, \ldots, n$) is a complex zero of $P$.

***Proof.*** From the fundamental theorem of algebra (Theorem 5.6.1), the polynomial function $P$ has at least one complex zero, $r_1$. That is, there exists a complex number $r_1$ such that $P(r_1) = 0$. Therefore, by the factor theorem

(Theorem 5.4.4), $(x - r_1)$ is a factor of $P(x)$. Thus

$$P(x) = (x - r_1)Q_1(x) \qquad (2)$$

where $Q_1(x)$ is the quotient obtained when $P(x)$ is divided by $(x - r_1)$, and $Q_1(x)$ is of degree $n - 1$. Applying the fundamental theorem of algebra, if $n - 1 \geq 1$, there exists a complex number, $r_2$, such that $Q_1(r_2) = 0$. Then by the factor theorem,

$$Q_1(x) = (x - r_2)Q_2(x) \qquad (3)$$

where $Q_2(x)$ is the quotient obtained when $Q_1(x)$ is divided by $(x - r_2)$. Substituting from equation (3) into equation (2), we get

$$P(x) = (x - r_1)(x - r_2)Q_2(x)$$

Because $Q_1(r_2) = 0$, it follows from equation (2) that $P(r_2) = 0$, and hence $r_2$ is a complex zero of $P$. We continue this procedure until the factoring has been performed $n$ times; then we have

$$P(x) = (x - r_1)(x - r_2) \cdots (x - r_n)Q_n(x)$$

where each $r_i (i = 1, 2, \ldots, n)$ is a complex zero of $P$. Because there are $n$ factors of the form $(x - r_i)$, the polynomial $Q_n(x)$ must be a constant and that constant must be the coefficient of $x^n$ in the expansion. Thus $Q_n(x) = a_0$, and therefore

$$P(x) = a_0(x - r_1)(x - r_2) \cdots (x - r_n)$$

where each $r_i$ is a complex zero of $P$.

If in equation (1), a factor $(x - r_i)$ occurs $k$ times, then $r_i$ is called a *zero of multiplicity k*. If a zero of multiplicity $k$ is counted as $k$ zeros, then it follows from Theorem 5.6.2 that a polynomial function $P$, for which $P(x)$ is of degree $n \geq 1$, has *at least n zeros* (some of which may be repeated). However, we can prove that such a polynomial function has *exactly n zeros*, and this fact is stated as the next theorem.

**5.6.3 THEOREM**   If $P(x)$ is a polynomial of degree $n \geq 1$, with complex coefficients, then $P$ has exactly $n$ complex zeros.

*Proof.* From Theorem 5.6.2, $P$ has at least $n$ complex zeros. If we now show that $P$ can not have more than $n$ zeros, then the theorem will be proved.

Let $r$ be any number other than $r_1, r_2, \ldots, r_n$, the $n$ complex zeros of $P$ given by Theorem 5.6.2. Because equation (1) is an identity (it is true for all values of $x$), it follows that

$$P(r) = a_0(r - r_1)(r - r_2) \cdots (r - r_n) \qquad (4)$$

None of the factors $(r - r_i)$ is zero because $r \neq r_i$ $(i = 1, 2, \ldots, n)$; furthermore, $a_0 \neq 0$. Therefore, the right member of equation (4) is not zero, and so $P(r) \neq 0$. Thus $r$ is not a zero of $P$. Hence $P$ has exactly $n$ complex zeros.

**ILLUSTRATION 2.** The function $P$ defined by

$$P(x) = (x - 4)^3(x + 1)^2(x - 3)$$

has six zeros, and they are 4, 4, 4, $-1$, $-1$, and 3. The number 4 is a zero of multiplicity three, and $-1$ is a zero of multiplicity two. •

**EXAMPLE 1**
Show that 3 is a zero of multiplicity two of the polynomial function $\{(x, P(x)) | P(x) = 2x^4 - 11x^3 + 11x^2 + 15x - 9\}$, and find the other two zeros.

**SOLUTION**
To show that 3 is a zero of multiplicity two of the given polynomial function, we show that $(x - 3)^2$ is a factor of the polynomial $P(x)$. We use synthetic division to divide $P(x)$ by $(x - 3)$.

$$\begin{array}{r|rrrr} 3 & 2 & -11 & 11 & 15 & -9 \\ & & 6 & -15 & -12 & 9 \\ \hline & 2 & -5 & -4 & 3 & 0 \end{array}$$

Hence

$$2x^4 - 11x^3 + 11x^2 + 15x - 9 = (x - 3)(2x^3 - 5x^2 - 4x + 3) \quad (5)$$

We now divide $(2x^3 - 5x^2 - 4x + 3)$ by $(x - 3)$.

$$\begin{array}{r|rrrr} 3 & 2 & -5 & -4 & 3 \\ & & 6 & 3 & -3 \\ \hline & 2 & 1 & -1 & 0 \end{array}$$

Therefore,

$$2x^3 - 5x^2 - 4x + 3 = (x - 3)(2x^2 + x - 1) \quad (6)$$

Substituting from equation (6) into equation (5), we obtain

$$2x^4 - 11x^3 + 11x^2 + 15x - 9 = (x - 3)^2(2x^2 + x - 1)$$

The quadratic factor can now be factored into two linear factors, and we have

$$2x^4 - 11x^3 + 11x^2 + 15x - 9 = (x - 3)^2(2x - 1)(x + 1)$$

Because $(2x - 1) = 2(x - \frac{1}{2})$, it follows that

$$2x^4 - 11x^3 + 11x^2 + 15x - 9 = 2(x - 3)^2(x - \frac{1}{2})(x + 1)$$

Thus the zeros of the given polynomial function are 3, 3, $\frac{1}{2}$, and $-1$.

For each theorem relating to the zeros of a polynomial function, we have a statement regarding the roots of a polynomial equation. For instance, from

Theorem 5.6.3, we have the statement that a polynomial equation of degree $n$, with complex coefficients, has exactly $n$ complex roots.

**ILLUSTRATION 3.** The polynomial equation corresponding to the polynomial function of Illustration 2 is

$$(x - 4)^3(x + 1)^2(x - 3) = 0$$

This is an equation of the sixth degree, and the six roots are 4, 4, 4, −1, −1, and 3. •

EXAMPLE 2
Given that 3 is a root of the equation

$$x^3 - 4x^2 + 5x - 6 = 0$$

find the other two roots.

SOLUTION
Because 3 is a root of the given equation, 3 is a zero of the polynomial function $\{(x,P(x)) | P(x) = x^3 - 4x^2 + 5x - 6\}$. Thus $(x - 3)$ is a factor of the polynomial $P(x)$, and another factor can be found by dividing $P(x)$ by $(x - 3)$. We use synthetic division.

$$\begin{array}{r|rrrr} 3 & 1 & -4 & 5 & -6 \\ & & 3 & -3 & 6 \\ \hline & 1 & -1 & 2 & 0 \end{array}$$

Therefore,

$$x^3 - 4x^2 + 5x - 6 = (x - 3)(x^2 - x + 2)$$

Hence the given equation can be written as

$$(x - 3)(x^2 - x + 2) = 0$$

Equating each factor to zero, we have

$$x - 3 = 0 \qquad x^2 - x + 2 = 0$$

$$x = 3 \qquad x = \frac{1 \pm \sqrt{(-1)^2 - 4(1)(2)}}{2(1)}$$

$$= \frac{1 \pm i\sqrt{7}}{2}$$

Thus the other two roots are $\dfrac{1 + i\sqrt{7}}{2}$ and $\dfrac{1 - i\sqrt{7}}{2}$.

EXAMPLE 3
Given that −1 and $2i$ are roots of the equation

$$x^4 + x^3 + 3x^2 + (3 + 2i)x + 2i = 0$$

find the other two roots.

SOLUTION
Because −1 and $2i$ are roots of the given equation, −1 and $2i$ are zeros of the polynomial function

$$\{(x, P(x)) | P(x) = x^4 + x^3 + 3x^2 + (3 + 2i)x + 2i\}.$$

We use synthetic division to divide $P(x)$ by $(x + 1)$, and then we divide the quotient by $(x - 2i)$.

```
-1 | 1   1    3    3+2i   2i
       -1    0   -3      -2i
  2i | 1    0    3    2i      0
            2i  -4   -2i
       1   2i   -1    0
```

Therefore,
$$x^4 + x^3 + 3x^2 + (3+2i)x + 2i = (x+1)(x-2i)(x^2+2ix-1)$$

Thus the given equation can be written as
$$(x+1)(x-2i)(x^2+2ix-1) = 0$$

Equating each factor to zero, we get

$x+1 = 0 \qquad x-2i = 0 \qquad x^2+2ix-1 = 0$

$x = -1 \qquad x = 2i \qquad x = \dfrac{-2i \pm \sqrt{(2i)^2 - 4(1)(-1)}}{2(1)}$

$$= \dfrac{-2i \pm \sqrt{-4+4}}{2}$$

$$= \dfrac{-2i \pm 0}{2}$$

Therefore, the other two roots are $-i$ and $-i$; that is, $-i$ is a root of multiplicity two.

From the results of Example 2, we see that the polynomial function $\{(x, P(x)) | P(x) = x^3 - 4x^2 + 5x - 6\}$ has the two zeros $\frac{1}{2}(1 + i\sqrt{7})$ and $\frac{1}{2}(1 - i\sqrt{7})$, and they are conjugates of each other. Furthermore, from the quadratic formula, it follows that if a quadratic function, having real coefficients, has the complex zero $a + bi$, then the other zero is its conjugate $a - bi$. These two situations are special cases of the following theorem.

**5.6.4 THEOREM** If $P(x)$ is a polynomial with real coefficients, and if $z$ is a complex zero of $P$, then the conjugate $\bar{z}$ is also a zero of $P$.

**Proof.** Let
$$P(x) = a_0 x^n + a_1 x^{n-1} + a_2 x^{n-2} + \cdots + a_{n-1} x + a_n$$
where all the coefficients $a_i$ are real numbers. Because $z$ is a zero of $P$, $P(z) = 0$; thus
$$a_0 z^n + a_1 z^{n-1} + a_2 z^{n-2} + \cdots + a_{n-1} z + a_n = 0$$
Hence
$$\overline{a_0 z^n + a_1 z^{n-1} + a_2 z^{n-2} + \cdots + a_{n-1} z + a_n} = \bar{0} \qquad (7)$$

From equation (8) in Section 5.1 (the conjugate of the sum of complex numbers equals the sum of the conjugates of the numbers) and the fact that $\overline{0} = 0$, we have from equation (7)

$$\overline{a_0 z^n} + \overline{a_1 z^{n-1}} + \overline{a_2 z^{n-2}} + \cdots + \overline{a_{n-1} z} + \overline{a_n} = 0$$

Applying Theorem 5.1.3(ii) in each term of the left member of this equation, we have

$$\overline{a_0}\,\overline{z^n} + \overline{a_1}\,\overline{z^{n-1}} + \overline{a_2}\,\overline{z^{n-2}} + \cdots + \overline{a_{n-1}}\,\overline{z} + \overline{a_n} = 0 \qquad (8)$$

Because each $a_i$ is a real number $\overline{a_i} = a_i$ ($i = 1, 2, \ldots, n$); furthermore, from Theorem 5.1.3(iii), $\overline{z^i} = \overline{z}^i$. Thus equation (8) can be written as

$$a_0 \overline{z}^n + a_1 \overline{z}^{n-1} + a_2 \overline{z}^{n-2} + \cdots + a_{n-1}\overline{z} + a_n = 0 \qquad (9)$$

The left member of equation (9) is $P(\overline{z})$. Therefore, $P(\overline{z}) = 0$, and hence $\overline{z}$ is a zero of $P$.

### EXAMPLE 4

Find a polynomial $P(x)$ of the fourth degree with real coefficients if $P$ has $1 - i$ and $-2i$ as zeros.

**SOLUTION**

From Theorem 5.6.4, if $1 - i$ and $-2i$ are zeros of $P$, then their conjugates, $1 + i$ and $2i$, are also zeros. Hence

$$\begin{aligned} P(x) &= [x - (1 - i)][x - (1 + i)][x - (-2i)][x - 2i] \\ &= (x^2 - 2x + 2)(x^2 + 4) \\ &= x^4 - 2x^3 + 6x^2 - 8x + 8 \end{aligned}$$

### EXAMPLE 5

Given that $i$ is a root of the equation

$$2x^4 - 5x^3 + 3x^2 - 5x + 1 = 0$$

find the solution set of the equation.

**SOLUTION**

Because $i$ is a root of the given equation with real coefficients, it follows from Theorem 5.6.4 that its conjugate, $-i$, is also a root. We use synthetic division to divide the polynomial $(2x^4 - 5x^3 + 3x^2 - 5x + 1)$ by $(x - i)$, and then we divide the quotient by $[x - (-i)]$.

$$\begin{array}{r|rrrrr} i & 2 & -5 & 3 & -5 & 1 \\ & & 2i & -2-5i & 5+i & -1 \\ \hline -i & 2 & -5+2i & 1-5i & i & 0 \\ & & -2i & 5i & -i & \\ \hline & 2 & -5 & 1 & 0 & \end{array}$$

Therefore, the given equation can be written as

$$(x - i)(x + i)(2x^2 - 5x + 1) = 0$$

Equating each factor to zero, we obtain

$$\begin{array}{ccc} x - i = 0 & x + i = 0 & 2x^2 - 5x + 1 = 0 \\ x = i & x = -i & x = \dfrac{5 \pm \sqrt{25 - 8}}{4} \\ & & = \dfrac{5 \pm \sqrt{17}}{4} \end{array}$$

Thus the solution set of the given equation is
$$\left\{i, -i, \frac{5 + \sqrt{17}}{4}, \frac{5 - \sqrt{17}}{4}\right\}.$$

Observe that if $P(x)$ is a polynomial with complex coefficients, the conclusion of Theorem 5.6.4 is not necessarily true. For instance, in Example 3
$$P(x) = x^4 + x^3 + 3x^2 + (3 + 2i)x + 2i$$
and the zeros of $P$ are $-1$, $2i$, $-i$, and $-i$. The coefficient of $x$ and the constant term in the polynomial are complex numbers but not real numbers; even though $2i$ and $-i$ are zeros of $P$, their conjugates are not.

We now state two interesting theorems that are immediate consequences of Theorem 5.6.4. Their proofs are left as exercises (see Exercises 37 and 38 in Exercises 5.6).

**5.6.5 THEOREM**  If $P(x)$ is a polynomial with real coefficients and the degree of $P$ is an odd number, then $P$ has at least one real zero.

**5.6.6 THEOREM**  If $P(x)$ is a polynomial with real coefficients, then $P(x)$ can be expressed as a product of factors which are either linear or quadratic.

## EXERCISES 5.6

*In Exercises 1 through 8, find the zeros of the given polynomial function. State the multiplicity of each zero.*

1. $\{(x, P(x)) | P(x) = (x - 4)^2(x^2 - 4)\}$
2. $\{(x, P(x)) | P(x) = x^3(x^2 - 5)\}$
3. $\{(x, P(x)) | P(x) = x^2(x + 1)^2(x^2 + 1)\}$
4. $\{(x, P(x)) | P(x) = (x^2 - 25)^2\}$
5. $\{(x, P(x)) | P(x) = (x + 7)^3(x^2 - 7)^2\}$
6. $\{(x, P(x)) | P(x) = (x^2 + 2)^2(x^2 - 2)(2x + 1)\}$
7. $\{(x, P(x)) | P(x) = (3x + 4)^3(4x^2 - 9)^2(4x^2 + 9)\}$
8. $\{(x, P(x)) | P(x) = (x^2 - 9)^2(5x^2 - 17x + 6)^2\}$

9. Show that $-2$ and $3$ are zeros of the polynomial function
$$\{(x, P(x)) | P(x) = x^4 - 4x^3 - 7x^2 + 22x + 24\}$$
and find the other two zeros.

10. Show that $5$ and $-1$ are zeros of the polynomial function
$$\{(x, P(x)) | P(x) = x^4 + x^3 - 31x^2 - x + 30\}$$
and find the other two zeros.

11. Show that $-4$ is a zero of multiplicity two of the polynomial function
$$\{(x, P(x)) | P(x) = x^4 + 9x^3 + 23x^2 + 8x - 16\}$$
and find the other two zeros.

12. Show that 2 is a zero of multiplicity three of the polynomial function
$$\{(x, P(x))|P(x) = 3x^5 - 18x^4 + 38x^3 - 36x^2 + 24x - 16\}$$
and find the other two zeros.

13. Show that both $i$ and $-i$ are zeros of multiplicity two of the polynomial function
$$\{(x, P(x))|P(x) = x^6 + 4x^4 + 5x^2 + 2\}$$
and find the other two zeros.

14. Show that 2 is a zero and $2i$ is a zero of multiplicity two of the polynomial function
$$\{(x, P(x))|P(x) = x^5 - 2x^4 + 8x^3 - 16x^2 + 16x - 32\}$$
and find the other two zeros.

15. Show that $2 + 3i$ and $2 - 3i$ are zeros of the polynomial function
$$\{(x, P(x))|P(x) = 2x^4 - 7x^3 + 21x^2 + 17x - 13\}$$
and find the other two zeros.

16. Show that $1 + i\sqrt{2}$ and $1 - i\sqrt{2}$ are zeros of the polynomial function
$$\{(x, P(x))|P(x) = x^4 - 2x^2 + 8x - 3\}$$
and find the other two zeros.

17. Given that $-2$ is a root of the equation
$$5x^3 + 3x^2 - 12x + 4 = 0$$
find the other two roots.

18. Given that $\frac{4}{3}$ is a root of the equation
$$3x^3 - 16x^2 + 28x - 16 = 0$$
find the other two roots.

19. Given that $\frac{1}{2}$ and $-\frac{2}{3}$ are roots of the equation
$$6x^4 + 25x^3 + 8x^2 - 7x - 2 = 0$$
find the other two roots.

20. Given that $\sqrt{3}$ and $-\sqrt{3}$ are roots of the equation
$$x^4 + 3x^3 - 5x^2 - 9x + 6 = 0$$
find the other two roots.

21. Given that $i\sqrt{2}$ and $-i\sqrt{2}$ are roots of the equation
$$x^4 - 3x^3 + 4x^2 - 6x + 4 = 0$$
find the other two roots.

22. Given that $-3i$ is a root of the equation
$$x^3 + 4ix^2 - x + 6i = 0$$
find the other two roots.

23. Given that $1 - i$ is a root of the equation
$$x^3 - x^2 + (-2 + i)x + 3 - 3i = 0$$
find the other two roots.

24. Given that $-2 + i$ and $-i$ are roots of the equation
$$x^4 + 2x^3 + 2ix^2 - 2x - 1 - 2i = 0$$
find the other two roots.

*In Exercises 25 through 32, find a polynomial $P(x)$ of the stated degree with real coefficients for which the given numbers are zeros of P.*

25. Second degree; $4 + 3i$ is a zero of $P$.
26. Second degree; $3 - i$ is a zero of $P$.
27. Third degree; $2 - i\sqrt{5}$ and $-4$ are zeros of $P$.
28. Third degree; $5 + i\sqrt{3}$ and $2$ are zeros of $P$.
29. Fourth degree; $-3i$ is a zero of multiplicity two of $P$.
30. Fourth degree; $2 + i$ and $1 - i\sqrt{2}$ are zeros of $P$.
31. Fifth degree; $3, 3 + i\sqrt{2}$ and $-i\sqrt{2}$ are zeros of $P$.
32. Fifth degree; $3 - i$ (multiplicity two) and $1$ are zeros of $P$.

*In Exercises 33 through 36, find the solution set of the given equation if the given number is a root.*

33. $3x^4 - 2x^3 + 2x^2 - 8x - 40 = 0$; $2i$ is a root.
34. $5x^4 - 2x^3 + 46x^2 - 18x + 9 = 0$; $-3i$ is a root.
35. $2x^4 + 6x^3 + 33x^2 - 36x + 20 = 0$; $-2 - 4i$ is a root.
36. $3x^4 + 4x^3 + 9x^2 - 6x + 4 = 0$; $-1 + i\sqrt{3}$ is a root.
37. Prove Theorem 5.6.5.
38. Prove Theorem 5.6.6.

## 5.7 Rational Zeros of Polynomial Functions

We stated in Section 5.6 that it is difficult to find the zeros of a polynomial function, except in special cases. One of these special cases occurs when the coefficients are integers; in such a situation, if there are any rational zeros, they can be found by applying the following theorem.

**5.7.1 THEOREM** Suppose that
$$P(x) = a_0x^n + a_1x^{n-1} + a_2x^{n-2} + \cdots + a_{n-1}x + a_n$$

where $a_0, a_1, a_2, \cdots, a_n$ are integers. If $\frac{p}{q}$, in lowest terms, is a rational number and a zero of $P$, then $p$ is an integer factor of $a_n$ and $q$ is an integer factor of $a_0$.

**Proof.** Because $\frac{p}{q}$ is a zero of $P$, it is a solution of the equation $P(x) = 0$. Therefore,

$$a_0 \left(\frac{p}{q}\right)^n + a_1 \left(\frac{p}{q}\right)^{n-1} + a_2 \left(\frac{p}{q}\right)^{n-2} + \cdots + a_{n-1} \left(\frac{p}{q}\right) + a_n = 0$$

Multiplying each member of this equation by $q^n$, we obtain

$$a_0 p^n + a_1 p^{n-1} q + a_2 p^{n-2} q^2 + \cdots + a_{n-1} p q^{n-1} + a_n q^n = 0 \qquad (1)$$

We now add $-a_n q^n$ to each member of equation (1) and factor $p$ from each term in the resulting left member. Thus we have the equivalent equation

$$p(a_0 p^{n-1} + a_1 p^{n-2} q + a_2 p^{n-3} q^2 + \cdots + a_{n-1} q^{n-1}) = -a_n q^n \qquad (2)$$

Because $a_i$ ($i = 1, 2, \cdots, n$), $p$, and $q$ are integers, and the set of integers is closed with respect to addition and multiplication, the expression in parentheses in the left member of equation (2) is an integer, which we represent by $t$. Therefore, equation (2) can be written as

$$pt = -a_n q^n \qquad (3)$$

The left member of equation (3) is an integer having $p$ as a factor. Therefore, $p$ must be a factor of the right member, $-a_n q^n$. Because $\frac{p}{q}$ is in lowest terms, $p$ has no factor in common with $q$. Thus $p$ must be a factor of $a_n$.

Equation (1) is also equivalent to the equation

$$q(a_1 p^{n-1} + a_2 p^{n-2} q + \cdots + a_{n-1} p q^{n-2} + a_n q^{n-1}) = -a_0 p^n \qquad (4)$$

Now, the left member of equation (4) is an integer having $q$ as a factor; hence $q$ must be a factor of the right member, $-a_0 p^n$. Because $q$ has no factor in common with $p$, it follows that $q$ must be a factor of $a_0$.

**ILLUSTRATION 1.** If

$$P(x) = 4x^3 + 14x^2 + 10x - 3$$

then by Theorem 5.7.1, any rational zero $\frac{p}{q}$ of $P$ must be such that $p$ is an integer factor of $-3$ and $q$ is an integer factor of 4. Thus the possible values of $p$ are $1, -1, 3,$ and $-3$; and the possible values of $q$ are $1, -1, 2, -2, 4,$

and $-4$. Hence the set of possible rational zeros of $P$ is

$$\left\{1, -1, 3, -3, \frac{1}{2}, -\frac{1}{2}, \frac{3}{2}, -\frac{3}{2}, \frac{1}{4}, -\frac{1}{4}, \frac{3}{4}, -\frac{3}{4}\right\}$$

It should be noted that Theorem 5.7.1 does not guarantee that a polynomial function with integer coefficients has a rational zero; however, the theorem provides the means of locating the numbers that could be rational zeros. We can then use synthetic division to determine if any of the possible zeros are indeed actual zeros.

**ILLUSTRATION 2.** In Illustration 1 we found the set of all possible rational zeros of the polynomial

$$P(x) = 4x^3 + 14x^2 + 10x - 3$$

Because the polynomial is of the third degree, not more than three of the possibilities can be zeros. We now use synthetic division to ascertain which of them, if any, are zeros.

$$\begin{array}{r|rrrr} 1 & 4 & 14 & 10 & -3 \\ & & 4 & 18 & 28 \\ \hline & 4 & 18 & 28 & 25 \end{array} \qquad \begin{array}{r|rrrr} -1 & 4 & 14 & 10 & -3 \\ & & -4 & -10 & 0 \\ \hline & 4 & 10 & 0 & -3 \end{array}$$

$$\begin{array}{r|rrrr} 3 & 4 & 14 & 10 & -3 \\ & & 12 & 78 & 264 \\ \hline & 4 & 26 & 88 & 261 \end{array} \qquad \begin{array}{r|rrrr} -3 & 4 & 14 & 10 & -3 \\ & & -12 & -6 & -12 \\ \hline & 4 & 2 & 4 & -15 \end{array}$$

$$\begin{array}{r|rrrr} \frac{1}{2} & 4 & 14 & 10 & -3 \\ & & 2 & 8 & 9 \\ \hline & 4 & 16 & 18 & 6 \end{array} \qquad \begin{array}{r|rrrr} -\frac{1}{2} & 4 & 14 & 10 & -3 \\ & & -2 & -6 & -2 \\ \hline & 4 & 12 & 4 & -5 \end{array}$$

So far, we have seen that $P(1) = 25$, $P(-1) = -3$, $P(3) = 261$, $P(-3) = -15$, $P(\frac{1}{2}) = 6$, and $P(-\frac{1}{2}) = -5$. We continue.

$$\begin{array}{r|rrrr} \frac{3}{2} & 4 & 14 & 10 & -3 \\ & & 6 & 30 & 60 \\ \hline & 4 & 20 & 40 & 57 \end{array} \qquad \begin{array}{r|rrrr} -\frac{3}{2} & 4 & 14 & 10 & -3 \\ & & -6 & -12 & 3 \\ \hline & 4 & 8 & -2 & 0 \end{array}$$

Because $P(-\frac{3}{2}) = 0$, it follows that $-\frac{3}{2}$ is a zero of $P$. Furthermore, $(x + \frac{3}{2})$ is a factor of $P(x)$, and

$$P(x) = \left(x + \frac{3}{2}\right)(4x^2 + 8x - 2)$$

$$= 2\left(x + \frac{3}{2}\right)(2x^2 + 4x - 1)$$

The other two zeros of $P$ can be found by setting the quadratic factor equal to zero and solving the equation.

$$2x^2 + 4x - 1 = 0$$

$$x = \frac{-4 \pm \sqrt{16 + 8}}{4}$$

$$= \frac{-4 \pm 2\sqrt{6}}{4}$$

$$= \frac{-2 \pm \sqrt{6}}{2}$$

Therefore, the three zeros of $P$ are $-\frac{3}{2}$, $\frac{-2 + \sqrt{6}}{2}$, and $\frac{-2 - \sqrt{6}}{2}$. ●

A special case of Theorem 5.7.1 occurs when $a_0$ (the coefficient of $x^n$) is 1. Then

$$P(x) = x^n + a_1 x^{n-1} + a_2 x^{n-2} + \cdots + a_{n-1} x + a_n$$

where $a_1, a_2, \ldots, a_n$ are integers. For such a polynomial, any rational zero of $P$ must be an integer and, furthermore, must be an integer factor of $a_n$. This follows from the fact that if $\frac{p}{q}$ is a rational zero of $P$, then $p$ must be a factor of $a_n$ and $q$ must be a factor of 1 (the coefficient of $x^n$).

EXAMPLE 1

Find all the rational zeros of the polynomial function $P$ for which

$P(x) = x^5 + 4x^4 - 4x^3 - 34x^2 - 45x - 18$

SOLUTION

The possible rational zeros are integer factors of $-18$. These numbers are 1, $-1$, 2, $-2$, 3, $-3$, 6, $-6$, 9, $-9$, 18, and $-18$. We use synthetic division to determine which of them, if any, are zeros.

$$\underline{1\,|}\ \ 1 \quad\ \ 4 \quad -4 \quad -34 \quad -45 \quad -18$$
$$\phantom{\underline{1\,|}\ \ 1}\quad\ \ 1 \quad\ \ 5 \quad\ \ 1 \quad -33 \quad -78$$
$$\phantom{\underline{1\,|}}\ \ 1 \quad\ \ 5 \quad\ \ 1 \quad -33 \quad -78 \quad -96$$

$$\underline{-1\,|}\ \ 1 \quad\ \ 4 \quad -4 \quad -34 \quad -45 \quad -18$$
$$\phantom{\underline{-1\,|}\ \ 1}\quad -1 \quad -3 \quad\ \ 7 \quad\ \ 27 \quad\ \ 18$$
$$\phantom{\underline{-1\,|}}\ \ 1 \quad\ \ 3 \quad -7 \quad -27 \quad -18 \quad\ \ 0$$

Because $P(-1) = 0$, it follows that $-1$ is a zero of $P$, and

$$P(x) = (x + 1)(x^4 + 3x^3 - 7x^2 - 27x - 18)$$

Any other rational zeros of $P$ must be zeros of the second factor. Because $-1$ is a possible zero of the second factor (and if it is, then $-1$ is a multiple zero of $P$), we divide that factor by $(x + 1)$ to determine if the remainder is zero.

$$\begin{array}{r|rrrrr} -1 & 1 & 3 & -7 & -27 & -18 \\ & & -1 & -2 & 9 & 18 \\ \hline & 1 & 2 & -9 & -18 & 0 \end{array}$$

Therefore,
$$x^4 + 3x^3 - 7x^2 - 27x - 18 = (x + 1)(x^3 + 2x^2 - 9x - 18)$$
so that
$$P(x) = (x + 1)(x + 1)(x^3 + 2x^2 - 9x - 18)$$

Continuing, we check for further rational zeros of $P$ by considering the polynomial $(x^3 + 2x^2 - 9x - 18)$.

$$\begin{array}{r|rrrr} -1 & 1 & 2 & -9 & -18 \\ & & -1 & -1 & 10 \\ \hline & 1 & 1 & -10 & -8 \end{array} \qquad \begin{array}{r|rrrr} 2 & 1 & 2 & -9 & -18 \\ & & 2 & 8 & -2 \\ \hline & 1 & 4 & -1 & -20 \end{array}$$

$$\begin{array}{r|rrrr} -2 & 1 & 2 & -9 & -18 \\ & & -2 & 0 & 18 \\ \hline & 1 & 0 & -9 & 0 \end{array}$$

Thus
$$x^3 + 2x^2 - 9x - 18 = (x + 2)(x^2 - 9)$$
and so
$$P(x) = (x + 1)(x + 1)(x + 2)(x^2 - 9)$$
Factoring the fourth factor, we obtain
$$P(x) = (x + 1)^2(x + 2)(x - 3)(x + 3)$$
Hence the zeros of $P$ are $-1$, $-1$, $-2$, $3$, and $-3$.

In searching for the zeros of a polynomial function with real coefficients, the procedure can often be shortened if "upper and lower bounds" of the zeros can be determined.

**5.7.2 DEFINITION** If $P$ is polynomial function, an *upper bound* of the real zeros of $P$ is any number that is greater than or equal to the largest real zero. A *lower bound* of the real zeros of $P$ is any number that is less than or equal to the smallest real zero.

**ILLUSTRATION 3.** In Example 1 the zeros of the given function are $-3$, $-2$, $-1$, $-1$, and 3. The number 4 is an upper bound of these zeros. Another upper bound is 3. Actually, any number that is greater than or equal to 3 is an upper bound of these zeros. The number $-3$ is a lower bound of these zeros; the numbers $-\frac{7}{2}$ and $-10$ are also lower bounds. Any number that is less than or equal to $-3$ will serve as a lower bound. ●

The following theorem states that upper and lower bounds of the real zeros of a polynomial function $P$ with real coefficients can be determined by observing the behavior of the signs of the numbers in the bottom row of the synthetic division of $P$.

**5.7.3 THEOREM**  Suppose that

$$P(x) = a_0 x^n + a_1 x^{n-1} + a_2 x^{n-2} + \cdots + a_{n-1} x + a_n$$

where $a_0, a_1, a_2, \ldots, a_n$ are real numbers and $a_0 > 0$. In the synthetic division of $P(x)$ by $(x - r)$,

(i) If $r > 0$, and there are no negative numbers in the bottom row, then $r$ is an upper bound of the real zeros of $P$.
(ii) If $r < 0$, and the signs of the numbers in the bottom row are alternately plus and minus (where zero can be appropriately denoted by either $+0$ or $-0$), then $r$ is a lower bound of the real zeros of $P$.

***Proof.*** To prove part (i), we consider the quotient $Q(x)$ and the remainder $R$ when $P(x)$ is divided by $(x - r)$, and we have

$$P(x) = (x - r)Q(x) + R \tag{5}$$

By hypothesis, the numbers in the bottom row of the synthetic division of $P(x)$ by $(x - r)$ are nonnegative. Therefore, $R$ and the coefficients in $Q(x)$ are positive or zero. Furthermore, we are given that the coefficient $a_0$ of $x^{n-1}$ in $Q(x)$ is positive.

To show that $r$ is an upper bound of the real zeros of $P$, we shall show that if $s$ is any positive number for which $s > r$, then $P(s) \neq 0$. If in equation (5), $s$ is substituted for $x$, we have

$$P(s) = (s - r)Q(s) + R \tag{6}$$

Because $s > r$, then $(s - r) > 0$. Furthermore, $s > 0$ and hence $Q(s) > 0$ (remember, the first coefficient in $Q(x)$ is positive and all the other coefficients are nonnegative). Also, $R \geq 0$. Thus, from equation (6), it follows that $P(s) > 0$. Therefore, $P(s) \neq 0$, and hence $r$ is an upper bound of the real zeros of $P$.

The proof of part (ii) is similar by considering

$$P(-x) = (-x - r)Q(-x) + R$$

However, here we must discuss separately the case when $n$ is odd and the case when $n$ is even. This proof is omitted.

**ILLUSTRATION 4.** Suppose that $P(x) = 2x^3 - x^2 + 4x - 2$. Then, if we use synthetic division to divide $P(x)$ by $(x - 1)$, we have

$$\begin{array}{r|rrrr} 1 & 2 & -1 & 4 & -2 \\ & & 2 & 1 & 5 \\ \hline & 2 & 1 & 5 & 3 \end{array}$$

Because all the numbers in the bottom row are positive, it follows from Theorem 5.7.3(i) that 1 is an upper bound of the real zeros of $P$. If we divide $P(x)$ by $(x + 1)$, we have

$$\begin{array}{r|rrrr} -1 & 2 & -1 & 4 & -2 \\ & & -2 & 3 & -7 \\ \hline & 2 & -3 & 7 & -9 \end{array}$$

Because the signs of the numbers in the bottom row are alternately plus and minus, it follows from Theorem 5.7.3(ii) that $-1$ is a lower bound of the real zeros of $P$. ●

Because finding the zeros of a polynomial function $P$ is equivalent to finding the solutions of the polynomial equation $P(x) = 0$, we can apply Theorem 5.7.1 to find the rational roots of a polynomial equation. If all but two of the roots of a polynomial equation are rational, then the nonrational roots can be found by the quadratic formula; this was demonstrated in Illustration 2.

**EXAMPLE 2**
Find the rational roots of the equation

$$\frac{5}{3}x^4 - \frac{2}{3}x^3 - \frac{11}{6}x^2 + 2x = -\frac{3}{2}$$

If possible, find all of the roots.

**SOLUTION**
We first multiply each member of the given equation by 6 in order that the coefficients are integers (recall that to apply Theorem 5.7.1 the coefficients must be integers). We obtain the equivalent equation

$$10x^4 - 4x^3 - 11x^2 + 12x + 9 = 0$$

Let $P(x) = 10x^4 - 4x^3 - 11x^2 + 12x + 9$. If $\frac{p}{q}$ is a rational root of the equation $P(x) = 0$, then $p$ must be an integer factor of 9 and $q$ must be an integer factor of 10. Therefore, the possible values of $p$ are $1, -1, 3, -3, 9$, and $-9$; the possible values of $q$ are $1, -1, 2, -2, 5, -5, 10$, and $-10$. The set of possible rational roots of the equation is, therefore,

$$\left\{1, -1, 3, -3, 9, -9, \frac{1}{2}, -\frac{1}{2}, \frac{3}{2}, -\frac{3}{2}, \frac{9}{2}, -\frac{9}{2}, \frac{1}{5}, -\frac{1}{5}, \frac{3}{5}, \right.$$
$$\left. -\frac{3}{5}, \frac{9}{5}, -\frac{9}{5}, \frac{1}{10}, -\frac{1}{10}, \frac{3}{10}, -\frac{3}{10}, \frac{9}{10}, -\frac{9}{10}\right\}$$

We apply synthetic division to test these possible roots one by one.

$$\begin{array}{r|rrrrr} 1 & 10 & -4 & -11 & 12 & 9 \\ & & 10 & 6 & -5 & 7 \\ \hline & 10 & 6 & -5 & 7 & 16 \end{array} \qquad \begin{array}{r|rrrrr} -1 & 10 & -4 & -11 & 12 & 9 \\ & & -10 & 14 & -3 & -9 \\ \hline & 10 & -14 & 3 & 9 & 0 \end{array}$$

Therefore, $P(-1) = 0$, and so $-1$ is a root of the equation $P(x) = 0$; furthermore, $(x + 1)$ is a factor of $P(x)$. We now apply synthetic division to the quotient $Q(x) = 10x^3 - 14x^2 + 3x + 9$. We test $-1$ by dividing $Q(x)$ by $(x + 1)$.

$$\begin{array}{r|rrrr} -1 & 10 & -14 & 3 & 9 \\ & & -10 & 24 & -27 \\ \hline & 10 & -24 & 27 & -18 \end{array}$$

Because the signs of the numbers in the bottom row are alternately plus and minus, it follows from Theorem 5.7.3(ii) that $-1$ is a lower bound of the real zeros of $Q$ (and hence, of $P$). Thus $-9$, $-\frac{9}{2}$, $-3$, $-\frac{9}{5}$ and $-\frac{3}{2}$ are eliminated as possible roots of the given equation. We test 3.

$$\begin{array}{r|rrrr} 3 & 10 & -14 & 3 & 9 \\ & & 30 & 48 & 153 \\ \hline & 10 & 16 & 51 & 162 \end{array}$$

Because all the numbers in the bottom row are positive, it follows from Theorem 5.7.3(i) that 3 is an upper bound of the real zeros of $Q$ (and hence, of $P$). Therefore, we eliminate 9 and $\frac{9}{2}$ as possible roots of the given equation. We now test $\frac{1}{2}$.

$$\begin{array}{r|rrrr} \frac{1}{2} & 10 & -14 & 3 & 9 \\ & & 5 & -\frac{9}{2} & \\ \hline & 10 & -9 & \text{stop} & \end{array}$$

The notation "stop" is indicated because once a fraction has appeared in the second row, each successive entry in the bottom row will also be a fraction. Thus the last number in the bottom row cannot be zero. A similar situation occurs when we test $-\frac{1}{2}$, $\frac{3}{2}$, $\pm\frac{1}{5}$, and $\frac{3}{5}$. We now test $-\frac{3}{5}$.

$$\begin{array}{r|rrrr} -\frac{3}{5} & 10 & -14 & 3 & 9 \\ & & -6 & 12 & -9 \\ \hline & 10 & -20 & 15 & 0 \end{array}$$

Thus $Q(-\frac{3}{5}) = 0$; consequently, $(x + \frac{3}{5})$ is a factor of $Q(x)$, and therefore, also a factor of $P(x)$. Hence $-\frac{3}{5}$ is a root of the equation $P(x) = 0$. The equation $P(x) = 0$ can be written as

$$(x + 1)\left(x + \frac{3}{5}\right)(10x^2 - 20x + 15) = 0$$

or, equivalently (dividing both members by 5),

$$(x + 1)\left(x + \frac{3}{5}\right)(2x^2 - 4x + 3) = 0$$

We can find the other two roots by solving the quadratic equation

$$2x^2 - 4x + 3 = 0$$

$$x = \frac{4 \pm \sqrt{16 - 24}}{4}$$

$$= \frac{4 \pm 2i\sqrt{2}}{4}$$

$$= \frac{2 \pm i\sqrt{2}}{2}$$

The roots of the given equation are, therefore, $-1$, $-\frac{3}{5}$, $\frac{2 + i\sqrt{2}}{2}$, and $\frac{2 - i\sqrt{2}}{2}$.

**EXAMPLE 3**
Prove that the following equation has no rational roots.

$$2x^3 - 2x^2 - 4x + 1 = 0$$

**SOLUTION**

Let $P(x) = 2x^3 - 2x^2 - 4x + 1$. If $\frac{p}{q}$ is a rational root of the equation $P(x) = 0$, then $p$ must be an integer factor of 1 and $q$ must be an integer factor of 2. The possible values of $p$ are, therefore, 1 and $-1$; the possible values of $q$ are 1, $-1$, 2, and $-2$. Thus the set of possible rational roots of the equation is $\{1, -1, \frac{1}{2}, -\frac{1}{2}\}$. We test each of these possible roots by synthetic division.

```
 1 | 2  -2  -4   1        -1 | 2  -2  -4   1
   |     2   0  -4           |    -2   4   0
   |_____         |_____
     2   0  -4  -3             2  -4   0   1


 1/2 | 2  -2  -4   1      -1/2 | 2  -2  -4   1
     |     1  -1/2              |    -1   3/2
     |_____          |_____
       2  -1  stop                2  -3  stop
```

Therefore, $P(1) \neq 0$, $P(-1) \neq 0$, $P(\frac{1}{2}) \neq 0$, and $P(-\frac{1}{2}) \neq 0$. Hence the equation has no rational roots.

## EXERCISES 5.7

*In Exercises 1 through 12, find all the rational zeros of the given polynomial function. If possible, find all the zeros of the function.*

1. $\{(x, P(x)) \mid P(x) = x^3 - 3x^2 - x + 3\}$
2. $\{(x, P(x)) \mid P(x) = x^3 - 4x^2 + x + 6\}$
3. $\{(x, P(x)) \mid P(x) = x^3 - 7x - 6\}$
4. $\{(x, P(x)) \mid P(x) = x^3 - x^2 - 8x + 12\}$
5. $\{(x, P(x)) \mid P(x) = x^4 + 3x^3 - 12x^2 - 13x - 15\}$
6. $\{(x, P(x)) \mid P(x) = x^4 - 3x^3 + x^2 + 7x - 30\}$
7. $\{(x, P(x)) \mid P(x) = 3x^3 + 8x^2 - 1\}$
8. $\{(x, P(x)) \mid P(x) = 4x^3 - 31x + 15\}$
9. $\{(x, P(x)) \mid P(x) = 6x^4 - 37x^3 + 63x^2 - 33x + 5\}$
10. $\{(x, P(x)) \mid P(x) = 8x^4 + 6x^3 - 13x^2 - x + 3\}$
11. $\{(x, P(x)) \mid P(x) = x^4 - 2x^3 - 9x^2 + 20x - 4\}$
12. $\{(x, P(x)) \mid P(x) = 2x^4 - x^3 + 2x^2 - 7x + 3\}$

*In Exercises 13 through 24, find all the rational roots of the given equation. If possible, find all the roots.*

13. $x^3 + 2x^2 - 7x + 4 = 0$
14. $x^3 - 3x^2 - 10x + 24 = 0$
15. $2x^3 - 13x^2 + 27x - 18 = 0$
16. $x^3 - 8x - 8 = 0$
17. $x^5 + 2x^4 - 13x^3 - 14x^2 + 24x = 0$
18. $9x^4 - 3x^3 + 7x^2 - 3x - 2 = 0$
19. $12x^4 - 5x^3 - 38x^2 + 15x + 6 = 0$
20. $2x^3 - \frac{25}{2}x^2 + \frac{7}{2}x - 3 = 0$
21. $x^3 + \frac{17}{3}x^2 - \frac{5}{3}x + 2 = 0$
22. $3x^4 + x^3 + 12x^2 - 5x - 3 = 0$
23. $\frac{1}{14}x^4 + \frac{1}{7}x^3 - \frac{1}{2}x^2 - \frac{1}{2}x + 1 = 0$
24. $18x^6 + 3x^5 - 25x^4 - 41x^3 - 15x^2 = 0$

*In Exercises 25 through 28, prove that the given equation has no rational roots.*

25. $x^3 - 9x - 6 = 0$
26. $2x^3 + 6x^2 - 3 = 0$
27. $3x^4 - x^3 + 4x^2 + 2x - 2 = 0$
28. $x^4 - x^3 - 4x^2 - 16 = 0$

29. A rectangular box is to be made from a piece of cardboard, 6 centimeters wide and 14 centimeters long, by cutting equal squares from the four corners and turning up the sides. If the volume of the box is to be 40 cubic centimeters, what should be the length of the side of the square to be cut out?
30. The area of a right triangle is 6 square centimeters. Find the lengths of the sides of the triangle if the length of one of the sides is 2 centimeters shorter than the length of the hypotenuse.
31. A slice of thickness 1 centimeter is cut off from one side of a cube. If the volume of the remaining figure is 180 cubic centimeters, how long is the edge of the original cube?

## 5.8 Graphs of Polynomial Functions

Before discussing the general polynomial fuction, we consider the power function defined by

$$f(x) = ax^n$$

where $n$ is a positive integer. Sketches of the graphs of the power function for $a = 1$ and $n$ having values 1, 2, 3, 4, 5, and 6 are shown in Figures 5.8.1(a) through (f), respectively. Sketches of the graphs of the power function for $a = -1$ and $n$ having values 1 through 6 are shown in Figures 5.8.2(a) through (f), respectively. They are mirror images in the $x$ axis of the corre-

5.8] Graphs of Polynomial Functions 313

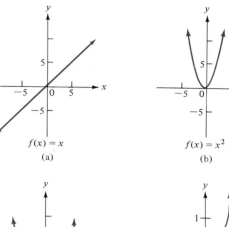

$f(x) = x$
(a)

$f(x) = x^2$
(b)

$f(x) = x^3$
(c)

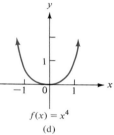
$f(x) = x^4$
(d)

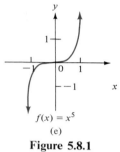
$f(x) = x^5$
(e)

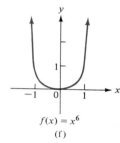
$f(x) = x^6$
(f)

**Figure 5.8.1**

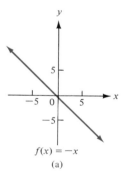
$f(x) = -x$
(a)

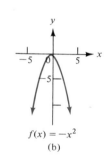
$f(x) = -x^2$
(b)

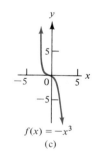
$f(x) = -x^3$
(c)

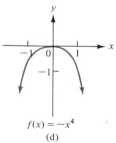
$f(x) = -x^4$
(d)

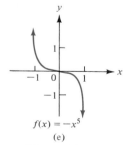
$f(x) = -x^5$
(e)

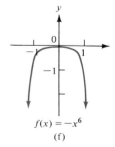
$f(x) = -x^6$
(f)

**Figure 5.8.2**

sponding graphs in Figures 5.8.1(a) through (f). Let us suppose that $a > 0$ and note that the graph of the function for $a < 0$ is the mirror image in the $x$ axis of the corresponding graph for $a > 0$. All of the graphs contain the origin, and this is the only intersection of the curve with either axis. If $n > 1$, the $x$ axis is tangent to the graph at the origin. If $n$ is a positive even integer, the graph is in the first and second quadrants and is symmetric with respect to the $y$ axis. If $n$ is a positive odd integer, the graph is in the first and third quadrants and is symmetric with respect to the origin. There are no asymptotes. As $|x|$ increases without bound, so does $|f(x)|$.

A polynomial function of the $n$th degree with real coefficients is defined by

$$P(x) = a_0 x^n + a_1 x^{n-1} + a_2 x^{n-2} + \cdots + a_n \qquad (1)$$

where $a_0 \neq 0$, $n$ is a positive integer, and $a_0, a_1, \ldots, a_n$ are real numbers. When $n = 1$, we have

$$P(x) = a_0 x + a_1$$

which is a linear function, and its graph is a straight line. When $n = 2$, we have

$$P(x) = a_0 x^2 + a_1 x + a_2$$

which is a quadratic function, and its graph is a parabola.

We now consider graphs of polynomial functions for which $n \geq 3$. If the right-hand side of equation (1) is considered as a fraction with a denominator of 1, we can conclude that there are no vertical asymptotes. As $|x|$ increases without bound, $|a_n x^n|$ increases without bound and will become larger than the sum of all the other terms in the polynomial. Therefore, the form of the graph for large values of $|x|$ will be determined by the values of the term $a_n x^n$. We can conclude then that the shape of the graph for large values of $|x|$ will be that of the graph of the power function of degree $n$. There are no horizontal asymptotes. If $a_n > 0$, the function will be increasing for large values of $x$, so the graph will be going up to the right as in Figures 5.8.3(a) and (b). If $a_n < 0$, the function will be decreasing for large values of $x$, and the graph will be going down to the right as in Figures 5.8.3(c) and (d). If $a_n > 0$ and $n$ is even, the graph goes up to the left as in Figure 5.8.3(a);

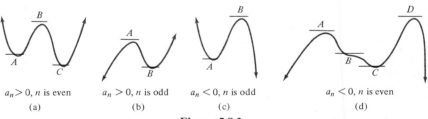

$a_n > 0$, $n$ is even  $a_n > 0$, $n$ is odd  $a_n < 0$, $n$ is odd  $a_n < 0$, $n$ is even
(a)　　　　　　　(b)　　　　　　　(c)　　　　　　　(d)

**Figure 5.8.3**

whereas if $n$ is odd, the graph goes down to the left as in Figure 5.8.3(b). If $a_n < 0$ and $n$ is even, the graph goes down to the left as in Figure 5.8.3(d), and if $n$ is odd, the graph goes up to the left as in Figure 5.8.3(c).

An important aid in drawing a sketch of the graph of a polynomial function is to determine any "relative extrema" of the function. A precise definition of relative extrema and the method of determining them require concepts of the calculus. However, it is worthwhile here to give an intuitive discussion. Relative extrema of a function consist of relative minimum and relative maximum values of the function. For a polynomial function, the graph will have a horizontal tangent line at a point where there is a relative extremum. In Figure 5.8.3(a) the function has one relative maximum value at the point $B$ and two relative minimum values at the points $A$ and $C$. Observe that there are horizontal tangent lines at these three points. In Figure 5.8.3(b) there are two relative extrema, a relative maximum value at $A$ and a relative minimum value at $B$. There are also two relative extrema in Figure 5.8.3(c); there is a relative minimum value at $A$ and a relative maximum value at $B$. There are three relative extrema for the function of Figure 5.8.3(d); relative maximum values occur at points $A$ and $D$, and a relative minimum value is at $C$. For the function of Figure 5.8.3(d), note that there is a horizontal tangent line at point $B$, but there is no relative extremum there. We state without proof a theorem giving the number of possible relative extrema for a polynomial function.

**5.8.1 THEOREM** A polynomial function of the $n$th degree with real coefficients has at most $n - 1$ relative extrema.

From Theorem 5.8.1 it follows that a polynomial function of the fourth degree with real coefficients has at most three relative extrema. Therefore, the graph in Figure 5.8.3(a) could be that of a fourth-degree polynomial. The graph in Figure 5.8.3(b) could be that of a polynomial of the third degree with real coefficients because such a polynomial has at most two relative extrema.

In the following examples we apply the above concepts to draw a sketch of the graph of a polynomial. When finding some arbitrary points on the graph by preparing a table of values of $x$ and $P(x)$, it is usually easier to compute $P(x)$ by synthetic division.

**EXAMPLE 1**
Draw a sketch of the graph of the function $P$ defined by

$$P(x) = x^3 - 6x^2 + 9x - 4$$

**SOLUTION**
Because $P(x)$ is a third-degree polynomial, there are at most two relative extrema. Furthermore, because the degree of the polynomial is odd and the coefficient of $x^3$ is positive, the graph goes up to the right and down to the left. Therefore, the graph is probably similar to that shown in Figure 5.8.3(b). We prepare Table 5.8.1 by using synthetic division. The computations for

$P(-2) = -54$, $P(-1) = -20$, and $P(1) = 0$ are shown, and the other computations are similar.

$$\begin{array}{r|rrrr} -2 & 1 & -6 & 9 & -4 \\ & & -2 & 16 & -50 \\ \hline & 1 & -8 & 25 & -54 \end{array} \quad \begin{array}{r|rrrr} -1 & 1 & -6 & 9 & -4 \\ & & -1 & 7 & -16 \\ \hline & 1 & -7 & 16 & -20 \end{array} \quad \begin{array}{r|rrrr} 1 & 1 & -6 & 9 & -4 \\ & & 1 & -5 & 4 \\ \hline & 1 & -5 & 4 & 0 \end{array}$$

**Table 5.8.1**

| $x$ | $-2$ | $-1$ | $0$ | $1$ | $2$ | $3$ | $4$ | $5$ |
|---|---|---|---|---|---|---|---|---|
| $P(x)$ | $-54$ | $-20$ | $-4$ | $0$ | $-2$ | $-4$ | $0$ | $16$ |

In Figure 5.8.4 we have plotted the points $(-2, -54)$, $(-1, -20)$, $(0, -4)$, $(1, 0)$, $(2, -2)$, $(3, -4)$, $(4, 0)$, and $(5, 16)$ given from Table 5.8.1. We have chosen different size units on the $y$ axis than on the $x$ axis. With these points and the previous information, we draw the sketch shown in Figure 5.8.5.

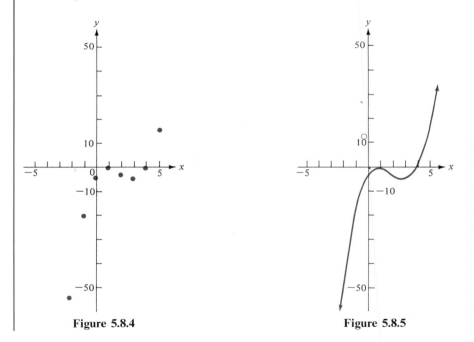

Figure 5.8.4    Figure 5.8.5

In Example 1, with our limited knowledge (that is, without using concepts of calculus) we cannot be absolutely sure that the relative extrema occur at $(1, 0)$ and $(3, -4)$, but we can assume that the graph has the general appearance given in Figure 5.8.5. A similar comment holds for the graph in Example 2.

## EXAMPLE 2

Draw a sketch of the graph of the function $P$ defined by

$$P(x) = 3x^4 - 4x^3 - 12x^2 + 12$$

**SOLUTION**

There are at most three relative extrema because $P(x)$ is a fourth-degree polynomial. The degree of the polynomial is even and the coefficient of $x^4$ is positive; therefore, the graph goes up to the right and up to the left. Because $P(0) = 12$, the graph intersects the $y$ axis at 12. Values of $P(x)$ are computed by synthetic division for some values of $x$ and these are given in Table 5.8.2.

**Table 5.8.2**

| $x$    | $-2$ | $-1$ | 0  | 1   | 2    | 3  |
|--------|------|------|----|-----|------|----|
| $P(x)$ | 44   | 7    | 12 | $-1$ | $-20$ | 39 |

In Figure 5.8.6 we have plotted the points $(-2, 44)$, $(-1, 7)$, $(0, 12)$, $(1, -1)$, $(2, -20)$, and $(3, 39)$ given from Table 5.8.2. Observe that we have chosen different size units on the $y$ axis than on the $x$ axis. With these points and the information found above, we draw the sketch of the graph shown in Figure 5.8.7.

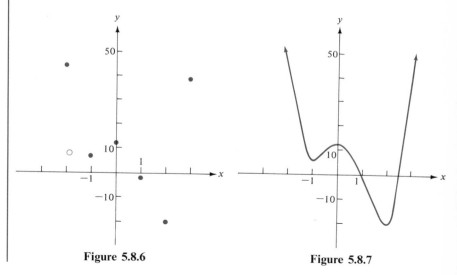

**Figure 5.8.6**    **Figure 5.8.7**

The following theorem gives us information about the real roots of a polynomial equation. We apply it in Section 5.9.

**5.8.2 THEOREM**  Let $P(x)$ be a polynomial with real coefficients. If $a, b \in R$ and $a < b$, then if $P(a)$ and $P(b)$ are opposite in sign, there is a real number $c$ between $a$ and $b$ such that $P(c) = 0$.

The proof of Theorem 5.8.2 is omitted. However, because the graph of a polynomial function is a continuous unbroken curve, the truth of the theorem should be intuitive to you because of the following reasoning: If $P(a)$ and $P(b)$ are opposite in sign, then the points $(a, P(a))$ and $(b, P(b))$ are on opposite sides of the $x$-axis; thus the graph of $y = P(x)$ must intersect the $x$-axis in at least one point $(c, 0)$, where $c$ is between $a$ and $b$.

We show this geometrically in Figures 5.8.8 and 5.8.9. In Figure 5.8.8 we have a portion of the graph of a polynomial function $P$ from the point $(a, P(a))$ to $(b, P(b))$, where $P(a) < 0$ and $P(b) > 0$. The graph intersects the $x$ axis at the point $(c, 0)$, where $a < c < b$. Figure 5.8.9 shows the situation when $P(a) > 0$ and $P(b) < 0$.

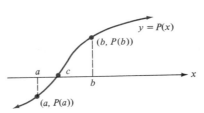

Figure 5.8.8

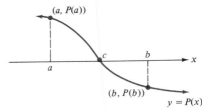

Figure 5.8.9

## EXERCISES 5.8

*In Exercises 1 through 24, draw a sketch of the graph of the given polynomial function.*

1. $f(x) = 2x^2$
2. $g(x) = 3(x + 1)^2$
3. $h(x) = -4(x - 2)^2$
4. $F(x) = -3x^3$
5. $G(x) = (x + 2)^3$
6. $H(x) = (2x - 1)^3$
7. $f(x) = (x - 2)^4$
8. $g(x) = -(x + 2)^4$
9. $h(x) = (x + 1)^5$
10. $f(x) = -(x - 1)^6$
11. $P(x) = x^3 - 2x^2 - 5x + 6$
12. $P(x) = x^3 - 6x^2 + 9x - 4$
13. $F(x) = x^3 - 3x^2 + 3$
14. $G(x) = x^3 + 4x^2 + 4x$
15. $P(x) = 3x^3 - 4x^2 - 5x + 2$
16. $H(x) = 6x^3 + 29x^2 + x - 6$
17. $f(x) = x^4 - 5x^3 + 2x^2 + 8x$
18. $g(x) = x^4 - 5x^2 + 4$
19. $P(x) = x^4 + x^3 - 7x^2 - x + 6$
20. $P(x) = x^4 - 6x^3 + 11x^2 - 6x$
21. $G(x) = 3x^4 + 5x^3 - 5x^2 - 5x + 2$
22. $H(x) = 2x^4 - x^3 - 6x^2 - x + 2$
23. $F(x) = x^5 + 3x^4 + 3x^3 + x^2$
24. $f(x) = x^6 - 3x^5 + 3x^4 - x^3$

## 5.9 Real Roots of Polynomial Equations

In Example 3 of Section 5.7 we showed that the equation

$$2x^3 - 2x^2 - 4x + 1 = 0 \qquad (1)$$

has no rational roots. Thus any real roots of this equation must be irrational. The equation is of the third degree and hence has three roots. However, because the imaginary roots must occur in pairs (as a result of Theorem

5.6.4), it follows that equation (1) has either three real roots, all of which are irrational, or two imaginary roots and one real root, which is irrational. Additional information regarding the real roots of a polynomial equation with real coefficients can be obtained by *Descartes' rule of signs*. Theorem 5.9.1 gives this rule, which involves the concept of "variation in sign" of a polynomial.

If the terms of a polynomial with real coefficients are written in descending powers of the variable (the terms having zero coefficients are omitted), then a *variation in sign* occurs if two successive coefficients are opposite in sign. For example, if

$$P(x) = 2x^3 - 2x^2 - 4x + 1$$

the coefficients have, successively, the signs $+, -, -, +$; thus there are two variations in sign.

**ILLUSTRATION 1.** If $Q(x) = x^4 - 6x^2 - 2x - 1$, then $Q(x)$ has one variation in sign. Furthermore, $Q(-x) = x^4 - 6x^2 + 2x - 1$, and $Q(-x)$ has three variations in sign. If $R(x) = x^3 + 2x + 5$, then $R(x)$ has no variations in sign. •

**5.9.1 THEOREM** *Descartes' Rule of Signs.* If $P(x)$ is a polynomial having real coefficients, the number of positive roots of the equation $P(x) = 0$ either is equal to the number of variations in sign of $P(x)$, or is less than this number by an even natural number. Furthermore, the number of negative roots of the equation is equal to the number of variations in sign of $P(-x)$, or is less than this number by an even natural number.

The proof of Descartes' rule of signs is omitted because it is beyond the scope of this book.

**ILLUSTRATION 2.** We apply Descartes' rule of signs to equation (1). If $P(x) = 2x^3 - 2x^2 - 4x + 1$, then $P(x)$ has two variations in sign. Therefore, by Descartes' rule of signs the number of positive roots of equation (1) is either two or zero. Furthermore, $P(-x) = -2x^3 - 2x^2 + 4x + 1$, and $P(-x)$ has one variation in sign. Thus equation (1) has one negative root. •

From Illustration 2 and the discussion in the first paragraph of this section it follows that the three roots of equation (1) are such that either two are positive irrational numbers and one is a negative irrational number, or else two are imaginary numbers and one is a negative irrational number.

Observe that Descartes' rule of signs states that if a polynomial $P(x)$ has an odd number of variations in sign, then the equation $P(x) = 0$ has an odd number of positive roots; thus we are certain of at least one positive root in such a case. Similarly, if $P(-x)$ has an odd number of variations in sign, then there is at least one negative root of the equation $P(x) = 0$.

### EXAMPLE 1

From Descartes' rule of signs determine information concerning the number of positive, negative, and imaginary roots of each of the following equations.

(a) $x^3 + 6x - 2 = 0$
(b) $x^4 + 2x^2 - 5 = 0$
(c) $6x^4 - x^3 + 2x - 3 = 0$

### SOLUTION

(a) Let $P(x) = x^3 + 6x - 2$. Because $P(x)$ has one variation in sign, there is one positive root of the equation. $P(-x) = -x^3 - 6x - 2$. Because $P(-x)$ has no variations in sign, there are no negative roots of the equation. Because the equation is of the third degree, it has three roots, and hence there are two imaginary roots.
(b) Let $Q(x) = x^4 + 2x^2 - 5$. There is one positive root of the equation because $Q(x)$ has one variation in sign. $Q(-x) = x^4 + 2x^2 - 5$. Thus $Q(-x)$ has one variation in sign; consequently, there is one negative root of the equation. The equation has four roots, and hence there are two imaginary roots.
(c) Let $R(x) = 6x^4 - x^3 + 2x - 3$. Because $R(x)$ has three variations in sign, the number of positive roots of the equation is either three or one. $R(-x) = 6x^4 + x^3 - 2x - 3$, and $R(-x)$ has one variation in sign. Therefore, the equation has one negative root. Thus the equation has either three positive roots, one negative root, and no imaginary roots; or else it has one positive root, one negative root, and two imaginary roots.

**ILLUSTRATION 3.** We have learned that the three roots of equation (1) are such that either two are positive irrational numbers and one is a negative irrational number, or else two are imaginary numbers and one is a negative irrational number. With $P(x) = 2x^3 - 2x^2 - 4x + 1$, we compute $P(1)$ and $P(2)$ by synthetic division.

```
1 | 2  -2  -4   1         2 | 2  -2  -4   1
        2   0  -4                 4   4   0
    ─────────────             ─────────────
    2   0  -4  -3             2   2   0   1
```

Therefore, $P(1) = -3$ and $P(2) = 1$. Because $P(1)$ and $P(2)$ are opposite in sign, it follows from Theorem 5.8.2 that there is a real number $c$ between 1 and 2 such that $P(c) = 0$; therefore, equation (1) has a positive root between 1 and 2. Thus we are certain that equation (1) has two positive irrational roots and one negative irrational root.

We can apply Theorem 5.8.2 to locate integers between which the other two roots lie. Because $P(0) = 1$, it follows that $P(0)$ and $P(1)$ are opposite in sign, and hence the equation has a positive root between 0 and 1. We compute $P(-1)$ and $P(-2)$ by synthetic division.

```
-1 | 2  -2  -4   1        -2 | 2  -2  -4    1
        -2   4   0                -4  12  -16
    ─────────────             ─────────────
    2  -4   0   1             2  -6   8  -15
```

Thus $P(-1)$ and $P(-2)$ are opposite in sign, and consequently a negative root is between $-2$ and $-1$. •

**EXAMPLE 2**
Determine all the information you can concerning the number of positive, negative, and imaginary roots of each of the following equations.

(a) $3x^4 + x^2 + 7x + 1 = 0$
(b) $x^5 + 5x^2 - 4 = 0$

**SOLUTION**
(a) Let $P(x) = 3x^4 + x^2 + 7x + 1$. Because $P(x)$ has no variations in sign, there is no positive root. $P(-x) = 3x^4 + x^2 - 7x + 1$. Because $P(-x)$ has two variations in sign, there are either two negative roots or no negative roots. Furthermore, $P(0) = 1$ and $P(-1) = -2$. Because $P(0)$ and $P(-1)$ have opposite signs, it follows from Theorem 5.8.2 that there is a number $c$ between $0$ and $-1$ such that $P(c) = 0$; therefore, the number $c$ is a negative root of the equation. Thus the equation has two negative roots and two imaginary roots.

(b) Let $Q(x) = x^5 + 5x^2 - 4$. Because $Q(x)$ has one variation in sign, there is one positive root. $Q(-x) = -x^5 + 5x^2 - 4$, and $Q(-x)$ has two variations in sign. Therefore, there are either two negative roots or no negative roots. We compute $Q(-1)$ by synthetic division

$$\begin{array}{r|rrrrrr} -1 & 1 & 0 & 0 & 5 & 0 & -4 \\ & & -1 & 1 & -1 & -4 & 4 \\ \hline & 1 & -1 & 1 & 4 & -4 & 0 \end{array}$$

Therefore, $Q(-1) = 0$ and thus $-1$ is a root of the equation. Hence the equation has one positive root, two negative roots, and two imaginary roots.

We have seen that there are polynomial equations having integer coefficients that have no rational roots. Equation (1) is such an equation. It has three irrational roots, and we have located these roots between $-2$ and $-1$, between $0$ and $1$, and between $1$ and $2$. These irrational roots can be approximated to any degree of accuracy by various techniques and using automatic computers. A standard procedure that is often used is the one called "Newton's method of approximation," which involves concepts of the calculus. Horner's and Graeffe's methods are two other procedures that appear in texts on the theory of equations. There is a graphical way of approximating roots of polynomial equations. This process involves finding the zeros of the corresponding polynomial function by approximating the $x$ intercepts of the graph of the function. Obviously, this procedure is not very accurate. There is an elementary method that involves repeated use of Theorem 5.8.2, and the following example demonstrates it.

**EXAMPLE 3**
Find the approximate value, to the nearest hundredth, of the smallest positive root of equation (1).

**SOLUTION**
The equation is

$$2x^3 - 2x^2 - 4x + 1 = 0$$

and we know from Illustration 3 that the smallest positive root is between 0 and 1. Let $P(x) = 2x^3 - 2x^2 - 4x + 1$. The interval between 0 and 1 is divided into ten equal subintervals to give the numbers $0, 0.1, 0.2, 0.3, \ldots, 0.9, 1$. We use synthetic division to find $P(x)$ at each of these numbers until we have a change in sign.

$$\begin{array}{r|rrrr} 0.1 & 2 & -2 & -4 & 1 \\ & & 0.2 & -0.18 & -0.418 \\ \hline & 2 & -1.8 & -4.18 & 0.582 \end{array} \qquad \begin{array}{r|rrrr} 0.2 & 2 & -2 & -4 & 1 \\ & & 0.4 & -0.32 & -0.864 \\ \hline & 2 & -1.6 & -4.32 & 0.136 \end{array}$$

$$\begin{array}{r|rrrr} 0.3 & 2 & -2 & -4 & 1 \\ & & 0.6 & -0.42 & -1.326 \\ \hline & 2 & -1.4 & -4.42 & -0.326 \end{array}$$

Because $P(0.2) = 0.136$ and $P(0.3) = -0.326$, the root lies between 0.2 and 0.3. We now divide the interval between 0.2 and 0.3 into ten equal subintervals to give the numbers $0.2, 0.21, 0.22, 0.23, \ldots, 0.29, 0.3$. We use synthetic division to find $P(x)$ at each of these numbers until we have a change in sign.

$$\begin{array}{r|rrrr} 0.22 & 2 & -2 & -4 & 1 \\ & & 0.44 & -0.3432 & -0.9555 \\ \hline & 2 & -1.56 & -4.3432 & 0.0445 \end{array}$$

$$\begin{array}{r|rrrr} 0.23 & 2 & -2 & -4 & 1 \\ & & 0.46 & -0.3542 & -1.0015 \\ \hline & 2 & -1.54 & -4.3542 & -0.0015 \end{array}$$

Thus $P(0.22) = 0.443$ and $P(0.23) = -0.0015$, and hence the root lies between 0.22 and 0.23. To obtain the root accurate to the nearest hundredth, we find $P(0.225)$.

$$\begin{array}{r|rrrr} 0.225 & 2 & -2 & -4 & 1 \\ & & 0.45 & -0.349 & -0.9785 \\ \hline & 2 & -1.55 & -4.349 & 0.0215 \end{array}$$

Therefore, $P(0.225) = 0.0215$. It follows that the root is greater than 0.225, and hence it is closer to 0.23 than to 0.22. Thus the smallest positive root to the nearest hundredth is 0.23.

We conclude by giving an example showing how Descartes' rule of signs is used to prove that certain numbers are irrational.

EXAMPLE 4
Prove that $\sqrt{3}$ is irrational.

SOLUTION
Let $x = \sqrt{3}$. Then $x^2 = 3$ or, equivalently,

$$x^2 - 3 = 0 \tag{2}$$

If $P(x) = x^2 - 3$, then $P(x)$ has one variation in sign. Therefore, by Descartes' rule of signs there is one positive root of equation (2). From

Theorem 5.7.1 the only possible positive rational roots are 1 and 3. By synthetic division we show that neither of these numbers is a root. Thus the positive root is an irrational number, and it is $\sqrt{3}$.

## EXERCISES 5.9

In Exercises 1 through 12, use Descartes' rule of signs to determine information concerning the number of positive, negative, and imaginary roots of the given equation.

1. $x^3 - 4x^2 - 2 = 0$
2. $5x^3 - 3x - 7 = 0$
3. $4x^3 + 6x^2 - 3x + 5 = 0$
4. $2x^3 + x + 1 = 0$
5. $3x^3 - 4x^2 + 2x - 5 = 0$
6. $x^3 + 2x^2 + 3x + 1 = 0$
7. $x^4 + 7x^3 + x - 8 = 0$
8. $3x^4 - 2x^3 + 8x^2 - x - 7 = 0$
9. $6x^4 + 8x^2 + x = 0$
10. $5x^4 - 3x^3 - 2 = 0$
11. $2x^5 - 6x^4 - x^2 + 4x - 1 = 0$
12. $x^6 + 3x^4 + 2x^3 - 4x^2 + 2x - 5 = 0$

In Exercises 13 through 20, determine all the information you can concerning the number of positive, negative, and imaginary roots of the given equation. Determine any rational roots and locate any irrational roots between two consecutive integers.

13. $x^3 - 6x + 3 = 0$
14. $x^3 + 3x - 20 = 0$
15. $x^4 + x^2 - 1 = 0$
16. $x^3 + 3x^2 - 2x - 5 = 0$
17. $4x^4 - 3x^3 + 2x - 5 = 0$
18. $2x^4 - 14x^3 + 24x^2 + x - 4 = 0$
19. $3x^4 - 21x^3 + 36x^2 + 2x - 8 = 0$
20. $x^4 + 2x^3 - 9x^2 - 8x + 14 = 0$

In Exercises 21 through 26, use the method of Example 3 to find the approximate value, to the nearest hundredth, of the indicated root.

21. $x^3 - 4x - 8 = 0$; the positive root
22. $x^3 - 2x + 7 = 0$; the negative root
23. $x^4 - 10x + 5 = 0$; the smallest positive root
24. $x^4 - 10x + 5 = 0$; the largest positive root
25. $2x^4 - 2x^3 + x^2 + 3x - 4 = 0$; the negative root
26. $x^4 + x^3 - 3x^2 - x - 4 = 0$; the positive root

In Exercises 27 through 32, prove that the given number is irrational.

27. $\sqrt{5}$
28. $2\sqrt{7}$
29. $\sqrt[3]{10}$
30. $\sqrt[4]{8}$
31. $2 + \sqrt{5}$
32. $3 + 2\sqrt{3}$

## REVIEW EXERCISES (CHAPTER 5)

In Exercises 1 through 10, perform the indicated operations and express the result in the form $a + bi$.

1. $(8 + 3i) + (10 - 2i)$
2. $(11 - 2i) - (-5 + 6i)$
3. $(\frac{1}{2} - i) - (\frac{1}{4} - \frac{1}{3}i)$
4. $(3 + \frac{2}{3}i) + (-1 - i)$
5. $\sqrt{-9}\sqrt{-49}$
6. $\sqrt{-8}\sqrt{-24}\sqrt{-48}$
7. $(-4 + 2i)(-3 + i)$
8. $(-5 - i)^2$
9. $(5 - 2i) \div (-4 - 3i)$
10. $i \div (-6 - i)$

*In Exercises 11 and 12, perform geometrically the addition.*

**11.** $(4 + 5i) + (6 - 3i)$

**12.** $(-3 + 2i) + (-7 - i)$

*In Exercises 13 and 14, write the given expression without absolute value bars.*

**13.** $|4 - 3i|$

**14.** $|-6 + 2i|$

*In Exercises 15 and 16, find an equation with real coefficients for which the given numbers are roots.*

**15.** $3, -2, -4i$

**16.** $1 + i, \frac{1}{2} - \frac{3}{2}i$

*In Exercises 17 and 18, find the solution set of the given equation, and write the solutions in standard form.*

**17.** $2ix^2 - 5x + 7i = 0$

**18.** $3ix^2 - 4x - 2i = 0$

**19.** Find the three cube roots of $-8$.

**20.** Find the four fourth roots of $81$.

*In Exercises 21 and 22, use the remainder theorem to find the remainder for the indicated division.*

**21.** $(3x^3 + 4x^2 - 3x - 5) \div (x + 2)$

**22.** $(2x^4 - 5x^2 - 2x + 1) \div (x - 1)$

*In Exercises 23 and 24, use the factor theorem to answer the question.*

**23.** Is $(x - 3)$ a factor of $x^3 + 2x^2 - 12x - 9$?

**24.** Is $(x + 4)$ a factor of $2x^3 + 9x^2 + 6x + 8$?

**25.** Find a value of $k$ so that $(x - i)$ is a factor of $2kx^3 - 5x^2 + 3kx$.

**26.** Find a value of $k$ so that $(x + 2)$ is a factor of $2x^4 + 2kx^3 - x^2 - 3kx - 8$.

*In Exercises 27 through 30, use synthetic division to find the quotient and remainder.*

**27.** $(2x^4 + 7x^3 - 4x + 5) \div (x + 3)$

**28.** $(x^3 - 6x^2 + 8x - 5) \div (x - 4)$

**29.** $(x^6 - 64) \div (x - 2)$

**30.** $(x^5 + 243) \div (x + 3)$

*In Exercises 31 and 32, find the indicated function values by synthetic division.*

**31.** $P(x) = 2x^4 - 8x^2 - 10x - 3$; $P(-2)$ and $P(3)$

**32.** $P(x) = 3x^4 + 10x^3 - 6x^2 + 1$; $P(-4)$ and $P(-\frac{1}{3})$

*In Exercises 33 and 34, find the zeros of the given polynomial function. State the multiplicity of each zero.*

**33.** $\{(x, P(x)) | P(x) = (x^2 - 9)(x^2 + 4)^2(6x^2 + x - 15)\}$

**34.** $\{(x, P(x)) | P(x) = (x - 5)^3(x^2 + 5)(x^2 + 2x - 1)^2\}$

**35.** Show that $-1 + i$ is a zero of the polynomial function $\{(x, P(x)) | P(x) = x^4 + 3x^3 + 3x^2 - 2\}$ and find the other three zeros.

**36.** Find the solution set of the equation $2x^4 - x^3 + 33x^2 - 16x + 16 = 0$ given that $4i$ is a root.

In Exercises 37 and 38, find a ploynomial P(x) of the stated degree with real coefficients for which the given numbers are zeros of P.

37. Fourth degree; $1 - i$ and $1 + i\sqrt{3}$ are zeros of P.
38. Fifth degree; $\sqrt{2} + i$ (multiplicity two) and $-1$ are zeros of P.

In Exercises 39 and 40, find all the rational zeros of the given polynomial function. If possible, find all the zeros of the function.

39. $\{(x, P(x))|P(x) = x^4 + x^3 - 4x^2 + 2x - 12\}$
40. $\{(x, P(x))|P(x) = 2x^4 + 3x^3 - 4x^2 + 13x - 6\}$

In Exercises 41 through 44, find all the rational roots of the given equation. If possible, find all the roots.

41. $x^3 - 2x^2 - 9 = 0$
43. $6x^4 - 25x^3 - 3x^2 + 5x + 1 = 0$
42. $x^4 - 3x^3 - 10x^2 + 28x - 16 = 0$
44. $3x^3 - x^2 + 16x + 12 = 0$

In Exercises 45 through 48, draw a sketch of the graph of the polynomial function.

45. $f(x) = (x - 3)^4$
47. $h(x) = x^3 - 6x^2 + 9x + 6$
46. $g(x) = (x + 4)^3$
48. $F(x) = x^4 - 8x^2 + 9$

In Exercises 49 and 50, prove that the given equation has no rational roots. Then use Descartes' rule of signs to determine information concerning the number of positive, negative, and imaginary roots.

49. $x^3 - 7x^2 + x + 3 = 0$
50. $x^4 + 2x^3 + 6x - 3 = 0$

In Exercises 51 and 52, determine all the information you can concerning the number of positive, negative, and imaginary roots of the given equation. Determine any rational roots and locate any irrational roots between two consecutive integers.

51. $x^4 - 2x^3 - 13x^2 + 33x - 14 = 0$
52. $3x^4 + 10x^3 - 11x^2 - 4x + 2 = 0$

In Exercises 53 and 54, use the method of Example 3 in Section 5.9 to find the approximate value, to the nearest hundredth, of the indicated root.

53. $x^3 - 3x^2 - 6x + 2 = 0$; the negative root
54. $x^3 + 9x^2 + 15x - 21 = 0$; the positive root

# 6
# Systems of Equations and Inequalities

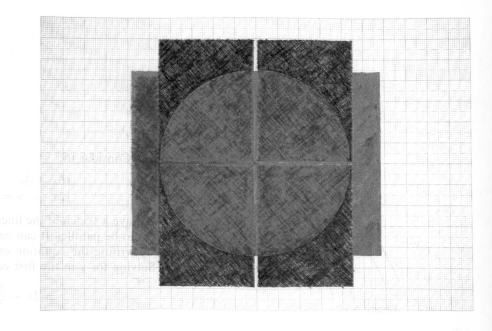

# 6.1 Systems of Linear Equations in Two Variables

It is often necessary to deal with more than one equation in several variables. In such situations the given equations are called a *system of equations*. By the solution set of a system of equations we mean the set of solutions that are common to the equations in the system.

We write a system of equations by using a left brace. In particular, a system of two linear equations in two variables $x$ and $y$ can be written as

$$\begin{cases} a_1 x + b_1 y = c_1 \\ a_2 x + b_2 y = c_2 \end{cases}$$

Where $a_1, b_1, c_1, a_2, b_2,$ and $c_2$ are real numbers.

**ILLUSTRATION 1.** The following pair of equations is a system of two linear equations in two variables.

(I) $\quad \begin{cases} 2x + y = 3 \\ 5x + 3y = 10 \end{cases}$

The solution set of each of the two equations in system (I) is an infinite set of ordered pairs of real numbers, and the graphs of these sets are straight lines. Figure 6.1.1 shows on the same coordinate system sketches of the graphs of the equations. From the figure it is apparent that the two lines intersect at one and only one point. This is the point $(-1, 5)$, which can be verified by substituting into the equations. Doing this, we have

$$2(-1) + 5 = 3$$

and

$$5(-1) + 3(5) = 10$$

We conclude that the ordered pair $(-1, 5)$ is the only ordered pair that is common to the solution sets of the two equations. Hence the solution set of system (I) is $\{(-1, 5)\}$. Using symbols, we may write

$$\{(x, y) | 2x + y = 3\} \cap \{(x, y) | 5x + 3y = 10\} = \{(-1, 5)\}$$

**ILLUSTRATION 2.** Consider the system

(II) $\quad \begin{cases} 6x - 3y = 5 \\ 2x - y = 4 \end{cases}$

In Figure 6.1.2 we have a sketch of the lines having the equations in system (II); the lines seem to be parallel. It can easily be proved that the lines are indeed parallel by writing the equation of each of the lines in the slope-intercept form. Solving for $y$ in the first equation, we have

$$y = 2x - \frac{5}{3}$$

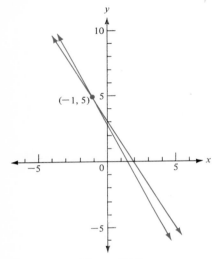

Figure 6.1.1

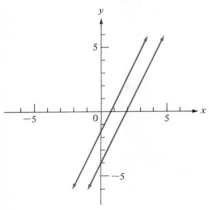

Figure 6.1.2

and solving for $y$ in the second equation, we obtain

$$y = 2x - 4$$

We see that each line has a slope 2. Therefore, the lines are parallel. They are not the same line because the $y$ intercepts, $-\frac{5}{3}$ and $-4$, are not equal. It follows that the two lines have no point in common and so the system (II) has no solution. Using symbols, we may write

$$\{(x, y) \mid 6x - 3y = 5\} \cap \{(x, y) \mid 2x - y = 4\} = \emptyset$$

●

**ILLUSTRATION 3.** For the system

(III) $$\begin{cases} 3x + 2y = 4 \\ 6x + 4y = 8 \end{cases}$$

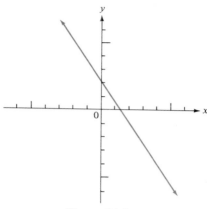

**Figure 6.1.3**

the graphs of the two equations are the same line (see Figure 6.1.3). This fact is evident when the equations are written in the slope-intercept form. Solving each of the equations for $y$, we obtain

$$y = -\frac{3}{2}x + 2$$

The solution sets of the two equations are equal, and their intersection consists of all the ordered pairs in either one of the two solution sets. It is an infinite set.

●

If we are given a system of two linear equations in two variables, three possibilities arise.

**POSSIBILITY 1:** The solution sets of the two equations are equal. This occurs in Illustration 3 with system (III). The graphs are the same line. The equations are said to be *dependent*.

**POSSIBILITY 2:** The intersection of the two solution sets is the empty set. This occurs in Illustration 2 with system (II). The graphs are distinct parallel lines. The equations are said to be *inconsistent*.

**POSSIBILITY 3:** The intersection of the two solution sets contains one and only one ordered pair. This occurs in Illustration 1 with system (I). The graphs intersect in one and only one point. The equations are said to be *consistent* and *independent*.

When two linear equations in two variables are consistent and independent, the solution obtained from the graphs (as in Illustration 1) is generally only an approximation because drawing the graphs depends upon measuring devices. To obtain exact solutions of systems of linear equations, we must use algebraic methods. These methods consist of replacing the given system by another system that has exactly the same solution set.

**6.1.1 DEFINITION** Two systems of equations that have the same solution set are said to be *equivalent systems*.

If any equation in a given system is replaced by an equivalent equation, the resulting system is equivalent to the given system. Furthermore, if any two equations of a given system are interchanged the resulting system is equivalent to the given system.

One method for finding the solution set of a system of two linear equations in two variables is called the *substitution method*. For any ordered pair $(x, y)$ in the solution set of a system of equations, the variables $x$ and $y$ in one equation represent the same numbers as the variables $x$ and $y$ in the other equation. Therefore, if we apply the substitution axiom and replace one of the variables in one of the equations by its equal from the other equation, we have an equivalent system. The next illustration shows the procedure.

**ILLUSTRATION 4.** We apply the method of substitution to find the solution set of system (I), which is

$$\begin{cases} 2x + y = 3 \\ 5x + 3y = 10 \end{cases}$$

If we replace the first equation by the equivalent equation $y = -2x + 3$, we obtain the equivalent system

$$\begin{cases} y = -2x + 3 \\ 5x + 3y = 10 \end{cases}$$

We apply the substitution axiom and replace $y$ in the second equation by its equal $(-2x + 3)$ from the first equation, and we have the equivalent system

$$\begin{cases} y = -2x + 3 \\ 5x + 3(-2x + 3) = 10 \end{cases}$$

Simplifying the second equation, we obtain

$$\begin{cases} y = -2x + 3 \\ -x + 9 = 10 \end{cases}$$

Solving the second equation for $x$, we have

$$\begin{cases} y = -2x + 3 \\ x = -1 \end{cases}$$

Substituting the value of $x$ from the second equation into the first equation, we have

$$\begin{cases} y = 5 \\ x = -1 \end{cases}$$

**EXAMPLE 1**
Use the substitution method to find the solution set of the system

(IV) $\quad \begin{cases} 8x + 3y = 7 \\ 4x + y = 3 \end{cases}$

**SOLUTION**
This latter system is equivalent to the given system; therefore, the solution set is $\{(-1, 5)\}$, which agrees with the result of Illustration 1. ●

We replace the second equation by the equivalent equation obtained by solving for $y$, and we have the equivalent system

$$\begin{cases} 8x + 3y = 7 \\ y = 3 - 4x \end{cases}$$

We apply the substitution axiom and replace $y$ in the first equation by its equal $(3 - 4x)$ from the second equation, and we have the equivalent system

$$\begin{cases} 8x + 3(3 - 4x) = 7 \\ y = 3 - 4x \end{cases}$$

Simplifying the first equation, we have

$$\begin{cases} 9 - 4x = 7 \\ y = 3 - 4x \end{cases}$$

Solving the first equation for $x$, we have

$$\begin{cases} x = \dfrac{1}{2} \\ y = 3 - 4x \end{cases}$$

We substitute the value of $x$ from the first equation into the second equation, and we have

(V) $\quad \begin{cases} x = \dfrac{1}{2} \\ y = 1 \end{cases}$

System (V) is equivalent to system (IV). Hence the solution set of system (IV) is $\{(\tfrac{1}{2}, 1)\}$.

The solution $(\tfrac{1}{2}, 1)$ can be checked by substituting into the equations of the given system. Doing this, we have

$$8\left(\frac{1}{2}\right) + 3(1) = 4 + 3$$
$$= 7$$

and

$$4\left(\frac{1}{2}\right) + 1 = 2 + 1$$
$$= 3$$

The next theorem gives another method for obtaining a system of two linear equations in two variables equivalent to a given one.

**6.1.2 THEOREM** The system of equations

(VI) $$\begin{cases} a_1x + b_1y = c_1 \\ a_2x + b_2y = c_2 \end{cases}$$

is equivalent to the system

(VII) $$\begin{cases} a_1x + b_1y = c_1 \\ k_1(a_1x + b_1y) + k_2(a_2x + b_2y) = k_1c_1 + k_2c_2 \end{cases}$$

where $k_1$ and $k_2$ are real numbers and $k_2 \neq 0$.

Furthermore, system (VI) is equivalent to the system

(VIII) $$\begin{cases} k_1(a_1x + b_1y) + k_2(a_2x + b_2y) = k_1c_1 + k_2c_2 \\ a_2x + b_2y = c_2 \end{cases}$$

where $k_1$ and $k_2$ are real numbers and $k_1 \neq 0$.

*Proof.* We prove that the system (VI) is equivalent to system (VII). To do this, we must show that any solution of system (VI) is a solution of system (VII) and, conversely, that any solution of system (VII) is a solution of system (VI).

If the ordered pair $(r, s)$ is a solution of system (VI), then

$$a_1r + b_1s = c_1$$

and

$$a_2r + b_2s = c_2$$

Therefore, by the substitution axiom,

$$k_1(a_1r + b_1s) + k_2(a_2r + b_2s) = k_1c_1 + k_2c_2$$

Thus any solution of system (VI) is a solution of system (VII).

If the ordered pair $(r, s)$ is a solution of system (VII), then

$$a_1r + b_1s = c_1$$

and

$$k_1(a_1r + b_1s) + k_2(a_2r + b_2s) = k_1c_1 + k_2c_2$$

Substituting $c_1$ for $a_1r + b_1s$ in the second equation, we have

$$k_1c_1 + k_2(a_2r + b_2s) = k_1c_1 + k_2c_2$$
$$k_2(a_2r + b_2s) = k_2c_2$$

or, equivalently (because $k_2 \neq 0$),

$$a_2r + b_2s = c_2$$

Thus any solution of system (VII) is a solution of system (VI).

The proof that system (VI) is equivalent to system (VIII) is identical.

When applying Theorem 6.1.2 to obtain the second equation of system (VII) from the equations of system (VI), we use the terminology "multiply the equation $a_1x + b_1y = c_1$ by $k_1$, multiply the equation $a_2x + b_2y = c_2$ by $k_2$, and add the resulting equations." It is understood that "to multiply an equation by a number" means to multiply each member of the equation by that number and "to add two equations" means to add corresponding members of the equations.

**ILLUSTRATION 5.** We apply Theorem 6.1.2 to find the solution set of system (I), which is

$$\begin{cases} 2x + y = 3 \\ 5x + 3y = 10 \end{cases}$$

When applying Theorem 6.1.2 we wish to choose the multipliers of the equations in such a way that we eliminate one of the variables. Because the coefficient of $y$ in the first equation is 1 and the coefficient of $y$ in the second equation is 3 to obtain an equation not involving $y$ we replace the second equation by the sum of $-3$ times the first equation and 1 times the second equation. We then have the equivalent system

$$\begin{cases} 2x + y = 3 \\ -3(2x + y) + 1(5x + 3y) = (-3)(3) + (1)(10) \end{cases}$$

Simplifying the second equation, we obtain

$$\begin{cases} 2x + y = 3 \\ -x = 1 \end{cases}$$

Solving the second equation for $x$, we have

$$\begin{cases} 2x + y = 3 \\ x = -1 \end{cases}$$

We substitute the value of $x$ from the second equation into the first equation and obtain

$$\begin{cases} 2(-1) + y = 3 \\ x = -1 \end{cases}$$

or, equivalently,

$$\begin{cases} y = 5 \\ x = -1 \end{cases}$$

Therefore, the solution set of the given system is $\{(-1, 5)\}$, which agrees with the results of Illustrations 1 and 4. ●

## EXAMPLE 2

Find the solution set of the system

$$\begin{cases} 4x + 3y = 6 \\ 3x + 5y = -1 \end{cases}$$

### SOLUTION

Because the coefficient of $x$ in the first equation is 4 and in the second equation it is 3, we can obtain an equation not involving $x$ by replacing the second equation by the sum of 3 times the first equation and $-4$ times the second equation. We then have the equivalent system

$$\begin{cases} 4x + 3y = 6 \\ 3(4x + 3y) - 4(3x + 5y) = 3(6) + (-4)(-1) \end{cases}$$

Simplifying the second equation, we obtain

$$\begin{cases} 4x + 3y = 6 \\ -11y = 22 \end{cases}$$

Solving the second equation for $y$, we have

$$\begin{cases} 4x + 3y = 6 \\ y = -2 \end{cases}$$

Substituting the value of $y$ from the second equation into the first equation, we have

$$\begin{cases} 4x + 3(-2) = 6 \\ y = -2 \end{cases}$$

or, equivalently,

$$\begin{cases} x = 3 \\ y = -2 \end{cases}$$

Hence the solution set of the given system is $\{(3, -2)\}$.

We check the solution by substituting into the original equations, and we have

$$4(3) + 3(-2) = 12 - 6 \\ = 6$$

and

$$3(3) + 5(-2) = 9 - 10 \\ = -1$$

In the next two illustrations we show what happens when Theorem 6.1.2 is used to find the solution set of a system where the two equations are either inconsistent or dependent.

**ILLUSTRATION 6.** System (II) of Illustration 2 is

$$\begin{cases} 6x - 3y = 5 \\ 2x - y = 4 \end{cases}$$

If we replace the second equation by the sum of 1 times the first equation and $-3$ times the second equation, we have the equivalent system

$$\begin{cases} 6x - 3y = 5 \\ 1(6x - 3y) - 3(2x - y) = 1(5) + (-3)(4) \end{cases}$$

Simplifying the second equation, we obtain

$$\begin{cases} 6x - 3y = 5 \\ 0 = -7 \end{cases}$$

The solution set of this latter system is the empty set, $\emptyset$, because there is no ordered pair $(x, y)$ for which the second equation is a true statement. Hence the solution set of the given system (II) is $\emptyset$. The two equations are inconsistent. •

**ILLUSTRATION 7.** System (III) of Illustration 3 is

$$\begin{cases} 3x + 2y = 4 \\ 6x + 4y = 8 \end{cases}$$

Replacing the second equation by the sum of $-2$ times the first equation and 1 times the second equation, we have the equivalent system

$$\begin{cases} 3x + 2y = 4 \\ -2(3x + 2y) + 1(6x + 4y) = (-2)(4) + (1)(8) \end{cases}$$

Simplifying the second equation, we obtain

$$\begin{cases} 3x + 2y = 4 \\ 0 = 0 \end{cases}$$

The second equation of this latter system is an identity, that is, it is a true statement for any ordered pair $(x, y)$. Therefore, the solution set of the system is the same as the solution set of the first equation. The solution set can be written as $\{(x, y) \mid 3x + 2y = 4\}$. The equations are dependent. •

**EXAMPLE 3**

Find the solution set of the system

(IX) $\begin{cases} \dfrac{x}{3} + \dfrac{y}{2} = 2 \\ 2(x - y) - y = 5 \end{cases}$

**SOLUTION**

We first replace each equation by an equivalent equation. We multiply each member of the first equation by 6 and obtain $2x + 3y = 12$. The second equation is equivalent to the equation $2x - 3y = 5$. Hence system (IX) is equivalent to the system

(X) $\begin{cases} 2x + 3y = 12 \\ 2x - 3y = 5 \end{cases}$

Replacing the second equation by the sum of 1 times the first equation and 1 times the second equation, we have the equivalent system

$$\begin{cases} 2x + 3y = 12 \\ 4x = 17 \end{cases}$$

We solve the second equation for $x$ and obtain

(XI) $$\begin{cases} 2x + 3y = 12 \\ x = \dfrac{17}{4} \end{cases}$$

Another system of equations that is equivalent to system (X) is obtained by replacing the first equation by the sum of 1 times the first equation and $-1$ times the second equation; this system is

$$\begin{cases} 6y = 7 \\ 2x - 3y = 5 \end{cases}$$

We solve the first equation for $y$ and obtain

(XII) $$\begin{cases} y = \dfrac{7}{6} \\ 2x - 3y = 5 \end{cases}$$

From systems (XI) and (XII) we see that the solution set of system (IX) is $\{(\tfrac{17}{4}, \tfrac{7}{6})\}$. The solution can be checked by substituting into the equations of system (IX).

Note in Example 3 that after obtaining system (XI), the value of $y$ can be found by substituting $\tfrac{17}{4}$ for $x$ into the first equation of system (XI). In the solution shown, the use of system (XII) gives an alternative procedure for finding the value of $y$.

EXAMPLE 4

Find the solution set of the system

$$\begin{cases} \dfrac{4}{x} + \dfrac{3}{y} = 4 \\ \dfrac{2}{x} - \dfrac{6}{y} = -3 \end{cases}$$

SOLUTION

We replace the first equation by the sum of 1 times the first equation and $-2$ times the second equation, and we have the equivalent system

$$\begin{cases} 1\left(\dfrac{4}{x} + \dfrac{3}{y}\right) - 2\left(\dfrac{2}{x} - \dfrac{6}{y}\right) = 1(4) + (-2)(-3) \\ \dfrac{2}{x} - \dfrac{6}{y} = -3 \end{cases}$$

Simplifying the first equation, we get

$$\begin{cases} \dfrac{15}{y} = 10 \\ \dfrac{2}{x} - \dfrac{6}{y} = -3 \end{cases}$$

Solving the first equation for $\dfrac{1}{y}$, we have

$$\begin{cases} \dfrac{1}{y} = \dfrac{2}{3} \\ \dfrac{2}{x} - \dfrac{6}{y} = -3 \end{cases}$$

In the second equation we substitute $\dfrac{2}{3}$ for $\dfrac{1}{y}$, and we have

$$\begin{cases} \dfrac{1}{y} = \dfrac{2}{3} \\ \dfrac{2}{x} - 4 = -3 \end{cases}$$

or equivalently,

$$\begin{cases} \dfrac{1}{y} = \dfrac{2}{3} \\ \dfrac{2}{x} = 1 \end{cases}$$

or, equivalently,

$$\begin{cases} y = \dfrac{3}{2} \\ x = 2 \end{cases}$$

Therefore, the solution set of the given system is $\{(2, \tfrac{3}{2})\}$.

In Example 4 we treated the variables as $\dfrac{1}{x}$ and $\dfrac{1}{y}$. If we multiply each of the given equations by $xy$ (the LCD), we obtain the system

$$\begin{cases} 4y + 3x = 4xy \\ 2y - 6x = -3xy \end{cases}$$

Note that each of the equations in this system is of the second degree (the terms $4xy$ and $-3xy$ are second-degree terms). To solve this system for $x$ and $y$ requires a more complicated procedure than that used in Example 4.

# EXERCISES 6.1

*In Exercises 1 through 10, draw a sketch of the graph of the system of equations. Classify the equations as (i) consistent and independent, (ii) consistent and dependent, or (iii) inconsistent. If the equations are consistent and independent, determine the solution set of the system from the graphs.*

1. $\begin{cases} x - y = 8 \\ 2x + y = 1 \end{cases}$

2. $\begin{cases} y = 8 + 2x \\ 6x + 3y = 0 \end{cases}$

3. $\begin{cases} 2x + y = 6 \\ 8x = 6y + 9 \end{cases}$

4. $\begin{cases} 9x - 3y = 7 \\ y = 3x - \dfrac{5}{2} \end{cases}$

5. $\begin{cases} y = 2x - 4 \\ 6x - 3y - 12 = 0 \end{cases}$

6. $\begin{cases} 2x - 3y = -1 \\ 5x - 4y = 8 \end{cases}$

7. $\begin{cases} 4x - 2y - 7 = 0 \\ x = \dfrac{1}{2}y + 5 \end{cases}$

8. $\begin{cases} 3x - y = 1 \\ 6x + 5y = 2 \end{cases}$

9. $\begin{cases} 2x + 6y = -11 \\ 4x - 3y = -2 \end{cases}$

10. $\begin{cases} y = 3x - 5 \\ 6x - 2y = 10 \end{cases}$

*In Exercises 11 through 26, find the solution set of the given system by using either Theorem 6.1.2 or the method of substitution.*

11. $\begin{cases} 5x - 2y - 5 = 0 \\ 3x + y - 3 = 0 \end{cases}$

12. $\begin{cases} 3x + 4y - 4 = 0 \\ 5x + 2y - 8 = 0 \end{cases}$

13. $\begin{cases} 4x + 3y + 6 = 0 \\ 3x - 2y - 4 = 0 \end{cases}$

14. $\begin{cases} 8x - 3y = 5 \\ 5x - 2y = 4 \end{cases}$

15. $\begin{cases} 5x + 3y = 3 \\ x + 9y = 2 \end{cases}$

16. $\begin{cases} 5x + 6y = -5 \\ 15x - 3y = 13 \end{cases}$

17. $\begin{cases} 3x + 4y - 4 = 0 \\ 6x - 2y - 3 = 0 \end{cases}$

18. $\begin{cases} 18x + 3y - 10 = 0 \\ 2x - 2y - 5 = 0 \end{cases}$

19. $\begin{cases} 8x + 5y = 3 \\ 7x + 3y = -7 \end{cases}$

20. $\begin{cases} 2x - 5y = -21 \\ 5x + 3y = -6 \end{cases}$

21. $\begin{cases} \dfrac{x}{3} + \dfrac{y}{2} = 1 \\ \dfrac{x}{4} - \dfrac{y}{3} = -1 \end{cases}$

22. $\begin{cases} \dfrac{x}{2} - \dfrac{y}{6} = 1 \\ \dfrac{x}{3} + \dfrac{y}{2} = -1 \end{cases}$

23. $\begin{cases} \dfrac{6}{x} + \dfrac{3}{y} = -2 \\ \dfrac{4}{x} + \dfrac{7}{y} = -2 \end{cases}$

24. $\begin{cases} \dfrac{2}{x} + \dfrac{3}{y} = 2 \\ \dfrac{4}{x} - \dfrac{3}{y} = 1 \end{cases}$

25. $\begin{cases} \dfrac{3}{x} - \dfrac{2}{y} = 14 \\ \dfrac{6}{x} + \dfrac{3}{y} = -7 \end{cases}$

26. $\begin{cases} \dfrac{1}{x} - \dfrac{10}{y} = 6 \\ \dfrac{2}{x} + \dfrac{5}{y} = 2 \end{cases}$

## 6.2 Systems of Linear Equations in Three Variables

Consider the equation
$$2x - y + 4z = 10 \tag{1}$$
for which the replacement set of each of the three variables $x$, $y$, and $z$, is $R$. Equation (1) is a linear (first-degree) equation in the three variables, and a solution of this equation is an ordered triple of real numbers. An ordered

triple of numbers $(r, s, t)$ is a *solution* of equation (1) if, when $x$ is replaced by $r$, $y$ is replaced by $s$, and $z$ is replaced by $t$, the resulting statement is true; in such a case the ordered triple is said to *satisfy* the equation. The set of all solutions is the *solution set* of the equation.

The ordered triple $(3, 4, 2)$ is a solution of equation (1) because

$$2(3) - (4) + 4(2) = 10$$

Some other ordered triples that satisfy equation (1) are $(1, 0, 2)$, $(0, 2, 3)$, $(5, 4, 1)$, $(2, 2, 2)$, $(-3, -4, 3)$, and $(-1, 4, 4)$. It appears that the solution set of equation (1) is an infinite set.

The graph of an equation in three variables is a set of points represented by ordered triples of real numbers; such points involve a three-dimensional coordinate system which we do not discuss in this book. However, it should be mentioned that the graph of a linear equation in three variables is a plane.

Suppose that we have the following system of linear equations in the variables $x, y,$ and $z$.

(I)
$$\begin{cases} a_1 x + b_1 y + c_1 z = d_1 \\ a_2 x + b_2 y + c_2 z = d_2 \\ a_3 x + b_3 y + c_3 z = d_3 \end{cases}$$

The solution set of system (I) is the intersection of the solution sets of the three equations. Because the graph of each of the equations in system (I) is a plane, the solution set can be interpreted graphically as the intersection of three planes. When this intersection consists of a single point, the equations of the system are said to be *consistent* and *independent*. Refer to Figure 6.2.1, which shows three planes whose intersection is a single point. As we proceed with the discussion, we shall show other possible relative positions of three planes.

Algebraic methods for finding the solution set of a system of three linear equations in three variables are analogous to those used to solve systems of linear equations in two variables.

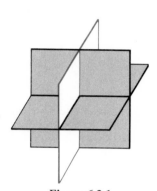

**Figure 6.2.1**

EXAMPLE 1

Find the solution set of the system

$$\begin{cases} x - y - 4z = 3 \\ 2x - 3y + 2z = 0 \\ 2x - y + 2z = 2 \end{cases}$$

by the method of substitution.

SOLUTION

We solve the first equation for $x$ and substitute the resulting expression for $x$ in the second and third equations; doing this, we have the equivalent system

$$\begin{cases} x = y + 4z + 3 \\ 2(y + 4z + 3) - 3y + 2z = 0 \\ 2(y + 4z + 3) - y + 2z = 2 \end{cases}$$

or, equivalently,

$$\begin{cases} x = y + 4z + 3 \\ -y + 10z = -6 \\ y + 10z = -4 \end{cases}$$

Solving the second equation for $y$ and substituting the resulting expression for $y$ in the third equation, we have the equivalent system

$$\begin{cases} x = y + 4z + 3 \\ y = 10z + 6 \\ (10z + 6) + 10z = -4 \end{cases}$$

or equivalently,

$$\begin{cases} x = y + 4z + 3 \\ y = 10z + 6 \\ 20z = -10 \end{cases}$$

or, equivalently,

$$\begin{cases} x = y + 4z + 3 \\ y = 10z + 6 \\ z = -\frac{1}{2} \end{cases}$$

Substituting the value of $z$ from the third equation into the second equation, we have

$$\begin{cases} x = y + 4z + 3 \\ y = 1 \\ z = -\frac{1}{2} \end{cases}$$

Substituting the values of $y$ and $z$ from the second and third equations into the first equation, we have

$$\begin{cases} x = 2 \\ y = 1 \\ z = -\frac{1}{2} \end{cases}$$

This latter system is equivalent to the given system. Hence the solution set of the given system is $\{(2, 1, -\frac{1}{2})\}$.

The solution can be checked by substituting into each of the given equations. Doing this, we have

$$\begin{cases} 2 - 1 + 2 = 3 \\ 4 - 3 - 1 = 0 \\ 4 - 1 - 1 = 2 \end{cases}$$

The equations of the given system are consistent and independent.

## EXAMPLE 2
Find the solution set of the system

(II) $\quad \begin{cases} 4x - 2y - 3z = 8 \\ 5x + 3y - 4z = 4 \\ 6x - 4y - 5z = 12 \end{cases}$

## SOLUTION
We first obtain an equivalent system in which the second and third equations do not involve the variable $x$. To eliminate $x$ between the first two equations, we observe that the coefficient of $x$ in the first equation is 4 and it is 5 in the second equation. Therefore, we replace the second equation by the sum of 5 times the first equation and $-4$ times the second equation. To eliminate $x$ between the first and third equations we replace the third equation by the sum of 6 times the first equation and $-4$ times the third equation. We have the equivalent system

$$\begin{cases} 4x - 2y - 3z = 8 \\ 5(4x - 2y - 3z) - 4(5x + 3y - 4z) = 5(8) + (-4)(4) \\ 6(4x - 2y - 3z) - 4(6x - 4y - 5z) = 6(8) + (-4)(12) \end{cases}$$

or, equivalently,

$$\begin{cases} 4x - 2y - 3z = 8 \\ -22y + z = 24 \\ 4y + 2z = 0 \end{cases}$$

We now wish to obtain an equivalent system in which we have an equation containing only one variable. Because the coefficient of $y$ in the second equation is $-22$ and in the third equation it is 4, we can obtain an equation not involving $y$ by replacing the third equation by the sum of 2 times the second equation and 11 times the third equation. We have

$$\begin{cases} 4x - 2y - 3z = 8 \\ -22y + z = 24 \\ 2(-22y + z) + 11(4y + 2z) = 2(24) + (11)(0) \end{cases}$$

or, equivalently,

(III) $\quad \begin{cases} 4x - 2y - 3z = 8 \\ -22y + z = 24 \\ z = 2 \end{cases}$

Substituting the value of $z$ from the third equation into the second equation, we have

$$\begin{cases} 4x - 2y - 3z = 8 \\ y = -1 \\ z = 2 \end{cases}$$

Substituting the values of $y$ and $z$ from the second and third equations of this latter system into the first equation, we have

(IV) $\quad \begin{cases} x = 3 \\ y = -1 \\ z = 2 \end{cases}$

System (IV) is equivalent to system (II). Hence the solution set of system (II) is $\{(3, -1, 2)\}$.

The solution set of system (II) can be found by working with combinations of equations other than those used in Example 2. The procedure followed in Example 2 is a methodical one, and it consists of obtaining an equivalent system in what is called *triangular form,* which is of the form of system (III). When a system is in triangular form, it is a simple matter to obtain an equivalent system like system (IV), which is the eventual goal.

In the next illustration we have a system of three linear equations in three variables where the equations are inconsistent. We show what happens when we try to solve such a system.

**ILLUSTRATION 1.** Suppose that we have the system

(V)
$$\begin{cases} 2x + y - z = 2 \\ x + 2y + 4z = 1 \\ 5x + y - 7z = 4 \end{cases}$$

We can replace this system by an equivalent one for which two of the equations do not involve $x$. Because the coefficient of $x$ in the first equation is 2 and it is 1 in the second equation, we replace the second equation by the sum of 1 times the first equation and $-2$ times the second equation. To eliminate $x$ between the first and third equations, we replace the third equation by the sum of 5 times the first equation and $-2$ times the third equation. We obtain the following system equivalent to system (V).

$$\begin{cases} 2x + y - z = 2 \\ 1(2x + y - z) - 2(x + 2y + 4z) = 1(2) + (-2)(1) \\ 5(2x + y - z) - 2(5x + y - 7z) = 5(2) + (-2)(4) \end{cases}$$

or, equivalently,

$$\begin{cases} 2x + y - z = 2 \\ -3y - 9z = 0 \\ 3y + 9z = 2 \end{cases}$$

Replacing the third equation by the sum of the second equation and the third equation, we have the equivalent system

$$\begin{cases} 2x + y - z = 2 \\ -3y - 9z = 0 \\ (-3y - 9z) + (3y + 9z) = 0 + 2 \end{cases}$$

or, equivalently,

(VI)
$$\begin{cases} 2x + y - z = 2 \\ -3y - 9z = 0 \\ 0 = 2 \end{cases}$$

We see that the solution set of system (VI) is the empty set, ∅, because there is no ordered triple $(x, y, z)$ for which the third equation is a true statement. Therefore, the solution set of system (V) is ∅, and the three equations are inconsistent. •

In Section 6.1, it was shown that when we have a system of two inconsistent linear equations in two variables, the graphs of the two equations are parallel lines. For a system of three inconsistent linear equations in three variables, the graphs of the three equations are planes that have no common intersection. The various possibilities for such a situation are shown in Figure 6.2.2 (a), (b), (c), and (d). In (a), the three planes are parallel. In (b), two of the planes are the same plane, and the third plane is parallel to it. In (c), two of the planes are parallel, the intersection of each of these planes with the third plane is a line, and the lines are parallel. In (d), no two planes are parallel, but two of the planes intersect in a line that is parallel to the third plane.

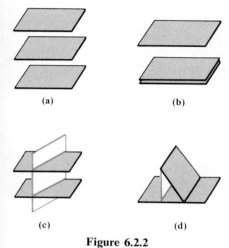

Figure 6.2.2

**ILLUSTRATION 2.** Consider the system

(VII) $\quad \begin{cases} 2x + 2y - z = 4 \\ 6x + 8y - 5z = 6 \\ 4x + 5y - 3z = 5 \end{cases}$

Replacing the second equation by the sum of 3 times the first equation and $-1$ times the second equation, and replacing the third equation by the sum of 2 times the first equation and $-1$ times the third equation, we have the equivalent system

$$\begin{cases} 2x + 2y - z = 4 \\ 3(2x + 2y - z) - 1(6x + 8y - 5z) = 3(4) + (-1)(6) \\ 2(2x + 2y - z) - 1(4x + 5y - 3z) = 2(4) + (-1)(5) \end{cases}$$

or, equivalently,

$$\begin{cases} 2x + 2y - z = 4 \\ -2y + 2z = 6 \\ -y + z = 3 \end{cases}$$

We now replace the third equation by the sum of 1 times the second equation and $-2$ times the third equation, and we obtain the equivalent system

$$\begin{cases} 2x + 2y - z = 4 \\ -2y + 2z = 6 \\ 1(-2y + 2z) - 2(-y + z) = 1(6) + (-2)(3) \end{cases}$$

or, equivalently,

(VIII) $\quad \begin{cases} 2x + 2y - z = 4 \\ y - z = -3 \\ 0 = 0 \end{cases}$

The third equation of system (VIII) is an identity because it is a true statement for any ordered triple $(x, y, z)$. In particular, it is true for the ordered triple $(0, 0, t)$ where $t$ is any real number. Therefore, if in system (VIII) we replace the third equation $0 = 0$ by the equation $z = t$, we will have an equivalent system in triangular form. We have then the system

(IX) $\quad \begin{cases} 2x + 2y - z = 4 \\ \phantom{2x + 2}y - z = -3 \\ \phantom{2x + 2y - }z = t \end{cases}$

We substitute the value of $z$ from the third equation into the second equation, and we obtain

$\begin{cases} 2x + 2y - z = 4 \\ \phantom{2x + 2}y = t - 3 \\ \phantom{2x + 2y - }z = t \end{cases}$

We now substitute the values of $y$ and $z$ from the second and third equations into the first equation, and we have

$\begin{cases} 2x + 2(t - 3) - t = 4 \\ \phantom{2x + 2(t - 3) - }y = t - 3 \\ \phantom{2x + 2(t - 3) - }z = t \end{cases}$

or, equivalently,

$\begin{cases} 2x + 2t - 6 - t = 4 \\ \phantom{2x + 2t - 6 - }y = t - 3 \\ \phantom{2x + 2t - 6 - }z = t \end{cases}$

or, equivalently,

$\begin{cases} x = -\tfrac{1}{2}t + 5 \\ y = t - 3 \\ z = t \end{cases}$

Therefore, any ordered triple of the form $(-\tfrac{1}{2}t + 5, t - 3, t)$ is a solution of system (IX), or, equivalently, system (VII). Therefore, the solution set of system (VII) is

$$\{(-\tfrac{1}{2}t + 5, t - 3, t) \mid t \in R\}$$

This is an infinite set and by assigning an arbitrary value to $t$ we get an ordered triple in the set. Assigning to $t$ the values $0, 1, 2, 4, -1, -2$, and $-4$, we obtain, respectively, these ordered triples: $(5, -3, 0)$, $(\tfrac{9}{2}, -2, 1)$, $(4, -1, 2)$, $(3, 1, 4)$, $(\tfrac{11}{2}, -4, -1)$, $(6, -5, -2)$, and $(7, -7, -4)$. ●

The solution set of system (VII) is an infinite set because the equations are *dependent*. Notice that the sum of 2 times the third equation and $-1$ times

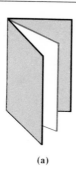
(a)

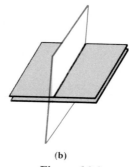
(b)

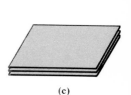
(c)

**Figure 6.2.3**

the first equation gives the equation

$$2(4x + 5y - 3z) - 1(2x + 2y - z) = 2(5) + (-1)(4)$$
$$6x + 8y - 5z = 6$$

which is the second equation.

For a system of three dependent linear equations in three variables, the graphs of the three equations are either three planes having a line in common, or else they are the same plane. The various possibilities are shown in Figures 6.2.3 (a), (b), and (c). In (a), the graphs are three distinct planes having a line in common. In (b), two of the planes are identical, and the third plane intersects them in a line. In (c), the three planes are identical.

## EXERCISES 6.2

*In Exercises 1 through 12, find the solution set of the given system of equations. If the equations are either inconsistent or dependent, then so indicate.*

1. $\begin{cases} 4x + 3y + z = 15 \\ x - y - 2z = 2 \\ 2x - 2y + z = 4 \end{cases}$

2. $\begin{cases} 2x + 3y + z = 8 \\ 5x + 2y - 3z = -13 \\ x - 2y + 5z = 15 \end{cases}$

3. $\begin{cases} x - y + 3z = 2 \\ 2x + 2y - z = 5 \\ 5x + 2z = 7 \end{cases}$

4. $\begin{cases} 3x + 2y - z = 4 \\ 3x + y + 3z = -2 \\ 6x - 3y - 2z = -6 \end{cases}$

5. $\begin{cases} x + \frac{1}{3}(y - z) = -1 \\ y - \frac{1}{2}(z - 2x) = 1 \\ z - \frac{1}{4}(2x - y) = -2 \end{cases}$

6. $\begin{cases} 3x - 2y + 4z = 4 \\ 7x + 5y - z = 9 \\ x + 9y - 9z = 1 \end{cases}$

7. $\begin{cases} 2x - 3y - 5z = 4 \\ x + 7y + 6z = -7 \\ 7x - 2y - 9z = 6 \end{cases}$

8. $\begin{cases} 3x - 5y + 2z = -2 \\ 2x + 3z = -3 \\ 4y - 3z = 8 \end{cases}$

9. $\begin{cases} x - y + 5z = 2 \\ 4x - 3y + 5z = 3 \\ 3x - 2y + 4z = 1 \end{cases}$

10. $\begin{cases} 5x - 4y + 5z = 6 \\ 6x + y - 2x = 4 \\ 4x - 9y + 12z = 5 \end{cases}$

11. $\begin{cases} \frac{1}{x} + \frac{1}{y} - \frac{1}{z} = 5 \\ \frac{3}{x} - \frac{1}{y} + \frac{2}{z} = 12 \\ \frac{1}{x} + \frac{2}{y} + \frac{1}{z} = 9 \end{cases}$

12. $\begin{cases} \frac{3}{x} - \frac{3}{y} + \frac{1}{z} = -1 \\ \frac{2}{x} + \frac{1}{y} - \frac{4}{z} = 0 \\ \frac{1}{x} + \frac{4}{y} + \frac{1}{z} = 5 \end{cases}$

*In Exercises 13 and 14, show that the given system has an infinite number of solutions in the solution set and use the variable t to express the solution set as an infinite set of ordered pairs. Then assign the values 0, 1, 2, −1, and −2 to t and find five ordered triples in the solution set.*

13. $\begin{cases} 5x - 4y + 3z = 1 \\ 3x - 5y + 7z = 11 \end{cases}$

14. $\begin{cases} 3x - 2y - 5z = 5 \\ 2x - y - 4z = 3 \end{cases}$

*In Exercises 15 through 18, determine if the equations are consistent or inconsistent. To show they are inconsistent, solve a system of three of the equations and show that no member of the solution set satisfies the fourth equation. To show that they are consistent, find the solution set.*

15. $\begin{cases} 2x + y + 3z = -4 \\ x - 4y - 2z = 3 \\ 4x - 2y + z = 4 \\ 5x + 3y + 4z = 5 \end{cases}$

16. $\begin{cases} 2x + 3y - z = 5 \\ 4x - 3y + 3z = 5 \\ 3x + y + 4z = 2 \\ x - 2y + z = 1 \end{cases}$

17. $\begin{cases} 2x + 4y + 3z = 5 \\ x - 4y - 2z = 7 \\ 4x - 3y + 5z = 2 \\ 3x + 2y + 4z = 8 \end{cases}$

18. $\begin{cases} x - y - 4z = 0 \\ x - y + 2z = 6 \\ 3x + y - 5z = -1 \\ x - 2y + z = 7 \end{cases}$

## 6.3  Word Problems

In our previous discussions of word problems it was necessary to represent each of the unknown numbers by symbols using only one variable. We now can apply systems of equations to solving word problems and represent each of the unknown numbers by a different variable. You will see that some problems which can be solved by using a system of linear equations in two or more variables can also be solved by using a single linear equation in one variable. By assigning a separate variable to represent each unknown number, it is usually easier to form an equation. However, remember that it is necessary to have as many independent equations as there are variables.

## EXAMPLE 1

Three pounds of tea and 8 pounds of coffee cost $26, and 5 pounds of tea and 6 pounds of coffee cost $27.20. What is the cost of 1 pound of tea and what is the cost of 1 pound of coffee?

### SOLUTION

Let $x$ represent the number of cents in the cost of 1 pound of tea and $y$ represent the number of cents in the cost of 1 pound of coffee.

Because the number of cents in the cost of 3 pounds of tea plus the number of cents in the cost of 8 pounds of coffee is 2600, we have the equation $3x + 8y = 2600$. Because the number of cents in the cost of 5 pounds of tea plus the number of cents in the cost of 6 pounds of coffee is 2720, we have the equation $5x + 6y = 2720$. Thus we have the system of equations

$$\begin{cases} 3x + 8y = 2600 \\ 5x + 6y = 2720 \end{cases}$$

Replacing the second equation by the sum of 5 times the first equation and $-3$ times the second equation, we have the equivalent system

$$\begin{cases} 3x + 8y = 2600 \\ 5(3x + 8y) - 3(5x + 6y) = 5(2600) + (-3)(2720) \end{cases}$$

Simplifying the second equation, we obtain

$$\begin{cases} 3x + 8y = 2600 \\ 22y = 4840 \end{cases}$$

Solving the second equation for $y$, we have

$$\begin{cases} 3x + 8y = 2600 \\ y = 220 \end{cases}$$

Substituting 220 for $y$ in the first equation, we have

$$\begin{cases} 3x + 1760 = 2600 \\ y = 220 \end{cases}$$

Solving the first equation for $x$, we obtain

$$\begin{cases} x = 280 \\ y = 220 \end{cases}$$

Therefore, $(x, y) = (280, 220)$. Hence the price per pound of tea is $2.80 and the price per pound of coffee is $2.20.

**CHECK:** The value of 3 pounds of tea at 280 cents per pound is 840 cents, and the value of 8 pounds of coffee at 220 cents per pound is 1760 cents; $840 + 1760 = 2600$. The value of 5 pounds of tea at 280 cents per pound is 1400 cents, and the value of 6 pounds of coffee at 220 cents per pound is 1320 cents; $1400 + 1320 = 2720$.

## EXAMPLE 2

A college rowing team can row 2 miles downstream in 8 minutes, but

### SOLUTION

Let $x$ represent the number of miles per hour in the rate of rowing in still water, and let $y$ represent the number of miles per hour in the rate of the it takes 12 minutes for the team to row the same distance upstream. How fast can the team row in still water, and what is the rate of the current?

Table 6.3.1

|  | Number of Hours in Time | × | Number of Miles per Hour in Effective Rate | = | Number of Miles in Distance |
|---|---|---|---|---|---|
| Downstream | $\frac{2}{15}$ |  | $x + y$ |  | $\frac{2}{15}(x + y)$ |
| Upstream | $\frac{1}{5}$ |  | $x - y$ |  | $\frac{1}{5}(x - y)$ |

current. Table 6.3.1 gives expressions involving $x$ and $y$ for the number of miles in the distance each way.

The number of miles in the distance traveled each way can be represented by either 2 or one of the entries in the last column of Table 6.3.1. Therefore, we have the system of equations

$$\begin{cases} \frac{2}{15}(x + y) = 2 \\ \frac{1}{5}(x - y) = 2 \end{cases}$$

or, equivalently,

$$\begin{cases} x + y = 15 \\ x - y = 10 \end{cases}$$

We replace the first equation by the sum of the first equation and the second equation. Also, we replace the second equation by the sum of the first equation and $-1$ times the second equation. We have then the equivalent system

$$\begin{cases} 2x = 25 \\ 2y = 5 \end{cases}$$

or, equivalently,

$$\begin{cases} x = 12\frac{1}{2} \\ y = 2\frac{1}{2} \end{cases}$$

Thus the team can row $12\frac{1}{2}$ miles per hour in still water, and the rate of the current is $2\frac{1}{2}$ miles per hour.

**CHECK:** The effective rate downstream is 15 miles per hour, and therefore the team can row 2 miles down the river in 8 minutes. The effective rate upstream is 10 miles per hour, and therefore the team can row 2 miles up the river in 12 minutes.

## EXAMPLE 3

Two people began a certain job on Monday and the work was to be completed on Tuesday. On Monday the first person worked for 10 hours and the second person worked for 8 hours, and one-half of the job was done. On Tuesday the first person had to stop working after only 5 hours, and the second person had to work a total of 12 hours to complete the job. How long would it have taken each person alone to do the whole job?

### SOLUTION

Let $x$ represent the number of hours it would have taken the first person alone to do the whole job, and $y$ represent the number of hours it would have taken the second person alone to do the whole job. The rate of work of the first person is $\frac{1}{x}$ of the job per hour, and the rate of work of the second person is $\frac{1}{y}$ of the job per hour. In Table 6.3.2 we obtain the fractional part of the work done by each person on each day.

Table 6.3.2

| | Rate of Work | × | Number of Hours Worked | = | Fractional Part of Job Done |
|---|---|---|---|---|---|
| First person on Monday | $\frac{1}{x}$ | | 10 | | $\frac{10}{x}$ |
| Second person on Monday | $\frac{1}{y}$ | | 8 | | $\frac{8}{y}$ |
| First person on Tuesday | $\frac{1}{x}$ | | 5 | | $\frac{5}{x}$ |
| Second person on Tuesday | $\frac{1}{y}$ | | 12 | | $\frac{12}{y}$ |

Because the two people do one-half of the job on Monday, the sum of the entries in the last column and the first two rows is $\frac{1}{2}$. Also, the two people do one-half of the job on Tuesday, and therefore the sum of the entries in the last column and the last two rows is also $\frac{1}{2}$. Hence we have the system of equations

$$\begin{cases} \frac{10}{x} + \frac{8}{y} = \frac{1}{2} \\ \frac{5}{x} + \frac{12}{y} = \frac{1}{2} \end{cases}$$

To solve this system, we treat the variables as $\frac{1}{x}$ and $\frac{1}{y}$, as we did in Example 4 of Section 6.1. We replace the first equation by the sum of 1 times the first equation and $-2$ times the second equation, and we have the equivalent system

$$\begin{cases} 1\left(\frac{10}{x} + \frac{8}{y}\right) - 2\left(\frac{5}{x} + \frac{12}{y}\right) = 1\left(\frac{1}{2}\right) + (-2)\left(\frac{1}{2}\right) \\ \frac{5}{x} + \frac{12}{y} = \frac{1}{2} \end{cases}$$

Simplifying the first equation, we have

$$\begin{cases} -\dfrac{16}{y} = -\dfrac{1}{2} \\ \dfrac{5}{x} + \dfrac{12}{y} = \dfrac{1}{2} \end{cases}$$

or, equivalently,

$$\begin{cases} \dfrac{1}{y} = \dfrac{1}{32} \\ \dfrac{5}{x} + \dfrac{12}{y} = \dfrac{1}{2} \end{cases}$$

Substituting $\dfrac{1}{32}$ for $\dfrac{1}{y}$ in the second equation, we have the equivalent system

$$\begin{cases} \dfrac{1}{y} = \dfrac{1}{32} \\ \dfrac{5}{x} + \dfrac{12}{32} = \dfrac{1}{2} \end{cases}$$

or, equivalently,

$$\begin{cases} \dfrac{1}{y} = \dfrac{1}{32} \\ \dfrac{5}{x} = \dfrac{1}{2} - \dfrac{3}{8} \end{cases}$$

or, equivalently,

$$\begin{cases} \dfrac{1}{y} = \dfrac{1}{32} \\ \dfrac{1}{x} = \dfrac{1}{40} \end{cases}$$

or, equivalently,

$$\begin{cases} y = 32 \\ x = 40 \end{cases}$$

Thus the first person can do the whole job alone in 40 hours, and the second person can do the whole job alone in 32 hours.

**CHECK:** On Monday the fractional part of the job done by the first person is $\tfrac{10}{40}$ or $\tfrac{1}{4}$, and the fractional part of the job done by the second person is $\tfrac{8}{32}$ or $\tfrac{1}{4}$; and $\tfrac{1}{4} + \tfrac{1}{4} = \tfrac{1}{2}$. On Tuesday the fractional part of the job done by the first person is $\frac{5}{40}$ or $\frac{1}{8}$, and the fractional part of the job done by the second person is $\frac{12}{32}$ or $\frac{3}{8}$; and $\frac{1}{8} + \frac{3}{8} = \frac{1}{2}$.

### EXAMPLE 4

A man has a total of $15,000 in three investments. One of the investments consists of bonds that pay 6 per cent annual interest, another is a savings account that pays 5 per cent annual interest, and the third is a business. Two years ago the business lost 3 per cent and his net income from the three investments was $550. Last year the business earned 9 per cent and his net income from the three investments was $910. How much does he have in each of the investments?

### SOLUTION

Let $x$ represent the number of dollars invested in bonds, let $y$ represent the number of dollars invested in the savings account, and let $z$ represent the number of dollars invested in the business.

Because there is a total of $15,000 invested, we have the equation $x + y + z = 15,000$. Each year the sum of the income from the bonds and the savings account is given by $0.06x + 0.05y$. However, because two years ago the business lost 3 per cent and the net income was $550, we have the equation $0.06x + 0.05y - 0.03z = 550$. Furthermore, because last year the business earned 9 per cent and the net income was $910, we have the equation $0.06x + 0.05y + 0.09z = 910$. We have then the following system of equations, where the second and third equations are obtained by multiplying both members of each of the corresponding equations by 100.

$$\begin{cases} x + y + z = 15{,}000 \\ 6x + 5y - 3z = 55{,}000 \\ 6x + 5y + 9z = 91{,}000 \end{cases}$$

Replacing the third equation by the sum of 1 times the third equation and $-1$ times the second equation, we have the equivalent system

$$\begin{cases} x + y + z = 15{,}000 \\ 6x + 5y - 3z = 55{,}000 \\ \phantom{6x + 5y +} 12z = 36{,}000 \end{cases}$$

Replacing the second equation by the sum of 6 times the first equation and $-1$ times the second equation, and solving the third equation for $z$, we have the equivalent system

$$\begin{cases} x + y + z = 15{,}000 \\ 6(x + y + z) - 1(6x + 5y - 3z) = 6(15{,}000) + (-1)(55{,}000) \\ z = 3000 \end{cases}$$

or, equivalently,

$$\begin{cases} x + y + z = 15{,}000 \\ y + 9z = 35{,}000 \\ z = 3000 \end{cases}$$

Substituting 3000 for $z$ in the second equation and solving for $y$, we have the equivalent system

$$\begin{cases} x + y + z = 15{,}000 \\ y = 8000 \\ z = 3000 \end{cases}$$

Substituting 8000 for $y$ and 3000 for $z$ in the first equation, and solving for $x$, we have the equivalent system
$$\begin{cases} x = 4000 \\ y = 8000 \\ z = 3000 \end{cases}$$

Therefore, $4000 is invested in bonds, $8000 is in the savings account, and $3000 is invested in the business.

CHECK: $4000 + $8000 + $3000 = $15,000. Two years ago the income from the bonds was $240, the income from the savings account was $400 and the loss from the business was $90; hence the net income was $240 + $400 − $90 = $550. Last year the income from the bonds was $240, the income from the savings account was $400, and the income from the business was $270; therefore, the net income was $240 + $400 + $270 = $910.

EXAMPLE 5
A group of 14 people spent $28 for admission tickets to Cinema One, charging $2.50 for adults, $1.50 for students, and $1 for children. If they had attended Cinema Two, charging $4 for adults, $2 for students, and $1 for children, they would have spent $42 for admission tickets. How many adults, how many students, and how many children were in the group?

SOLUTION
Let $a$, $s$, and $c$ represent, respectively, the number of adults, the number of students, and the number of children in the group. Then because there were 14 people in the group, $a + s + c = 14$. Because Cinema One charges $2.50 for adults, $1.50 for students, and $1 for children, and the total for admission tickets to Cinema One was $28, we have the equation $2.5a + 1.5s + c = 28$ or, equivalently (if we multiply by 2 to eliminate the decimals), $5a + 3s + 2c = 56$. Because Cinema Two charges $4 for adults, $2 for students, and $1 for children, and the total for admission tickets to Cinema Two was $42, we have the equation $4a + 2s + c = 42$. We have then the system of equations

(I) $$\begin{cases} a + s + c = 14 \\ 5a + 3s + 2c = 56 \\ 4a + 2s + c = 42 \end{cases}$$

Replacing the second equation by the sum of 5 times the first equation and −1 times the second equation, and replacing the third equation by the sum of 4 times the first equation and −1 times the third equation, we have the equivalent system
$$\begin{cases} a + s + c = 14 \\ 2s + 3c = 14 \\ 2s + 3c = 14 \end{cases}$$

Replacing the third equation by the sum of 1 times the second equation and −1 times the third equation, we have the equivalent system

(II) $$\begin{cases} a + s + c = 14 \\ 2s + 3c = 14 \\ 0 = 0 \end{cases}$$

The third equation of this system is an identity because it is a true statement for any ordered triple $(a, s, c)$. In particular it is true for the ordered triple $(0, 0, t)$ and so in system (II) we replace the third equation by the equation $c = t$ and we have the equivalent system

$$\begin{cases} a + s + c = 14 \\ \phantom{a + {}} 2s + 3c = 14 \\ \phantom{a + s + {}} c = t \end{cases}$$

Substituting the value of $c$ from the third equation into the second, and solving for $s$, we have

$$\begin{cases} a + s + c = 14 \\ s = 7 - \tfrac{3}{2}t \\ c = t \end{cases}$$

We now substitute into the first equation the values of $s$ and $c$, and we have

$$\begin{cases} a + (7 - \tfrac{3}{2}t) + t = 14 \\ s = 7 - \tfrac{3}{2}t \\ c = t \end{cases}$$

or, equivalently,

(III) $$\begin{cases} a = 7 + \tfrac{1}{2}t \\ s = 7 - \tfrac{3}{2}t \\ c = t \end{cases}$$

System (III) is equivalent to system (I), and so the equations of system (I) are dependent. It follows that a solution of system (I) is an ordered triple of the form $(7 + \tfrac{1}{2}t, 7 - \tfrac{3}{2}t, t)$. Becase $a$, $s$, and $c$ must represent nonnegative integers, each of the numbers $t$, $7 - \tfrac{3}{2}t$, and $7 + \tfrac{1}{2}t$ must be nonnegative integers. If $t = 0$, $7 - \tfrac{3}{2}t = 7$, and $7 + \tfrac{1}{2}t = 7$. Hence $(7, 7, 0)$ is a solution. If $t = 1$, $7 - \tfrac{3}{2}t = \tfrac{11}{2}$ and $7 + \tfrac{1}{2}t = \tfrac{15}{2}$; thus $t = 1$ does not give a solution to the problem. If $t = 2$, $7 - \tfrac{3}{2}t = 4$, and $7 + \tfrac{1}{2}t = 8$. Therefore, $(8, 4, 2)$ is a solution. If $t = 3$, both $7 - \tfrac{3}{2}t$ and $7 + \tfrac{1}{2}t$ are not integers and so $t = 3$ does not give a solution. If $t = 4$, $7 - \tfrac{3}{2}t = 1$, and $7 + \tfrac{1}{2}t = 9$. Hence $(9, 1, 4)$ is a solution. If $t$ is an integer greater than 4, $7 - \tfrac{3}{2}t$ is a negative number. Therefore, the solution set is $\{(7, 7, 0), (8, 4, 2), (9, 1, 4)\}$. Thus there are three possible combinations of people in the group: seven adults, seven students, and no children; eight adults, four students, and two children; or nine adults, one student, and four children.

**CHECK:** If seven adults, seven students, and no children are in the group, the number of dollars in the total cost of admission tickets to Cinema One is $(2.5)(7) + (1.5)7 = 28$, and the number of dollars in the total cost of admission tickets to Cinema Two is $(4)(7) + (2)(7) = 42$. If eight adults, four students, and two children are in the group, the number of dollars in the total cost of admission tickets to Cinema One is $(2.5)(8) + (1.5)(4) + (1)(2) = 28$, and the number of dollars in the total cost of admission tickets to Cinema Two is $(4)(8) + (2)(4) + (1)(2) = 42$. If nine adults, one student, and four children are in the group, the number of dollars in the total cost of admission tickets to Cinema One is $(2.5)(9) + (1.5)(1) + (1)(4) = 28$, and the number of dollars in the total cost of admission tickets to Cinema Two is $(4)(9) + (2)(1) + (1)(4) = 42$.

## EXERCISES 6.3

1. Two pounds of India tea and 5 pounds of China tea can be purchased for $16.72, and 3 pounds of India tea and 4 pounds of China tea cost $16.96. What is the price per pound of each kind of tea?

2. The cost of sending a telegram is based on a flat rate for the first 10 words and a fixed charge for each additional word. If a telegram of 15 words costs $4 and a telegram of 19 words cost $5.08, what is the flat rate and what is the fixed charge for each additional word?

3. A group of women decided to contribute equal amounts toward obtaining a speaker for a book review. If there were ten more women, each would have paid $2 less. However, if there were five less women, each would have paid $2 more. How many women were in the group and how much was the speaker paid?

4. A woman has a certain amount of money invested. If she had $6000 more invested at a rate of 1 per cent lower, she would have the same yearly income from the investment. Furthermore, if she had $4500 less invested at a rate 1 per cent higher, her yearly income from the investment would also be the same. How much does she have invested, and at what rate is it invested?

5. An airplane flying at its normal speed in still air against a wind of 40 kilometers per hour covers a certain distance in 1 hour and 30 minutes. On the return trip the plane flies at its normal speed in still air, but with the aid of a tail wind of 30 kilometers per hour, the plane covers the same distance in 1 hour and 20 minutes. What is the normal speed of the plane in still air and what is the distance traveled each way?

6. If the numerator and denominator of a fraction are each increased by 3, the resulting fraction is equivalent to $\frac{3}{5}$. However, if the numerator and denominator are each decreased by 3, the resulting fraction is equivalent to $\frac{3}{7}$. What is the fraction?

7. If either 4 is added to the denominator of a fraction or 2 is subtracted from the numerator of the fraction, the resulting fraction is equivalent to $\frac{1}{2}$. What is the fraction?

8. A rowing team can row 4 kilometers down a river in 8 minutes, but it takes the team 12 minutes to row the same distance up the river. How fast can the team row in still water and what is the rate of the current?

9. It takes a man 23 minutes longer to jog 5 miles than it takes his son to run the same distance. However, if the man doubles his speed, he can run the distance in 1 minute less than his son. What is the man's rate of jogging and what is the son's rate of running?

10. The distance between two automobiles is 140 kilometers. If the cars are driven toward each other, they will meet in 48 minutes. However, if they are driven in the same direction they will meet in 4 hours. What is the rate at which each car is driven?

11. If a girl works for 8 minutes and her brother works for 15 minutes, they can wash the front windows of their house. Also, if the girl works for 12 minutes and her brother works for 10 minutes, they can wash the same windows. How long will it take each person working alone to wash the windows?

12. A large pipe and a small pipe are used to fill a tank and a third pipe is used to drain the tank. If all three pipes are open, it takes 2 hours to fill the

tank. If the large pipe and the drain are open and the small pipe is closed, it takes 4 hours to fill the tank. If the small pipe and the drain are open and the large pipe is closed, it takes 6 hours to fill the tank. How long will it take each pipe to fill the tank alone if the drain is closed, and how long will it take the drain to empty a full tank?

13. One alloy contains 6 grams of copper, 4 grams of zinc, and 10 grams of lead. A second alloy contains 12 grams of copper, 5 grams of zinc, and 3 grams of lead. A third alloy contains 8 grams of copper, 6 grams of zinc, and 6 grams of lead. How many grams of each alloy should be combined in order to make a new alloy containing 34 grams of copper, 17 grams of zinc, and 19 grams of lead?

14. Part of $25,000 is invested at 5 per cent, another part is invested at 6 per cent, and a third part is invested at 8 per cent. The total yearly income from the three investments is $1600. Furthermore, the income from the 8 per cent investment yields the same amount as the sum of the incomes from the other two investments. How much is invested at each rate?

15. An equation of a parabola, whose axis is parallel to the $y$ axis, has the form $y = ax^2 + bx + c$. Find an equation of such a parabola if it contains the points $(1, -1)$, $(2, 3)$, and $(3, 15)$.

16. An equation of a circle is of the form $x^2 + y^2 + ax + by + c = 0$. Find an equation of the circle that contains the points $(-2, 8)$, $(2, 6)$, and $(-7, 3)$.

17. On a store counter, there was a supply of three sizes of Christmas cards. The large cards cost 25 cents each; the medium cards cost 20 cents each; and the small cards cost 15 cents each. A woman purchased ten cards, which consisted of one-fourth of the available large cards, one-third of the available medium cards, and one-half of the available small cards. The total cost of her cards was $2.05. If there were 21 cards remaining on the counter after her purchase, how many of each kind of card did she buy?

## 6.4 Systems Involving Quadratic Equations

Until now, our discussion of systems of equations has been confined to systems of linear equations. Here we consider systems of two equations in two variables in which at least one of the equations is quadratic. Perhaps the simplest such system is one that contains a linear equation and a quadratic equation. In this case the system can be solved by the method of substitution. The linear equation can be solved for one variable in terms of the other and the resulting expression can be substituted into the quadratic equation, as shown in the following example.

**EXAMPLE 1**
Find the solution set of the system

(I) $\begin{cases} x^2 - 2y = 2 \\ x - 3y = -1 \end{cases}$

**SOLUTION**
We solve the second equation for $y$, and we obtain the equivalent system

$\begin{cases} x^2 - 2y = 2 \\ y = \dfrac{x+1}{3} \end{cases}$

Replacing $y$ in the first equation by its equal from the second equation, we have the equivalent system

$$\begin{cases} x^2 - \dfrac{2x + 2}{3} = 2 \\ y = \dfrac{x + 1}{3} \end{cases}$$

or, equivalently,

(II) $\quad \begin{cases} 3x^2 - 2x - 8 = 0 \\ y = \dfrac{x + 1}{3} \end{cases}$

We now solve the first equation.

$$(3x + 4)(x - 2) = 0$$
$$3x + 4 = 0 \qquad x - 2 = 0$$
$$x = -\dfrac{4}{3} \qquad x = 2$$

Because the first equation of system (II) is equivalent to the two equations, $x = -\frac{4}{3}$ and $x = 2$, system (II) is equivalent to the two systems

$$\begin{cases} x = -\dfrac{4}{3} \\ y = \dfrac{x + 1}{3} \end{cases} \quad \text{and} \quad \begin{cases} x = 2 \\ y = \dfrac{x + 1}{3} \end{cases}$$

In each of the latter two systems we substitute into the second equation the value of $x$ from the first equation, and we have

(III) $\quad \begin{cases} x = -\dfrac{4}{3} \\ y = -\dfrac{1}{9} \end{cases} \quad \text{and} \quad \begin{cases} x = 2 \\ y = 1 \end{cases}$

System (I) is equivalent to systems (III). Thus the solution set of system (I) is $\{(-\frac{4}{3}, -\frac{1}{9}), (2, 1)\}$.

Figure 6.4.1 shows sketches of the graphs of the two equations of system (I) on the same coordinate system. The graph of the first equation is a parabola, and the graph of the second equation is a line. The graphs intersect at the points $(-\frac{4}{3}, -\frac{1}{9})$ and $(2, 1)$.

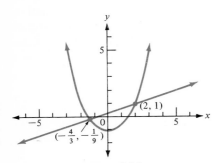

Figure 6.4.1

## EXAMPLE 2

Find the solution set of the system

(IV) $\quad \begin{cases} x^2 + y^2 = 25 \\ 3x + 4y = 25 \end{cases}$

Draw sketches of the graphs of the two equations on the same coordinate system.

### SOLUTION

Solving the second equation for $y$ and replacing $y$ in the first equation by the resulting expression, we have the equivalent system

(V) $\quad \begin{cases} x^2 + \left(\dfrac{25 - 3x}{4}\right)^2 = 25 \\ y = \dfrac{25 - 3x}{4} \end{cases}$

We solve the first equation by first multiplying each member by 16.

$$16x^2 + (625 - 150x + 9x^2) = 400$$
$$25x^2 - 150x + 225 = 0$$
$$x^2 - 6x + 9 = 0$$
$$(x - 3)^2 = 0$$

Hence the roots of this quadratic equation are 3 and 3; that is, 3 is a double root. Therefore, system (V) is equivalent to the system

$$\begin{cases} x = 3 \\ y = \dfrac{25 - 3x}{4} \end{cases}$$

Substituting 3 for $x$ in the second equation, we obtain

$$\begin{cases} x = 3 \\ y = 4 \end{cases}$$

Thus the solution set of the given system (IV) is $\{(3, 4)\}$.

Sketches of the graphs of the two equations of system (IV) are shown in Figure 6.4.2. We see that the line is tangent to the circle at the point (3, 4); this is the geometric significance of the double root. The point of tangency can be considered as two intersections of the line and the circle.

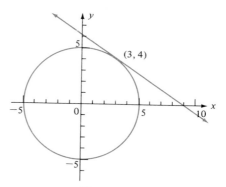

**Figure 6.4.2**

## EXAMPLE 3

Find the solution set of the system

(VI) $\quad \begin{cases} x^2 + y^2 = 2 \\ x - y = 4 \end{cases}$

Draw sketches of the graphs of the two equations on the same coordinate system.

## SOLUTION

Solving the second equation for $x$ and replacing $x$ in the first equation by the resulting expression, we have the equivalent system

(VII) $\quad \begin{cases} (y + 4)^2 + y^2 = 2 \\ x = y + 4 \end{cases}$

We solve the first equation for $y$.

$$y^2 + 8y + 16 + y^2 = 2$$
$$2y^2 + 8y + 14 = 0$$
$$y^2 + 4y + 7 = 0$$
$$y = \frac{-4 \pm \sqrt{4^2 - 4(1)(7)}}{2(1)}$$
$$= \frac{-4 \pm \sqrt{-12}}{2}$$
$$= \frac{-4 \pm 2i\sqrt{3}}{2}$$
$$= -2 \pm i\sqrt{3}$$

Hence the first equation of system (VII) is equivalent to the two equations $y = -2 + i\sqrt{3}$ and $y = -2 - i\sqrt{3}$. Therefore, system (VII) is equivalent to the two systems

$$\begin{cases} y = -2 + i\sqrt{3} \\ x = y + 4 \end{cases} \quad \text{and} \quad \begin{cases} y = -2 - i\sqrt{3} \\ x = y + 4 \end{cases}$$

In each of these two systems we substitute the value of $y$ from the first equation into the second equation and we have

(VIII) $\quad \begin{cases} y = -2 + i\sqrt{3} \\ x = 2 + i\sqrt{3} \end{cases} \quad \text{and} \quad \begin{cases} y = -2 - i\sqrt{3} \\ x = 2 - i\sqrt{3} \end{cases}$

System (VI) is equivalent to systems (VIII). Thus the solution set of system (VI) is $\{(2 + i\sqrt{3}, -2 + i\sqrt{3}), (2 - i\sqrt{3}, -2 - i\sqrt{3})\}$.

Sketches of the graphs of the two equations of system (VI) are shown in Figure 6.4.3. The line and the circle have no points of intersection.

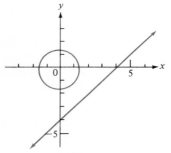

**Figure 6.4.3**

As in Example 3, when the solutions of a system of equations are ordered pairs of imaginary numbers, there are no points of intersection of the graphs that correspond to such solutions. This situation occurs because the coordinates of points in the real plane are real numbers.

Theorem 6.1.2 provides us with a method for solving a system of linear equations. There is a corresponding theorem for systems of higher-degree equations, for which the proof is similar; but the theorem is not stated here. However, we apply this theorem in the following example, involving a system of two quadratic equations.

## EXAMPLE 4

Find the solution set of the system

(IX) $\quad \begin{cases} 2x^2 - 3y^2 = 6 \\ 6x^2 + y^2 = 58 \end{cases}$

Draw sketches of the graphs of the two equations on the same coordinate system.

### SOLUTION

We wish to replace the given system (IX) by one that has an equation containing only one variable. Because the coefficients of $y^2$ are $-3$ in the first equation and $1$ in the second equation, we obtain an equation not involving $y$ by replacing the second equation by the sum of $1$ times the first equation and $3$ times the second equation. We have the equivalent system

$$\begin{cases} 2x^2 - 3y^2 = 6 \\ 1(2x^2 - 3y^2) + 3(6x^2 + y^2) = (1)(6) + (3)(58) \end{cases}$$

Simplifying the second equation, we obtain

$$\begin{cases} 2x^2 - 3y^2 = 6 \\ 20x^2 = 180 \end{cases}$$

or, equivalently,

(X) $\quad \begin{cases} 2x^2 - 3y^2 = 6 \\ x^2 = 9 \end{cases}$

The second equation of system (X) is equivalent to the two equations $x = 3$ and $x = -3$. Therefore, system (X) is equivalent to the two systems

$$\begin{cases} 2x^2 - 3y^2 = 6 \\ x = 3 \end{cases} \quad \text{and} \quad \begin{cases} 2x^2 - 3y^2 = 6 \\ x = -3 \end{cases}$$

In each of these two systems we substitute into the first equation the value of $x$ from the second equation, and we have

$$\begin{cases} 18 - 3y^2 = 6 \\ x = 3 \end{cases} \quad \text{and} \quad \begin{cases} 18 - 3y^2 = 6 \\ x = -3 \end{cases}$$

or, equivalently,

(XI) $\quad \begin{cases} y^2 = 4 \\ x = 3 \end{cases} \quad \text{and} \quad \begin{cases} y^2 = 4 \\ x = -3 \end{cases}$

The first equation in each of systems (XI) is equivalent to the two equations $y = 2$ and $y = -2$. Hence the two systems (XI) are equivalent to the four systems

(XII) $\quad \begin{cases} y = 2 \\ x = 3 \end{cases} \quad \begin{cases} y = -2 \\ x = 3 \end{cases} \quad \begin{cases} y = 2 \\ x = -3 \end{cases} \quad \begin{cases} y = -2 \\ x = -3 \end{cases}$

The given system (IX) is equivalent to systems (XII). Therefore, the solution set of system (IX) is $\{(3, 2), (3, -2), (-3, 2), (-3, -2)\}$.

Sketches of the graphs of the two equations of system (IX) are shown in Figure 6.4.4. The graph of the first equation is a hyperbola and the graph of the second equation is an ellipse. The two graphs intersect at the four points $(3, 2)$, $(3, -2)$, $(-3, 2)$, and $(-3, -2)$.

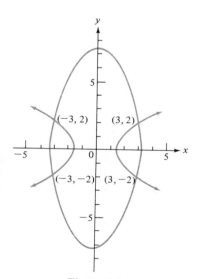

Figure 6.4.4

In the next example we have a system of two quadratic equations in which the second equation involves three second-degree terms. Because the right member of the second equation is zero and the left member can be factored, the second equation is equivalent to two linear equations. Therefore, the given system is equivalent to two systems each consisting of a quadratic equation and a linear equation.

EXAMPLE 5
Find the solution set of the system

(XIII) $\begin{cases} x^2 + y^2 = 16 \\ 2x^2 - 3xy + y^2 = 0 \end{cases}$

SOLUTION
We factor the left member of the second equation, and we obtain the equivalent system

$$\begin{cases} x^2 + y^2 = 16 \\ (2x - y)(x - y) = 0 \end{cases}$$

Because the second equation is equivalent to the two equations $2x - y = 0$ and $x - y = 0$, the given system is equivalent to the two systems

$$\begin{cases} x^2 + y^2 = 16 \\ 2x - y = 0 \end{cases} \quad \text{and} \quad \begin{cases} x^2 + y^2 = 16 \\ x - y = 0 \end{cases}$$

or, equivalently,

$$\begin{cases} x^2 + y^2 = 16 \\ y = 2x \end{cases} \quad \text{and} \quad \begin{cases} x^2 + y^2 = 16 \\ y = x \end{cases}$$

In each of the latter two systems we substitute into the first equation the value of $y$ from the second equation, and we have

$$\begin{cases} x^2 + 4x^2 = 16 \\ y = 2x \end{cases} \quad \text{and} \quad \begin{cases} x^2 + x^2 = 16 \\ y = x \end{cases}$$

or, equivalently,

(XIV) $\begin{cases} x^2 = \dfrac{16}{5} \\ y = 2x \end{cases} \quad \text{and} \quad \begin{cases} x^2 = 8 \\ y = x \end{cases}$

The equation $x^2 = \frac{16}{5}$ is equivalent to the two equations $x = \frac{4}{5}\sqrt{5}$ and $x = -\frac{4}{5}\sqrt{5}$. The equation $x^2 = 8$ is equivalent to the two equations $x = 2\sqrt{2}$ and $x = -2\sqrt{2}$. Therefore, the two systems (XIV) are equivalent to the four systems

$$\begin{cases} x = \dfrac{4}{5}\sqrt{5} \\ y = 2x \end{cases} \quad \begin{cases} x = -\dfrac{4}{5}\sqrt{5} \\ y = 2x \end{cases} \quad \begin{cases} x = 2\sqrt{2} \\ y = x \end{cases} \quad \begin{cases} x = -2\sqrt{2} \\ y = x \end{cases}$$

In each of these latter systems we substitute the value of $x$ from the first equation into the second equation, and we have

$$\begin{cases} x = \frac{4}{5}\sqrt{5} \\ y = \frac{8}{5}\sqrt{5} \end{cases} \begin{cases} x = -\frac{4}{5}\sqrt{5} \\ y = -\frac{8}{5}\sqrt{5} \end{cases} \begin{cases} x = 2\sqrt{2} \\ y = 2\sqrt{2} \end{cases} \begin{cases} x = -2\sqrt{2} \\ y = -2\sqrt{2} \end{cases}$$

The solution set of the given system (XIII) is

$$\left\{ \left(\frac{4}{5}\sqrt{5}, \frac{8}{5}\sqrt{5}\right), \left(-\frac{4}{5}\sqrt{5}, -\frac{8}{5}\sqrt{5}\right), \left(2\sqrt{2}, 2\sqrt{2}\right), \left(-2\sqrt{2}, -2\sqrt{2}\right) \right\}$$

The system in the next example consists of two quadratic equations in which all the terms containing variables are of the second degree; that is, there are no first-degree terms. If one of the equations is replaced by a combination of the two equations in which no constant term appears, the system that results can be solved by the method used in Example 5.

**EXAMPLE 6**

Find the solution set of the system

(XV) $\quad \begin{cases} 2x^2 + xy = 16 \\ 12x^2 - y^2 = 32 \end{cases}$

**SOLUTION**

We replace the second equation by the sum of $-2$ times the first equation and 1 times the second equation, and we have the equivalent system

$$\begin{cases} 2x^2 + xy = 16 \\ -2(2x^2 + xy) + 1(12x^2 - y^2) = (-2)(16) + (1)(32) \end{cases}$$

Simplifying the second equation, we have

$$\begin{cases} 2x^2 + xy = 16 \\ 8x^2 - 2xy - y^2 = 0 \end{cases}$$

Factoring the left member of the second equation, we have

$$\begin{cases} 2x^2 + xy = 16 \\ (2x - y)(4x + y) = 0 \end{cases}$$

The second equation is equivalent to the two equations $2x - y = 0$ and $4x + y = 0$. Hence this latter system is equivalent to the two systems

$$\begin{cases} 2x^2 + xy = 16 \\ 2x - y = 0 \end{cases} \quad \text{and} \quad \begin{cases} 2x^2 + xy = 16 \\ 4x + y = 0 \end{cases}$$

or, equivalently,

$$\begin{cases} 2x^2 + xy = 16 \\ y = 2x \end{cases} \quad \text{and} \quad \begin{cases} 2x^2 + xy = 16 \\ y = -4x \end{cases}$$

In each of these two systems, if we substitute into the first equation the value of $y$ from the second equation, we have

$$\begin{cases} 2x^2 + 2x^2 = 16 \\ y = 2x \end{cases} \quad \text{and} \quad \begin{cases} 2x^2 - 4x^2 = 16 \\ y = -4x \end{cases}$$

or, equivalently,

(XVI) $\quad \begin{cases} x^2 = 4 \\ y = 2x \end{cases} \quad \text{and} \quad \begin{cases} x^2 = -8 \\ y = -4x \end{cases}$

The equation $x^2 = 4$ is equivalent to the two equations $x = 2$ and $x = -2$, and the equation $x^2 = -8$ is equivalent to the two equations $x = 2i\sqrt{2}$ and $x = -2i\sqrt{2}$. Thus the two systems (XVI) are equivalent to the four systems

$$\begin{cases} x = 2 \\ y = 2x \end{cases} \quad \begin{cases} x = -2 \\ y = 2x \end{cases} \quad \begin{cases} x = 2i\sqrt{2} \\ y = -4x \end{cases} \quad \begin{cases} x = -2i\sqrt{2} \\ y = -4x \end{cases}$$

If in each of these systems we substitute in the second equation the value of $x$ from the first equation, we have the equivalent systems

(XVII) $\quad \begin{cases} x = 2 \\ y = 4 \end{cases} \quad \begin{cases} x = -2 \\ y = -4 \end{cases} \quad \begin{cases} x = 2i\sqrt{2} \\ y = -8i\sqrt{2} \end{cases} \quad \begin{cases} x = -2i\sqrt{2} \\ y = 8i\sqrt{2} \end{cases}$

The given system (XV) is equivalent to systems (XVII). Therefore, the solution set of system (XV) is $\{(2, 4), (-2, -4), (2i\sqrt{2}, -8i\sqrt{2}), (-2i\sqrt{2}, 8i\sqrt{2})\}$.

# EXERCISES 6.4

*In Exercises 1 through 24, find the solution set of the given system of equations. In Exercises 1 through 8 and 11 through 14, draw sketches of the graphs of the equations and obtain approximate values for the coordinates of the points of intersection.*

1. $\begin{cases} x^2 + y^2 = 25 \\ x - y + 1 = 0 \end{cases}$

2. $\begin{cases} x^2 + y^2 = 25 \\ x - 2y = -2 \end{cases}$

3. $\begin{cases} x^2 - y = 1 \\ x - 2y = -1 \end{cases}$

4. $\begin{cases} x^2 - y^2 = 9 \\ 2x + y = 6 \end{cases}$

5. $\begin{cases} x^2 - y^2 = 9 \\ x + y - 5 = 0 \end{cases}$

6. $\begin{cases} 4x^2 + y^2 = 25 \\ 2x + y + 1 = 0 \end{cases}$

7. $\begin{cases} x^2 - y - 4 = 0 \\ x - y - 3 = 0 \end{cases}$

8. $\begin{cases} 4x^2 + y - 3 = 0 \\ 8x + y - 7 = 0 \end{cases}$

9. $\begin{cases} x^2 - 2y^2 = 2 \\ x + 2y = 2 \end{cases}$

10. $\begin{cases} x^2 + xy + y^2 = 3 \\ x + y + 1 = 0 \end{cases}$

11. $\begin{cases} x^2 + y^2 = 4 \\ x^2 + 2y = 4 \end{cases}$

12. $\begin{cases} 4x^2 + y^2 = 17 \\ x^2 + y = 5 \end{cases}$

13. $\begin{cases} x^2 - y^2 = 15 \\ xy = 4 \end{cases}$

14. $\begin{cases} x^2 + y^2 = 25 \\ xy = 12 \end{cases}$

15. $\begin{cases} x^2 + y^2 = 25 \\ x^2 + 4y^2 = 64 \end{cases}$

16. $\begin{cases} 3x^2 + 2y^2 = 59 \\ 2x^2 + y^2 = 34 \end{cases}$

17. $\begin{cases} x^2 - y^2 = 9 \\ y^2 - 2x = 6 \end{cases}$

18. $\begin{cases} x^2 + y^2 = 16 \\ 9x^2 - 4y^2 = 36 \end{cases}$

19. $\begin{cases} x^2 - xy + 4 = 0 \\ 2x^2 - 2xy + y^2 = 8 \end{cases}$

20. $\begin{cases} 2x^2 - xy - y^2 = 0 \\ xy = 9 \end{cases}$

21. $\begin{cases} 10x^2 - xy + 4y^2 = 28 \\ 2x^2 - 3xy - 2y^2 = 0 \end{cases}$

22. $\begin{cases} 4x^2 - 5xy + 3y^2 = 24 \\ 2x^2 - 3xy + 2y^2 = 16 \end{cases}$

23. $\begin{cases} \dfrac{4}{x^2} + \dfrac{1}{y^2} = 15 \\ \dfrac{3}{x^2} - \dfrac{2}{y^2} = 14 \end{cases}$

24. $\begin{cases} \dfrac{3}{x^2} + \dfrac{1}{y^2} = 7 \\ \dfrac{5}{x^2} - \dfrac{2}{y^2} = -3 \end{cases}$

25. The sum of the reciprocals of two numbers is $\frac{4}{15}$ and their product is 60. Find the numbers.

26. The sum of the squares of two numbers is $\frac{5}{18}$ and the sum of 6 times the smaller number and 4 times the larger number is 3. Find the numbers.

27. A group of students planned a field trip and agreed to contribute equal amounts toward the transportation costs of $150. Later five more students decided to go on the trip and the transportation cost for each student was reduced by $1.50. Find the number of students who actually made the trip and the amount each paid for transportation.

28. Find the dimensions of a rectangle whose perimeter is 40 centimeters and whose area is 96 square centimeters.

29. A cyclist traveled a certain distance at his usual speed. If his speed had been 2 miles per hour faster, he would have traveled the distance in 1 hour less time. If his speed had been 2 miles per hour slower, he would have taken 2 hours longer. Find the distance traveled and his usual speed.

30. A piece of tin is in the form of a rectangle whose area is 486 square centimeters. A square of side 3 centimeters is cut from each corner, and an open box is made by turning up the ends and sides. If the volume of the box is 504 cubic centimeters, what are the dimensions of the piece of tin?

## 6.5 Systems of Inequalities and Introduction to Linear Programming

Two intersecting lines divide the points of the plane into four regions. Each of these regions is the intersection of two half planes and is defined by a pair of linear inequalities.

**ILLUSTRATION 1.** The two inequalities

$$6x - y - 5 > 0 \quad \text{and} \quad 4x + 3y - 7 > 0$$

define the region that is the intersection of the half plane below the line $y = 6x - 5$ and the half plane above the line $y = -\frac{4}{3}x + \frac{7}{3}$. This region is shaded in Figure 6.5.1. ●

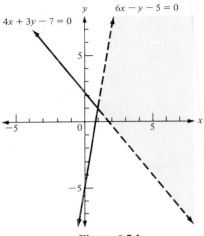

**Figure 6.5.1**

The next example gives a region defined by a system of three linear inequalities.

**EXAMPLE 1**
Draw a sketch of the region defined by the following inequalities.

$$2x - y + 6 \geq 0$$
$$x + 6y - 23 < 0$$
$$5x - 9y + 2 \leq 0$$

**SOLUTION**
Each of the inequalities defines a half plane, two of which are closed. Let $L_1$ be the line having the equation $2x - y + 6 = 0$, $L_2$ be the line having the equation $x + 6y - 23 = 0$, and $L_3$ the line having the equation $5x - 9y + 2 = 0$. The lines $L_1$ and $L_2$ intersect at the point $P(-1, 4)$, $L_1$ and $L_3$ intersect at $Q(-4, -2)$, and $L_2$ and $L_3$ intersect at $R(5, 3)$. Refer to Figure 6.5.2. The first inequality $y \leq 2x + 6$ defines the half plane below line $L_1$ together with $L_1$. The inequality $y < -\frac{1}{6}x + \frac{23}{6}$ defines the half plane below $L_2$, and the inequality $y \geq \frac{5}{9}x + \frac{2}{9}$ defines the half plane above $L_3$ together with $L_3$. The region, whose points satisfy all three inequalities, consists of the interior of the triangle $PQR$ and the points on the line segments $QP$ and $QR$, excluding the points $P$ and $R$.

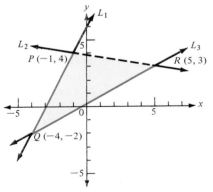

**Figure 6.5.2**

In the following example we have a region defined by a system of four inequalities in which two of the inequalities are quadratic and two are linear.

**EXAMPLE 2**
Draw a sketch of the region defined by the following system of inequalities.

$$x^2 - 5 \leq y \leq 5 - x^2$$
$$|x| > 1$$

**SOLUTION**
The given system of inequalities is equivalent to the system of the following four inequalities.

$$x^2 - 5 \leq y \qquad y \leq 5 - x^2 \qquad x > 1 \qquad x < -1$$

The graph of the inequality $x^2 - 5 \leq y$ consists of the points on the parabola $y = x^2 - 5$ and the points above the parabola. The graph of the inequality $y \leq 5 - x^2$ consists of the points on the parabola $y = 5 - x^2$ and the points below the parabola. The graph of the inequality $x > 1$ is the half plane lying to the right of the line $x = 1$, and the graph of the inequality $x < -1$ is the half plane lying to the left of the line $x = -1$. The graph of each of these inequalities is shaded in Figure 6.5.3, and the triple-shaded region in the figure is the one required.

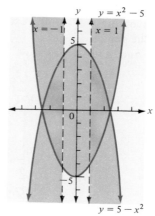

**Figure 6.5.3**

**EXAMPLE 3**
When flying economy class from the United States to Europe, a passenger is allowed to check, without additional cost, two pieces of luggage with the following restrictions: the sum of the three dimensions (length, width, and height) of the larger piece of luggage must not exceed 62

**SOLUTION**
(a) Both $x$ and $y$ must be positive, and $x$ must be greater than $y$ and less than or equal to $2y$. Therefore, we have the inequalities

$$x > 0 \qquad (1)$$
$$y > 0 \qquad (2)$$
$$x > y \qquad (3)$$
$$x \leq 2y \qquad (4)$$

inches, and the total dimensions of the two pieces must not exceed 106 inches. (a) Suppose that the height of the larger piece of luggage is 20 inches, and the length is greater than the width but not more than twice the width. If the length is $x$ inches and the width is $y$ inches, what are the inequalities involving $x$ and $y$? Show on a graph the region of permissible values of $x$ and $y$. (b) Suppose that the sum of the three dimensions of the larger piece of luggage is 58 inches. Furthermore, the height of the smaller piece is 12 inches and the length is to be greater than the width but not more than 10 inches greater. If $x$ inches is the length and $y$ inches is the width, what are the inequalities involving $x$ and $y$? Show on a graph the region of permissible values of $x$ and $y$.

Because $x + y + 20$ must be less than or equal to 62, we have the inequality

$$x + y \leq 42 \tag{5}$$

The lines associated with inequalities (1) through (5) are designated by $L_1$, $L_2$, $L_3$, $L_4$, and $L_5$, respectively. These lines have the equations

$$L_1: \quad x = 0, \qquad L_2: \quad y = 0 \qquad L_3: \quad x = y$$
$$L_4: \quad x = 2y \qquad L_5: \quad x + y = 42$$

Lines $L_1$, $L_2$, $L_3$, and $L_4$ intersect at the origin. The $x$ intercept of $L_5$ is 42 and the $y$ intercept of $L_5$ is also 42. If we solve simultaneously the equations for $L_3$ and $L_5$, we obtain the point of intersection $P(21, 21)$. The point of intersection of $L_4$ and $L_5$ is $Q(28, 14)$. Figure 6.5.4 shows the five lines.

Inequality (3) defines the half plane below line $L_3$ and inequality (4) defines the closed half plane above line $L_4$. The inequality $y \leq 42 - x$, which is equivalent to inequality (5), defines the closed half plane below $L_5$. The region, whose points satisfy all five inequalities, consists of the interior of the triangle $OPQ$ and the points on the line segments $OQ$ and $QP$ excluding the origin and the point $P$.

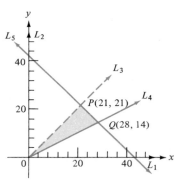

Figure 6.5.4

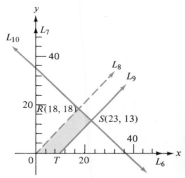

Figure 6.5.5

(b) The length and width must be positive, $x$ must be greater than $y$, and $x$ must be less than or equal to $y + 10$. We have then the inequalities

$$x > 0 \tag{6}$$
$$y > 0 \tag{7}$$
$$x > y \tag{8}$$
$$x \leq y + 10 \tag{9}$$

Because the sum of the three dimensions of the larger piece of luggage is 58 inches and the height of the smaller piece is 12 inches, it follows that $x + y$ must be less than or equal to $106 - (58 + 12)$. Therefore, we have the inequality

$$x + y \leq 36 \tag{10}$$

The lines $L_6$, $L_7$, $L_8$, $L_9$, and $L_{10}$ are associated with inequalities (6) through (10), respectively, and they have the equations

$$L_6: \ x = 0 \qquad L_7: \ y = 0 \qquad L_8: \ x = y$$
$$L_9: \ x = y + 10 \qquad L_{10}: \ x + y = 36$$

Lines $L_6$, $L_7$, and $L_8$ intersect at the origin and lines $L_8$ and $L_{10}$ intersect at the point $R(18, 18)$. Lines $L_9$ and $L_{10}$ intersect at the point $S(23, 13)$. Line $L_9$ intersects the $x$ axis at the point $T(10, 0)$. See Figure 6.5.5. Inequality (8) defines the half plane below line $L_8$. Inequality (9) defines the closed half plane above line $L_9$. Inequality (10) defines the closed half plane below line $L_{10}$. The region whose points satisfy the five inequalities (6) through (10) is the interior of the trapezoid $ORST$ and the points on the line segments $SR$ and $TS$ excluding the points $R$ and $T$.

The following example demonstrates a procedure that can be extended to an arbitrary number of linear inequalities.

EXAMPLE 4
Find the region, if there is one, whose points satisfy each of the following inequalities.

$$x \geq 1 \tag{11}$$
$$y \geq 0 \tag{12}$$
$$4x - 3y - 12 \leq 0 \tag{13}$$
$$2x - y - 4 \leq 0 \tag{14}$$
$$x + y - 7 \leq 0 \tag{15}$$

SOLUTION
Let lines $L_1$, $L_2$, $L_3$, $L_4$, and $L_5$ have the following equations, which are obtained from inequalities (11) through (15), respectively.

$$L_1: x = 1$$
$$L_2: y = 0$$
$$L_3: 4x - 3y - 12 = 0$$
$$L_4: 2x - y - 4 = 0$$
$$L_5: x + y - 7 = 0$$

We now determine if there are any points that satisfy all five of the inequalities. Solving simultaneously the equations for $L_1$ and $L_2$, we obtain the point of intersection $P(1, 0)$. We check to see if the coordinates of this point satisfy inequalities (13), (14), and (15), and we see that they do. The coordinates of $P$ then satisfy each of the five inequalities, and therefore $P$ is in the required region. We find the point of intersection of $L_1$ and $L_3$ to be $Q(1, -\frac{8}{3})$. The coordinates of $Q$ satisfy inequality (15) but not (14) and (12). Hence the point $Q$ is not in the required region.

Continuing, we see that $L_1$ and $L_4$ intersect at the point $R(1, -2)$, and the coordinates of $R$ satisfy inequalities (13) and (15) but not (12). Lines $L_1$ and $L_5$ intersect at the point $S(1, 6)$ and the coordinates of this point satisfy the three inequalities (12), (13), and (14). Therefore, the point $S$ is in the required region. Taking the equations in pairs, we have ten points of intersection. We see that the point of intersection of $L_2$ and $L_4$, which is $T(2, 0)$, and the point of intersection of $L_4$ and $L_5$, which is $U(\frac{11}{3}, \frac{10}{3})$ are in the required region. The remaining four points are not. Therefore, we have four points $P$, $S$, $T$, and $U$, whose coordinates satisfy all five inequalities. We plot these four points and the five lines, as shown in Figure 6.5.6. Inequalities (11) through (15) define half planes, and the interior and sides of the quadrilateral $PSTU$ give us all the points whose coordinates satisfy the given system.

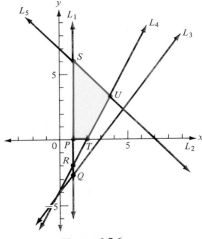

**Figure 6.5.6**

Systems of linear inequalities, like the one in Example 4, occur in the fields of economics, business, statistics, science, and engineering, among others. In practice, many unknowns are usually involved, as well as a large number of inequalities, and electronic computers perform most of the computation. Before discussing an application of a system of linear inequalities, we define what is meant by a convex region.

**6.5.1 DEFINITION**  A region is said to be *convex* if and only if for every pair of points $P$ and $Q$ in the region, the line segment $PQ$ lies entirely in the region.

**ILLUSTRATION 2.** The shaded region in Figure 6.5.6 is convex. The region in Figure 6.5.7 is not convex because every point of the line segment $AB$ shown in the figure is not in the region. ●

Suppose a merchant has a fixed amount of floor space available and a certain amount of money to invest in two different kinds of merchandise.

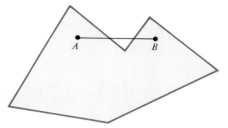

**Figure 6.5.7**

Knowing the cost and profit per unit of each kind of merchandise and the amount of floor space required for each unit, the dealer would be interested in determining the number of units of each kind he should stock, in order for his profit to be a maximum. Or suppose three customers, each having specific requirements, are to be serviced with goods from two warehouses whose capacities are fixed. Knowing the costs per unit of servicing each customer from each warehouse, it is desired to determine the amount of goods to be shipped to each customer from each warehouse, in order for the total cost of servicing to be a minimum. These are two decision problems that can be solved by methods referred to as *mathematical programming*. The independent variables in such decision problems are called *primary variables,* and they are subject to restrictions which are called *constraints*. The problem involves maximizing or minimizing a value of an algebraic expression involving the primary variables. When the constraints can be expressed as a system of linear inequalities, and the algebraic expression is linear, we have a problem in *linear programming*.

We can solve a problem in linear programming involving two primary variables by geometric methods. We shall demonstrate this geometric method by an example because this will appeal to your intuition and aid you in understanding more complicated techniques. The geometric method involves first finding the set of all *feasible solutions to the problem*. By a feasible solution, we mean one which satisfies all the constraints. Each of the constraints is an inequality, and the region common to the graphs of these inequalities is the *set of all feasible solutions*. In the case of two primary variables, this region is a polygon. To determine which feasible solution is the optimum (the "best") solution, we make use of the following theorem, which we state without proof.

**6.5.2 THEOREM**   If $a, b, c \in R$ and

$$z = ax + by + c$$

where $(x, y)$ corresponds to a point in a closed convex polygonal region, then the values of $x$ and $y$ which maximize and minimize $z$ occur at the vertices of the polygon.

**ILLUSTRATION 3.**   Suppose

$$z = 9x - 3y + 10$$

where $x$ and $y$ satisfy the five inequalities of Example 4. The set of all feasible solutions to the system of inequalities (11) through (15) consists of the points that are either in the interior or on the sides of the quadrilateral *PSTU* in

Figure 6.5.6. By Theorem 6.5.2 the maximum and minimum values of $z$ must occur at a vertex. We compute the value of $z$ at each vertex.

At $(0, 1)$, $z = 9(0) - 3(1) + 10 = 7$

At $(1, 6)$, $z = 9(1) - 3(6) + 10 = 1$

At $(2, 0)$, $z = 9(2) - 3(0) + 10 = 28$

At $\left(\frac{11}{3}, \frac{10}{3}\right)$, $z = 9\left(\frac{11}{3}\right) - 3\left(\frac{10}{3}\right) + 10 = 33$

The maximum value of $z$ occurs at $(\frac{11}{3}, \frac{10}{3})$, and it is 33. The minimum value of $z$ occurs at $(1, 6)$, and it is 1. ●

EXAMPLE 5

A company manufactures two products, A and B, and each of these products must be processed on two different machines. Product A requires 1 minute of work time per unit on machine 1 and 4 minutes of work time per unit on machine 2. Product B requires 2 minutes of work time per unit on machine 1 and 3 minutes of work time per unit on machine 2. Each day 100 minutes are available on machine 1 and 200 minutes are available on machine 2. To satisfy certain customers, the company must produce at least 6 units per day of product A and at least 12 units per day of product B. If the profit of each unit of product A is \$5 and the profit of each unit of product B is \$6, how many units of each product should be produced daily in order to maximize the company's profits?

SOLUTION

Let $x$ represent the number of units of product A to be produced daily, and let $y$ represent the number of units of product B to be produced daily. If $P$ dollars is the company's daily profit, then

$$P = 5x + 6y$$

We wish to maximize $P$ subject to the following constraints.

$$x + 2y \leq 100 \qquad 4x + 3y \leq 200 \qquad x \geq 6 \qquad y \geq 12 \qquad (6)$$

The set of all feasible solutions to the system of inequalities (6) consists of the points that are either in the interior or on the sides of the quadrilateral shown in Figure 6.5.8. The vertices of this quadrilateral are at the points $(6, 12)$, $(41, 12)$, $(20, 40)$, and $(6, 47)$. The maximum solution must occur at one of the vertices. We compute the value of $P$ at each vertex.

At $(6, 12)$, $P = 5(6) + 6(12) = 102$

At $(41, 12)$, $P = 5(41) + 6(12) = 277$

At $(20, 40)$, $P = 5(20) + 6(40) = 340$

At $(6, 47)$, $P = 5(6) + 6(47) = 312$

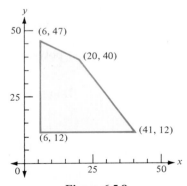

Figure 6.5.8

Hence the maximum solution occurs at (20, 40). The company should manufacture 20 units of product A and 40 units of product B daily to realize a maximum daily profit of $340.

## EXERCISES 6.5

In Exercises 1 through 18, draw a sketch of the region (if any) defined by the given system of inequalities.

1. $y - 2x \leq 4$, $2x + y \leq 4$
2. $4x + y \geq 6$, $6x - 2y \leq 7$
3. $2x + y \geq 4$, $y - 2x \geq 4$
4. $4x + 3y - 7 > 0$, $6x - y - 5 < 0$
5. $3 + y \leq x \leq y - 4$
6. $-1 < x - y \leq 2$
7. $1 < x^2 + y^2 < 16$
8. $x^2 + y^2 \leq 25$, $x \geq 1$, $y \leq 4$
9. $x - y - 1 \leq 0$, $x + y + 1 \geq 0$, $x - y + 1 \geq 0$, $x + y - 1 \leq 0$
10. $x - 2y < 4$, $11 - 6x < y$, $4x + 5y < 29$
11. $x^2 - y + 4 > 0$, $x - y \leq 4$
12. $x^2 - y + 1 < 0$, $x + y - 4 > 0$
13. $x^2 + y^2 \geq 8$, $x < y$
14. $x^2 - 4 \leq y \leq 4 - y^2$
15. $4 \leq x^2 + y^2 \leq 9$, $|x| > 1$
16. $16 \leq x^2 + y^2 \leq 36$, $|x| > 3$
17. $x \geq 0$, $y \geq 0$, $y \leq 3$, $x + y - 5 \leq 0$, $2x + y - 8 \leq 0$
18. $x \geq 0$, $y \geq 0$, $y - x + 1 \geq 0$, $x + y - 5 \leq 0$, $x + 3y - 8 \leq 0$

19. If $z = 2x + 3.5y$, maximize $z$ subject to the following constraints.

$$x \geq 0 \quad y \geq 0 \quad x + y \leq 6$$
$$3x + 5y \leq 20 \quad x - y \geq 1$$

20. If $z = 4x + 3y$, maximize $z$ subject to the following constraints.

$$x + 2y \leq 6 \quad 3x + y \leq 8 \quad x \geq 0 \quad y \geq 0$$

21. A company manufactures two products, A and B, and it requires three different machines to process each product. Product A requires 10 hours of time on machine $M_1$, 6 hours of time on machine $M_2$, and 12 hours of time on machine $M_3$. Product B requires 10 hours of time on $M_1$, 12 hours on $M_2$, and 4 hours on $M_3$. If the profit of each unit of product A is $200 and the profit of each unit of product B is $360, how many units of each product should be produced in each two-week period if there are 240 hours of time available on each machine and the company wishes to maximize the profit?

22. A wholesaler has 24,000 square feet of storage space available and $20,000 that can be spent for merchandise of types A and B. Each unit of type A costs $4 and requires 6 square feet of storage space, and each unit of type B costs $8 and requires 8 square feet of storage space. If the wholesaler expects a profit of $2 per unit on type A and $4.50 per unit on type B, how many units of each should be bought and stocked in order to maximize the profit?

23. A distributor of refrigerators has two warehouses which supply three different retailers. To deliver a refrigerator to retailer $R_1$ costs $27 from warehouse $W_1$ and $36 from warehouse $W_2$. It costs $9 to deliver a refrigerator from warehouse $W_1$ to retailer $R_2$ and $6 to deliver one from $W_2$ to $R_2$. For retailer $R_3$, it costs $15 if the refrigerator comes from $W_1$, and $30 if it comes from $W_2$. Suppose that retailer $R_1$ orders three refrigerators, retailer $R_2$ orders four refrigerators, and two refrigerators are ordered from retailer $R_3$. If the distributor has five refrigerators in stock in warehouse $W_1$ and four in warehouse $W_2$, how many refrigerators should be shipped from each warehouse to each retailer in order for the distributor to minimize the delivery costs?

## REVIEW EXERCISES (CHAPTER 6)

In Exercises 1 through 4, draw a sketch of the graph of the system of equations. Classify the equations as (i) consistent and independent, (ii) consistent and dependent, or (iii) inconsistent. If the equations are consistent and independent, determine the solution set of the system from the graphs.

1. $\begin{cases} 4x + 3y = 6 \\ 2x + y = 4 \end{cases}$
2. $\begin{cases} 3x - 2y = 4 \\ 9x - 6y = 8 \end{cases}$
3. $\begin{cases} 2y = 4x - 6 \\ 6x = 3y + 9 \end{cases}$
4. $\begin{cases} 4x + 2y = 5 \\ 8x - 2y = 1 \end{cases}$

In Exercises 5 through 12, classify the equations of the given system as (i) consistent and independent, (ii) consistent and dependent, or (iii) inconsistent. If the equations are consistent, find the solution set of the given system.

5. $\begin{cases} 2x + y + 1 = 0 \\ 3x + 2y + 4 = 0 \end{cases}$
6. $\begin{cases} 3x - 2y + 7 = 0 \\ 2x - 3y + 8 = 0 \end{cases}$
7. $\begin{cases} 2x - 5y = 7 \\ 6x - 15y = 14 \end{cases}$

8. $\begin{cases} \dfrac{3}{x} - \dfrac{2}{y} = 8 \\ \dfrac{9}{x} + \dfrac{4}{y} = -6 \end{cases}$
9. $\begin{cases} x + 2y + 2z = 1 \\ x - 3y - 2z = 4 \\ 6x + y - z = 21 \end{cases}$
10. $\begin{cases} 6x + 4y + 5z = 14 \\ 4x - 3y - z = 2 \\ 14x - 10y - 9z = 10 \end{cases}$

11. $\begin{cases} 4x - 3y - 6z = 7 \\ 2x - y - 4z = 3 \\ 3x - 2y - 5z = 5 \end{cases}$
12. $\begin{cases} 3x + 3y - 5z = 4 \\ 6x + 2y - 3z = 7 \\ 3x - 5y + 9z = 5 \end{cases}$

In Exercises 13 and 14, determine if the equations are consistent or inconsistent. To show they are inconsistent, solve a system of three of the equations and show that no member of the solution set satisfies the fourth equation. To show that they are consistent, find the solution set.

13. $\begin{cases} 2x + 4y + 3z = 8 \\ x - 2y + 3z = 1 \\ 2x - 6y - 6z = 9 \\ x + 2y - 6z = 9 \end{cases}$
14. $\begin{cases} 4x - 4y + z = 5 \\ 2x + y + 3z = 6 \\ 6x + 3y + 4z = 8 \\ 3x - 2y + 3z = 5 \end{cases}$

15. Show that the following system has an infinite number of solutions in the solution set and use the variable $t$ to express the solution set as an infinite set of ordered pairs. Then assign the values 0, 1, 2, $-1$, and $-2$ to $t$ and find five ordered triples in the solution set.

$$\begin{cases} 2x - 3y + 2z = -1 \\ -3x + 4y - 4z = 3 \end{cases}$$

In Exercises 16 through 19, find the solution set of the given system of equations.

16. $\begin{cases} 7x + 3y = 9 \\ x^2 + 2y = 1 \end{cases}$
17. $\begin{cases} 3x^2 + 2y^2 = 7 \\ 5x^2 - y^2 = 3 \end{cases}$
18. $\begin{cases} 4x^2 - 3y^2 = -8 \\ y^2 + 2xy = 8 \end{cases}$
19. $\begin{cases} 2x^2 - xy + y^2 = 8 \\ 4x^2 + xy + y^2 = 6 \end{cases}$

In Exercises 20 through 23, draw a sketch of the region (if any) defined by the given system of inequalities.

20. $x^2 + y^2 > 25$, $y < 4x + 3$
21. $3x + 2y \leq 12$, $2y - 4 > x$, $x \geq 0$
22. $4 \leq x^2 + y^2 < 16$
23. $x^2 + y^2 \geq 4$, $|x| < 25$

24. If $z = x + 4y$, minimize $z$ subject to the following constraints:

$$x + 4y \geq 8 \qquad x - y \leq 4 \qquad x \geq 2$$

25. A pilot makes a check flight in an airplane. The pilot flies a distance of 80 kilometers against the wind in 10 minutes, and then flies back the same distance with the wind in 8 minutes. If the plane's speed in still air is the same in both directions, what is the rate of the wind and what is the plane's speed in still air?

26. A drain at the bottom of a tank is always open, and there are two identical pipes that bring water into the tank. If only one pipe is open, it takes 12 hours to fill the tank, but if both pipes are open, it only takes 3 hours to fill the tank. How long does it take the drain to empty a full tank?

27. An investment yields an annual interest of $750. If $500 more is invested and the rate is 1 per cent less, the annual interest is $650. What is the amount of the investment and the rate of interest?

28. A woman bought 100 stamps in denominations of 10, 15, and 31 cents, and the number of 15-cent stamps purchased was 10 less than the combined total of the other two denominations. If she paid $17.50 for the stamps, how many stamps of each denomination did she buy?

# 7
# Matrices and Determinants

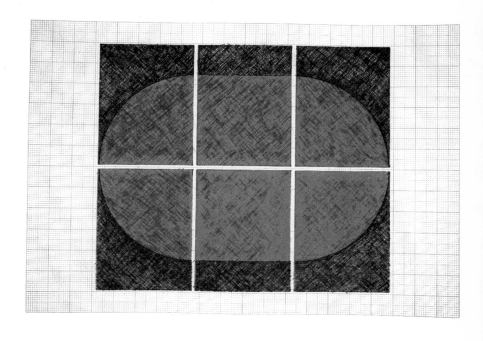

# 7.1 Matrices

In Example 2 of Section 6.2 we have the system of equations

(I) $\begin{cases} 4x - 2y - 3z = 8 \\ 5x + 3y - 4z = 4 \\ 6x - 4y - 5z = 12 \end{cases}$

We solved this system by obtaining the equivalent system in triangular form

$$\begin{cases} 4x - 2y - 3z = 8 \\ -22y + z = 24 \\ z = 2 \end{cases}$$

By using a zero as the coefficient of a variable that does not appear in an equation, this latter system can be written as

(II) $\begin{cases} 4x - 2y - 3z = 8 \\ 0x - 22y + z = 24 \\ 0x + 0y + z = 2 \end{cases}$

The procedure used to obtain system (II) involves operations on the equations of system (I) and a series of equivalent systems until an equivalent system in triangular form is found. These operations cause changes in the coefficients of the variables and changes in the constant terms in the right members of the equations. Thus we are concerned essentially with these numbers (the coefficients and the constant terms) that appear in each of the equivalent systems. Therefore, to simplify the calculations, we introduce a notation for recording the coefficients and constant terms so that the variables do not have to be written. The numbers involved in system (I) are listed in the following way.

(III) $\begin{bmatrix} 4 & -2 & -3 & 8 \\ 5 & 3 & -4 & 4 \\ 6 & -4 & -5 & 12 \end{bmatrix}$

and the numbers in system (II) are listed as

(IV) $\begin{bmatrix} 4 & -2 & -3 & 8 \\ 0 & -22 & 1 & 24 \\ 0 & 0 & 1 & 2 \end{bmatrix}$

Each of the arrays, (III) and (IV), is called a *matrix*. The numbers in the matrix are called the *elements* of the matrix. The elements that appear next to each other horizontally form a *row,* and the elements that appear vertically form a *column.* Hence in matrix (III), the elements in the first row are 4, −2, −3, and 8, and the elements in the second column are −2, 3, and −4. The number of rows and the number of columns in a matrix determine the *order of a matrix.* If there are $m$ rows and $n$ columns, then the matrix is of order $m \times n$ (read "$m$ by $n$"). Matrices (III) and (IV) are of order $3 \times 4$ (read "three by four"). If $m \neq n$, the matrix is called a *rectangular matrix,* and if $m = n$, the matrix is referred to as a *square matrix of order n.*

We use a double subscript notation for the elements of a matrix. The symbol $a_{ij}$ denotes the element appearing in the $i$th row and $j$th column. The symbol $i$ is called the *row subscript* and $j$ is called the *column subscript of $a_{ij}$.* Then if $m$ and $n$ are positive integers, the matrix of order $m \times n$ is an array of the form

$$\begin{bmatrix} a_{11} & a_{12} & a_{13} & \cdots & a_{1n} \\ a_{21} & a_{22} & a_{23} & \cdots & a_{2n} \\ a_{31} & a_{32} & a_{33} & \cdots & a_{3n} \\ \cdot & \cdot & \cdot & \cdots & \cdot \\ \cdot & \cdot & \cdot & \cdots & \cdot \\ \cdot & \cdot & \cdot & \cdots & \cdot \\ a_{m1} & a_{m2} & a_{m3} & \cdots & a_{mn} \end{bmatrix}$$

where each $a_{ij}$ represents a real number.

Suppose that we have a system of linear equations where the terms involving the variables are in the left member and the constants are in the right member. Such a system of three linear equations is system (I). The matrix whose only elements are the coefficients of the variables, listed as they appear in the equations, is called the *coefficient matrix.* For example, the coefficient matrix of system (I) is

$$\begin{bmatrix} 4 & -2 & -3 \\ 5 & 3 & -4 \\ 6 & -4 & -5 \end{bmatrix}$$

which is a square matrix of order three. The matrix obtained from the coefficient matrix by listing the constants in an additional column on the right is called the *augmented matrix.* For system (I) the augmented matrix is matrix (III).

To solve a system of equations by using matrices, we start with the augmented matrix and perform operations on the rows to obtain a matrix of an equivalent system. We continue the process until we obtain a matrix of a system in triangular form. The rules for performing the operations are called *elementary row operations,* and they are given in the following theorem.

**7.1.1 THEOREM** If we have an augmented matrix of a system of linear equations, each of the following operations produces a matrix of an equivalent system of linear equations:

(i) Interchanging any two rows.
(ii) Multiplying each element of a row by the same nonzero number.
(iii) Replacing a given row by a new row whose elements are the sum of $k_1$ times the elements of the given row and $k_2$ times the corresponding elements of any other row, where $k_1$ and $k_2$ are real numbers, and $k_1 \neq 0$.

The proof of Theorem 7.1.1 utilizes the corresponding operations on the equations of the system and it is omitted.

When applying Theorem 7.1.1 (ii) we indicate that we are multiplying a row by a certain real number, it being understood that to multiply a row by a number means to multiply each element of the row by the number. In a similar manner, when applying Theorem 7.1.1 (iii), we state that we are adding one row to another row which means that we are adding corresponding elements of the two rows.

The following illustration shows how Theorem 7.1.1 is used to solve system (I). You should compare the computation with that of Example 2 in Section 6.2, which involves the same system of equations.

**ILLUSTRATION 1.** System (I) is

$$\begin{cases} 4x - 2y - 3z = 8 \\ 5x + 3y - 4z = 4 \\ 6x - 4y - 5z = 12 \end{cases}$$

and the augmented matrix is

$$\begin{bmatrix} 4 & -2 & -3 & 8 \\ 5 & 3 & -4 & 4 \\ 6 & -4 & -5 & 12 \end{bmatrix}$$

We replace the second row by the sum of 5 times the first row and $-4$ times the second row, and we replace the third row by the sum of 6 times the first row and $-4$ times the third row. Thus we have the matrix

$$\begin{bmatrix} 4 & -2 & -3 & 8 \\ 0 & -22 & 1 & 24 \\ 0 & 4 & 2 & 0 \end{bmatrix}$$

In the latter matrix we replace the third row by the sum of 2 times the second row and 11 times the third row, and we obtain

$$\begin{bmatrix} 4 & -2 & -3 & 8 \\ 0 & -22 & 1 & 24 \\ 0 & 0 & 24 & 48 \end{bmatrix}$$

We now multiply the third row by $\frac{1}{24}$ and we get

$$\begin{bmatrix} 4 & -2 & -3 & 8 \\ 0 & -22 & 1 & 24 \\ 0 & 0 & 1 & 2 \end{bmatrix}$$

which is matrix (IV), the augmented matrix for the system of equations (II). System (II) is in triangular form and is equivalent to the given system (I). The solution set is then found as in Example 2, Section 6.2. ●

## EXAMPLE 1

Use matrices to find the solution set of the system

$$\begin{cases} \frac{3}{8}x + \frac{1}{12}y - \frac{5}{4} = 0 \\ \frac{2}{3}x - \frac{1}{6}y - \frac{1}{3} = 0 \end{cases}$$

## SOLUTION

We first replace each equation by an equivalent equation in which the coefficients and constants are integers and the constants are in the right members. We multiply each member of the first equation by 24 and each member of the second equaton by 6; we obtain the equivalent system

$$\begin{cases} 9x + 2y = 30 \\ 4x - y = 2 \end{cases}$$

The augmented matrix of this system is

$$\begin{bmatrix} 9 & 2 & 30 \\ 4 & -1 & 2 \end{bmatrix}$$

Replacing the second row by the sum of 4 times the first row and $-9$ times the second row, we obtain

$$\begin{bmatrix} 9 & 2 & 30 \\ 0 & 17 & 102 \end{bmatrix}$$

We now multiply the second row by $\frac{1}{17}$ and we have

$$\begin{bmatrix} 9 & 2 & 30 \\ 0 & 1 & 6 \end{bmatrix}$$

which is the augmented matrix of the system

$$\begin{cases} 9x + 2y = 30 \\ y = 6 \end{cases}$$

Substituting into the first equation the value of $y$ from the second equation, we have the system

$$\begin{cases} x = 2 \\ y = 6 \end{cases}$$

which is equivalent to the given system. Hence, the solution set of the given system is $\{(2, 6)\}$.

In the next example we solve a system of four linear equations in four variables by the matrix method. By a solution of the system we mean an ordered four-tuple $(r, s, t, u)$ such that each of the equations of the system is satisfied if $w$ is replaced by $r$, $x$ is replaced by $s$, $y$ is replaced by $t$, and $z$ is replaced by $u$.

## EXAMPLE 2

Use matrices to find the solution set of the system

$$\begin{cases} w + x + y + z = 5 \\ 3z + x = w - 2 \\ 2x + 2y = 3w + 2 \\ y = w + z \end{cases}$$

## SOLUTION

We replace each equation by an equivalent equation in which the terms involving variables are in the left member and the constant terms are in the right member. We have the following equivalent system in which the equations are written so that terms involving the same variable are in a vertical column.

$$\begin{cases} w + x + y + z = 5 \\ -w + x \phantom{+ y} + 3z = -2 \\ -3w + 2x + 2y \phantom{+ z} = 2 \\ -w \phantom{+ 2x} + y - z = 0 \end{cases}$$

The augmented matrix of this system is

$$\begin{bmatrix} 1 & 1 & 1 & 1 & 5 \\ -1 & 1 & 0 & 3 & -2 \\ -3 & 2 & 2 & 0 & 2 \\ -1 & 0 & 1 & -1 & 0 \end{bmatrix}$$

We replace the second row by the sum of the first and second rows; we replace the third row by the sum of the third row and 3 times the first row; we replace the fourth row by the sum of the first and fourth rows. We have then the matrix

$$\begin{bmatrix} 1 & 1 & 1 & 1 & 5 \\ 0 & 2 & 1 & 4 & 3 \\ 0 & 5 & 5 & 3 & 17 \\ 0 & 1 & 2 & 0 & 5 \end{bmatrix}$$

We now interchange the second and fourth rows because then we will have 1 as an element in the second row and second column; this will make it easier to obtain 0's in the third and fourth rows of the second column. We have then the matrix

$$\begin{bmatrix} 1 & 1 & 1 & 1 & 5 \\ 0 & 1 & 2 & 0 & 5 \\ 0 & 5 & 5 & 3 & 17 \\ 0 & 2 & 1 & 4 & 3 \end{bmatrix}$$

We replace the third row by the sum of the third row and $-5$ times the second row, and we replace the fourth row by the sum of the fourth row and $-2$ times the second row. We obtain the matrix

$$\begin{bmatrix} 1 & 1 & 1 & 1 & 5 \\ 0 & 1 & 2 & 0 & 5 \\ 0 & 0 & -5 & 3 & -8 \\ 0 & 0 & -3 & 4 & -7 \end{bmatrix}$$

Replacing the fourth row by the sum of 3 times the third row and $-5$ times the fourth row, we have the matrix

$$\begin{bmatrix} 1 & 1 & 1 & 1 & 5 \\ 0 & 1 & 2 & 0 & 5 \\ 0 & 0 & -5 & 3 & -8 \\ 0 & 0 & 0 & -11 & 11 \end{bmatrix}$$

We now multiply the fourth row by $-\frac{1}{11}$ and we have the matrix

$$\begin{bmatrix} 1 & 1 & 1 & 1 & 5 \\ 0 & 1 & 2 & 0 & 5 \\ 0 & 0 & -5 & 3 & -8 \\ 0 & 0 & 0 & 1 & -1 \end{bmatrix}$$

This latter matrix is the augmented matrix of the system in triangular form

$$\begin{cases} w + x + y + z = 5 \\ \quad\; x + 2y \qquad\;\; = 5 \\ \qquad\quad -5y + 3z = -8 \\ \qquad\qquad\qquad\;\; z = -1 \end{cases}$$

We substitute into the third equation the value of $z$ from the fourth equation, and we obtain $y = 1$; then substituting 1 for $y$ in the second equation, we have $x = 3$; by substituting in the first equation 3 for $x$, 1 for $y$, and $-1$ for $z$, we obtain $w = 2$. We have then the following system, which is equivalent to the given system

$$\begin{cases} w = 2 \\ x = 3 \\ y = 1 \\ z = -1 \end{cases}$$

Thus the solution set of the given system is $\{(2, 3, 1, -1)\}$.

### EXAMPLE 3

Use matrices to find the solution set of the system

(V) $\begin{cases} 2x + y - 3z = 0 \\ 3x + 2y - 4z = 2 \\ x - y - 3z = -6 \end{cases}$

### SOLUTION

The augmented matrix of the system is

$$\begin{bmatrix} 2 & 1 & -3 & 0 \\ 3 & 2 & -4 & 2 \\ 1 & -1 & -3 & -6 \end{bmatrix}$$

Replacing the second row by the sum of 3 times the first row and $-2$ times the second row, and replacing the third row by the sum of the first row and $-2$ times the third row, we obtain

$$\begin{bmatrix} 2 & 1 & -3 & 0 \\ 0 & -1 & -1 & -4 \\ 0 & 3 & 3 & 12 \end{bmatrix}$$

We replace the third row by the sum of 3 times the second row and the third row, and we have

$$\begin{bmatrix} 2 & 1 & -3 & 0 \\ 0 & -1 & -1 & -4 \\ 0 & 0 & 0 & 0 \end{bmatrix}$$

This matrix is the augmented matrix of the system

(V) $\begin{cases} 2x + y - 3z = 0 \\ -y - z = -4 \\ 0 = 0 \end{cases}$

The third equation is an identity for any ordered triple $(x, y, z)$ and in particular for $(0, 0, t)$. So we replace the equation $0 = 0$ by the equation $z = t$, and we have the equivalent system

(VI) $\begin{cases} 2x + y - 3z = 0 \\ -y - z = -4 \\ z = t \end{cases}$

System (VI) is equivalent to system (V), and the equations of system (V) are dependent. In system (VI) we substitute the value of $z$ from the third equation into the second and solve for $y$; we have the equivalent system

$$\begin{cases} 2x + y - 3z = 0 \\ y = 4 - t \\ z = t \end{cases}$$

Substituting the values of $y$ and $z$ from the second and third equations into the first, we obtain

$$\begin{cases} 2x + (4 - t) - 3t = 0 \\ y = 4 - t \\ z = t \end{cases}$$

or equivalently,

$$\begin{cases} x = 2t - 2 \\ y = 4 - t \\ z = t \end{cases}$$

Therefore, the solution set of system (V) is $(2t - 2, 4 - t, t)$.

## EXERCISES 7.1

Use matrices to find the solution set of the given system.

1. $\begin{cases} 4x - 3y = 5 \\ 6x + 2y = 1 \end{cases}$

2. $\begin{cases} 4x + 3y = 2 \\ 5x + 4y = 1 \end{cases}$

3. $\begin{cases} \dfrac{x}{2} - \dfrac{y}{6} = 1 \\ \dfrac{x}{4} + \dfrac{y}{3} = 4 \end{cases}$

4. $\begin{cases} \dfrac{x}{3} + \dfrac{y}{2} = 0 \\ \dfrac{x}{6} + \dfrac{y}{8} = \dfrac{1}{2} \end{cases}$

5. $\begin{cases} x - 2y - z = 3 \\ x + y + z = 4 \\ x - 3y - z = 4 \end{cases}$

6. $\begin{cases} x + y - 2z = 5 \\ 3x + 2y = 4 \\ 2x + z = 2 \end{cases}$

7. $\begin{cases} \dfrac{1}{2}x + \dfrac{1}{3}y + \dfrac{1}{4}z = 2 \\ \dfrac{1}{4}x - \dfrac{2}{3}y - \dfrac{1}{4}z = \dfrac{1}{2} \\ \dfrac{1}{6}x + \dfrac{1}{9}y + \dfrac{1}{2}z = 4 \end{cases}$

8. $\begin{cases} \dfrac{x}{3} + \dfrac{y}{2} - \dfrac{z}{2} = 0 \\ -\dfrac{x}{6} + \dfrac{3y}{4} + \dfrac{z}{3} = \dfrac{5}{8} \\ \dfrac{3x}{2} + \dfrac{y}{4} + \dfrac{z}{6} = -\dfrac{1}{8} \end{cases}$

9. $\begin{cases} 2x + 3y - 4z = 4 \\ x + 2y - 5z = 6 \\ 4x + 5y - 2z = 0 \end{cases}$

10. $\begin{cases} w + 3x + y + z = 3 \\ -w - 3y - z = 0 \\ -2w + 3x - 4y + z = 0 \\ -w - 6x - 2y + 2z = -4 \end{cases}$

11. $\begin{cases} 2w + x - 3y - 3z = 4 \\ x + 2y + z = -3 \\ 2w - x + 3z = -3 \\ 5w - y + z = -6 \end{cases}$

12. $\begin{cases} 2w - 4x + y - 2x = 3 \\ 3w + x + 2y + 3z = 12 \\ w - 4x + 2y - 6z = 1 \\ 5w + x + 11z = 16 \end{cases}$

13. $\begin{cases} 2w + 3x - 4y - z = 3 \\ 3w + x + y + 2z = 1 \\ w - 2x + 3y - z = 0 \\ w - 2x - y - 9z = 5 \end{cases}$

14. $\begin{cases} 4w + x - 2y + z = 4 \\ 2w + 3x + 4y - 3z = -2 \\ 3w - 2x + 5y + 3z = 2 \\ w + 4x + 3y + z = 5 \end{cases}$

## 7.2 Properties of Matrices

In Section 7.1 we showed how matrices can be used to solve a system of linear equations. We defined a matrix of order $m \times n$ as one having $m$ rows and $n$ columns, where the number of rows is stated first. If $m = n$, we have a square matrix of order $n$. If a matrix has only one column it is called a *column matrix*, and if it has only one row it is called a *row matrix*.

**ILLUSTRATION 1.** The following matrices have the order indicated below the matrix.

$\begin{bmatrix} 5 & 4 \\ -2 & 7 \\ 0 & -4 \end{bmatrix}$   $\begin{bmatrix} 9 & -1 & 6 \\ 2 & 10 & -8 \end{bmatrix}$   $\begin{bmatrix} 7 \\ -3 \\ 0 \\ 4 \end{bmatrix}$   $[0 \ \ 2 \ \ -5]$   $\begin{bmatrix} -4 & 1 & 8 \\ 0 & -2 & 11 \\ 6 & -3 & 0 \end{bmatrix}$

$3 \times 2$   $2 \times 3$   $4 \times 1$   $1 \times 3$   $3 \times 3$

         Column matrix   Row matrix   Square matrix   ●

**7.2.1 DEFINITION** Two matrices are said to be *equal* if and only if their orders are equal and the corresponding elements are equal; that is,

$$\begin{bmatrix} r & s \\ t & u \end{bmatrix} = \begin{bmatrix} a & b \\ c & d \end{bmatrix}$$

if and only if

$$r = a \quad s = b \quad t = c \quad u = d$$

**ILLUSTRATION 2.** If

$$A = \begin{bmatrix} 4 & -3 & 0 \\ 1 & 3 & 2 \end{bmatrix} \quad B = \begin{bmatrix} 1 & 3 & 2 \\ 4 & -3 & 0 \end{bmatrix} \quad C = \begin{bmatrix} 2 \times 2 & -3 & 0 \\ 3 - 2 & \frac{6}{2} & \frac{2}{1} \end{bmatrix}$$

then $A \neq B$ and $B \neq C$, but $A = C$. •

We now define addition of two matrices that have the same number of rows and columns.

**7.2.2 DEFINITION** If $A$ and $B$ are two matrices having the same order, the *sum* of $A$ and $B$, denoted by $A + B$, is the matrix for which each of its elements is the sum of the corresponding elements of $A$ and $B$.

**ILLUSTRATION 3.** If

$$A = \begin{bmatrix} 5 & -2 & 1 \\ 3 & 0 & -4 \end{bmatrix} \quad \text{and} \quad B = \begin{bmatrix} 4 & 7 & 6 \\ -1 & 8 & -3 \end{bmatrix}$$

then

$$A + B = \begin{bmatrix} 5+4 & -2+7 & 1+6 \\ 3+(-1) & 0+8 & -4+(-3) \end{bmatrix}$$

$$= \begin{bmatrix} 9 & 5 & 7 \\ 2 & 8 & -7 \end{bmatrix}$$

•

Observe that addition of matrices having different orders is not defined. There is an "identity element" associated with the operation of addition of matrices. It is called a "zero matrix."

**7.2.3 DEFINITION** A *zero matrix* is one whose elements are all zeros. It is denoted by **0**.

**ILLUSTRATION 4.** The following matrices are zero matrices.

$$\begin{bmatrix} 0 & 0 \\ 0 & 0 \\ 0 & 0 \end{bmatrix} \quad \begin{bmatrix} 0 & 0 & 0 & 0 \end{bmatrix} \quad \begin{bmatrix} 0 \\ 0 \end{bmatrix}$$

•

**7.2.4 THEOREM** If $A$ is any matrix and $\mathbf{0}$ is the zero matrix of the same order as $A$, then

$$A + \mathbf{0} = A$$

and

$$\mathbf{0} + A = A$$

The proof of Theorem 7.2.4 follows immediately from the definitions of the zero matrix and the addition of matrices. For instance, if $A$ is a matrix of order $2 \times 3$, then

$$\begin{bmatrix} a_{11} & a_{12} & a_{13} \\ a_{21} & a_{22} & a_{23} \end{bmatrix} + \begin{bmatrix} 0 & 0 & 0 \\ 0 & 0 & 0 \end{bmatrix} = \begin{bmatrix} a_{11} & a_{12} & a_{13} \\ a_{21} & a_{22} & a_{23} \end{bmatrix}$$

**7.2.5 DEFINITION** If $A$ is any matrix, then the *negative of $A$*, denoted by $-A$, is the matrix for which each of the elements is the negative of the corresponding element of $A$.

**ILLUSTRATION 5.** If

$$A = \begin{bmatrix} 5 & -2 & 1 \\ 3 & 0 & -4 \end{bmatrix}, \quad \text{then} \quad -A = \begin{bmatrix} -5 & 2 & -1 \\ -3 & 0 & 4 \end{bmatrix} \quad \bullet$$

**7.2.6 THEOREM** The sum of a matrix $A$ and the negative of $A$ is a zero matrix; that is, if $A$ is any matrix, then

$$A + (-A) = \mathbf{0}$$

The proof of Theorem 7.2.6 is left as an exercise (see Exercise 41). Subtraction of matrices can be defined in terms of addition.

**7.2.7 DEFINITION** If $A$ and $B$ are matrices of the same order, then the *difference of $A$ and $B$*, denoted by $A - B$, is defined by

$$A - B = A + (-B)$$

**ILLUSTRATION 6**

$$\begin{bmatrix} 7 & -3 \\ 2 & 1 \\ -6 & 0 \end{bmatrix} - \begin{bmatrix} 4 & -2 \\ -5 & 0 \\ 3 & -8 \end{bmatrix} = \begin{bmatrix} 7 & -3 \\ 2 & 1 \\ -6 & 0 \end{bmatrix} + \begin{bmatrix} -4 & 2 \\ 5 & 0 \\ -3 & 8 \end{bmatrix}$$

$$= \begin{bmatrix} 3 & -1 \\ 7 & 1 \\ -9 & 8 \end{bmatrix} \quad \bullet$$

Matrix addition is commutative and associative, as given in the next theorem.

**7.2.8 THEOREM** If $A$, $B$, and $C$ are matrices of the same order, then
(i) $A + B = B + A$
(ii) $A + (B + C) = (A + B) + C$

The proof of Theorem 7.2.8 follows from the definition of matrix addition and the commutativity and associativity of addition of real numbers.

**EXAMPLE 1**
If
$$A = \begin{bmatrix} 2 & -1 & 0 \\ 3 & 4 & -2 \end{bmatrix}$$
$$B = \begin{bmatrix} -3 & 5 & 1 \\ -6 & 0 & 2 \end{bmatrix}$$
$$C = \begin{bmatrix} 1 & -4 & -5 \\ -3 & -1 & 7 \end{bmatrix}$$
find
(a) $A + B$
(b) $B + C$
(c) $(A + B) + C$
(d) $A + (B + C)$

**SOLUTION**

(a) $A + B = \begin{bmatrix} 2 & -1 & 0 \\ 3 & 4 & -2 \end{bmatrix} + \begin{bmatrix} -3 & 5 & 1 \\ -6 & 0 & 2 \end{bmatrix}$

$= \begin{bmatrix} 2 + (-3) & -1 + 5 & 0 + 1 \\ 3 + (-6) & 4 + 0 & -2 + 2 \end{bmatrix}$

$= \begin{bmatrix} -1 & 4 & 1 \\ -3 & 4 & 0 \end{bmatrix}$

(b) $B + C = \begin{bmatrix} -3 & 5 & 1 \\ -6 & 0 & 2 \end{bmatrix} + \begin{bmatrix} 1 & -4 & -5 \\ -3 & -1 & 7 \end{bmatrix}$

$= \begin{bmatrix} -3 + 1 & 5 + (-4) & 1 + (-5) \\ -6 + (-3) & 0 + (-1) & 2 + 7 \end{bmatrix}$

$= \begin{bmatrix} -2 & 1 & -4 \\ -9 & -1 & 9 \end{bmatrix}$

(c) $(A + B) + C = \begin{bmatrix} -1 & 4 & 1 \\ -3 & 4 & 0 \end{bmatrix} + \begin{bmatrix} 1 & -4 & -5 \\ -3 & -1 & 7 \end{bmatrix}$

$= \begin{bmatrix} -1 + 1 & 4 + (-4) & 1 + (-5) \\ -3 + (-3) & 4 + (-1) & 0 + 7 \end{bmatrix}$

$= \begin{bmatrix} 0 & 0 & -4 \\ -6 & 3 & 7 \end{bmatrix}$

(d) $A + (B + C) = \begin{bmatrix} 2 & -1 & 0 \\ 3 & 4 & -2 \end{bmatrix} + \begin{bmatrix} -2 & 1 & -4 \\ -9 & -1 & 9 \end{bmatrix}$

$= \begin{bmatrix} 2 + (-2) & -1 + 1 & 0 + (-4) \\ 3 + (-9) & 4 + (-1) & -2 + 9 \end{bmatrix}$

$= \begin{bmatrix} 0 & 0 & -4 \\ -6 & 3 & 7 \end{bmatrix}$

We define multiplication of a matrix by a scalar, where in this book a scalar denotes a real number.

**7.2.9 DEFINITION** The *product of a scalar k and a matrix A*, denoted by $kA$, is the matrix in which each element is obtained by multiplying $k$ times the corresponding element of $A$.

**ILLUSTRATION 7**

$$2\begin{bmatrix} 4 & -1 & 6 \\ -3 & 0 & \frac{1}{2} \end{bmatrix} = \begin{bmatrix} 8 & -2 & 12 \\ -6 & 0 & 1 \end{bmatrix}$$

and

$$-3\begin{bmatrix} -5 & -3 \\ 2 & 1 \\ 0 & -4 \end{bmatrix} = \begin{bmatrix} 15 & 9 \\ -6 & -3 \\ 0 & 12 \end{bmatrix}$$

●

The next definition involves the product of two matrices $A$ and $B$ where it is necessary that the number of columns of $A$ is the same as the number of rows of $B$.

**7.2.10 DEFINITION**  Suppose that $A$ is a matrix of order $m \times p$ and $B$ is a matrix of order $p \times n$. Then the *product of A and B*, denoted by $AB$, is the $m \times n$ matrix for which the element in the $i$th row and $j$th column is the sum of the products formed by multiplying each element in the $i$th row of $A$ by the corresponding element in the $j$th column of $B$.

**ILLUSTRATION 8.**  If

$$A = \begin{bmatrix} 2 & -3 \\ 4 & -1 \\ 1 & 5 \end{bmatrix} \quad \text{and} \quad B = \begin{bmatrix} 3 & 0 & -4 \\ -2 & 2 & -1 \end{bmatrix}$$

then

$$AB = \begin{bmatrix} 2 & -3 \\ 4 & -1 \\ 1 & 5 \end{bmatrix} \begin{bmatrix} 3 & 0 & -4 \\ -2 & 2 & -1 \end{bmatrix}$$

$$= \begin{bmatrix} (2)(3) + (-3)(-2) & (2)(0) + (-3)(2) & (2)(-4) + (-3)(-1) \\ (4)(3) + (-1)(-2) & (4)(0) + (-1)(2) & (4)(-4) + (-1)(-1) \\ (1)(3) + (5)(-2) & (1)(0) + (5)(2) & (1)(-4) + (5)(-1) \end{bmatrix}$$

$$= \begin{bmatrix} 12 & -6 & -5 \\ 14 & -2 & -15 \\ -7 & 10 & -9 \end{bmatrix}$$

and

$$BA = \begin{bmatrix} 3 & 0 & -4 \\ -2 & 2 & -1 \end{bmatrix} \begin{bmatrix} 2 & -3 \\ 4 & -1 \\ 1 & 5 \end{bmatrix}$$

$$= \begin{bmatrix} (3)(2) + (0)(4) + (-4)(1) & (3)(-3) + (0)(-1) + (-4)(5) \\ (-2)(2) + (2)(4) + (-1)(1) & (-2)(-3) + (2)(-1) + (-1)(5) \end{bmatrix}$$

$$= \begin{bmatrix} 2 & -29 \\ 3 & -1 \end{bmatrix}$$

●

Observe in Illustration 8 that $AB$ is not equal to $BA$; that is, matrix multiplication is not commutative.

If $A$ is a matrix of order $r \times s$ and $B$ is a matrix of order $u \times v$, then in order for $AB$ to be defined, $s$ must equal $u$, and then the order of $AB$ is $r \times v$. For instance, in Illustration 8, $AB$ is the product of a $3 \times 2$ matrix and a $2 \times 3$ matrix and the order of the product is $3 \times 3$. Furthermore, $BA$ is formed by multiplying a $2 \times 3$ matrix by a $3 \times 2$ matrix from which we obtain a $2 \times 2$ matrix.

If $A$ is a square matrix, then $A^2$ denotes the product $AA$, $A^3$ denotes the product $(AA)A$, and so on.

**EXAMPLE 2**

If
$$A = \begin{bmatrix} 2 & -1 \\ 1 & 3 \end{bmatrix}$$

$$B = \begin{bmatrix} 4 & 3 \\ -1 & -2 \end{bmatrix}$$

$$C = \begin{bmatrix} -3 & 0 \\ 2 & 1 \\ 3 & -2 \end{bmatrix}$$

find

(a) $AB$  (b) $BA$  (c) $CA$
(d) $C(AB)$

**SOLUTION**

(a) $AB = \begin{bmatrix} 2 & -1 \\ 1 & 3 \end{bmatrix} \begin{bmatrix} 4 & 3 \\ -1 & -2 \end{bmatrix}$

$= \begin{bmatrix} (2)(4) + (-1)(-1) & (2)(3) + (-1)(-2) \\ (1)(4) + (3)(-1) & (1)(3) + (3)(-2) \end{bmatrix}$

$= \begin{bmatrix} 9 & 8 \\ 1 & -3 \end{bmatrix}$

(b) $BA = \begin{bmatrix} 4 & 3 \\ -1 & -2 \end{bmatrix} \begin{bmatrix} 2 & -1 \\ 1 & 3 \end{bmatrix}$

$= \begin{bmatrix} (4)(2) + (3)(1) & (4)(-1) + (3)(3) \\ (-1)(2) + (-2)(1) & (-1)(-1) + (-2)(3) \end{bmatrix}$

$= \begin{bmatrix} 11 & 5 \\ -4 & -5 \end{bmatrix}$

(c) $CA = \begin{bmatrix} -3 & 0 \\ 2 & 1 \\ 3 & -2 \end{bmatrix} \begin{bmatrix} 2 & -1 \\ 1 & 3 \end{bmatrix}$

$= \begin{bmatrix} (-3)(2) + (0)(1) & (-3)(-1) + (0)(3) \\ (2)(2) + (1)(1) & (2)(-1) + (1)(3) \\ (3)(2) + (-2)(1) & (3)(-1) + (-2)(3) \end{bmatrix}$

$= \begin{bmatrix} -6 & 3 \\ 5 & 1 \\ 4 & -9 \end{bmatrix}$

(d) $C(AB) = \begin{bmatrix} -3 & 0 \\ 2 & 1 \\ 3 & -2 \end{bmatrix} \begin{bmatrix} 9 & 8 \\ 1 & -3 \end{bmatrix}$

$= \begin{bmatrix} (-3)(9) + (0)(1) & (-3)(8) + (0)(-3) \\ (2)(9) + (1)(1) & (2)(8) + (1)(-3) \\ (3)(9) + (-2)(1) & (3)(8) + (-2)(-3) \end{bmatrix}$

$$= \begin{bmatrix} -27 & -24 \\ 19 & 13 \\ 25 & 30 \end{bmatrix}$$

Even though matrix multiplication is not commutative, it is associative and distributive for square matrices of the same order (recall that a square matrix of order $n$ is one having $n$ rows and $n$ columns). The following theorem gives these properties.

**7.2.11 THEOREM**  If $A$, $B$, and $C$ are square matrices of order $n$, then

(i) $A(BC) = (AB)C$
(ii) $A(B + C) = AB + AC$
(iii) $(B + C)A = BA + CA$

The proof of Theorem 7.2.11 is omitted; however, in the following example we verify parts (i) and (ii) for three particular square matrices of order two. You are asked to verify part (iii) in Exercise 33 in Exercises 7.2.

**EXAMPLE 3**
If

$$A = \begin{bmatrix} -3 & 2 \\ 0 & 4 \end{bmatrix}$$

$$B = \begin{bmatrix} 1 & -4 \\ -2 & 3 \end{bmatrix}$$

$$C = \begin{bmatrix} -1 & 5 \\ 4 & 1 \end{bmatrix}$$

show that

(a) $A(BC) = (AB)C$
(b) $A(B + C) = AB + AC$

**SOLUTION**

(a) $A(BC) = \begin{bmatrix} -3 & 2 \\ 0 & 4 \end{bmatrix} \left( \begin{bmatrix} 1 & -4 \\ -2 & 3 \end{bmatrix} \begin{bmatrix} -1 & 5 \\ 4 & 1 \end{bmatrix} \right)$

$= \begin{bmatrix} -3 & 2 \\ 0 & 4 \end{bmatrix} \begin{bmatrix} -1 - 16 & 5 - 4 \\ 2 + 12 & -10 + 3 \end{bmatrix}$

$= \begin{bmatrix} -3 & 2 \\ 0 & 4 \end{bmatrix} \begin{bmatrix} -17 & 1 \\ 14 & -7 \end{bmatrix}$

$= \begin{bmatrix} 51 + 28 & -3 - 14 \\ 0 + 56 & 0 - 28 \end{bmatrix}$

$= \begin{bmatrix} 79 & -17 \\ 56 & -28 \end{bmatrix}$

$(AB)C = \left( \begin{bmatrix} -3 & 2 \\ 0 & 4 \end{bmatrix} \begin{bmatrix} 1 & -4 \\ -2 & 3 \end{bmatrix} \right) \begin{bmatrix} -1 & 5 \\ 4 & 1 \end{bmatrix}$

$= \begin{bmatrix} -3 - 4 & 12 + 6 \\ 0 - 8 & 0 + 12 \end{bmatrix} \begin{bmatrix} -1 & 5 \\ 4 & 1 \end{bmatrix}$

$= \begin{bmatrix} -7 & 18 \\ -8 & 12 \end{bmatrix} \begin{bmatrix} -1 & 5 \\ 4 & 1 \end{bmatrix}$

$= \begin{bmatrix} 7 + 72 & -35 + 18 \\ 8 + 48 & -40 + 12 \end{bmatrix}$

$= \begin{bmatrix} 79 & -17 \\ 56 & -28 \end{bmatrix}$

Therefore, $A(BC) = (AB)C$

(b) $A(B + C) = \begin{bmatrix} -3 & 2 \\ 0 & 4 \end{bmatrix} \left( \begin{bmatrix} 1 & -4 \\ -2 & 3 \end{bmatrix} + \begin{bmatrix} -1 & 5 \\ 4 & 1 \end{bmatrix} \right)$

$= \begin{bmatrix} -3 & 2 \\ 0 & 4 \end{bmatrix} \begin{bmatrix} 0 & 1 \\ 2 & 4 \end{bmatrix}$

$= \begin{bmatrix} 0 + 4 & -3 + 8 \\ 0 + 8 & 0 + 16 \end{bmatrix}$

$= \begin{bmatrix} 4 & 5 \\ 8 & 16 \end{bmatrix}$

$AB + AC = \begin{bmatrix} -3 & 2 \\ 0 & 4 \end{bmatrix} \begin{bmatrix} 1 & -4 \\ -2 & 3 \end{bmatrix} + \begin{bmatrix} -3 & 2 \\ 0 & 4 \end{bmatrix} \begin{bmatrix} -1 & 5 \\ 4 & 1 \end{bmatrix}$

$= \begin{bmatrix} -3 - 4 & 12 + 6 \\ 0 - 8 & 0 + 12 \end{bmatrix} + \begin{bmatrix} 3 + 8 & -15 + 2 \\ 0 + 16 & 0 + 4 \end{bmatrix}$

$= \begin{bmatrix} -7 & 18 \\ -8 & 12 \end{bmatrix} + \begin{bmatrix} 11 & -13 \\ 16 & 4 \end{bmatrix}$

$= \begin{bmatrix} 4 & 5 \\ 8 & 16 \end{bmatrix}$

Hence $A(B + C) = AB + AC$.

## EXERCISES 7.2

**1.** Given the matrices

$$A = \begin{bmatrix} 3 & -2 & 7 & 0 \\ 4 & 5 & -1 & 8 \end{bmatrix} \quad B = \begin{bmatrix} -6 \\ 2 \\ 5 \end{bmatrix}$$

(a) What is the order of matrix $A$?
(b) What is the order of matrix $B$?
(c) What is the element in the first row and second column of matrix $A$?
(d) What is the negative of matrix $A$?
(e) What is the negative of matrix $B$?
(f) What is the product of 3 and matrix $A$?
(g) What is the product of $-2$ and matrix $B$?

**2.** Given the matrices

$$A = \begin{bmatrix} 5 & -1 & -3 & 0 \\ 1 & 2 & -1 & -4 \\ 0 & -6 & 2 & 3 \\ 4 & -1 & 7 & 2 \end{bmatrix} \quad B = [2 \quad -1 \quad 5 \quad -6]$$

(a) What is the order of matrix $A$?
(b) What is the order of matrix $B$?
(c) What is the element in the third row and second column of matrix $A$?
(d) What is the negative of matrix $A$?
(e) What is the negative of matrix $B$?
(f) What is the product of 2 and matrix $A$?
(g) What is the product of $-3$ and matrix $B$?

3. Given the matrices

$$A = \begin{bmatrix} 7 & 2 & -1 \\ -3 & -4 & 0 \\ -2 & 1 & -5 \\ 0 & -4 & -1 \end{bmatrix} \quad B = \begin{bmatrix} -3 & 5 \\ 1 & -2 \end{bmatrix}$$

(a) What is the order of matrix $A$?
(b) What is the order of matrix $B$?
(c) What is the element in the second row and third column of matrix $A$?
(d) What is the zero matrix of the same order as matrix $A$?
(e) What is the zero matrix of the same order as matrix $B$?
(f) What is the product of $-4$ and matrix $A$?
(g) What is the product of 6 and matrix $B$?

4. Given the matrices

$$A = \begin{bmatrix} 9 & -4 & 10 \\ 2 & -5 & 1 \\ -7 & -2 & -3 \end{bmatrix} \quad B = \begin{bmatrix} 0 & -6 \\ -2 & 3 \\ -3 & 1 \end{bmatrix}$$

(a) What is the order of matrix $A$?
(b) What is the order of matrix $B$?
(c) What is the element in the third row and first column of matrix $A$?
(d) What is the zero matrix of the same order as matrix $A$?
(e) What is the zero matrix of the same order as matrix $B$?
(f) What is the product of $-1$ and matrix $A$?
(g) What is the negative of matrix $B$?

*In Exercises 5 through 10, perform the indicated operations on the matrices.*

5. $\begin{bmatrix} 3 & -5 \\ -1 & 2 \end{bmatrix} + \begin{bmatrix} -4 & 1 \\ -8 & 6 \end{bmatrix}$

6. $\begin{bmatrix} 4 & -1 \\ -5 & 6 \\ 0 & -2 \end{bmatrix} + \begin{bmatrix} -5 & -1 \\ 3 & -6 \\ -2 & -2 \end{bmatrix}$

7. $\begin{bmatrix} -8 & 3 & -1 \\ 9 & -2 & -5 \end{bmatrix} - \begin{bmatrix} 3 & -2 & 5 \\ -4 & -7 & 3 \end{bmatrix}$

8. $\begin{bmatrix} -4 & -3 & 5 \\ 2 & -4 & 1 \\ 0 & 6 & -7 \end{bmatrix} - \begin{bmatrix} -1 & 4 & -2 \\ 5 & 4 & 0 \\ -7 & 6 & -4 \end{bmatrix}$

9. $\begin{bmatrix} 2 & 6 & -2 \\ 1 & -1 & 3 \\ -5 & -7 & 2 \end{bmatrix} + \begin{bmatrix} -7 & -6 & 5 \\ 4 & -8 & -3 \\ 0 & 2 & -3 \end{bmatrix}$

10. $\begin{bmatrix} 9 & -10 \\ -6 & 2 \end{bmatrix} - \begin{bmatrix} -3 & -6 \\ 8 & -1 \end{bmatrix}$

11. Find $a$, $b$, $c$, and $d$ so that

$$\begin{bmatrix} 7 & -2 \\ -4 & 3 \end{bmatrix} - \begin{bmatrix} a & b \\ c & d \end{bmatrix} = \begin{bmatrix} -6 & 5 \\ -1 & 4 \end{bmatrix}$$

12. Find $a$, $b$, $c$, $d$, $e$, and $f$ so that

$$\begin{bmatrix} a & b \\ c & d \\ e & f \end{bmatrix} + \begin{bmatrix} -2 & -1 \\ 5 & 7 \\ -5 & 0 \end{bmatrix} = \begin{bmatrix} 6 & -1 \\ -5 & 3 \\ -8 & -3 \end{bmatrix}$$

*In Exercises 13 through 16, find the product of the scalar and the matrix.*

13. $4 \begin{bmatrix} -1 & 2 \\ 3 & 5 \\ -6 & 1 \end{bmatrix}$

14. $-5 \begin{bmatrix} -7 & 0 \\ -4 & 6 \end{bmatrix}$

15. $-2 \begin{bmatrix} -3 & 6 & -1 \\ 4 & -2 & 0 \\ -5 & 1 & -8 \end{bmatrix}$

16. $3 \begin{bmatrix} 4 & 3 & -3 \\ 1 & -2 & 7 \end{bmatrix}$

*In Exercises 17 through 24, find the product of the two matrices.*

17. $\begin{bmatrix} 2 & 3 \\ -1 & 5 \end{bmatrix} \begin{bmatrix} 2 & -1 \\ 0 & 3 \end{bmatrix}$

18. $\begin{bmatrix} 4 & -5 \\ 7 & 3 \end{bmatrix} \begin{bmatrix} 5 & -1 \\ -2 & 7 \end{bmatrix}$

19. $\begin{bmatrix} 1 & 2 & 3 \\ 4 & 5 & 7 \end{bmatrix} \begin{bmatrix} 1 & -1 \\ 2 & 0 \\ -1 & 1 \end{bmatrix}$

20. $\begin{bmatrix} 2 & 1 \end{bmatrix} \begin{bmatrix} 3 \\ 2 \end{bmatrix}$

21. $\begin{bmatrix} 1 & 2 & -3 \end{bmatrix} \begin{bmatrix} 2 \\ 1 \\ -1 \end{bmatrix}$

22. $\begin{bmatrix} 3 \\ 2 \end{bmatrix} \begin{bmatrix} 2 & 1 \end{bmatrix}$

23. $\begin{bmatrix} 2 \\ 1 \\ -1 \end{bmatrix} \begin{bmatrix} 1 & 2 & -3 \end{bmatrix}$

24. $\begin{bmatrix} -1 & 3 & -2 \end{bmatrix} \begin{bmatrix} 2 & 0 \\ 1 & 4 \\ -1 & -2 \end{bmatrix}$

*In Exercises 25 through 36, the matrices are as follows:*

$$A = \begin{bmatrix} -2 & 3 \\ 2 & -1 \end{bmatrix} \quad B = \begin{bmatrix} 2 & -1 \\ 3 & -2 \end{bmatrix} \quad C = \begin{bmatrix} -3 & 0 \\ 1 & -2 \end{bmatrix} \quad D = \begin{bmatrix} -1 & 4 \\ 0 & 0 \end{bmatrix} \quad I = \begin{bmatrix} 1 & 0 \\ 0 & 1 \end{bmatrix}$$

25. Verify that $A + B = B + A$.
26. Verify that $B + C = C + B$.
27. Verify that $A + (B + C) = (A + B) + C$.
28. Verify that $A + \mathbf{0} = A$ and $A + (-A) = \mathbf{0}$.
29. Verify that $A\mathbf{0} = \mathbf{0}$ and $\mathbf{0}A = \mathbf{0}$.
30. Verify that $AI = A$ and $IA = A$.
31. Verify that $AB \neq BA$.
32. Verify that $A(B + C) = AB + AC$.
33. Verify that $(B + C)A = BA + CA$.
34. Verify that $A(BC) = (AB)C$.
35. Verify that $DA = DB$ even though $D \neq \mathbf{0}$ and $A \neq B$.
36. Verify that $(B + C)(B - C) \neq B^2 - C^2$.

*In Exercises 37 through 40, find the matrix X that satisfies the given equation.*

37. $X - 3\begin{bmatrix} 2 & -1 \\ 3 & 1 \end{bmatrix} = \begin{bmatrix} -2 & 5 \\ 1 & 4 \end{bmatrix}$

38. $2\begin{bmatrix} -4 & -1 & 3 & 0 \\ 2 & -5 & -2 & 6 \end{bmatrix} + 3X = \begin{bmatrix} -5 & 10 & 3 & -6 \\ 7 & -7 & 11 & 0 \end{bmatrix}$

39. $-2\begin{bmatrix} 5 & -1 \\ 6 & 0 \\ -2 & -5 \end{bmatrix} - 4X = \begin{bmatrix} -2 & 6 \\ 4 & 4 \\ 0 & 2 \end{bmatrix}$

40. $-3X - 4\begin{bmatrix} 6 \\ 3 \\ -1 \\ 2 \end{bmatrix} = \begin{bmatrix} -9 \\ -6 \\ 7 \\ 1 \end{bmatrix}$

41. Prove Theorem 7.2.6.

## 7.3 Determinants

Associated with each square matrix is a number called the determinant of the matrix. In Section 7.6 we apply determinants to solve systems of linear equations. But in this and the next section we discuss methods of computing the value of a determinant.

**7.3.1 DEFINITION**  If $A$ is the square matrix of order two,

$$\begin{bmatrix} a_{11} & a_{12} \\ a_{21} & a_{22} \end{bmatrix}$$

then the *determinant* of $A$, denoted by "det $A$," is defined by

$$\det A = a_{11}a_{22} - a_{12}a_{21} \tag{1}$$

The determinant of a square matrix of order two is called a *second-order determinant*.

The usual notation for the determinant of a matrix is to write the elements of the matrix and replace the brackets by vertical lines. Thus the second-order determinant of the matrix of Definition 7.3.1 is written

$$\begin{vmatrix} a_{11} & a_{12} \\ a_{21} & a_{22} \end{vmatrix}$$

and equation (1) is equivalent to

$$\begin{vmatrix} a_{11} & a_{12} \\ a_{21} & a_{22} \end{vmatrix} = a_{11}a_{22} - a_{12}a_{21} \qquad (2)$$

Observe that a determinant is a number and the determinant notation in the left member of equation (2) is simply a way of denoting the number in the right member. The two products in the right member of equation (2) can be remembered as the products of the elements in the diagonals in the determinant notation in the left member.

**ILLUSTRATION 1.** If

$$A = \begin{vmatrix} 5 & -3 \\ 2 & -4 \end{vmatrix}$$

then

$$\det A = (5)(-4) - (-3)(2) \\ = -14$$

•

We now define a "third-order determinant."

**7.3.2 DEFINITION** If $A$ is the square matrix of order three,

$$\begin{bmatrix} a_{11} & a_{12} & a_{13} \\ a_{21} & a_{22} & a_{23} \\ a_{31} & a_{32} & a_{33} \end{bmatrix}$$

then the *determinant* of $A$, denoted by $\det A$, is defined by

$$\det A = a_{11}a_{22}a_{33} - a_{11}a_{23}a_{32} + a_{12}a_{23}a_{31} - a_{12}a_{21}a_{33} \\ + a_{13}a_{21}a_{32} - a_{13}a_{22}a_{31} \qquad (3)$$

The determinant of a square matrix of order three is called a *third-order determinant*.

As with second-order determinants, a third-order determinant is a number. It also can be represented by writing the elements of the matrix and replacing the brackets by vertical lines. Hence equation (3) can be written as

$$\begin{vmatrix} a_{11} & a_{12} & a_{13} \\ a_{21} & a_{22} & a_{23} \\ a_{31} & a_{32} & a_{33} \end{vmatrix} = a_{11}a_{22}a_{33} - a_{11}a_{23}a_{32} + a_{12}a_{23}a_{31} - a_{12}a_{21}a_{33} \\ + a_{13}a_{21}a_{32} - a_{13}a_{22}a_{31} \quad (4)$$

In the right member of equation (4) observe that there are six terms and each term is a product of three factors. Also note that each product consists of one element from each row and one element from each column such that no two factors in any one term are elements of the same row or column; furthermore, all such possible products appear.

Before computing a third-order determinant, we define what is meant by the "minor" and the "cofactor" of an element in a matrix.

**7.3.3 DEFINITION** If $a_{ij}$ is an element of a square matrix $A$ of order three, then the *minor* of $a_{ij}$, denoted by $M_{ij}$, is the determinant of the matrix of order two, obtained by deleting row $i$ and column $j$.

**ILLUSTRATION 2.** Suppose that $A$ is the matrix

$$\begin{bmatrix} 2 & 3 & -1 \\ -4 & 1 & -2 \\ 5 & 0 & -3 \end{bmatrix}$$

The elements of the first row are $a_{11} = 2$, $a_{12} = 3$, and $a_{13} = -1$. $M_{11}$ is the minor of 2 and is obtained by deleting the first row and first column of $A$.

$$M_{11} = \begin{vmatrix} 1 & -2 \\ 0 & -3 \end{vmatrix}$$
$$= (1)(-3) - (-2)(0)$$
$$= -3$$

$M_{12}$ is the minor of 3 and is obtained by deleting the first row and second column of $A$.

$$M_{12} = \begin{vmatrix} -4 & -2 \\ 5 & -3 \end{vmatrix}$$
$$= (-4)(-3) - (-2)(5)$$
$$= 22$$

$M_{13}$ is the minor of $-1$ and is obtained by deleting the first row and third column of $A$.

$$M_{13} = \begin{vmatrix} -4 & 1 \\ 5 & 0 \end{vmatrix}$$
$$= (-4)(0) - (1)(5)$$
$$= -5$$

●

**ILLUSTRATION 3.** Suppose that $A$ is the matrix

$$\begin{bmatrix} a_{11} & a_{12} & a_{13} \\ a_{21} & a_{22} & a_{23} \\ a_{31} & a_{32} & a_{33} \end{bmatrix}$$

The minor of $a_{22}$ is $M_{22}$ and

$$M_{22} = \begin{vmatrix} a_{11} & a_{13} \\ a_{31} & a_{33} \end{vmatrix}$$
$$= a_{11}a_{33} - a_{13}a_{31}$$

The minor of $a_{23}$ is $M_{23}$ and

$$M_{23} = \begin{vmatrix} a_{11} & a_{12} \\ a_{31} & a_{32} \end{vmatrix}$$
$$= a_{11}a_{32} - a_{12}a_{31}$$ •

**7.3.4 DEFINITION** If $a_{ij}$ is an element of a square matrix $A$ of order three, and $M_{ij}$ is the minor of $a_{ij}$, then the *cofactor* of the element $a_{ij}$, denoted by $A_{ij}$, is defined by

$$A_{ij} = (-1)^{i+j} M_{ij}$$

A mnemonic device for determining whether $(-1)^{i+j}$ is $+1$ or $-1$ is the so-called "checkerboard rule," which alternates $+$ and $-$ signs over the determinant starting with a $+$ in the upper left corner as follows:

$$\begin{vmatrix} + & - & + \\ - & + & - \\ + & - & + \end{vmatrix}$$

**ILLUSTRATION 4.** Suppose that $A$ is the matrix of Illustration 2 and consider the cofactors of the elements of the first row. The cofactor of 2 is $A_{11}$ and

$$A_{11} = (-1)^{1+1} \begin{vmatrix} 1 & -2 \\ 0 & -3 \end{vmatrix}$$
$$= (-1)^2(-3)$$
$$= -3$$

The cofactor of 3 is $A_{12}$ and

$$A_{12} = (-1)^{1+2} \begin{vmatrix} -4 & -2 \\ 5 & -3 \end{vmatrix}$$
$$= (-1)^3(22)$$
$$= -22$$

The cofactor of $-1$ is $A_{13}$ and

$$A_{13} = (-1)^{1+3} \begin{vmatrix} -4 & 1 \\ 5 & 0 \end{vmatrix}$$
$$= (-1)^4(-5)$$
$$= -5$$

**ILLUSTRATION 5.** Suppose that $A$ is the matrix

$$\begin{bmatrix} a_{11} & a_{12} & a_{13} \\ a_{21} & a_{22} & a_{23} \\ a_{31} & a_{32} & a_{33} \end{bmatrix}$$

The product of $a_{11}$ and its cofactor is $a_{11} \cdot A_{11}$, and

$$a_{11} \cdot A_{11} = a_{11} \cdot (-1)^{1+1} \begin{vmatrix} a_{22} & a_{23} \\ a_{32} & a_{33} \end{vmatrix}$$
$$= a_{11} \begin{vmatrix} a_{22} & a_{23} \\ a_{32} & a_{33} \end{vmatrix}$$

The product of $a_{12}$ and its cofactor is $a_{12} \cdot A_{12}$, and

$$a_{12} \cdot A_{12} = a_{12} \cdot (-1)^{1+2} \begin{vmatrix} a_{21} & a_{23} \\ a_{31} & a_{33} \end{vmatrix}$$
$$= -a_{12} \begin{vmatrix} a_{21} & a_{23} \\ a_{31} & a_{33} \end{vmatrix}$$

The product of $a_{13}$ and its cofactor is $a_{13} \cdot A_{13}$, and

$$a_{13} \cdot A_{13} = a_{13} \cdot (-1)^{1+3} \begin{vmatrix} a_{21} & a_{22} \\ a_{31} & a_{32} \end{vmatrix}$$
$$= a_{13} \begin{vmatrix} a_{21} & a_{22} \\ a_{31} & a_{32} \end{vmatrix}$$

Referring back to equation (4) which gives a formula for computing a third-order determinant, we group pairs of terms in the right member and we see that equation (4) is equivalent to the equation

$$\begin{vmatrix} a_{11} & a_{12} & a_{13} \\ a_{21} & a_{22} & a_{23} \\ a_{31} & a_{32} & a_{33} \end{vmatrix} = a_{11}(a_{22}a_{33} - a_{23}a_{32}) - a_{12}(a_{21}a_{33} - a_{23}a_{31}) \\ + a_{13}(a_{21}a_{32} - a_{22}a_{31}) \quad (5)$$

Each of the expressions in parentheses in the right member of equation (5) is a second-order determinant. Hence by using determinant notation, equation (5) can be written as

$$\begin{vmatrix} a_{11} & a_{12} & a_{13} \\ a_{21} & a_{22} & a_{23} \\ a_{31} & a_{32} & a_{33} \end{vmatrix} = a_{11} \begin{vmatrix} a_{22} & a_{23} \\ a_{32} & a_{33} \end{vmatrix} - a_{12} \begin{vmatrix} a_{21} & a_{23} \\ a_{31} & a_{33} \end{vmatrix} + a_{13} \begin{vmatrix} a_{21} & a_{22} \\ a_{31} & a_{32} \end{vmatrix}$$

or, equivalently,

$$\det A = a_{11}A_{11} + a_{12}A_{12} + a_{13}A_{13} \tag{6}$$

Notice in the right member of equation (6) that the terms are the products of the elements of the first row and their cofactors (refer to Illustration 5). Equation (6) is a special case of the following theorem.

**7.3.5 THEOREM**    Let $A$ be a square matrix of order three. Then $\det A$ is the sum of the three products obtained by multiplying each element in any row or column by its cofactor.

We omit the proof of Theorem 7.3.5. However, we derived equation (6) from Definition 7.3.2. Equation (6) is the special case of Theorem 7.3.5, where the elements are those in the first row of the matrix $A$ of Definition 7.3.2.

We now verify Theorem 7.3.5 for another special case. If in the right member of equation (4), we group together the first two terms, the fourth and fifth terms, and the third and sixth terms, we see that equation (4) is equivalent to the equation

$$\begin{vmatrix} a_{11} & a_{12} & a_{13} \\ a_{21} & a_{22} & a_{23} \\ a_{31} & a_{32} & a_{33} \end{vmatrix} = a_{11}(a_{22}a_{33} - a_{23}a_{32}) - a_{21}(a_{12}a_{33} - a_{13}a_{32}) + a_{31}(a_{12}a_{23} - a_{13}a_{22})$$

which, by using determinant notation, is equivalent to the equation

$$\begin{vmatrix} a_{11} & a_{12} & a_{13} \\ a_{21} & a_{22} & a_{23} \\ a_{31} & a_{32} & a_{33} \end{vmatrix} = a_{11}\begin{vmatrix} a_{22} & a_{23} \\ a_{32} & a_{33} \end{vmatrix} - a_{21}\begin{vmatrix} a_{12} & a_{13} \\ a_{32} & a_{33} \end{vmatrix} + a_{31}\begin{vmatrix} a_{12} & a_{13} \\ a_{22} & a_{23} \end{vmatrix}$$

or, equivalently,

$$\det A = a_{11}A_{11} + a_{21}A_{21} + a_{31}A_{31} \tag{7}$$

In the right member of equation (7) the terms are the products of the elements of the first column and their cofactors. Thus equation (7) is another special case of Theorem 7.3.5. In Exercises 29 and 30 of Exercises 7.3 you are asked to verify Theorem 7.3.5 for two other special cases.

When Theorem 7.3.5 is used to find the value of a determinant, we say that the determinant is evaluated by the cofactors of the elements of the row or column in which the elements appear.

## EXAMPLE 1

If $A$ is the matrix of Illustration 2, evaluate $\det A$ in two ways: (a) by the cofactors of the elements of the first row; and (b) by the cofactors of the elements of the second column.

### SOLUTION

(a) $\begin{vmatrix} 2 & 3 & -1 \\ -4 & 1 & -2 \\ 5 & 0 & -3 \end{vmatrix} = 2 \cdot (-1)^{1+1} \begin{vmatrix} 1 & -2 \\ 0 & -3 \end{vmatrix} + 3 \cdot (-1)^{1+2} \begin{vmatrix} -4 & -2 \\ 5 & -3 \end{vmatrix}$
$+ (-1) \cdot (-1)^{1+3} \begin{vmatrix} -4 & 1 \\ 5 & 0 \end{vmatrix}$
$= 2(1)(-3 - 0) + 3(-1)(12 + 10) + (-1)(1)(0 - 5)$
$= 2(-3) - 3(22) - 1(-5)$
$= -67$

(b) $\begin{vmatrix} 2 & 3 & -1 \\ -4 & 1 & -2 \\ 5 & 0 & -3 \end{vmatrix} = 3 \cdot (-1)^{1+2} \begin{vmatrix} -4 & -2 \\ 5 & -3 \end{vmatrix} + 1 \cdot (-1)^{2+2} \begin{vmatrix} 2 & -1 \\ 5 & -3 \end{vmatrix}$
$+ 0 \cdot (-1)^{3+2} \begin{vmatrix} 2 & -1 \\ -4 & -2 \end{vmatrix}$
$= 3(-1)(12 + 10) + 1(1)(-6 + 5) + 0$
$= -3(22) + 1(-1)$
$= -67$

Observe in the solution of Example 1 that because 0 is an element in the second column, the computation involved in part (b) is less than that involved in part (a). It is apparent, then, that if any elements are zero, the determinant should be evaluated by the cofactors of the row or column containing the most zeros.

Because we have defined a third order determinant, we may define the minor and cofactor of an element in a square matrix of order four just as we did in Definitions 7.3.3 and 7.3.4 for a square matrix of order three. However, the minors are now third order determinants.

**ILLUSTRATION 6.** Suppose that $A$ is the matrix

$$\begin{bmatrix} 5 & -2 & -4 & -1 \\ 4 & 1 & 2 & -2 \\ 0 & 3 & -1 & 0 \\ 1 & -1 & 0 & -3 \end{bmatrix}$$

The elements of the second row are $a_{21} = 4$, $a_{22} = 1$, $a_{23} = 2$, and $a_{24} = -2$. The minor of $a_{23}$, denoted by $M_{23}$, is the third order determinant obtained by deleting the second row and third column of matrix $A$. Therefore,

$$M_{23} = \begin{vmatrix} 5 & -2 & -1 \\ 0 & 3 & 0 \\ 1 & -1 & -3 \end{vmatrix}$$

The cofactor of $a_{23}$ is denoted by $A_{23}$, and

$$A_{23} = (-1)^{2+3} M_{23}$$

$$= (-1) \begin{vmatrix} 5 & -2 & -1 \\ 0 & 3 & 0 \\ 1 & -1 & -3 \end{vmatrix}$$

We evaluate the third order determinant by the cofactors of the elements of the second row because there are two zeros there. We have then

$$A_{23} = (-1)\left[ 0 + (3)(-1)^{2+2} \begin{vmatrix} 5 & -1 \\ 1 & -3 \end{vmatrix} + 0 \right]$$

$$= (-1)[3(-15 + 1)]$$
$$= (-1)[3(-14)]$$
$$= 42 \qquad \bullet$$

The determinant of a square matrix of order four can be evaluated by cofactors in a manner similar to the way third order determinants are evaluated. The method is based on an extension of Theorem 7.3.5 to fourth order determinants. The following example shows the procedure.

EXAMPLE 2
Find det $A$ if

$$A = \begin{bmatrix} 2 & 4 & 0 & 1 \\ -3 & 0 & 3 & 2 \\ 5 & 1 & 0 & -1 \\ -4 & 0 & 0 & 1 \end{bmatrix}$$

SOLUTION
Because there are three zeros in the third column, we expand by the cofactors of the elements of that column.

$$\det A = \begin{vmatrix} 2 & 4 & 0 & 1 \\ -3 & 0 & 3 & 2 \\ 5 & 1 & 0 & -1 \\ -4 & 0 & 0 & 1 \end{vmatrix}$$

$$= 0 \cdot A_{13} + 3 \cdot (-1)^{2+3} \cdot A_{23} + 0 \cdot A_{33} + 0 \cdot A_{43}$$

$$= (-3) \begin{vmatrix} 2 & 4 & 1 \\ 5 & 1 & -1 \\ -4 & 0 & 1 \end{vmatrix}$$

We expand this third-order determinant by the cofactors of the elements of the third row and we obtain

$$\det A = (-3)\left[ (-4)(-1)^{3+1} \begin{vmatrix} 4 & 1 \\ 1 & -1 \end{vmatrix} + 0 + (1)(-1)^{3+3} \begin{vmatrix} 2 & 4 \\ 5 & 1 \end{vmatrix} \right]$$

$$= (-3)[-4(-5) + 1(-18)]$$
$$= (-3)[20 - 18]$$
$$= -6$$

We showed how third order determinants can be evaluated by using cofactors that are second order determinants and then we expressed fourth order determinants in terms of third order cofactors. We use this pattern to

**7.3.6 DEFINITION** If $A$ is a square matrix of order $n$, then $\det A$ is the sum of the $n$ products obtained by multiplying each element of the first row by its cofactor.

Following is an important theorem that allows us to evaluate an $n$th order determinant by using the cofactors of the elements of any row or column. It is an extension of Theorem 7.3.5.

**7.3.7 THEOREM** If $A$ is a square matrix of order $n$ then $\det A$ is the sum of the $n$ products obtained by multiplying each element of any row or column by its cofactor.

The proof of Theorem 7.3.7 is complicated and is omitted.

In the next example we show how Theorem 7.3.7 is used to evaluate a fifth order determinant. We purposely use a determinant having some zeros as elements in order to facilitate the computation.

**EXAMPLE 3**

Evaluate the determinant

$$\begin{vmatrix} 0 & 4 & 0 & 3 & 0 \\ 1 & 3 & 1 & 4 & 3 \\ 0 & -2 & 0 & 0 & 0 \\ 2 & 1 & 2 & -3 & -1 \\ -1 & -5 & 0 & -4 & 0 \end{vmatrix}$$

**SOLUTION**

There are four zeros in the third row. Therefore, we expand by the cofactors of the elements of that row.

$$\begin{vmatrix} 0 & 4 & 0 & 3 & 0 \\ 1 & 3 & 1 & 4 & 3 \\ 0 & -2 & 0 & 0 & 0 \\ 2 & 1 & 2 & -3 & -1 \\ -1 & -5 & 0 & -4 & 0 \end{vmatrix} = (-2)(-1)^{3+2} \begin{vmatrix} 0 & 0 & 3 & 0 \\ 1 & 1 & 4 & 3 \\ 2 & 2 & -3 & -1 \\ -1 & 0 & -4 & 0 \end{vmatrix}$$

In the resulting fourth order determinant, the first row has three zeros, and so we expand by the cofactors of the elements of that row. We have

$$(-2)(-1) \begin{vmatrix} 0 & 0 & 3 & 0 \\ 1 & 1 & 4 & 3 \\ 2 & 2 & -3 & -1 \\ -1 & 0 & -4 & 0 \end{vmatrix} = 2 \left[ 3(-1)^{1+3} \begin{vmatrix} 1 & 1 & 3 \\ 2 & 2 & -1 \\ -1 & 0 & 0 \end{vmatrix} \right]$$

Because there are two zeros in the third row, we expand the third order determinant by the cofactors of the elements of that row. We have

$$6 \begin{vmatrix} 1 & 1 & 3 \\ 2 & 2 & -1 \\ -1 & 0 & 0 \end{vmatrix} = 6 \left[ (-1)(-1)^{3+1} \begin{vmatrix} 1 & 3 \\ 2 & -1 \end{vmatrix} \right]$$

$$= -6(-1 - 6)$$
$$= 42$$

## EXERCISES 7.3

*In Exercises 1 through 4, let*

$$A = \begin{vmatrix} -3 & 4 & 0 \\ 1 & 2 & -5 \\ -1 & 3 & 6 \end{vmatrix}$$

1. Find the following minors of elements of $A$: $M_{11}$, $M_{23}$, $M_{32}$.
2. Find the following minors of elements of $A$: $M_{21}$, $M_{12}$, $M_{33}$.
3. Find the following cofactors of elements of $A$: $A_{21}$, $A_{31}$, $A_{13}$.
4. Find the following cofactors of elements of $A$: $A_{11}$, $A_{23}$, $A_{32}$.

*In Exercises 5 through 8, let*

$$A = \begin{vmatrix} 5 & -2 & 0 & 3 \\ -1 & 4 & -1 & 2 \\ 3 & -1 & 6 & 0 \\ 1 & 3 & 4 & 1 \end{vmatrix}$$

5. Find the following minors of elements of $A$: $M_{12}$, $M_{22}$, $M_{32}$, $M_{42}$.
6. Find the following minors of elements of $A$: $M_{14}$, $M_{24}$, $M_{34}$, $M_{44}$.
7. Find the following cofactors of elements of $A$: $A_{31}$, $A_{32}$, $A_{33}$, $A_{34}$.
8. Find the following cofactors of elements of $A$: $A_{21}$, $A_{22}$, $A_{23}$, $A_{24}$.

*In Exercises 9 through 20, evaluate the given determinant.*

9. $\begin{vmatrix} 2 & -3 \\ 4 & 5 \end{vmatrix}$

10. $\begin{vmatrix} -3 & 7 \\ 6 & 2 \end{vmatrix}$

11. $\begin{vmatrix} 4 & -5 \\ -2 & 1 \end{vmatrix}$

12. $\begin{vmatrix} -2 & 6 \\ -3 & -2 \end{vmatrix}$

13. $\begin{vmatrix} 1 & 2 & 0 \\ 2 & -1 & 3 \\ 1 & 5 & -4 \end{vmatrix}$

14. $\begin{vmatrix} -6 & -3 & 2 \\ 3 & 2 & 0 \\ 4 & 2 & -1 \end{vmatrix}$

15. $\begin{vmatrix} 5 & 0 & -2 & 0 \\ 0 & 1 & 0 & -1 \\ 0 & 2 & -4 & 0 \\ -3 & 0 & 0 & -5 \end{vmatrix}$

16. $\begin{vmatrix} 2 & -3 & 0 & -2 \\ 6 & 0 & 0 & 0 \\ 1 & -1 & -4 & 2 \\ 0 & 2 & 3 & 4 \end{vmatrix}$

17. $\begin{vmatrix} 1 & 2 & 3 & 4 \\ 0 & 1 & 2 & 3 \\ 0 & 0 & 1 & 2 \\ 0 & 0 & 0 & 1 \end{vmatrix}$

18. $\begin{vmatrix} 0 & x & 0 & 0 \\ 0 & 0 & y & 0 \\ 0 & 0 & 0 & z \\ w & 0 & 0 & 0 \end{vmatrix}$

19. $\begin{vmatrix} 1 & 3 & 0 & -1 & 0 \\ -2 & -1 & 3 & 2 & -3 \\ 0 & 2 & 0 & 4 & 0 \\ -1 & 1 & 0 & 3 & -2 \\ 0 & -4 & 0 & 1 & 0 \end{vmatrix}$

20. $\begin{vmatrix} 2 & 0 & 0 & -3 & 0 \\ 3 & 4 & 1 & -2 & 1 \\ 0 & 0 & 0 & -1 & 0 \\ -1 & -3 & 1 & 2 & 1 \\ -4 & 0 & 3 & -4 & 0 \end{vmatrix}$

*In Exercises 21 through 28, solve for x.*

21. $\begin{vmatrix} x & 2 \\ -1 & 5 \end{vmatrix} = 7$

22. $\begin{vmatrix} 4 & -2 \\ x & 3 \end{vmatrix} = 6$

23. $\begin{vmatrix} x & -2 \\ 5 & x \end{vmatrix} = 14$

24. $\begin{vmatrix} x^2 & -3 \\ x & 1 \end{vmatrix} = 4$

25. $\begin{vmatrix} x & 0 & 0 \\ 3 & -1 & 5 \\ 2 & 2 & -6 \end{vmatrix} = 8$   26. $\begin{vmatrix} x & 1 & 0 \\ 3 & -2 & -2 \\ -4 & 4 & 5 \end{vmatrix} = 5$   27. $\begin{vmatrix} x & 2 & 5 \\ 1 & 0 & 4 \\ x^2 & -1 & 2 \end{vmatrix} = 3$   28. $\begin{vmatrix} x & 0 & -1 \\ 3 & x & 4 \\ 2 & 2 & 1 \end{vmatrix} = 10$

29. Use equation (4) to verify Theorem 7.3.5 where det $A$ is the sum of the products of the elements of the third row and their cofactors.

30. Use equation (4) to verify Theorem 7.3.5 where det $A$ is the sum of the products of the elements of the second column and their cofactors.

*In Exercises 31 through 34, let*

$$A = \begin{bmatrix} a_{11} & a_{12} & a_{13} & a_{14} & a_{15} \\ a_{21} & a_{22} & a_{23} & a_{24} & a_{25} \\ a_{31} & a_{32} & a_{33} & a_{34} & a_{35} \\ a_{41} & a_{42} & a_{43} & a_{44} & a_{45} \\ a_{51} & a_{52} & a_{53} & a_{54} & a_{55} \end{bmatrix}$$

31. Find all of the terms of det $A$ that contain the three factors $a_{14}$, $a_{32}$, and $a_{55}$.

32. Find all of the terms of det $A$ that contain the three factors $a_{22}$, $a_{35}$, and $a_{43}$.

33. Find all of the terms of det $A$ that contain the two factors $a_{23}$ and $a_{41}$.

34. Find all of the terms of det $A$ that contain the two factors $a_{15}$ and $a_{51}$.

## 7.4 Properties of Determinants

When applying Theorem 7.3.7 to evaluate the determinant of a matrix of order $n$ if $n$ is large, the numerical computation is usually quite tedious. For instance, if the matrix is of order five, then, when evaluating the determinant by the cofactors of any row or column, a sum of five terms is obtained and each term contains the determinant of a matrix of order four; to evaluate each of these determinants, we obtain a sum of four terms each of which contains the determinant of a matrix of order three. Therefore, there are $5 \cdot 4$, or 20, determinants of matrices of order three to be evaluated. If the determinant of a matrix of order six is evaluated by Theorem 7.3.7, a sum of six terms, each containing the determinant of a matrix of order five, is obtained; thus, there are $6 \cdot 20$, or 120, determinants of matrices of order three to be evaluated. Of course, the computation is simplified if many of the elements of the original matrix are zero.

We now discuss some properties of determinants that are useful to reduce the amount of computation involved in their evaluation. These properties enable us to obtain one matrix from another, having the same determinant, and in which some of the elements are zero.

**7.4.1 THEOREM**  If $A$ is a matrix such that every element of a row or column is zero, then det $A = 0$.

*Proof.* We apply Theorem 7.3.7 and evaluate det $A$ by the cofactors of the elements of the row or column in which the zeros appear. Hence det $A$ is the sum of $n$ products, each of which has a factor of zero. Therefore, det $A = 0$.

**ILLUSTRATION 1.** Each of the following results is a consequence of Theorem 7.4.1.

$$\begin{vmatrix} 1 & 0 \\ 4 & 0 \end{vmatrix} = 0 \qquad \begin{vmatrix} -2 & 3 & 1 \\ 0 & 0 & 0 \\ 4 & -7 & 2 \end{vmatrix} = 0 \qquad \begin{vmatrix} 2 & 6 & 0 & -1 \\ 0 & -4 & 0 & 3 \\ 4 & 3 & 0 & 0 \\ 0 & 1 & 0 & -5 \end{vmatrix} = 0 \quad \bullet$$

We do not give a general proof for the next three theorems. Instead we follow the statement of each theorem with an illustration which demonstrates the theorem for a matrix of order three.

**7.4.2 THEOREM** If a matrix $B$ is obtained from a matrix $A$ by interchanging two rows (or columns), then $\det B = -\det A$.

**ILLUSTRATION 2**

(a) Suppose that

$$A = \begin{bmatrix} a_{11} & a_{12} & a_{13} \\ a_{21} & a_{22} & a_{23} \\ a_{31} & a_{32} & a_{33} \end{bmatrix}$$

and $B$ is the matrix obtained from $A$ by interchanging the second and third rows. That is,

$$B = \begin{bmatrix} a_{11} & a_{12} & a_{13} \\ a_{31} & a_{32} & a_{33} \\ a_{21} & a_{22} & a_{23} \end{bmatrix}$$

If $\det A$ is evaluated by the cofactors of the elements of the second row, then

$$\det A = -a_{21} \begin{vmatrix} a_{12} & a_{13} \\ a_{32} & a_{33} \end{vmatrix} + a_{22} \begin{vmatrix} a_{11} & a_{13} \\ a_{31} & a_{33} \end{vmatrix} - a_{23} \begin{vmatrix} a_{11} & a_{12} \\ a_{31} & a_{32} \end{vmatrix} \quad (1)$$

If $\det B$ is evaluated by the cofactors of the elements of the third row, then

$$\det B = a_{21} \begin{vmatrix} a_{12} & a_{13} \\ a_{32} & a_{33} \end{vmatrix} - a_{22} \begin{vmatrix} a_{11} & a_{13} \\ a_{31} & a_{33} \end{vmatrix} + a_{23} \begin{vmatrix} a_{11} & a_{12} \\ a_{31} & a_{32} \end{vmatrix} \quad (2)$$

Comparing equations (1) and (2), we see that $\det B = -\det A$.

(b) Let $A$ be the matrix defined in part (a) and let $C$ be the matrix obtained from $A$ by interchanging the first and third columns. Then

$$C = \begin{bmatrix} a_{13} & a_{12} & a_{11} \\ a_{23} & a_{22} & a_{21} \\ a_{33} & a_{32} & a_{31} \end{bmatrix}$$

If det $A$ is evaluated by the cofactors of the elements of the first column, then

$$\det A = a_{11} \begin{vmatrix} a_{22} & a_{23} \\ a_{32} & a_{33} \end{vmatrix} - a_{21} \begin{vmatrix} a_{12} & a_{13} \\ a_{32} & a_{33} \end{vmatrix} + a_{31} \begin{vmatrix} a_{12} & a_{13} \\ a_{22} & a_{23} \end{vmatrix}$$

or, equivalently,

$$\det A = a_{11}(a_{22}a_{33} - a_{23}a_{32}) - a_{21}(a_{12}a_{33} - a_{13}a_{32})$$
$$+ a_{31}(a_{12}a_{23} - a_{13}a_{22}) \quad (3)$$

If det $C$ is evaluated by the cofactors of the elements of the third column, then

$$\det C = a_{11} \begin{vmatrix} a_{23} & a_{22} \\ a_{33} & a_{32} \end{vmatrix} - a_{21} \begin{vmatrix} a_{13} & a_{12} \\ a_{33} & a_{32} \end{vmatrix} + a_{31} \begin{vmatrix} a_{13} & a_{12} \\ a_{23} & a_{22} \end{vmatrix}$$

or, equivalently,

$$\det C = a_{11}(a_{23}a_{32} - a_{22}a_{33}) - a_{21}(a_{13}a_{32} - a_{12}a_{33})$$
$$+ a_{31}(a_{13}a_{22} - a_{12}a_{23})$$

or, equivalently,

$$\det C = -a_{11}(a_{22}a_{33} - a_{23}a_{32}) + a_{21}(a_{12}a_{33} - a_{13}a_{32})$$
$$- a_{31}(a_{12}a_{23} - a_{13}a_{22}) \quad (4)$$

From Equations (3) and (4) it follows that $\det C = -\det A$. ●

**7.4.3 THEOREM**  If a matrix $B$ is obtained from a matrix $A$ by multiplying every element of one row (or column) by a real number $k$, then $\det B = k(\det A)$.

**ILLUSTRATION 3.** Let $A$ be the matrix defined in part (a) of Illustration 2 and let $B$ be the matrix obtained from $A$ by multiplying the elements of the second column by $k$; then

$$B = \begin{bmatrix} a_{11} & ka_{12} & a_{13} \\ a_{21} & ka_{22} & a_{23} \\ a_{31} & ka_{32} & a_{33} \end{bmatrix}$$

If det $A$ is evaluated by the cofactors of the elements of the second column, then

$$\det A = -a_{12} \begin{vmatrix} a_{21} & a_{23} \\ a_{31} & a_{33} \end{vmatrix} + a_{22} \begin{vmatrix} a_{11} & a_{13} \\ a_{31} & a_{33} \end{vmatrix} - a_{32} \begin{vmatrix} a_{11} & a_{13} \\ a_{21} & a_{23} \end{vmatrix} \quad (5)$$

If det $B$ is evaluated by the cofactors of the elements of the second column, then

$$\det B = -ka_{12} \begin{vmatrix} a_{21} & a_{23} \\ a_{31} & a_{33} \end{vmatrix} + ka_{22} \begin{vmatrix} a_{11} & a_{13} \\ a_{31} & a_{33} \end{vmatrix} - ka_{32} \begin{vmatrix} a_{11} & a_{13} \\ a_{21} & a_{23} \end{vmatrix} \quad (6)$$

Comparing equations (5) and (6), we see that $\det B = k(\det A)$. ●

**7.4.4 THEOREM** If a matrix $B$ is obtained from a matrix $A$ by replacing any row (or column) of $A$ by the sum of that row (or column) and $k$ times another row (or column), then $\det B = \det A$.

**ILLUSTRATION 4.** Suppose that $A$ is the matrix defined in part (a) of Illustration 2, and let $B$ be the matrix obtained from $A$ by adding to the third row the product of $k$ times the first row. Therefore,

$$B = \begin{bmatrix} a_{11} & a_{12} & a_{13} \\ a_{21} & a_{22} & a_{23} \\ a_{31} + ka_{11} & a_{32} + ka_{12} & a_{33} + ka_{13} \end{bmatrix}$$

If $\det B$ is evaluated by the cofactors of the elements of the third row, then

$$\det B = (a_{31} + ka_{11}) \begin{vmatrix} a_{12} & a_{13} \\ a_{22} & a_{23} \end{vmatrix} + (a_{32} + ka_{12}) \left( - \begin{vmatrix} a_{11} & a_{13} \\ a_{21} & a_{23} \end{vmatrix} \right)$$
$$+ (a_{33} + ka_{13}) \begin{vmatrix} a_{11} & a_{12} \\ a_{21} & a_{22} \end{vmatrix} \quad (7)$$

Because the cofactors $A_{31}$, $A_{32}$, and $A_{33}$ are defined by

$$A_{31} = \begin{vmatrix} a_{12} & a_{13} \\ a_{22} & a_{23} \end{vmatrix}, \quad A_{32} = - \begin{vmatrix} a_{11} & a_{13} \\ a_{21} & a_{23} \end{vmatrix}, \quad \text{and} \quad A_{33} = \begin{vmatrix} a_{11} & a_{12} \\ a_{21} & a_{22} \end{vmatrix}$$

it follows from equation (7) that

$$\det B = (a_{31} + ka_{11})A_{31} + (a_{32} + ka_{12})A_{32} + (a_{33} + ka_{13})A_{33}$$

or, equivalently,

$$\det B = (a_{31}A_{31} + a_{32}A_{32} + a_{33}A_{33}) + k(a_{11}A_{31} + a_{12}A_{32} + a_{13}A_{33}) \quad (8)$$

We compute the second expression in parentheses in the right member of equation (8).

$$a_{11}A_{31} + a_{12}A_{32} + a_{13}A_{33} = a_{11}(a_{12}a_{23} - a_{13}a_{22}) - a_{12}(a_{11}a_{23} - a_{13}a_{21})$$
$$+ a_{13}(a_{11}a_{22} - a_{12}a_{21})$$
$$= a_{11}a_{12}a_{23} - a_{11}a_{13}a_{22} - a_{12}a_{11}a_{23} + a_{12}a_{13}a_{21}$$
$$+ a_{13}a_{11}a_{22} - a_{13}a_{12}a_{21}$$

Therefore,

$$a_{11}A_{31} + a_{12}A_{32} + a_{13}A_{33} = 0 \quad (9)$$

From Theorem 7.3.5 it follows that the first expression in parentheses in the right member of equation (8) is $\det A$; that is,

$$a_{31}A_{31} + a_{32}A_{32} + a_{33}A_{33} = \det A \quad (10)$$

Substituting from equations (9) and (10) into equation (8), we obtain

$$\det B = \det A + k(0)$$

or, equivalently,
$$\det B = \det A$$

**ILLUSTRATION 5.** Let
$$A = \begin{bmatrix} -3 & 5 & -1 \\ 1 & 0 & -3 \\ -2 & 4 & 1 \end{bmatrix}$$

We evaluate $\det A$ by the cofactors of the elements of the second row.

$$\det A = 1(-1)^{2+1}\begin{vmatrix} 5 & -1 \\ 4 & 1 \end{vmatrix} + 0 + (-3)(-1)^{2+3}\begin{vmatrix} -3 & 5 \\ -2 & 4 \end{vmatrix}$$
$$= (-1)(5+4) + 0 + 3(-12+10)$$
$$= (-1)(9) + 3(-2)$$
$$= -15$$

(a) Let $B$ be the matrix obtained from $A$ by interchanging the first and third columns. Then by Theorem 7.4.2, $\det B = -\det A$. We verify this fact and evaluate $\det B$ by the cofactors of the elements of the second row.

$$\det B = \begin{vmatrix} -1 & 5 & -3 \\ -3 & 0 & 1 \\ 1 & 4 & -2 \end{vmatrix}$$

$$= (-3)(-1)^{2+1}\begin{vmatrix} 5 & -3 \\ 4 & -2 \end{vmatrix} + 0 + 1(-1)^{2+3}\begin{vmatrix} -1 & 5 \\ 1 & 4 \end{vmatrix}$$
$$= 3(-10+12) + (-1)(-4-5)$$
$$= 3(2) - (-9)$$
$$= 15$$
$$= -\det A$$

(b) Let $C$ be the matrix obtained from $A$ by multiplying every element of the third row by 3. Then by Theorem 7.4.3, $\det C = 3(\det A)$. We verify this fact and evaluate $\det C$ by the cofactors of the elements of the second row.

$$\det C = \begin{vmatrix} -3 & 5 & -1 \\ 1 & 0 & -3 \\ -6 & 12 & 3 \end{vmatrix}$$

$$= 1(-1)^{2+1}\begin{vmatrix} 5 & -1 \\ 12 & 3 \end{vmatrix} + 0 + (-3)(-1)^{2+3}\begin{vmatrix} -3 & 5 \\ -6 & 12 \end{vmatrix}$$
$$= (-1)(15+12) + 3(-36+30)$$
$$= (-1)(27) + 3(-6)$$
$$= -45$$
$$= 3(\det A)$$

(c) Let $D$ be the matrix obtained from $A$ by replacing the third column by the sum of the third column and 3 times the first column. Then by Theorem 7.4.4, $\det D = \det A$. We verify this fact and evaluate $\det D$ by the cofactors of the elements of the second row.

$$\det D = \begin{vmatrix} -3 & 5 & -10 \\ 1 & 0 & 0 \\ -2 & 4 & -5 \end{vmatrix}$$

$$= 1(-1)^{2+1} \begin{vmatrix} 5 & -10 \\ 4 & -5 \end{vmatrix} + 0 + 0$$

$$= (-1)(-25 + 40)$$

$$= -15$$

$$= \det A \qquad \bullet$$

Theorem 7.4.4 is used to obtain a row (or column) for which all but one of the elements is zero. If there is an element of the matrix that is 1 or $-1$, then the procedure is similar to that used in part (c) of Illustration 5. If there is no element that is 1 or $-1$ in the given matrix, then the number 1 can be obtained by first applying Theorem 7.4.4 or Theorem 7.4.3 in some manner (see the next example).

EXAMPLE 1
Evaluate the determinant

$$\begin{vmatrix} 2 & 6 & 5 \\ 5 & -3 & -7 \\ 3 & 4 & -2 \end{vmatrix}$$

by first obtaining two zeros in a row or column.

SOLUTION
There is no 1 or $-1$ appearing as an element; however, if we replace the second column by the sum of the second column and $-1$ times the third column, we have

$$\begin{vmatrix} 2 & 6 & 5 \\ 5 & -3 & -7 \\ 3 & 4 & -2 \end{vmatrix} = \begin{vmatrix} 2 & 1 & 5 \\ 5 & 4 & -7 \\ 3 & 6 & -2 \end{vmatrix}$$

$$= \begin{vmatrix} 0 & 1 & 0 \\ -3 & 4 & -27 \\ -9 & 6 & -32 \end{vmatrix} \qquad \text{(replacing the first column by the sum of the first column and } -2 \text{ times the second column; and replacing the third column by the sum of the third column and } -5 \text{ times the second column)}$$

$$= 1(-1)^{1+2} \begin{vmatrix} -3 & -27 \\ -9 & -32 \end{vmatrix} \qquad \text{(evaluating by the minors of the elements of the first row)}$$

$$= (-1)(96 - 243)$$

$$= 147$$

## EXAMPLE 2

Evaluate the determinant by first obtaining three zeros in a row or column.

$$\begin{vmatrix} 5 & -3 & 3 & -5 \\ -1 & 8 & -3 & 8 \\ 4 & -2 & 2 & -4 \\ 2 & 6 & -1 & 4 \end{vmatrix}$$

### SOLUTION

Because 2 is a factor of each element of the third row, it follows from Theorem 7.4.3 that

$$\begin{vmatrix} 5 & -3 & 3 & -5 \\ -1 & 8 & -3 & 8 \\ 4 & -2 & 2 & -4 \\ 2 & 6 & -1 & 4 \end{vmatrix} = 2 \begin{vmatrix} 5 & -3 & 3 & -5 \\ -1 & 8 & -3 & 8 \\ 2 & -1 & 1 & -2 \\ 2 & 6 & -1 & 4 \end{vmatrix}$$

We now apply Theorem 7.4.4 and obtain three zeros in the third row. We replace the first column by the sum of the first column and $-2$ times the third column; we replace the second column by the sum of the second column and the third column; and we replace the fourth column by the sum of the fourth column and 2 times the third column. Hence

$$2 \begin{vmatrix} 5 & -3 & 3 & -5 \\ -1 & 8 & -3 & 8 \\ 2 & -1 & 1 & -2 \\ 2 & 6 & -1 & 4 \end{vmatrix} = 2 \begin{vmatrix} -1 & 0 & 3 & 1 \\ 5 & 5 & -3 & 2 \\ 0 & 0 & 1 & 0 \\ 4 & 5 & -1 & 2 \end{vmatrix}$$

$$= 2(1)(-1)^{3+3} \begin{vmatrix} -1 & 0 & 1 \\ 5 & 5 & 2 \\ 4 & 5 & 2 \end{vmatrix} \quad \text{(evaluating by the minors of the elements of the third row)}$$

$$= 2 \begin{vmatrix} -1 & 0 & 0 \\ 5 & 5 & 7 \\ 4 & 5 & 6 \end{vmatrix} \quad \text{(replacing the third column by the sum of the third column and the first column)}$$

$$= 2(-1)(-1)^{1+1} \begin{vmatrix} 5 & 7 \\ 5 & 6 \end{vmatrix} \quad \text{(evaluating by the minors of the elements of the first row)}$$

$$= -2(30 - 35)$$

$$= 10$$

**7.4.5 THEOREM** If $A$ is a square matrix having two identical rows (or columns), then $\det A = 0$.

*Proof.* Let $B$ be the matrix obtained from $A$ by interchanging the two identical rows (or columns). Then $A$ and $B$ are the same matrix and so

$$\det B = \det A \tag{11}$$

Furthermore, from Theorem 7.4.2, it follows that

$$\det B = -\det A \tag{12}$$

Substituting from equation (11) into equation (12), we obtain

$$\det A = -\det A$$
$$2 \det A = 0$$
$$\det A = 0$$

**ILLUSTRATION 6.** If

$$A = \begin{bmatrix} 2 & -3 & 2 \\ 4 & 2 & 4 \\ -1 & 5 & -1 \end{bmatrix}$$

then because the first and third columns are identical, it follows from Theorem 7.4.5 that $\det A = 0$. ●

## EXERCISES 7.4

*In Exercises 1 through 14, give the reason that the equation is true.*

1. $\begin{vmatrix} 3 & 0 & -2 \\ 5 & 0 & -4 \\ 2 & 0 & 6 \end{vmatrix} = 0$

2. $\begin{vmatrix} -4 & 3 & -4 \\ 5 & 6 & 5 \\ 0 & -1 & 0 \end{vmatrix} = 0$

3. $\begin{vmatrix} 1 & -3 & 2 & 1 \\ 0 & 5 & 6 & 0 \\ 1 & -3 & 2 & 1 \\ 3 & 5 & 4 & 7 \end{vmatrix} = 0$

4. $\begin{vmatrix} 2 & 3 & -1 & 5 \\ 7 & 1 & 0 & 4 \\ 0 & 0 & 0 & 0 \\ -1 & -2 & 5 & 1 \end{vmatrix} = 0$

5. $\begin{vmatrix} -2 & 3 & 0 \\ 3 & 4 & 2 \\ -3 & 2 & 5 \end{vmatrix} = -\begin{vmatrix} -2 & 3 & 0 \\ -3 & 2 & 5 \\ 3 & 4 & 2 \end{vmatrix}$

6. $\begin{vmatrix} 1 & 2 & -3 \\ 0 & -3 & -1 \\ 4 & 0 & -2 \end{vmatrix} = -\begin{vmatrix} 1 & 2 & -3 \\ 0 & 3 & 1 \\ 4 & 0 & -2 \end{vmatrix}$

7. $\begin{vmatrix} 2 & -2 & 3 & 0 \\ 0 & -3 & 5 & -4 \\ 4 & 0 & -5 & -1 \\ -2 & -1 & 1 & 5 \end{vmatrix} = -\begin{vmatrix} 2 & 2 & 3 & 0 \\ 0 & 3 & 5 & -4 \\ 4 & 0 & -5 & -1 \\ -2 & 1 & 1 & 5 \end{vmatrix}$

8. $\begin{vmatrix} 3 & -2 & 1 & -2 \\ -4 & 1 & 6 & -1 \\ 2 & 3 & 0 & 5 \\ 0 & 1 & -2 & 2 \end{vmatrix} = -\begin{vmatrix} 3 & -2 & 1 & -2 \\ -4 & -1 & 6 & 1 \\ 2 & 5 & 0 & 3 \\ 0 & 2 & -2 & 1 \end{vmatrix}$

9. $\begin{vmatrix} 1 & 4 & 3 \\ -4 & 0 & -1 \\ 0 & -2 & -3 \end{vmatrix} = 2 \begin{vmatrix} 1 & 2 & 3 \\ -4 & 0 & -1 \\ 0 & -1 & -3 \end{vmatrix}$

10. $\begin{vmatrix} 5 & 3 \\ -4 & 1 \end{vmatrix} = -\begin{vmatrix} 5 & 3 \\ 4 & -1 \end{vmatrix}$

11. $\begin{vmatrix} 1 & 2 \\ 3 & -2 \end{vmatrix} = \begin{vmatrix} 3 & 2 \\ 1 & -2 \end{vmatrix}$

12. $\begin{vmatrix} -1 & 0 & -3 & 2 \\ 2 & 1 & 0 & 5 \\ 0 & 3 & -6 & 1 \\ -4 & -2 & 0 & 3 \end{vmatrix} = -3 \begin{vmatrix} -1 & 0 & 1 & 2 \\ 2 & 1 & 0 & 5 \\ 0 & 3 & 2 & 1 \\ -4 & -2 & 0 & 3 \end{vmatrix}$

13. $\begin{vmatrix} 1 & 0 & -4 \\ 5 & -1 & 0 \\ 1 & 3 & 0 \end{vmatrix} = \begin{vmatrix} 1 & 0 & -4 \\ 5 & -1 & 0 \\ 0 & 3 & 4 \end{vmatrix}$

14. $\begin{vmatrix} 2 & 5 & 4 \\ 1 & 2 & 0 \\ -3 & 1 & -2 \end{vmatrix} = \begin{vmatrix} 2 & 1 & 4 \\ 1 & 0 & 0 \\ -3 & 7 & -2 \end{vmatrix}$

In Exercises 15 through 18, evaluate the given determinant by first obtaining two zeros in a row or column.

15. $\begin{vmatrix} -1 & 6 & 4 \\ -1 & 2 & 1 \\ -2 & 7 & 4 \end{vmatrix}$

16. $\begin{vmatrix} 1 & -1 & 5 \\ 2 & 4 & 3 \\ 4 & 1 & 7 \end{vmatrix}$

17. $\begin{vmatrix} 4 & -3 & 3 \\ 6 & -4 & 5 \\ -3 & 3 & -2 \end{vmatrix}$

18. $\begin{vmatrix} -5 & 6 & 9 \\ 3 & 0 & 2 \\ 1 & 4 & 6 \end{vmatrix}$

In Exercises 19 through 22, evaluate the given determinant by first obtaining three zeros in a row or column.

19. $\begin{vmatrix} 2 & 2 & -2 & -2 \\ 1 & -3 & 1 & 0 \\ 2 & 3 & -1 & 1 \\ 2 & -6 & 3 & -2 \end{vmatrix}$

20. $\begin{vmatrix} 3 & -1 & 2 & -3 \\ 0 & 1 & -1 & 3 \\ -1 & 2 & 1 & -4 \\ -2 & 3 & 4 & 3 \end{vmatrix}$

21. $\begin{vmatrix} 2 & 1 & -1 & 5 \\ 4 & 5 & 2 & 4 \\ 5 & 8 & 2 & 6 \\ -2 & 4 & -2 & 1 \end{vmatrix}$

22. $\begin{vmatrix} -4 & 1 & -5 & -5 \\ 1 & -2 & -4 & -3 \\ -3 & -3 & 3 & -4 \\ -5 & 2 & -7 & 8 \end{vmatrix}$

In Exercises 23 and 24, evaluate the given determinant by first obtaining four zeros in a row or column.

23. $\begin{vmatrix} -1 & 0 & -3 & 0 & 2 \\ 2 & -2 & 0 & 1 & -1 \\ 1 & 0 & 0 & 0 & 3 \\ 0 & 2 & 1 & -1 & 4 \\ -1 & -1 & 0 & 2 & 0 \end{vmatrix}$

24. $\begin{vmatrix} 1 & 4 & 4 & -2 & 1 \\ 1 & 1 & 2 & 1 & 1 \\ -1 & -1 & -3 & 1 & -1 \\ -2 & -4 & -6 & 4 & -1 \\ 1 & 1 & 3 & 2 & 0 \end{vmatrix}$

25. Prove that an equation of the line through the two points $P_1(x_1, y_1)$ and $P_2(x_2, y_2)$ is
$$\begin{vmatrix} x & y & 1 \\ x_1 & y_1 & 1 \\ x_2 & y_2 & 1 \end{vmatrix} = 0$$

In Exercises 26 and 27, use the result of Exercise 25 to find an equation of the line through the two given points.

26. $(4, -1)$ and $(-3, 2)$

27. $(-2, 5)$ and $(3, -4)$

28. Show that $\begin{vmatrix} a & b & 1 \\ a^2 & b^2 & 1 \\ a^3 & b^3 & 1 \end{vmatrix} = ab(b-a)(a-1)(b-1)$.

29. Show that $\begin{vmatrix} a^2 & a & 1 \\ b^2 & b & 1 \\ c^2 & c & 1 \end{vmatrix} = (a-b)(a-c)(b-c)$.

## 7.5 The Inverse of a Square Matrix

In the discussion in this section we need the concept of "transpose of a matrix," which we now define.

**7.5.1 DEFINITION** The *transpose of a matrix A*, denoted by $A^t$, is the matrix for which the $i$th row of $A^t$ is the $i$th column of $A$ and the $j$th column of $A^t$ is the $j$th row of $A$; that is, corresponding rows and columns are interchanged.

**ILLUSTRATION 1.** If

$$A = \begin{bmatrix} 3 & 1 \\ 4 & -2 \\ 0 & 5 \end{bmatrix} \quad \text{and} \quad B = \begin{bmatrix} -1 & 5 \\ -6 & 2 \end{bmatrix}$$

then

$$A^t = \begin{bmatrix} 3 & 4 & 0 \\ 1 & -2 & 5 \end{bmatrix} \quad \text{and} \quad B^t = \begin{bmatrix} -1 & -6 \\ 5 & 2 \end{bmatrix} \quad \bullet$$

In Section 1.2 we had Axiom 1.2.5, which states the existence of the number 1 such that for any real number $a$,

$$a \cdot 1 = a$$

The number 1 is the multiplicative identity for multiplication of real numbers. Does the set of matrices of a given order have a multiplicative identity? The answer, in general, is no. However, there is a multiplicative identity for the set of all square matrices of a given order. The definition uses the terminology *main diagonal,* which is the diagonal of elements from the upper left corner of the matrix to the lower right corner.

**7.5.2 DEFINITION** The *multiplicative identity* for the set of square matrices of order $n$, denoted by $I$, is the square matrix of order $n$ whose elements on the main diagonal are ones and whose other elements are all zeros.

**ILLUSTRATION 2.** The identity matrix for multiplication for the set of square matrices of order three is the matrix $I$, defined by

$$I = \begin{bmatrix} 1 & 0 & 0 \\ 0 & 1 & 0 \\ 0 & 0 & 1 \end{bmatrix}$$

If $A$ is any square matrix of order three; that is

$$A = \begin{bmatrix} a_{11} & a_{12} & a_{13} \\ a_{21} & a_{22} & a_{23} \\ a_{31} & a_{32} & a_{33} \end{bmatrix}$$

then

$$AI = \begin{bmatrix} a_{11} & a_{12} & a_{13} \\ a_{21} & a_{22} & a_{23} \\ a_{31} & a_{32} & a_{33} \end{bmatrix} \begin{bmatrix} 1 & 0 & 0 \\ 0 & 1 & 0 \\ 0 & 0 & 1 \end{bmatrix} \quad \text{and} \quad IA = \begin{bmatrix} 1 & 0 & 0 \\ 0 & 1 & 0 \\ 0 & 0 & 1 \end{bmatrix} \begin{bmatrix} a_{11} & a_{12} & a_{13} \\ a_{21} & a_{22} & a_{23} \\ a_{31} & a_{32} & a_{33} \end{bmatrix}$$

$$= \begin{bmatrix} a_{11} & a_{12} & a_{13} \\ a_{21} & a_{22} & a_{23} \\ a_{31} & a_{32} & a_{33} \end{bmatrix} \qquad\qquad\qquad = \begin{bmatrix} a_{11} & a_{12} & a_{13} \\ a_{21} & a_{22} & a_{23} \\ a_{31} & a_{32} & a_{33} \end{bmatrix}$$

$$= A \qquad\qquad\qquad\qquad\qquad\qquad = A \qquad\qquad \bullet$$

**7.5.3 THEOREM**    If $M$ is a square matrix of order $n$ and if $I$ is the multiplicative identity of order $n$, then

$$MI = M \quad \text{and} \quad IM = M$$

The proof of this theorem for square matrices of order three appears in Illustration 2. A similar proof can be given for square matrices of any order.

We know that for every real number $a$, except $0$, there exists a real number called the multiplicative inverse of $a$, denoted by $a^{-1}$, such that

$$a \cdot a^{-1} = 1 \quad \text{and} \quad a^{-1} \cdot a = 1$$

Does every square matrix $A$ have a multiplicative inverse $A^{-1}$ such that

$$AA^{-1} = I \quad \text{and} \quad A^{-1}A = I?$$

Before answering this question we give an illustration that shows there is a multiplicative inverse for a particular square matrix of order two.

**ILLUSTRATION 3.** Suppose that

$$A = \begin{bmatrix} 5 & -2 \\ -2 & 1 \end{bmatrix} \quad \text{and} \quad B = \begin{bmatrix} 1 & 2 \\ 2 & 5 \end{bmatrix}$$

then

$$AB = \begin{bmatrix} (5)(1) + (-2)(2) & (5)(2) + (-2)(5) \\ (-2)(1) + (1)(2) & (-2)(2) + (1)(5) \end{bmatrix}$$

$$= \begin{bmatrix} 1 & 0 \\ 0 & 1 \end{bmatrix}$$

$$= I$$

and

$$BA = \begin{bmatrix} (1)(5) + (2)(-2) & (1)(-2) + (2)(1) \\ (2)(5) + (5)(-2) & (2)(-2) + (5)(1) \end{bmatrix}$$

$$= \begin{bmatrix} 1 & 0 \\ 0 & 1 \end{bmatrix}$$

$$= I$$

Because $AB = I$ and $BA = I$, $B$ is the multiplicative inverse of $A$ and $A$ is the multiplicative inverse of $B$. •

The answer to the question posed before Illustration 3 is yes, provided that $\det A \neq 0$. The next theorem states this fact and gives a method for computing the multiplicative inverse of a square matrix when it exists.

**7.5.4 THEOREM**  Suppose that $A$ is the square matrix of order $n$

$$A = \begin{bmatrix} a_{11} & a_{12} & a_{13} & \cdots & a_{1n} \\ a_{21} & a_{22} & a_{23} & \cdots & a_{2n} \\ a_{31} & a_{32} & a_{33} & \cdots & a_{3n} \\ \vdots & \vdots & \vdots & & \vdots \\ a_{n1} & a_{n2} & a_{n3} & \cdots & a_{nn} \end{bmatrix}$$

and that $A_{ij}$ is the cofactor of the element $a_{ij}$. Then if $\det A \neq 0$,

$$A^{-1} = \frac{1}{\det A} \begin{bmatrix} A_{11} & A_{12} & A_{13} & \cdots & A_{1n} \\ A_{21} & A_{22} & A_{23} & \cdots & A_{2n} \\ A_{31} & A_{32} & A_{33} & \cdots & A_{3n} \\ \vdots & \vdots & \vdots & & \vdots \\ A_{n1} & A_{n2} & A_{n3} & \cdots & A_{nn} \end{bmatrix}^t$$

with

$$AA^{-1} = I \quad \text{and} \quad A^{-1}A = I$$

The proof of Theorem 7.5.4 for the general case is beyond the scope of this book. However, following is the proof that $AA^{-1} = I$ for square matrices of order two. If

$$A = \begin{bmatrix} a_{11} & a_{12} \\ a_{21} & a_{22} \end{bmatrix} \quad \text{then} \quad A^{-1} = \frac{1}{\det A} \begin{bmatrix} A_{11} & A_{12} \\ A_{21} & A_{22} \end{bmatrix}^t$$

Because $A_{ij}$ is the cofactor of $a_{ij}$, we have

$$A_{11} = a_{22} \quad A_{12} = -a_{21} \quad A_{21} = -a_{12} \quad A_{22} = a_{11}$$

Substituting these values into the matrix for $A^{-1}$, we have

$$A^{-1} = \frac{1}{\det A} \begin{bmatrix} a_{22} & -a_{21} \\ -a_{12} & a_{11} \end{bmatrix}^t$$

$$= \frac{1}{\det A} \begin{bmatrix} a_{22} & -a_{12} \\ -a_{21} & a_{11} \end{bmatrix}$$

418 Matrices and Determinants [Ch. 7

Let $A = \begin{bmatrix} 5 & 2 \\ 3 & -1 \end{bmatrix}$. Then equation (1) can be written as

$$A \begin{bmatrix} x \\ y \end{bmatrix} = \begin{bmatrix} -1 \\ 6 \end{bmatrix}$$

We multiply on the left both sides of this equation by $A^{-1}$, because $\det A \neq 0$, and we have

$$A^{-1}\left(A \begin{bmatrix} x \\ y \end{bmatrix}\right) = A^{-1} \begin{bmatrix} -1 \\ 6 \end{bmatrix}$$

$$(A^{-1}A) \begin{bmatrix} x \\ y \end{bmatrix} = A^{-1} \begin{bmatrix} -1 \\ 6 \end{bmatrix}$$

$$I \begin{bmatrix} x \\ y \end{bmatrix} = A^{-1} \begin{bmatrix} -1 \\ 6 \end{bmatrix}$$

$$\begin{bmatrix} x \\ y \end{bmatrix} = A^{-1} \begin{bmatrix} -1 \\ 6 \end{bmatrix} \qquad (2)$$

We compute $A^{-1}$.

$$A^{-1} = \frac{1}{\det A} \begin{bmatrix} A_{11} & A_{12} \\ A_{21} & A_{22} \end{bmatrix}^t$$

$$= \frac{1}{-11} \begin{bmatrix} -1 & -3 \\ -2 & 5 \end{bmatrix}^t$$

$$= -\frac{1}{11} \begin{bmatrix} -1 & -2 \\ -3 & 5 \end{bmatrix}$$

$$= \begin{bmatrix} \frac{1}{11} & \frac{2}{11} \\ \frac{3}{11} & -\frac{5}{11} \end{bmatrix}$$

We substitute the matrix $A^{-1}$ in equation (2) and we obtain

$$\begin{bmatrix} x \\ y \end{bmatrix} = \begin{bmatrix} \frac{1}{11} & \frac{2}{11} \\ \frac{3}{11} & -\frac{5}{11} \end{bmatrix} \begin{bmatrix} -1 \\ 6 \end{bmatrix}$$

$$= \begin{bmatrix} -\frac{1}{11} + \frac{12}{11} \\ -\frac{3}{11} - \frac{30}{11} \end{bmatrix}$$

$$= \begin{bmatrix} 1 \\ -3 \end{bmatrix}$$

Therefore, $x = 1$ and $y = -3$.

[7.5]   The Inverse of a Square Matrix   417

Thus

$$AA^{-1} = \begin{bmatrix} 1 & 2 & 1 \\ 1 & 3 & 0 \\ 4 & 0 & 2 \end{bmatrix} \begin{bmatrix} -\frac{3}{5} & \frac{2}{5} & \frac{3}{10} \\ \frac{1}{5} & \frac{1}{5} & -\frac{1}{10} \\ \frac{6}{5} & -\frac{4}{5} & -\frac{1}{10} \end{bmatrix}$$

$$= \begin{bmatrix} -\frac{3}{5} + \frac{2}{5} + \frac{6}{5} & \frac{2}{5} + \frac{2}{5} - \frac{4}{5} & \frac{3}{10} - \frac{2}{10} - \frac{1}{10} \\ -\frac{3}{5} + \frac{3}{5} + 0 & \frac{2}{5} + \frac{3}{5} + 0 & \frac{3}{10} - \frac{3}{10} + 0 \\ -\frac{12}{5} + 0 + \frac{12}{5} & \frac{8}{5} + 0 - \frac{8}{5} & \frac{12}{10} + 0 - \frac{2}{10} \end{bmatrix}$$

$$= \begin{bmatrix} 1 & 0 & 0 \\ 0 & 1 & 0 \\ 0 & 0 & 1 \end{bmatrix}$$

$$= I$$

It is apparent that computation of the inverse of a matrix having an order greater than three is rather cumbersome. There are other systematic and more efficient methods for computing the inverse of a matrix but they involve concepts that we do not discuss in this book.

A very important application of the multiplicative inverse of a square matrix is its use in solving a system of linear equations. In the following illustration we show the method for a system of two linear equations.

**ILLUSTRATION 4.** Suppose that we have the system of equations

$$\begin{cases} 5x + 2y = -1 \\ 3x - y = 6 \end{cases}$$

This system can be written in matrix form as

$$\begin{bmatrix} 5 & 2 \\ 3 & -1 \end{bmatrix} \begin{bmatrix} x \\ y \end{bmatrix} = \begin{bmatrix} -1 \\ 6 \end{bmatrix} \quad (1)$$

We wish to show that $AA^{-1} = I$. We have

$$AA^{-1} = \begin{bmatrix} a_{11} & a_{12} \\ a_{21} & a_{22} \end{bmatrix} \begin{bmatrix} \dfrac{a_{22}}{\det A} & \dfrac{-a_{12}}{\det A} \\ \dfrac{-a_{21}}{\det A} & \dfrac{a_{11}}{\det A} \end{bmatrix}$$

$$= \begin{bmatrix} \dfrac{a_{11}a_{22} - a_{12}a_{21}}{\det A} & \dfrac{-a_{11}a_{12} + a_{12}a_{11}}{\det A} \\ \dfrac{a_{21}a_{22} - a_{22}a_{21}}{\det A} & \dfrac{-a_{21}a_{12} + a_{22}a_{11}}{\det A} \end{bmatrix}$$

$$= \begin{bmatrix} \dfrac{\det A}{\det A} & \dfrac{0}{\det A} \\ \dfrac{0}{\det A} & \dfrac{\det A}{\det A} \end{bmatrix}$$

$$= \begin{bmatrix} 1 & 0 \\ 0 & 1 \end{bmatrix}$$

$$= I$$

In a similar way we can show that $A^{-1}A = I$ (see Exercise 18). Observe that because $AA^{-1} = I$ and $A^{-1}A = I$, the commutative property for multiplication of matrices holds for matrices $A$ and $A^{-1}$ even though it does not hold in general for square matrices of the same order.

The existence of $A^{-1}$ depends upon $\det A \neq 0$. If a matrix $A$ has an inverse, $A$ is said to be *nonsingular*. If a matrix $A$ does not have an inverse (that is, if $\det A = 0$), then $A$ is said to be *singular*.

**EXAMPLE 1**
If
$$A = \begin{bmatrix} 2 & -4 \\ 3 & -5 \end{bmatrix}$$
find $A^{-1}$ and show that
$$AA^{-1} = I$$
and
$$A^{-1}A = I$$

**SOLUTION**
$$\det A = 2(-5) - 3(-4)$$
$$= 2$$
Because $\det A \neq 0$, $A^{-1}$ exists and
$$A^{-1} = \frac{1}{\det A} \begin{bmatrix} A_{11} & A_{12} \\ A_{21} & A_{22} \end{bmatrix}^t$$
$$= \frac{1}{2} \begin{bmatrix} -5 & -3 \\ 4 & 2 \end{bmatrix}^t$$
$$= \frac{1}{2} \begin{bmatrix} -5 & 4 \\ -3 & 2 \end{bmatrix}$$
$$= \begin{bmatrix} -\dfrac{5}{2} & 2 \\ -\dfrac{3}{2} & 1 \end{bmatrix}$$

Therefore,

$$AA^{-1} = \begin{bmatrix} 2 & -4 \\ 3 & -5 \end{bmatrix} \begin{bmatrix} -\frac{5}{2} & 2 \\ -\frac{3}{2} & 1 \end{bmatrix} \qquad A^{-1}A = \begin{bmatrix} -\frac{5}{2} & 2 \\ -\frac{3}{2} & 1 \end{bmatrix} \begin{bmatrix} 2 & -4 \\ 3 & -5 \end{bmatrix}$$

$$= \begin{bmatrix} -5+6 & 4-4 \\ -\frac{15}{2}+\frac{15}{2} & 6-5 \end{bmatrix} \qquad\qquad = \begin{bmatrix} -5+6 & 10-10 \\ -3+3 & 6-5 \end{bmatrix}$$

$$= \begin{bmatrix} 1 & 0 \\ 0 & 1 \end{bmatrix} \qquad\qquad\qquad\qquad = \begin{bmatrix} 1 & 0 \\ 0 & 1 \end{bmatrix}$$

$$= I \qquad\qquad\qquad\qquad\qquad = I$$

## EXAMPLE 2

If

$$A = \begin{bmatrix} 1 & 2 & 1 \\ 1 & 3 & 0 \\ 4 & 0 & 2 \end{bmatrix}$$

find $A^{-1}$ and show that $AA^{-1} = I$.

**SOLUTION**

We first compute det $A$. We expand by elements of the third column, and we have

$$\det A = 1 \begin{vmatrix} 1 & 3 \\ 4 & 0 \end{vmatrix} + 2 \begin{vmatrix} 1 & 2 \\ 1 & 3 \end{vmatrix}$$

$$= -12 + 2(1)$$

$$= -10$$

Because det $A \neq 0$, $A^{-1}$ exists. We now compute the cofactors of the elements of $A$.

$$A_{11} = \begin{vmatrix} 3 & 0 \\ 0 & 2 \end{vmatrix} \qquad A_{12} = (-1)\begin{vmatrix} 1 & 0 \\ 4 & 2 \end{vmatrix} \qquad A_{13} = \begin{vmatrix} 1 & 3 \\ 4 & 0 \end{vmatrix}$$

$$= 6 \qquad\qquad\qquad = -2 \qquad\qquad\qquad = -12$$

$$A_{21} = (-1)\begin{vmatrix} 2 & 1 \\ 0 & 2 \end{vmatrix} \qquad A_{22} = \begin{vmatrix} 1 & 1 \\ 4 & 2 \end{vmatrix} \qquad A_{23} = (-1)\begin{vmatrix} 1 & 2 \\ 4 & 0 \end{vmatrix}$$

$$= -4 \qquad\qquad\qquad = -2 \qquad\qquad\qquad = 8$$

$$A_{31} = \begin{vmatrix} 2 & 1 \\ 3 & 0 \end{vmatrix} \qquad A_{32} = (-1)\begin{vmatrix} 1 & 1 \\ 1 & 0 \end{vmatrix} \qquad A_{33} = \begin{vmatrix} 1 & 2 \\ 1 & 3 \end{vmatrix}$$

$$= -3 \qquad\qquad\qquad = 1 \qquad\qquad\qquad = 1$$

Therefore,

$$A^{-1} = -\frac{1}{10}\begin{bmatrix} 6 & -2 & -12 \\ -4 & -2 & 8 \\ -3 & 1 & 1 \end{bmatrix}^t$$

$$= -\frac{1}{10}\begin{bmatrix} 6 & -4 & -3 \\ -2 & -2 & 1 \\ -12 & 8 & 1 \end{bmatrix}$$

## EXAMPLE 3

Use the matrix method of Illustration 4 to solve the following system of equations.

$$\begin{cases} x + 2y + z = 2 \\ x + 3y = 1 \\ 4x + 2z = -4 \end{cases}$$

## SOLUTION

We write the given system in matrix form.

$$\begin{bmatrix} 1 & 2 & 1 \\ 1 & 3 & 0 \\ 4 & 0 & 2 \end{bmatrix} \begin{bmatrix} x \\ y \\ z \end{bmatrix} = \begin{bmatrix} 2 \\ 1 \\ -4 \end{bmatrix} \quad (3)$$

Let

$$A = \begin{bmatrix} 1 & 2 & 1 \\ 1 & 3 & 0 \\ 4 & 0 & 2 \end{bmatrix}$$

Then equation (3) can be written as

$$A \begin{bmatrix} x \\ y \\ z \end{bmatrix} = \begin{bmatrix} 2 \\ 1 \\ -4 \end{bmatrix} \quad (4)$$

Because $A$ is the matrix of Example 2, $A^{-1}$ exists. Therefore, we can multiply on the left, both sides of equation (4) by $A^{-1}$, and we have

$$\begin{bmatrix} x \\ y \\ z \end{bmatrix} = A^{-1} \begin{bmatrix} 2 \\ 1 \\ -4 \end{bmatrix}$$

We use the result of Example 2 for $A^{-1}$ and we have

$$\begin{bmatrix} x \\ y \\ z \end{bmatrix} = \begin{bmatrix} -\frac{3}{5} & \frac{2}{5} & \frac{3}{10} \\ \frac{1}{5} & \frac{1}{5} & -\frac{1}{10} \\ \frac{6}{5} & -\frac{4}{5} & -\frac{1}{10} \end{bmatrix} \begin{bmatrix} 2 \\ 1 \\ -4 \end{bmatrix}$$

$$= \begin{bmatrix} -\frac{6}{5} + \frac{2}{5} - \frac{6}{5} \\ \frac{2}{5} + \frac{1}{5} + \frac{2}{5} \\ \frac{12}{5} - \frac{4}{5} + \frac{2}{5} \end{bmatrix}$$

$$= \begin{bmatrix} -2 \\ 1 \\ 2 \end{bmatrix}$$

Therefore, $x = -2$, $y = 1$, and $z = 2$.

## EXERCISES 7.5

*In Exercises 1 through 4, find the transpose of the given matrix.*

1. $\begin{bmatrix} 2 & -4 & 1 \\ -3 & 5 & 0 \end{bmatrix}$
2. $\begin{bmatrix} 9 & -3 \\ -7 & 2 \end{bmatrix}$
3. $\begin{bmatrix} -5 & 0 & 2 \\ 3 & -6 & 8 \\ 0 & -1 & 4 \end{bmatrix}$
4. $\begin{bmatrix} 4 & -3 & 6 & 1 \\ 0 & -2 & -5 & -1 \\ 7 & 0 & 3 & 4 \\ 2 & -6 & 0 & 5 \end{bmatrix}$

*In Exercises 5 through 14, the given matrix is A. Find $A^{-1}$, if it exists, and show that $AA^{-1} = I$. If A is singular (that is, $A^{-1}$ does not exist), then state this fact.*

5. $\begin{bmatrix} 5 & 2 \\ 2 & 1 \end{bmatrix}$
6. $\begin{bmatrix} 2 & -1 \\ -5 & 3 \end{bmatrix}$
7. $\begin{bmatrix} -2 & 3 \\ 4 & -6 \end{bmatrix}$
8. $\begin{bmatrix} 7 & -1 \\ -5 & 1 \end{bmatrix}$
9. $\begin{bmatrix} -4 & 1 \\ -5 & 2 \end{bmatrix}$
10. $\begin{bmatrix} 4 & -1 & 2 \\ -2 & 0 & -2 \\ 3 & 1 & 5 \end{bmatrix}$
11. $\begin{bmatrix} 3 & 3 & 1 \\ 1 & 4 & 1 \\ 2 & 3 & 1 \end{bmatrix}$
12. $\begin{bmatrix} 0 & 1 & 2 \\ -3 & 4 & 0 \\ 0 & -2 & -1 \end{bmatrix}$
13. $\begin{bmatrix} 1 & -2 & 1 \\ 2 & 2 & -1 \\ 1 & 1 & 0 \end{bmatrix}$
14. $\begin{bmatrix} 1 & 0 & 0 \\ 2 & 1 & 0 \\ 4 & 0 & 1 \end{bmatrix}$

*In Exercises 15 and 16, find $A^{-1}$ for the given matrix A without using Theorem 7.5.4. (Hint: Let $A^{-1} = \begin{bmatrix} a & b \\ c & d \end{bmatrix}$, and from the fact $AA^{-1} = I$, obtain four equations in a, b, c, and d.)*

15. $\begin{bmatrix} 3 & -4 \\ -2 & 3 \end{bmatrix}$
16. $\begin{bmatrix} -3 & 1 \\ 2 & -2 \end{bmatrix}$

17. If $I$ is the multiplicative identity for the set of square matrices of order two, prove that $II^{-1} = I$.

18. Prove that if $A$ is a square matrix of order two, then $A^{-1}A = I$.

*In Exercises 19 through 28, let*

$$A = \begin{bmatrix} 3 & 1 \\ 7 & 3 \end{bmatrix} \quad \text{and} \quad B = \begin{bmatrix} -2 & 1 \\ -3 & 2 \end{bmatrix}$$

19. Show that $(AB)^{-1} = B^{-1}A^{-1}$.
20. Show that $(BA)^{-1} = A^{-1}B^{-1}$.
21. Show that $(A^{-1})^{-1} = A$.
22. Show that $(B^{-1})^{-1} = B$.
23. Show that $(A^{-1})^t = (A^t)^{-1}$.
24. Show that $(B^{-1})^t = (B^t)^{-1}$.
25. Show that $(A^{-1})^2 = (A^2)^{-1}$.
26. Show that $(B^{-1})^2 = (B^2)^{-1}$.
27. Show that $\det(A^{-1}) = \dfrac{1}{\det A}$.
28. Show that $\det(B^{-1}) = \dfrac{1}{\det B}$.

In Exercises 29 through 36, use the matrix method of Illustration 4 and Example 3 to solve the system of equations. In Exercises 29 and 30, use the result that

$$\begin{bmatrix} 2 & -1 \\ -5 & 3 \end{bmatrix}^{-1} = \begin{bmatrix} 3 & 1 \\ 5 & 2 \end{bmatrix}$$

and in Exercises 33 and 34, use the result that

$$\begin{bmatrix} 1 & 0 & 2 \\ 0 & -1 & 1 \\ 1 & 3 & 0 \end{bmatrix}^{-1} = \begin{bmatrix} 3 & -6 & -2 \\ -1 & 2 & 1 \\ -1 & 3 & 1 \end{bmatrix}$$

29. $\begin{cases} 2x - y = -2 \\ -5x + 3y = 7 \end{cases}$

30. $\begin{cases} 2x - y = 1 \\ -5x + 3y = -5 \end{cases}$

31. $\begin{cases} 3x + 2y = -1 \\ 4x + 3y = 0 \end{cases}$

32. $\begin{cases} 3x + 2y = 3 \\ 6x - 6y = 1 \end{cases}$

33. $\begin{cases} x + 2z = 5 \\ -y + z = 3 \\ x + 3y = 0 \end{cases}$

34. $\begin{cases} x + 2z = 12 \\ -y + z = 7 \\ x + 3y = -4 \end{cases}$

35. $\begin{cases} 4x + 2y + 3z = 4 \\ 5x - y + 4z = 12 \\ x - 3y + 2z = 0 \end{cases}$

36. $\begin{cases} 2x + y + 3z = 5 \\ x + y + z = 0 \\ 4x + y - 2z = -15 \end{cases}$

## 7.6 Cramer's Rule

Consider the system of two linear equations in two variables

(I) $\qquad \begin{cases} a_{11}x + a_{12}y = k_1 \\ a_{21}x + a_{22}y = k_2 \end{cases}$

where in each equation at least one of the coefficients of the variables is nonzero. Without loss of generality, we assume in the first equation that the coefficient of $x$ is nonzero; because if the coefficient of $x$ is zero, then the coefficient of $y$ is nonzero and we can interchange the variables. To solve system (I), we use the matrix method to find an equivalent system in triangular form. The augmented matrix of system (I) is

$$\begin{bmatrix} a_{11} & a_{12} & k_1 \\ a_{21} & a_{22} & k_2 \end{bmatrix}$$

We obtain a matrix of an equivalent system by replacing the second row with the sum of $-a_{21}$ times the first row and $a_{11}$ times the second row (here we are applying Theorem 7.1.1(iii) in which it is necessary that $a_{11} \neq 0$). We have then

$$\begin{bmatrix} a_{11} & a_{12} & k_1 \\ 0 & (a_{11}a_{22} - a_{12}a_{21}) & (a_{11}k_2 - a_{21}k_1) \end{bmatrix}$$

which is the augmented matrix of the system

$$\begin{cases} a_{11}x + a_{12}y = k_1 \\ (a_{11}a_{22} - a_{12}a_{21})y = a_{11}k_2 - a_{21}k_1 \end{cases}$$

In the second equation of this system, the coefficient of $y$ and the right member can be written as determinants; that is, the system is

(II) $$\begin{cases} a_{11}x + a_{12}y = k_1 \\ \begin{vmatrix} a_{11} & a_{12} \\ a_{21} & a_{22} \end{vmatrix} y = \begin{vmatrix} a_{11} & k_1 \\ a_{21} & k_2 \end{vmatrix} \end{cases}$$

If

$$\begin{vmatrix} a_{11} & a_{12} \\ a_{21} & a_{22} \end{vmatrix} \neq 0$$

we can solve the second equation for $y$, and we have

$$y = \frac{\begin{vmatrix} a_{11} & k_1 \\ a_{21} & k_2 \end{vmatrix}}{\begin{vmatrix} a_{11} & a_{12} \\ a_{21} & a_{22} \end{vmatrix}} \quad \text{if} \quad \begin{vmatrix} a_{11} & a_{12} \\ a_{21} & a_{22} \end{vmatrix} \neq 0 \quad (1)$$

The value of $x$ can be found by substituting the value of $y$ from equation (1) into the first equation of system (II); or, alternatively, we can solve for $x$ by using the same procedure as we used to solve for $y$. We obtain

$$x = \frac{\begin{vmatrix} k_1 & a_{12} \\ k_2 & a_{22} \end{vmatrix}}{\begin{vmatrix} a_{11} & a_{12} \\ a_{21} & a_{22} \end{vmatrix}} \quad \text{if} \quad \begin{vmatrix} a_{11} & a_{12} \\ a_{21} & a_{22} \end{vmatrix} \neq 0 \quad (2)$$

Formulas (1) and (2) are known as *Cramer's rule* for the solution of a system of two linear equations in two variables. Note that in each of the denominators, we have the determinant of the coefficient matrix of the given system. In the numerator for the value of $x$, we have the determinant of the matrix obtained from the coefficient matrix by replacing the coefficients, $a_{11}$ and $a_{21}$, of $x$ by the numbers $k_1$ and $k_2$, respectively; and in the numerator for the value of $y$ we have the determinant of the matrix obtained from the coefficient matrix by replacing the coefficients, $a_{12}$ and $a_{22}$, of $y$ by $k_1$ and $k_2$, respectively.

EXAMPLE 1

Use Cramer's rule to find the solution set of the system

$$\begin{cases} 4x + 3y = 6 \\ 2x - 5y = 16 \end{cases}$$

SOLUTION

The determinant of the coefficient matrix is

$$\begin{vmatrix} 4 & 3 \\ 2 & -5 \end{vmatrix} = -20 - 6 = -26$$

From formulas (2) and (1), we have, respectively,

$$x = \frac{\begin{vmatrix} 6 & 3 \\ 16 & -5 \end{vmatrix}}{-26} \qquad y = \frac{\begin{vmatrix} 4 & 6 \\ 2 & 16 \end{vmatrix}}{-26}$$

$$= \frac{-30 - 48}{-26} \qquad\qquad = \frac{64 - 12}{-26}$$

$$= \frac{-78}{-26} \qquad\qquad\qquad = \frac{52}{-26}$$

$$= 3 \qquad\qquad\qquad\qquad = -2$$

Therefore, the solution set is $\{(3, -2)\}$.

When applying formulas (1) and (2) to solve system (I), it is necessary that the determinant of the coefficient matrix is not zero. If the determinant of the coefficient matrix is zero, then system (II) is

(III) $\qquad \begin{cases} a_{11}x + a_{12}y = k_1 \\ \phantom{a_{11}x + a_{12}y} 0 = \begin{vmatrix} a_{11} & k_1 \\ a_{21} & k_2 \end{vmatrix} \end{cases}$

If

$$\begin{vmatrix} a_{11} & k_1 \\ a_{21} & k_2 \end{vmatrix} \neq 0$$

then there is no ordered pair $(x, y)$ for which the second equation of system (III) is a true statement, and thus the equations of system (I) are inconsistent. If

$$\begin{vmatrix} a_{11} & k_1 \\ a_{21} & k_2 \end{vmatrix} = 0$$

then system (III) is

$$\begin{cases} a_{11}x + a_{12}y = k_1 \\ \phantom{a_{11}x + a_{12}y} 0 = 0 \end{cases}$$

and hence the equations of system (I) are dependent.

The results of the preceding discussion are summarized in the following theorem.

**7.6.1 THEOREM** Suppose that we have the system of two linear equations in two variables

$$\begin{cases} a_{11}x + a_{12}y = k_1 \\ a_{21}x + a_{22}y = k_2 \end{cases}$$

where either $a_{11} \neq 0$ or $a_{12} \neq 0$, and either $a_{21} \neq 0$ or $a_{22} \neq 0$.

(i) If
$$\begin{vmatrix} a_{11} & a_{12} \\ a_{21} & a_{22} \end{vmatrix} \neq 0$$
then the system has a unique solution given by

$$x = \frac{\begin{vmatrix} k_1 & a_{12} \\ k_2 & a_{22} \end{vmatrix}}{\begin{vmatrix} a_{11} & a_{12} \\ a_{21} & a_{22} \end{vmatrix}} \quad \text{and} \quad y = \frac{\begin{vmatrix} a_{11} & k_1 \\ a_{21} & k_2 \end{vmatrix}}{\begin{vmatrix} a_{11} & a_{12} \\ a_{21} & a_{22} \end{vmatrix}} \tag{3}$$

(ii) If
$$\begin{vmatrix} a_{11} & a_{12} \\ a_{21} & a_{22} \end{vmatrix} = 0 \quad \text{and} \quad \begin{vmatrix} a_{11} & k_1 \\ a_{21} & k_2 \end{vmatrix} \neq 0$$
then the equations of the system are inconsistent.

(iii) If
$$\begin{vmatrix} a_{11} & a_{12} \\ a_{21} & a_{22} \end{vmatrix} = 0 \quad \text{and} \quad \begin{vmatrix} a_{11} & k_1 \\ a_{21} & k_2 \end{vmatrix} = 0$$
then the equations of the system are dependent.

In parts (ii) and (iii), the determinant
$$\begin{vmatrix} a_{11} & k_1 \\ a_{21} & k_2 \end{vmatrix}$$
can be replaced by the determinant
$$\begin{vmatrix} k_1 & a_{12} \\ k_2 & a_{22} \end{vmatrix}$$

**ILLUSTRATION 1.** If we attempt to use Cramer's rule to solve the system
$$\begin{cases} 6x - 3y = 5 \\ 2x - y = 4 \end{cases}$$
we first find the determinant of the coefficient matrix. It is
$$\begin{vmatrix} 6 & -3 \\ 2 & -1 \end{vmatrix} = -6 + 6$$
$$= 0$$

Therefore, the system does not have a unique solution. Because
$$\begin{vmatrix} 6 & 5 \\ 2 & 4 \end{vmatrix} = 24 - 10$$
$$= 14$$
$$\neq 0$$

it follows from Theorem 7.6.1 (ii) that the equations of the given system are inconsistent.

Note that the given system is the same as the one in Illustrations 2 and 6 of Section 6.1. •

**ILLUSTRATION 2.** Consider the system
$$\begin{cases} 3x + 2y = 4 \\ 6x + 4y = 8 \end{cases}$$

The determinant of the coefficient matrix is
$$\begin{vmatrix} 3 & 2 \\ 6 & 4 \end{vmatrix} = 12 - 12 \\ = 0$$

Furthermore,
$$\begin{vmatrix} 3 & 4 \\ 6 & 8 \end{vmatrix} = 24 - 24 \\ = 0$$

Therefore, from Theorem 7.6.1 (iii) we conclude that the equations of the given system are dependent.

The given system is the same as the one in Illustrations 3 and 7 of Section 6.1. •

The following theorem is a generalization of Cramer's rule to a system of $n$ linear equations in $n$ variables.

**7.6.2 THEOREM**  Suppose that we have the system
$$\begin{cases} a_{11}x_1 + a_{12}x_2 + \cdots + a_{1n}x_n = k_1 \\ a_{21}x_1 + a_{22}x_2 + \cdots + a_{2n}x_n = k_2 \\ \vdots \qquad \vdots \qquad \vdots \qquad \vdots \\ a_{n1}x_1 + a_{n2}x_2 + \cdots + a_{nn}x_n = k_n \end{cases}$$

with
$$A = \begin{bmatrix} a_{11} & a_{12} & \cdots & a_{1n} \\ a_{21} & a_{22} & \cdots & a_{2n} \\ \vdots & \vdots & \vdots & \vdots \\ a_{n1} & a_{n2} & \cdots & a_{nn} \end{bmatrix} \quad \text{and} \quad \det A \neq 0$$

then
$$x_i = \frac{\begin{vmatrix} a_{11} & a_{12} & \cdots & k_1 & \cdots & a_{1n} \\ a_{21} & a_{22} & \cdots & k_2 & \cdots & a_{2n} \\ \vdots & \vdots & \vdots & \vdots & \vdots & \vdots \\ a_{n1} & a_{n2} & \cdots & k_n & \cdots & a_{nn} \end{vmatrix}}{\det A}$$

where the $i$th column is the one containing $k_1, k_2, \ldots, k_n$.

The proof of Theorem 7.6.2 for the general case is omitted. However, we use the matrix method of Section 7.5 to prove it for $n = 3$. We have then a system of three linear equations in three variables $x_1$, $x_2$, and $x_3$, which can be written as

$$A \begin{bmatrix} x_1 \\ x_2 \\ x_3 \end{bmatrix} = \begin{bmatrix} k_1 \\ k_2 \\ k_3 \end{bmatrix} \qquad (4)$$

where

$$A = \begin{bmatrix} a_{11} & a_{12} & a_{13} \\ a_{21} & a_{22} & a_{23} \\ a_{31} & a_{32} & a_{33} \end{bmatrix}$$

We multiply on both sides of equation (4) by $A^{-1}$ and obtain

$$\begin{bmatrix} x_1 \\ x_2 \\ x_3 \end{bmatrix} = A^{-1} \begin{bmatrix} k_1 \\ k_2 \\ k_3 \end{bmatrix}$$

Replacing $A^{-1}$ by its equivalent from Theorem 7.5.4, we have

$$\begin{bmatrix} x_1 \\ x_2 \\ x_3 \end{bmatrix} = \frac{1}{\det A} \begin{bmatrix} A_{11} & A_{12} & A_{13} \\ A_{21} & A_{22} & A_{23} \\ A_{31} & A_{32} & A_{33} \end{bmatrix}^t \begin{bmatrix} k_1 \\ k_2 \\ k_3 \end{bmatrix}$$

$$= \frac{1}{\det A} \begin{bmatrix} A_{11} & A_{21} & A_{31} \\ A_{12} & A_{22} & A_{32} \\ A_{13} & A_{23} & A_{33} \end{bmatrix} \begin{bmatrix} k_1 \\ k_2 \\ k_3 \end{bmatrix}$$

$$= \begin{bmatrix} \dfrac{k_1 A_{11} + k_2 A_{21} + k_3 A_{31}}{\det A} \\ \dfrac{k_1 A_{12} + k_2 A_{22} + k_3 A_{32}}{\det A} \\ \dfrac{k_1 A_{13} + k_2 A_{23} + k_3 A_{33}}{\det A} \end{bmatrix} \qquad (5)$$

In matrix (5) the elements are the evaluations of $\det A$ by the cofactors of the elements of the first, second, and third columns, respectively; except that the elements in the columns involved in each evaluation are replaced by $k_1$, $k_2$, and $k_3$. Therefore,

$$x_1 = \frac{\begin{vmatrix} k_1 & a_{12} & a_{13} \\ k_2 & a_{22} & a_{23} \\ k_3 & a_{32} & a_{33} \end{vmatrix}}{\begin{vmatrix} a_{11} & a_{12} & a_{13} \\ a_{21} & a_{22} & a_{23} \\ a_{31} & a_{32} & a_{33} \end{vmatrix}}, \quad x_2 = \frac{\begin{vmatrix} a_{11} & k_1 & a_{13} \\ a_{21} & k_2 & a_{23} \\ a_{31} & k_3 & a_{33} \end{vmatrix}}{\begin{vmatrix} a_{11} & a_{12} & a_{13} \\ a_{21} & a_{22} & a_{23} \\ a_{31} & a_{32} & a_{33} \end{vmatrix}}, \quad x_3 = \frac{\begin{vmatrix} a_{11} & a_{12} & k_1 \\ a_{21} & a_{22} & k_2 \\ a_{31} & a_{32} & k_3 \end{vmatrix}}{\begin{vmatrix} a_{11} & a_{12} & a_{13} \\ a_{21} & a_{22} & a_{23} \\ a_{31} & a_{32} & a_{33} \end{vmatrix}} \qquad (6)$$

## EXAMPLE 2

Use Cramer's rule to find the solution set of the system

$$\begin{cases} 4x - y + 2z = 8 \\ 2x + 3y + 2z = -4 \\ 3x + y - 2z = 6 \end{cases}$$

## SOLUTION

We first evaluate the determinant of the coefficient matrix.

$$\begin{vmatrix} 4 & -1 & 2 \\ 2 & 3 & 2 \\ 3 & 1 & -2 \end{vmatrix} = \begin{vmatrix} 7 & 0 & 0 \\ 2 & 3 & 2 \\ 3 & 1 & -2 \end{vmatrix}$$ (replacing the first row by the sum of the first row and the third row)

$$= 7 \begin{vmatrix} 3 & 2 \\ 1 & -2 \end{vmatrix}$$ (evaluating by the minors of the elements of the first row)

$$= -56$$

From formulas (6), we have

$$x = \frac{\begin{vmatrix} 8 & -1 & 2 \\ -4 & 3 & 2 \\ 6 & 1 & -2 \end{vmatrix}}{-56}, \quad y = \frac{\begin{vmatrix} 4 & 8 & 2 \\ 2 & -4 & 2 \\ 3 & 6 & -2 \end{vmatrix}}{-56}, \quad z = \frac{\begin{vmatrix} 4 & -1 & 8 \\ 2 & 3 & -4 \\ 3 & 1 & 6 \end{vmatrix}}{-56} \quad (7)$$

We evaluate each of the determinants in the numerators.

$$\begin{vmatrix} 8 & -1 & 2 \\ -4 & 3 & 2 \\ 6 & 1 & -2 \end{vmatrix} = \begin{vmatrix} 14 & 0 & 0 \\ -4 & 3 & 2 \\ 6 & 1 & -2 \end{vmatrix} \qquad \begin{vmatrix} 4 & 8 & 2 \\ 2 & -4 & 2 \\ 3 & 6 & -2 \end{vmatrix} = \begin{vmatrix} 4 & 0 & 2 \\ 2 & -8 & 2 \\ 3 & 0 & -2 \end{vmatrix}$$

$$= 14 \begin{vmatrix} 3 & 2 \\ 1 & -2 \end{vmatrix} \qquad\qquad = -8 \begin{vmatrix} 4 & 2 \\ 3 & -2 \end{vmatrix}$$

$$= -112 \qquad\qquad\qquad = 112$$

$$\begin{vmatrix} 4 & -1 & 8 \\ 2 & 3 & -4 \\ 3 & 1 & 6 \end{vmatrix} = \begin{vmatrix} 4 & -1 & 0 \\ 2 & 3 & -8 \\ 3 & 1 & 0 \end{vmatrix}$$

$$= 8 \begin{vmatrix} 4 & -1 \\ 3 & 1 \end{vmatrix}$$

$$= 56$$

Substituting into formulas (5), we get

$$x = \frac{-112}{-56} \qquad y = \frac{112}{-56} \qquad z = \frac{56}{-56}$$

$$= 2 \qquad\qquad = -2 \qquad\qquad = -1$$

Therefore, the solution set is $\{(2, -2, -1)\}$.

Even though Cramer's rule can be used to solve a system of $n$ equations in $n$ unknowns, it is tedious to apply for a system of more than three equations because of the many determinants that have to be evaluated. However, Cramer's rule as well as the matrix methods of Sections 7.1 and 7.5 can be programmed so that electronic computers do the calculations. In Exercises 13 through 18 in Exercises 7.6, you are asked to apply Cramer's rule to solve a system of four linear equations in four variables.

## EXERCISES 7.6

Find the solution set of the given system by Cramer's rule. If the equations of the system are either inconsistent or dependent, so indicate.

1. $\begin{cases} 2x - 3y = 4 \\ 3x - y = 1 \end{cases}$

2. $\begin{cases} x + 4y = 6 \\ 3x - 5y = -16 \end{cases}$

3. $\begin{cases} 2x + 6y = 3 \\ 8x - 3y = -6 \end{cases}$

4. $\begin{cases} 6x + y = 6 \\ 2x - 2y = -5 \end{cases}$

5. $\begin{cases} 2x - 2y + z = 0 \\ x + 5y - 7z = 3 \\ x - y - 3z = -7 \end{cases}$

6. $\begin{cases} x + 4y + z = 5 \\ 2x + y - 4z = 0 \\ 3x - 3y + z = -1 \end{cases}$

7. $\begin{cases} 4x + 3y + 5z = 0 \\ 2x - 4y - 3z = 0 \\ 6x - 2y + z = 0 \end{cases}$

8. $\begin{cases} 3x + y + 2z = 2 \\ 3x + 2y + 4z = 3 \\ 3x - y - 2z = 0 \end{cases}$

9. $\begin{cases} 6x + y - 2x = 0 \\ 2x + 3y - z = 0 \\ 4x + 2y - 3z = 2 \end{cases}$

10. $\begin{cases} x - y - z = 0 \\ 8x - 2y + z = 0 \\ x + 3y + 5z = 0 \end{cases}$

11. $\begin{cases} 2x - y + 3z = 5 \\ 3x - 2y + 7z = 3 \\ 3x - y + 2z = 10 \end{cases}$

12. $\begin{cases} 5x - 6y = 1 \\ 2x + 3z = 3 \\ 4y + 6z = 5 \end{cases}$

13. $\begin{cases} w - 2x + y - 2z = -4 \\ 2w + 2x - 3y - z = -1 \\ -w + x + 2y + 3z = 5 \\ x - 2y + z = 0 \end{cases}$

14. $\begin{cases} w - x - y - z = 2 \\ 2w + 4x - 3y = 6 \\ 3x - 4y - 2z = -1 \\ 2w + 4y + 3z = 5 \end{cases}$

15. $\begin{cases} 4w + 2x + 3y = 5 \\ 2w + 3x + z = 2 \\ 3w + 5x + 2z = 3 \\ 4x + 3y + 6z = 1 \end{cases}$

16. $\begin{cases} w - 2x - y + 2z = 4 \\ 3w - 3x + 2y - 2z = 8 \\ 2w + 2x - 3y + z = -2 \\ 4w + 7x - y - 6z = -5 \end{cases}$

17. $\begin{cases} 2w - 3x + y - 8z = -2 \\ w + 3x + 2y - z = 5 \\ -w + 2x + y + 3z = 3 \\ 3w + 2x + 3y - 7z = 5 \end{cases}$

18. $\begin{cases} 4w - 2x + y - z = 3 \\ 2w - 4x + y + z = -1 \\ 4w - 3x - 2z = 5 \\ 3w + x - y = 2 \end{cases}$

## REVIEW EXERCISES (CHAPTER 7)

In Exercises 1 through 6, use the matrix method of Section 7.1 to find the solution set of the given system of equations.

1. $\begin{cases} 5x - 7y = 4 \\ 2x + 3y = 19 \end{cases}$

2. $\begin{cases} 2x + 3y = 0 \\ 3x + 4y = 2 \end{cases}$

3. $\begin{cases} 3x - 5y + 2z = 4 \\ 4x + 2y + 7z = 1 \\ 5x - 9y - 3z = -11 \end{cases}$

Review Exercises 429

4. $\begin{cases} 3x - 2y - 2z = 3 \\ 6x + 4y + 3z = 3 \\ 3x - 6y + z = -2 \end{cases}$
5. $\begin{cases} 6w - 2x + 3y + 2z = 4 \\ 4w - 3x + 6y + 3z = 3 \\ 2w - 5x - 3y + 4z = 3 \\ 8w + 4x + 9y - 2z = 5 \end{cases}$
6. $\begin{cases} 2w + 4x + 2y - 3z = 0 \\ -4w + 2x + y + 3z = 4 \\ w + 3x - 2y - 6z = 6 \\ 3w - x + 3y + 9z = -6 \end{cases}$

*In Exercises 7 through 20, perform the indicated operations with matrices.*

7. $\begin{bmatrix} 3 & -4 \\ -7 & 1 \end{bmatrix} + \begin{bmatrix} -6 & 7 \\ -3 & 5 \end{bmatrix}$

8. $\begin{bmatrix} 1 & 2 & 8 \\ 3 & -4 & -1 \end{bmatrix} - \begin{bmatrix} 5 & -6 & 9 \\ 3 & 7 & -3 \end{bmatrix}$

9. $2\begin{bmatrix} -4 & 0 \\ 1 & -3 \\ 8 & -2 \end{bmatrix} - 3\begin{bmatrix} -2 & 2 \\ 3 & 5 \\ -1 & -3 \end{bmatrix}$

10. $\begin{bmatrix} 1 & 1 \\ 0 & 1 \end{bmatrix} \begin{bmatrix} 1 & 1 \\ -1 & 1 \end{bmatrix}$

11. $\begin{bmatrix} 1 & -1 \end{bmatrix} \begin{bmatrix} 1 \\ -1 \end{bmatrix}$

12. $-4\begin{bmatrix} 2 & -1 & 0 \\ 3 & 1 & -2 \end{bmatrix} + 3\begin{bmatrix} -2 & 1 & -1 \\ -3 & 0 & 2 \end{bmatrix}$

13. $\begin{bmatrix} 1 & 2 \\ -1 & 3 \end{bmatrix} \begin{bmatrix} 2 & 1 & 3 \\ 0 & -1 & 0 \end{bmatrix}$

14. $\begin{bmatrix} 5 & -1 \\ -2 & 0 \end{bmatrix} \begin{bmatrix} 1 \\ -2 \end{bmatrix}$

15. $\begin{bmatrix} a & b \\ c & d \end{bmatrix}^t$

16. $\begin{bmatrix} -7 & 2 \\ 0 & 1 \end{bmatrix}^{-1}$

17. $\begin{bmatrix} 1 & 3 & 0 \\ 0 & 1 & 1 \\ 2 & 0 & 0 \end{bmatrix}^{-1}$

18. $\begin{bmatrix} 3 & -1 & 0 \\ -2 & 1 & 2 \\ 0 & -3 & 4 \end{bmatrix}^t$

19. $\begin{bmatrix} -1 & 2 & 0 \end{bmatrix} \begin{bmatrix} -2 & 1 & 0 \\ 0 & -1 & 1 \\ 3 & 0 & 2 \end{bmatrix}$

20. $\begin{bmatrix} 1 \\ 2 \\ -3 \end{bmatrix} \begin{bmatrix} 0 & -2 & 1 \end{bmatrix}$

*In Exercises 21 through 24, find the matrix X that satisfies the given equation.*

21. $2X + 3\begin{bmatrix} 2 & -4 \\ 1 & 0 \end{bmatrix} = \begin{bmatrix} 4 & -6 \\ 3 & 4 \end{bmatrix}$

22. $4\begin{bmatrix} 1 & -2 \\ 0 & -1 \\ 3 & 2 \end{bmatrix} - X = \begin{bmatrix} 1 & -7 \\ -5 & -4 \\ 9 & 7 \end{bmatrix}$

23. $5\begin{bmatrix} 1 \\ -2 \\ -1 \end{bmatrix} - 4X = \begin{bmatrix} -3 \\ -6 \\ 7 \end{bmatrix}$

24. $\begin{bmatrix} -5 & 4 & 9 & -4 \end{bmatrix} = 3X - 2\begin{bmatrix} 1 & -2 & 0 & 5 \end{bmatrix}$

*In Exercises 25 through 30, evaluate the given determinant. In Exercises 25 and 26, first obtain two zeros in a row or column; and in Exercises 27 and 28, first obtain three zeros in a row or column.*

25. $\begin{vmatrix} 5 & -7 \\ 3 & -4 \end{vmatrix}$

26. $\begin{vmatrix} 6 & 3 \\ -2 & -4 \end{vmatrix}$

27. $\begin{vmatrix} 2 & 3 & -1 \\ 3 & 1 & 5 \\ 2 & 3 & 4 \end{vmatrix}$

**28.** $\begin{vmatrix} 1 & 1 & -1 \\ 0 & 1 & 1 \\ -4 & 2 & -3 \end{vmatrix}$

**29.** $\begin{vmatrix} 4 & 5 & 4 & 5 \\ 3 & 1 & 3 & 4 \\ 1 & 1 & 3 & 4 \\ 1 & 2 & 2 & 1 \end{vmatrix}$

**30.** $\begin{vmatrix} 2 & 5 & 4 & 1 \\ 1 & 2 & 3 & 2 \\ 3 & -1 & 5 & 2 \\ 4 & 2 & 1 & 1 \end{vmatrix}$

*In Exercises 31 and 32, verify the given equality.*

**31.** $\begin{vmatrix} a & b \\ c & d \end{vmatrix} = - \begin{vmatrix} b & a \\ d & c \end{vmatrix}$

**32.** $\begin{vmatrix} a & b \\ c & d \end{vmatrix} = \begin{vmatrix} a & b + ka \\ c & d + kc \end{vmatrix}$

*In Exercises 33 and 34, find $A^{-1}$ and show that $AA^{-1} = I$.*

**33.** $A = \begin{bmatrix} -2 & 1 \\ 4 & -3 \end{bmatrix}$

**34.** $A = \begin{bmatrix} 1 & 0 & 1 \\ 1 & 2 & 0 \\ 0 & 0 & -1 \end{bmatrix}$

*In Exercises 35 and 36, solve the system of equations by using the matrix method of Illustration 4 and Example 3 in Section 7.5.*

**35.** $\begin{cases} x + y - z = 0 \\ 2x - y + 3z = -1 \\ 2y - 3z = 1 \end{cases}$

**36.** $\begin{cases} 2x + 5y = 4 \\ 3x + 4y = -1 \end{cases}$

*In Exercises 37 through 39, use Cramer's rule to find the solution set of the given system of equations.*

**37.** $\begin{cases} 6x + 3y = 1 \\ 3x - 2y = 4 \end{cases}$

**38.** $\begin{cases} 4x + 2y - 3z = 10 \\ 2x - 3y - 4z = 8 \\ 6x - 5y - 2z = 6 \end{cases}$

**39.** $\begin{cases} 3x - 5y - 6z = 2 \\ 2x + 2y + 5z = -2 \\ 5x + 3y - z = 3 \\ 4x - 4y - 5z = 2 \end{cases}$

# 8 Sequences, Series, and Mathematical Induction

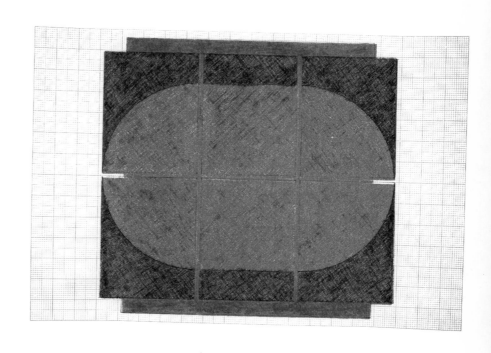

## 8.1 Sequences and Series

"Sequences" of numbers are often encountered in mathematics. For instance, the numbers

$$2, 4, 6, 8, 10 \tag{1}$$

form a sequence. This sequence is said to be *finite* because there is a first and last number. If the set of numbers which forms a sequence does not have both a first and last number, the sequence is said to be *infinite*. For instance, the sequence

$$\frac{1}{1}, \frac{3}{4}, \frac{5}{9}, \frac{7}{16}, \ldots \tag{2}$$

is infinite because the three dots with no number following indicate that there is no last number.

We now give a formal definition of a "sequence function."

**8.1.1 DEFINITION** (i) A *finite sequence function* is a function whose domain is the set

$$\{1, 2, 3, \ldots, n\}$$

of the first $n$ positive integers.

(ii) An *infinite sequence function* is a function whose domain is the set

$$\{1, 2, 3, \ldots, n, \ldots\}$$

of all positive integers.

The numbers in the range of a sequence function are called *elements*. We refer to the elements of a sequence function, listed in order, as *the sequence*.

**ILLUSTRATION 1.** Let $f$ be the function defined by

$$f(n) = 2n, \quad n \in \{1, 2, 3, 4, 5\}$$

Then $f$ is a finite sequence function, and

$$f(1) = 2 \quad f(2) = 4 \quad f(3) = 6 \quad f(4) = 8 \quad f(5) = 10$$

The elements of the sequence defined by $f$ are then 2, 4, 6, 8, and 10, and the sequence is

$$2, 4, 6, 8, 10$$

which is sequence (1). The ordered pairs in $f$ are (1, 2), (2, 4), (3, 6), (4, 8), and (5, 10). ●

**ILLUSTRATION 2.** Let $g$ be the function defined by

$$g(n) = \frac{2n - 1}{n^2}, \quad n \in \{1, 2, 3, \ldots\}$$

The function $g$ is an infinite sequence function. The elements of $g$ are

$$g(1) = \frac{1}{1} \quad g(2) = \frac{3}{4} \quad g(3) = \frac{5}{9} \quad g(4) = \frac{7}{16}$$

and so on. The sequence defined by $g$ is, therefore,

$$\frac{1}{1}, \frac{3}{4}, \frac{5}{9}, \frac{7}{16}, \ldots$$

which is sequence (2). Some of the ordered pairs in $g$ are $(1, 1)$, $(2, \frac{3}{4})$, $(3, \frac{5}{9})$, $(4, \frac{7}{16})$, and $(5, \frac{9}{25})$. •

The $n$th element of a sequence is called the *general element* of the sequence. We use the notation $a_n$ to denote the general element of a sequence. For sequence (1), $a_n = 2n$ and, therefore, $a_1 = 2, a_2 = 4, a_3 = 6, a_4 = 8$, and $a_5 = 10$. For sequence (2), $a_n = \frac{2n-1}{n^2}$, and hence $a_1 = \frac{1}{1}$, $a_2 = \frac{3}{4}$, $a_3 = \frac{5}{9}$, and so on. Often the general element of a sequence is stated when the elements are listed in order. Thus, for the elements of sequence (2) we would write

$$\frac{1}{1}, \frac{3}{4}, \frac{5}{9}, \frac{7}{16}, \ldots, \frac{2n-1}{n^2}, \ldots$$

You should distinguish between the elements of a sequence and the sequence itself. For instance, the sequence, for which $a_n = \frac{1}{n}$, has as its elements the reciprocals of the positive integers, and the sequence is

$$1, \frac{1}{2}, \frac{1}{3}, \frac{1}{4}, \ldots, \frac{1}{n}, \ldots \qquad (3)$$

The sequence for which

$$a_n = \begin{cases} 1, & \text{if } n \text{ is odd} \\ \frac{2}{n+2}, & \text{if } n \text{ is even} \end{cases}$$

has as its elements

$$a_1 = 1 \quad a_2 = \frac{1}{2} \quad a_3 = 1 \quad a_4 = \frac{1}{3} \quad a_5 = 1 \quad a_6 = \frac{1}{4}$$

and so on. The sequence is

$$1, \frac{1}{2}, 1, \frac{1}{3}, 1, \frac{1}{4}, \ldots \qquad (4)$$

The elements of sequences (3) and (4) are the same; however, the sequences are different.

You should realize that knowing several elements of a sequence does not determine a unique general element, as shown in the following illustration.

**ILLUSTRATION 3**

(a) The sequence, for which $a_n = 2n$, is

$$2, 4, 6, 8, 10, 12, \ldots, 2n, \ldots \qquad (5)$$

(b) The sequence for which

$$a_n = \begin{cases} 2n, & \text{if } n \text{ is odd} \\ 2a_{n-1}, & \text{if } n \text{ is even} \end{cases}$$

is

$$2, 4, 6, 12, 10, 20, \ldots, a_n, \ldots \qquad (6)$$

(c) The sequence, for which $a_n = 2n + (n-1)(n-2)(n-3)$, is

$$2, 4, 6, 14, 34, 72, \ldots, 2n + (n-1)(n-2)(n-3), \ldots \qquad (7)$$

Observe in Illustration 3 that sequences (5), (6), and (7) all have 2, 4, and 6 as the first three elements but the general element for each sequence is different. Therefore, to uniquely determine a sequence either the general element must be known or we must have a method for obtaining any element from the preceding one. It is not always possible that an equation can be given to determine the general element of a sequence. For instance, the sequence of prime numbers can be written as

$$2, 3, 5, 7, 11, 13, 17, 19, \ldots, a_n, \ldots$$

where $a_n$ is the $n$th prime number. For this sequence we cannot write an equation which defines $a_n$; however, $a_n$ can be determined for every positive integer $n$.

**EXAMPLE 1**
Write the first five elements of each of the following sequences for which the general element is given

(a) $a_n = \dfrac{n+2}{n(n+1)}$

(b) $a_n = (-1)^n \dfrac{1}{3^{n-1}}$

(c) $a_n = (-1)^{n-1} x^{2n+1}$

**SOLUTION**

(a) $a_1 = \dfrac{3}{1 \cdot 2}$ $\qquad a_2 = \dfrac{4}{2 \cdot 3}$ $\qquad a_3 = \dfrac{5}{3 \cdot 4}$ $\qquad a_4 = \dfrac{6}{4 \cdot 5}$ $\qquad a_5 = \dfrac{7}{5 \cdot 6}$

$\phantom{a_1} = \dfrac{3}{2}$ $\qquad \phantom{a_2} = \dfrac{2}{3}$ $\qquad \phantom{a_3} = \dfrac{5}{12}$ $\qquad \phantom{a_4} = \dfrac{3}{10}$ $\qquad \phantom{a_5} = \dfrac{7}{30}$

Hence the first five elements are

$$\dfrac{3}{2}, \dfrac{2}{3}, \dfrac{5}{12}, \dfrac{3}{10}, \dfrac{7}{30}$$

(b) $a_1 = (-1)^1 \dfrac{1}{3^0}$;  $a_2 = (-1)^2 \dfrac{1}{3^1}$;  $a_3 = (-1)^3 \dfrac{1}{3^2}$;  $a_4 = (-1)^4 \dfrac{1}{3^3}$;  $a_5 = (-1)^5 \dfrac{1}{3^4}$. Therefore, the first five elements are

$$-1, \frac{1}{3}, -\frac{1}{9}, \frac{1}{27}, -\frac{1}{81}$$

(c) $a_1 = (-1)^0 x^3$;  $a_2 = (-1)^1 x^5$;  $a_3 = (-1)^2 x^7$;  $a_4 = (-1)^3 x^9$;  $a_5 = (-1)^4 x^{11}$. Thus the first five elements are

$$x^3, -x^5, x^7, -x^9, x^{11}$$

**EXAMPLE 2**
Write the first twelve elements of the sequence for which

$$a_n = \begin{cases} n, & \text{if } n \text{ is odd} \\ n, & \text{if } n \text{ is even and not exactly divisible by 4} \\ \dfrac{1}{2}(n + a_{n-2}), & \text{if } n \text{ is even and exactly divisible by 4} \end{cases}$$

**SOLUTION**
$a_1 = 1$;  $a_2 = 2$;  $a_3 = 3$;  $a_4 = \tfrac{1}{2}(4 + 2)$;  $a_5 = 5$;  $a_6 = 6$;  $a_7 = 7$; $a_8 = \tfrac{1}{2}(8 + 6)$, $a_9 = 9$; $a_{10} = 10$; $a_{11} = 11$; $a_{12} = \tfrac{1}{2}(12 + 10)$. Therefore, the first twelve elements are

$$1, 2, 3, 3, 5, 6, 7, 7, 9, 10, 11, 11$$

The indicated sum of the elements of a sequence is called a *series*. Because the familiar operation of addition applies only to a finite set of numbers, we restrict the series (with one exception) in this text to those associated with finite sequences. The one exception, which is discussed intuitively in Section 8.5, is an infinite geometric series. The formal definition of the "sum" of an infinite series requires the concept of "limit," which is studied in the calculus.

Associated with the finite sequence

$$a_1, a_2, a_3, \ldots, a_n$$

is the series

$$a_1 + a_2 + a_3 + \cdots + a_n \tag{8}$$

**ILLUSTRATION 4**

(a) Associated with the sequence 1, 3, 5, 7, 9 is the series

$$1 + 3 + 5 + 7 + 9 \tag{9}$$

(b) The sequence having ten elements and whose general element is $(-1)^{n-1} x^{2n}$, is

$$x^2, -x^4, x^6, -x^8, x^{10}, -x^{12}, x^{14}, -x^{16}, x^{18}, -x^{20}$$

The series associated with this sequence is

$$x^2 - x^4 + x^6 - x^8 + x^{10} - x^{12} + x^{14} - x^{16} + x^{18} - x^{20} \qquad (10)$$

The *terms* in a series are the same as the corresponding elements in the associated sequence. Hence we refer to the *general term* of a series just as we refer to the general element of the associated sequence.

A notation called the *sigma notation* is now introduced to facilitate writing a series whose general term is known. This notation involves the use of the symbol $\Sigma$, the capital sigma of the Greek alphabet, which corresponds to our letter S. The following illustration gives some examples of the sigma notation.

**ILLUSTRATION 5**

(a) $\displaystyle\sum_{i=1}^{7} i^2 = 1^2 + 2^2 + 3^2 + 4^2 + 5^2 + 6^2 + 7^2$

(b) $\displaystyle\sum_{i=2}^{5} (2i + 3) = [2(2) + 3] + [2(3) + 3] + [2(4) + 3] + [2(5) + 3]$

$\qquad\qquad\qquad = 7 + 9 + 11 + 13$

(c) $\displaystyle\sum_{i=3}^{8} \frac{1}{i} = \frac{1}{3} + \frac{1}{4} + \frac{1}{5} + \frac{1}{6} + \frac{1}{7} + \frac{1}{8}$

The sigma notation can be defined by the equation

$$\sum_{i=m}^{n} F(i) = F(m) + F(m+1) + F(m+2) + \cdots + F(n) \qquad (11)$$

where $m$ and $n$ are integers and $m < n$.

The right side of formula (11) consists of the sum of $(n - m + 1)$ terms, the first of which is obtained by replacing $i$ by $m$ in $F(i)$, the second by replacing $i$ by $(m + 1)$ in $F(i)$, and so on, until the last term is obtained by replacing $i$ by $n$ in $F(i)$. The number $m$ is called the *lower limit* of the sum, and $n$ is called the *upper limit*. The symbol $i$ is called the *index of summation*. It is a "dummy" symbol because any other symbol can be used. For example,

$$\sum_{i=4}^{6} i^3 = 4^3 + 5^3 + 6^3$$

can be written also as

$$\sum_{j=4}^{6} j^3 = 4^3 + 5^3 + 6^3$$

Sometimes, the terms of the sum involve subscripts, as in series (8), for which we write

$$\sum_{i=1}^{n} a_i = a_1 + a_2 + a_3 + \cdots + a_n$$

**EXAMPLE 3**
Write series (9) and (10) with sigma notation.

**SOLUTION**
Series (9) is

$$1 + 3 + 5 + 7 + 9$$

With sigma notation, this series is

$$\sum_{i=1}^{5} (2i - 1)$$

Series (10) is

$$x^2 - x^4 + x^6 - x^8 + x^{10} - x^{12} + x^{14} - x^{16} + x^{18} - x^{20}$$

With sigma notation, this series is

$$\sum_{i=1}^{10} (-1)^{i-1} x^{2i}$$

## EXERCISES 8.1

*In Exercises 1 through 16, write the first eight elements of the sequence whose general element is given.*

1. $a_n = 2n + 3$
2. $a_n = \dfrac{3n - 1}{2}$
3. $a_n = \dfrac{n^2 + 1}{n}$
4. $a_n = \dfrac{1}{n^2 + 2}$
5. $a_n = (-1)^{n-1} \dfrac{n + 1}{2n - 1}$
6. $a_n = (-1)^{n+1} \dfrac{n}{2^n}$
7. $a_n = (-1)^n \dfrac{2^n}{1 + 2^n}$
8. $a_n = \dfrac{(-1)^{n-1}}{n(n + 1)}$
9. $a_n = \dfrac{(-1)^{n+1}}{n + 2} x^n$
10. $a_n = \dfrac{(-1)^n}{n^2} x^{2n-1}$
11. $a_n = n + (-1)^n n$
12. $a_n = \dfrac{1}{2n} - \dfrac{1}{3n}$
13. $a_n = \begin{cases} \dfrac{2}{n + 1}, & \text{if } n \text{ is odd} \\ 2, & \text{if } n \text{ is even} \end{cases}$
14. $a_n = \begin{cases} 1, & \text{if } n \text{ is odd} \\ \dfrac{4}{(n + 2)^2}, & \text{if } n \text{ is even} \end{cases}$
15. $a_n = \begin{cases} \dfrac{n + 1}{2}, & \text{if } n \text{ is odd} \\ a_{n-1}, & \text{if } n \text{ is even} \end{cases}$
16. $a_n = \begin{cases} n, & \text{if } n \text{ is odd} \\ \dfrac{n}{2}, & \text{if } n \text{ is even and not exactly divisible by 4} \\ \dfrac{1}{2}(a_{n-2} + a_{n-1}), & \text{if } n \text{ is even and exactly divisible by 4.} \end{cases}$

In Exercises 17 through 28, write the given series. In Exercises 17 through 24, find the sum of the series.

17. $\sum_{i=1}^{5} (4i - 3)$

18. $\sum_{i=1}^{7} (i + 1)^2$

19. $\sum_{j=2}^{6} \frac{j}{j-1}$

20. $\sum_{k=1}^{4} \frac{(-1)^{k+1}}{k}$

21. $\sum_{i=1}^{100} 5$

22. $\sum_{i=1}^{8} \frac{3i - 6}{2}$

23. $\sum_{k=0}^{5} \frac{1}{2^k}$

24. $\sum_{j=0}^{3} \frac{(-1)^j}{2^j + 1}$

25. $\sum_{i=1}^{8} (-1)^{i-1} x^{2i-1}$

26. $\sum_{i=1}^{5} 2^i x^{3i}$

27. $\sum_{i=1}^{n} f(x_i)$

28. $\sum_{i=1}^{n} f(x_{i-1}) h$

In Exercises 29 through 36, write the given series with sigma notation (there is no unique solution).

29. $1 + 3 + 5 + 7 + 9 + 11$

30. $2 + 4 + 6 + 8 + 10$

31. $4 - 7 + 10 - 13 + 16$

32. $1 + \frac{1}{3} + \frac{1}{9} + \frac{1}{27} + \frac{1}{81} + \frac{1}{243}$

33. $1 + \frac{3}{4} + \frac{5}{9} + \frac{7}{16} + \frac{9}{25} + \frac{11}{36}$

34. $1 - \frac{1}{4} + \frac{1}{16} - \frac{1}{64}$

35. $\frac{1}{2} - \frac{x^2}{4} + \frac{x^4}{6} - \frac{x^6}{8} + \frac{x^8}{10}$

36. $x - \frac{1}{2} x^3 + \frac{1}{3} x^5 - \frac{1}{4} x^7 + \frac{1}{5} x^9$

37. Write the general element and the first six elements of three different sequences each having as the first three elements 1, 3, and 5.

38. Write the general element of a sequence whose first three elements are 2, 4, and 6, and whose fourth element is $x$, where $x$ can be any real number.

## 8.2 Mathematical Induction

Mathematical induction is a certain general technique of proof in mathematics. This technique is used to prove that a theorem involving a variable is true for all positive-integer values of the variable. A proof by mathematical induction is based upon the following theorem.

**8.2.1 THEOREM** *Principle of Mathematical Induction.* Suppose that $P_n$ is a statement involving the positive integer $n$. Furthermore, suppose that the following two conditions are satisfied:

(i) $P_1$ is true (that is, the statement is true for $n = 1$).
(ii) If $k$ is an arbitrary positive integer for which $P_k$ is true, then $P_{k+1}$ is also true (that is, whenever the statement is true for $n = k$, it is also true for $n = k + 1$, where $k$ is an arbitrary positive integer).

Then the statement $P_n$ is true for all positive-integer values of $n$.

***Proof.*** Assume that conditions (i) and (ii) are satisfied but that the statement is not true for some positive-integer value of $n$. We shall show that this assumption leads to a contradiction. Let $M + 1$ be the smallest positive-integer value of $n$ for which the statement is not true. That is, we are assuming that the statement is true for $n = 1, n = 2, n = 3, \ldots, n = M$, where $M$ is a positive integer; and we are assuming that the statement is not true for $n = M + 1$. By condition (i) we know that the statement is true for $n = 1$. Therefore, $M \geq 1$. However, from condition (ii), because the statement is true for $n = M$, the statement must also be true for $n = M + 1$. Thus we have a contradiction to our assumption that there is some positive-integer value of $n$ for which the statement is not true. Hence the statement must be true for all positive-integer values of $n$.

**ILLUSTRATION 1.** We shall prove by mathematical induction that the sum of the first $n$ positive odd integers is equal to $n^2$. That is, we shall prove that

$$1 + 3 + 5 + \cdots + (2n - 1) = n^2 \qquad (1)$$

for all positive-integer values of $n$. The proof consists of part 1 (verifying condition (i) of Theorem 8.2.1), part 2 (verifying condition (ii) of Theorem 8.2.1), and conclusion.

**PART 1:** We first verify that formula (1) is true for $n = 1$. If $n = 1$, formula (1) becomes

$$1 = 1^2$$

which is true.

**PART 2:** We now show that if formula (1) is true for $n = k$, it is also true for $n = k + 1$, where $k$ is an arbitrary positive integer. That is, we assume

$$1 + 3 + 5 + \cdots + (2k - 1) = k^2 \qquad (2)$$

We wish to prove that if equation (2) is true, then the equation

$$1 + 3 + 5 + \cdots + (2k - 1) + [2(k + 1) - 1] = (k + 1)^2 \qquad (3)$$

is also true. The last term in the left member of equation (3) can be written as $2k + 2 - 1$; hence, $[2(k + 1) - 1] = (2k + 1)$. We add $[2(k + 1) - 1]$ to the left member of equation (2) and its equivalent $(2k + 1)$ to the right member, and we obtain

$$1 + 3 + 5 + \cdots + (2k - 1) + [2(k + 1) - 1] = k^2 + (2k + 1)$$

or, equivalently,

$$1 + 3 + 5 + \cdots + (2k - 1) + [2(k + 1) - 1] = (k + 1)^2$$

which is equation (3).

**CONCLUSION:** From part 1 we know that formula (1) is true for $n = 1$. Because it is true for $n = 1$, it follows from part 2 that the formula is true for $n = 1 + 1$ or 2; because it is true for $n = 2$, it is true for $n = 2 + 1$ or 3; because it is true for $n = 3$, it is true for $n = 3 + 1$ or 4; and so on. By the principle of mathematical induction the formula is true for all positive-integer values of $n$. ●

Observe in part 1 of Illustration 1 that it is only necessary to verify that the theorem is true for the smallest positive-integer value of $n$ ($n = 1$). However, often we verify a theorem for other successive values of $n$ (for instance, $n = 2$, $n = 3$, and $n = 4$) to convince us of a theorem's validity before proceeding to part 2. Verifying formula (1) for $n = 2$, $n = 3$, and $n = 4$, we have

If $n = 2$, $\qquad 1 + 3 = 2^2$; that is, $4 = 4$

If $n = 3$, $\qquad 1 + 3 + 5 = 3^2$; that is, $9 = 9$

If $n = 4$, $\qquad 1 + 3 + 5 + 7 = 4^2$; that is, $16 = 16$

### EXAMPLE 1

Prove the following formula by mathematical induction.

$$\sum_{i=1}^{n} i^2 = \frac{n(n+1)(2n+1)}{6} \quad (4)$$

### SOLUTION

**PART 1:** First formula (4) is verified for $n = 1$. When $n = 1$ the left side of formula (4) is

$$\sum_{i=1}^{1} i^2 = 1$$

and the right side of formula (4) is

$$\frac{1(1+1)(2+1)}{6} = \frac{1 \cdot 2 \cdot 3}{6}$$

$$= 1$$

Hence formula (4) is true when $n = 1$.

**PART 2:** We assume that formula (4) is true when $n = k$, where $k$ is any positive integer; and with this assumption we wish to prove that the formula is also true when $n = k + 1$. If the formula is true when $n = k$, then

$$\sum_{i=1}^{k} i^2 = \frac{k(k+1)(2k+1)}{6} \quad (5)$$

When $n = k + 1$, we have

$$\sum_{i=1}^{k+1} i^2 = 1^2 + 2^2 + 3^2 + \cdots + k^2 + (k+1)^2$$

$$= \sum_{i=1}^{k} i^2 + (k+1)^2$$

$$= \frac{k(k+1)(2k+1)}{6} + (k+1)^2 \quad \text{(by applying equation (5))}$$

$$= \frac{k(k+1)(2k+1) + 6(k+1)^2}{6}$$

$$= \frac{(k+1)[k(2k+1) + 6(k+1)]}{6}$$

$$= \frac{(k+1)(2k^2 + 7k + 6)}{6}$$

$$= \frac{(k+1)(k+2)(2k+3)}{6}$$

$$= \frac{(k+1)[(k+1)+1][2(k+1)+1]}{6}$$

Therefore, formula (4) is true when $n = k + 1$.

**CONCLUSION:** We have proved that the formula is true when $n = 1$, and we have also proved that when the formula is true for $n = k$, it is also true for $n = k + 1$. Therefore, by the principle of mathematical induction, formula (4) is true when $n$ is any positive integer.

EXAMPLE 2
Prove the following statement by mathematical induction.
$(x - y)$ is a factor of $(x^n - y^n)$ for all positive-integer values of $n$.

SOLUTION
**PART 1:** When $n = 1$, $(x^n - y^n)$ becomes $(x - y)$ which certainly has $(x - y)$ as a factor.

**PART 2:** We assume that $(x - y)$ is a factor of $(x^k - y^k)$, where $k$ is any positive integer; and with this assumption we wish to prove that $(x - y)$ is also a factor of $(x^{k+1} - y^{k+1})$. If we subtract and add $xy^k$ to $(x^{k+1} - y^{k+1})$, we obtain

$$x^{k+1} - y^{k+1} = x^{k+1} - xy^k + xy^k - y^{k+1}$$

or, equivalently,

$$x^{k+1} - y^{k+1} = x(x^k - y^k) + y^k(x - y) \tag{6}$$

We have assumed that $(x - y)$ is a factor of $x^k - y^k$; furthermore, $(x - y)$ is a factor of $y^k(x - y)$. Hence $(x - y)$ is a factor of each of the two terms in the right member of equation (6). Therefore, $(x - y)$ is a factor of $(x^{k+1} - y^{k+1})$.

**CONCLUSION** We have proved that the statement is true when $n = 1$, and we have also proved that when the statement is true for $n = k$, it is also true for $n = k + 1$. Therefore, by the principle of mathematical induction, the statement is true for all positive-integer values of $n$.

In Section 4.1 we stated that certain laws of exponents can be proved by mathematical induction. In Example 3 we prove Theorem 4.1.1(i) by mathematical induction and the following definition of positive-integer exponents.

**8.2.2 DEFINITION** Let $a$ be any real number. Then

(i) $a^1 = a$.
(ii) If $k$ is any positive integer such that $a^k$ is defined, let $a^{k+1} = a^k \cdot a$.

**EXAMPLE 3**
Prove Theorem 4.1.1(i): If $m$ and $n$ are positive integers and $a \in R$, then

$$a^m \cdot a^n = a^{m+n} \qquad (7)$$

**SOLUTION**
Let $m$ be an arbitrary positive integer. We wish to prove that the formula (7) is true for all positive integers $n$.

**PART 1:** We verify that the formula is true when $n = 1$. From part (i) of Definition 8.2.2 it follows that

$$a^m \cdot a^1 = a^m \cdot a$$

Applying Definition 8.2.2(ii) in the right member of this equation, we have

$$a^m \cdot a^1 = a^{m+1}$$

Hence formula (7) is true when $n = 1$.

**PART 2:** We assume that formula (7) is true when $n = k$, where $k$ is any positive integer; that is, we assume that

$$a^m \cdot a^k = a^{m+k} \qquad (8)$$

With this assumption we wish to prove that

$$a^m \cdot a^{k+1} = a^{m+(k+1)} \qquad (9)$$

To prove this, we start with the left member of equation (9) and replace $a^{k+1}$ by $a^k \cdot a$ (which follows from Definition 8.2.2(ii)). Thus

$$a^m \cdot a^{k+1} = a^m \cdot (a^k \cdot a) \qquad \text{(i)}$$

Applying the associative law for multiplication in the right member of equation (i), we obtain

$$a^m \cdot a^{k+1} = (a^m \cdot a^k) \cdot a \qquad \text{(ii)}$$

Substituting from equation (8) into the right member of equation (ii), we have

$$a^m \cdot a^{k+1} = a^{m+k} \cdot a \quad \text{(iii)}$$

We apply Definition 8.2.2(ii) in the right member of equation (iii) and we get

$$a^m \cdot a^{k+1} = a^{(m+k)+1} \quad \text{(iv)}$$

In the exponent of the right member of equation (iv), we use the associative law for addition, and we obtain

$$a^m \cdot a^{k+1} = a^{m+(k+1)}$$

which is equation (9) that we wished to prove.

**CONCLUSION:** From part 1 we know that formula (7) is true when $n = 1$. From part 2 we know that when formula (7) is true for $n = k$, it is also true for $n = k + 1$, where $k$ is any positive integer. Hence, by the principle of mathematical induction, formula (7) is true when $n$ is any positive integer.

## EXERCISES 8.2

*In Exercises 1 through 14, use mathematical induction to prove that the given formula is true for all positive-integer values of n.*

1. $\sum_{i=1}^{n} i = \dfrac{n(n+1)}{2}$

2. $\sum_{i=1}^{n} 2i = n(n+1)$

3. $\sum_{i=1}^{n} (3i-2) = \dfrac{n(3n-1)}{2}$

4. $\sum_{i=1}^{n} (3i-1) = \dfrac{n(3n+1)}{2}$

5. $\sum_{i=1}^{n} \dfrac{i(i+1)}{2} = \dfrac{n(n+1)(n+2)}{6}$

6. $\sum_{i=1}^{n} (2i-1)^2 = \dfrac{n(2n-1)(2n+1)}{3}$

7. $\sum_{i=1}^{n} 2^i = 2(2^n - 1)$

8. $\sum_{i=1}^{n} 3^i = \dfrac{3}{2}(3^n - 1)$

9. $\sum_{i=1}^{n} i^3 = \dfrac{n^2(n+1)^2}{4}$

10. $\sum_{i=1}^{n} (2i-1)^3 = n^2(2n^2 - 1)$

11. $\sum_{i=1}^{n} \dfrac{1}{i(i+1)} = \dfrac{n}{n+1}$

12. $\sum_{i=1}^{n} \dfrac{1}{(2i-1)(2i+1)} = \dfrac{n}{2n+1}$

13. $\displaystyle\sum_{i=1}^{n} \frac{1}{(3i-1)(3i+2)} = \frac{n}{2(3n+2)}$

14. $\displaystyle\sum_{i=1}^{n} ar^{i-1} = \frac{a - ar^n}{1-r}$

15. Prove by mathematical induction Theorem 4.1.1(ii): If $m$ and $n$ are positive integers and $a \in R$, then $(a^m)^n = a^{mn}$.

16. Prove by mathematical induction Theorem 4.1.1(iii): If $n$ is a positive integer and $a, b \in R$, then $(ab)^n = a^n b^n$.

17. Prove by mathematical induction that $(x+y)$ is a factor of $(x^{2n} - y^{2n})$ for all positive-integer values of $n$.

18. Prove by mathematical induction that $(x+y)$ is a factor of $(x^{2n-1} + y^{2n-1})$ for all positive-integer values of $n$.

## 8.3 Arithmetic Sequences and Series

The sequence

$$2, 5, 8, 11, 14, 17, 20 \qquad (1)$$

is one for which each element, except the first, can be obtained by adding 3 to the preceding element. This sequence is an example of an "arithmetic sequence," which we now define.

**8.3.1 DEFINITION** An *arithmetic sequence* is a sequence for which any element, except the first, can be obtained by adding to the preceding element a constant addend.

An arithmetic sequence is sometimes called an *arithmetic progression*. We can ascertain if a given sequence is an arithmetic sequence by subtracting each element from the succeeding one, and determine if there is a "common difference." Hence the constant addend in an arithmetic sequence is called the *common difference*, and it is denoted by $d$.

**ILLUSTRATION 1.** For the sequence

$$9, 5, 1, -3, -7, -11 \qquad (2)$$

$5 - 9 = -4$; $1 - 5 = -4$; $-3 - 1 = -4$; $-7 - (-3) = -4$; and $-11 - (-7) = -4$. Therefore, sequence (2) is an arithmetic sequence where the common difference $d$ is $-4$. •

In an arithmetic sequence, the number of elements is denoted by $N$, the first element is denoted by $a_1$, and the last element is denoted by $a_N$. In sequence (2), $N = 6$, $a_1 = 9$, and $a_6 = -11$.

The definition of an arithmetic sequence can be stated symbolically by giving the value of the first element $a_1$, the number of elements $N$, and the formula

$$a_{n+1} = a_n + d \qquad (3)$$

from which every element after the first can be obtained from the preceding one.

**ILLUSTRATION 2.** If $a_1 = 4$, $N = 8$, and
$$a_{n+1} = a_n + 3$$
the arithmetic sequence is
$$4, 7, 10, 13, 16, 19, 22, 25 \qquad \bullet$$

Formula (3) is called a *recursive formula*. From this formula we can write the general arithmetic sequence, for which the first element is $a_1$, the common difference is $d$, and the number of elements is $N$. We start with the element $a_1$ and each successive element is obtained from the preceding one by adding $d$ to it. Hence we have
$$a_1, a_1 + d, a_1 + 2d, a_1 + 3d, a_1 + 4d, \ldots, a_N$$
Refer to the first five elements and observe that each element is $a_1$ plus a multiple of $d$, where the coefficient of $d$ is one less than the number of the element. Intuitively, it appears that $a_N = a_1 + (N - 1)d$. We state this formally as a theorem and prove it by mathematical induction.

**8.3.2 THEOREM** The $N$th element of an arithmetic sequence is given by
$$a_N = a_1 + (N - 1)d \tag{4}$$

**Proof.** We first show that formula (4) is true if $N = 1$ by substituting 1 for $N$ in (4). We have
$$a_1 = a_1 + (1 - 1)d$$
That is,
$$a_1 = a_1$$
We now assume that formula (4) is true if $N = k$; thus we assume
$$a_k = a_1 + (k - 1)d \tag{5}$$
We wish to show that formula (4) is true if $N = k + 1$. By the definition of an arithmetic sequence
$$a_{k+1} = a_k + d \tag{6}$$
Substituting the value of $a_k$ from equation (5) into equation (6), we obtain
$$a_{k+1} = [a_1 + (k - 1)d] + d$$
$$a_{k+1} = a_1 + kd - d + d$$
$$a_{k+1} = a_1 + [(k + 1) - 1]d \tag{7}$$

Equation (7) is formula (4) with $N$ replaced by $k + 1$. We have shown that whenever equation (5) is true, then equation (7) is true. Therefore, by the principle of mathematical induction, formula (4) is valid for all natural numbers.

EXAMPLE 1

Find the thirtieth element of the arithmetic sequence for which the first element is 5 and the second element is 9.

SOLUTION

Let $a_{30}$ be the thirtieth element of the arithmetic sequence

$$5, 9, \ldots, a_{30}$$

Then $d = 9 - 5$ or $4$; $a_1 = 5$, and $N = 30$. From formula (4) we have

$$\begin{aligned} a_{30} &= a_1 + (30 - 1)d \\ &= 5 + 29 \cdot 4 \\ &= 121 \end{aligned}$$

EXAMPLE 2

Find the first element of an arithmetic sequence whose common difference is $-5$ and whose eighteenth element is $-21$.

SOLUTION

From formula (4) with $N = 18$, we have

$$a_{18} = a_1 + (18 - 1)d$$

Substituting $-21$ for $a_{18}$ and $-5$ for $d$, we have

$$-21 = a_1 + (17)(-5)$$

$$a_1 = 64$$

EXAMPLE 3

If the twelfth element of an arithmetic sequence is $-21$ and the twenty-fifth element is 18, what is the fourth element?

SOLUTION

We can consider the arithmetic sequence consisting of the first twenty-five elements; it is

$$a_1, \ldots, a_4, \ldots, a_{12}, \ldots, a_{25} \qquad (8)$$

Using formula (4) with $N = 25$ and $a_{25} = 18$, we have

$$18 = a_1 + (25 - 1)d$$

$$18 = a_1 + 24d \qquad (9)$$

The first twelve elements of sequence (8) also form an arithmetic sequence with $N = 12$ and $a_{12} = -21$. Using formula (4) with these values of $N$ and $a_{12}$, we have

$$-21 = a_1 + (12 - 1)d$$

$$-21 = a_1 + 11d \qquad (10)$$

We solve the system of equations (9) and (10) by replacing equation (10) by the sum of 1 times (9) and $-1$ times (10), and we have the system

$$\begin{cases} 18 = a_1 + 24d \\ 39 = 13d \end{cases}$$

which is equivalent to the system

$$\begin{cases} a_1 = -54 \\ d = 3 \end{cases}$$

Again using formula (4), with $N = 4$, $a_1 = -54$, and $d = 3$, we have

$$a_4 = -54 + (4 - 1)(3)$$
$$a_4 = -45$$

**8.3.3 DEFINITION**  In any arithmetic sequence, the elements between the first and last elements are called the *arithmetic means* between these two elements.

**ILLUSTRATION 3.**  Arithmetic sequence (1) is

$$2, 5, 8, 11, 14, 17, 20$$

Hence from Definition 8.3.3, it follows that 5, 8, 11, 14, and 17 are five arithmetic means between 2 and 20.  •

**EXAMPLE 4**
Insert three arithmetic means between 11 and 14.

**SOLUTION**
If $a_2$, $a_3$, and $a_4$ are the three arithmetic means, then we have the arithmetic sequence

$$11, a_2, a_3, a_4, 14$$

With $N = 5$ in formula (4), we obtain

$$a_5 = a_1 + (5 - 1)d$$

Because $a_1 = 11$ and $a_5 = 14$, we have

$$14 = 11 + 4d$$
$$\tfrac{3}{4} = d$$

Therefore,

$$a_2 = 11 + \tfrac{3}{4} \qquad a_3 = 11\tfrac{3}{4} + \tfrac{3}{4} \qquad a_4 = 12\tfrac{1}{2} + \tfrac{3}{4}$$
$$= 11\tfrac{3}{4} \qquad\qquad = 12\tfrac{1}{2} \qquad\qquad = 13\tfrac{1}{4}$$

The three arithmetic means are, therefore, $11\tfrac{3}{4}$, $12\tfrac{1}{2}$, and $13\tfrac{1}{4}$.

**ILLUSTRATION 4.**  To insert one arithmetic mean between the numbers $x$ and $y$, we let $M$ be the arithmetic mean and we have the arithmetic sequence

$$x, M, y$$

The common difference can be represented by either $M - x$ or $y - M$.

Therefore,
$$M - x = y - M$$
$$2M = x + y$$
$$M = \frac{x+y}{2} \qquad \bullet$$

The number $M$ obtained in Illustration 4 is called "the arithmetic mean" (or "average") of the numbers $x$ and $y$. We can generalize this concept and refer to "the arithmetic mean" of a set of numbers. We have the following formal definition.

**8.3.4 DEFINITION**  (i) The *arithmetic mean* (or *average*) of the numbers $x$ and $y$ is the number
$$\frac{x+y}{2}$$

(ii) The *arithmetic mean* (or *average*) of a set of numbers $x_1, x_2, x_3, \ldots, x_n$ is the number
$$\frac{x_1 + x_2 + x_3 + \cdots + x_n}{n}$$

**EXAMPLE 5**
On five separate examinations, a student received the following test scores: 78, 89, 62, 75, and 84. Find the arithmetic mean of these scores.

**SOLUTION**
Let $M$ be the arithmetic mean. Then from Definition 8.3.4(ii),
$$M = \frac{78 + 89 + 62 + 75 + 84}{5}$$
$$= \frac{388}{5}$$
$$= 77.6$$

An *arithmetic series* is the indicated sum of the elements of an arithmetic sequence.

**ILLUSTRATION 5.** The arithmetic series associated with the arithmetic sequence of Illustration 2 is
$$4 + 7 + 10 + 13 + 16 + 19 + 22 + 25$$
This arithmetic series can be written with the sigma notation as
$$\sum_{i=1}^{8} (3i + 1) \qquad \bullet$$

The arithmetic series associated with the general arithmetic sequence is
$$a_1 + (a_1 + d) + (a_1 + 2d) + \cdots + [a_1 + (N-1)d] \qquad (11)$$

To obtain a formula for the sum of $N$ terms of an arithmetic series, we denote the sum by $S_N$ and equate it to series (11). Hence

$$S_N = a_1 + (a_1 + d) + (a_1 + 2d) + \cdots + [a_1 + (N-1)d] \tag{12}$$

The series on the right side of equation (12) can be written in the reverse order with the $N$th term being written as $a_N$, the $(N-1)$th term being written as $(a_N - d)$, and so on, and the first term being written as $a_N - (n-1)d$. Therefore, we have the equation

$$S_N = a_N + (a_N - d) + (a_N - 2d) + \cdots + [a_N - (N-1)d] \tag{13}$$

If we add the equations (12) and (13) term-by-term, it follows that

$$S_N + S_N = (a_1 + a_N) + (a_1 + a_N) + (a_1 + a_N) + \cdots + (a_1 + a_N) \tag{14}$$

where on the right side the term $(a_1 + a_N)$ occurs $N$ times. Hence equation (14) is equivalent to

$$2S_N = N(a_1 + a_N)$$

$$S_N = \frac{N}{2}(a_1 + a_N) \tag{15}$$

Formula (15) can be written as

$$S_N = N\left(\frac{a_1 + a_N}{2}\right)$$

from which it follows that $S_N$ is the product of the number of terms and the arithmetic mean of the first and last terms.

If we substitute the value of $a_N$ from equation (4) into formula (15), we have

$$S_N = \frac{N}{2}(a_1 + [a_1 + (N-1)d])$$

$$S_N = \frac{N}{2}[2a_1 + (N-1)d] \tag{16}$$

**EXAMPLE 6**
Find the sum of the positive even integers less than 100.

**SOLUTION**
The positive even integers less than 100 form the arithmetic sequence

$$2, 4, 6, \ldots, 96, 98$$

We wish to find the sum of the associated arithmetic series, which is

$$2 + 4 + 6 + \cdots + 96 + 98$$

For this series, $a_1 = 2$, $d = 2$, $N = 49$, $a_{49} = 98$. From formula (15)

$$S_{49} = \frac{49}{2}(a_1 + a_{49})$$

$$= \frac{49}{2}(2 + 98)$$

$$= 2450$$

### EXAMPLE 7

Which of the following salary schedules offers more money to the employee over a 10-year period and how much more?

Schedule I: The starting salary is $12,000 per year and there is an annual increase of $900 thereafter.

Schedule II: The starting salary is $12,000 per year and there is a semiannual increase of $225 thereafter.

### SOLUTION

According to schedule I, the number of dollars received by the employee over a period of 10 years is the sum of the following arithmetic series of ten terms.

$$12{,}000 + 12{,}900 + \cdots + a_{10} \qquad (17)$$

Let $S_{10}$ be this sum, and from formula (16), we have

$$S_{10} = \frac{10}{2}[2a_1 + (10 - 1)d]$$

For series (17), $a_1 = 12{,}000$ and $d = 900$; hence

$$S_{10} = 5[2(12{,}000) + 9(900)]$$
$$= 160{,}500$$

According to schedule II, the number of dollars received by the employee over a period of 10 years is the sum of the following arithmetic series of twenty terms.

$$6000 + 6225 + 6450 + \cdots + a_{20} \qquad (18)$$

Denoting this sum by $S_{20}$ and using formula (16) with $a_1 = 6000$ and $d = 225$, we have

$$S_{20} = \frac{20}{2}[2a_1 + (20 - 1)d]$$

$$= 10[2(6000) + 19(225)]$$
$$= 162{,}750$$

Therefore, schedule II offers the employee $2250 more money over a 10-year period.

## EXERCISES 8.3

In Exercises 1 through 8, write the first five elements of an arithmetic sequence whose first element is a and whose common difference is d.

1. $a = 5$; $d = 3$
2. $a = -3$; $d = 2$
3. $a = 10$; $d = -4$
4. $a = 16$; $d = -5$
5. $a = -5$; $d = -7$
6. $a = 20$; $d = 10$
7. $a = x$; $d = 2y$
8. $a = u + v$; $d = -3v$

In Exercises 9 through 16, determine if the given elements form an arithmetic sequence. If they do, write the next two elements of the arithmetic sequence.

9. $3, -1, -5, -9$
10. $12, 7, 2, -3$
11. $2, -6, 10, -14$
12. $-1, -\frac{1}{3}, \frac{1}{3}, 1$
13. $\frac{1}{3}, \frac{1}{4}, \frac{1}{6}, \frac{1}{12}$
14. $\frac{1}{2}, \frac{3}{4}, \frac{7}{8}, \frac{15}{16}$
15. $x, (2x + y), (3x + 2y)$
16. $s, t, (2t - s)$

17. Find the twelfth element of an arithmetic sequence whose first element is 2 and whose second element is 5.
18. Find the tenth element of an arithmetic sequence whose first element is 8 and whose third element is 2.
19. Find the first element of an arithmetic sequence whose eighth element is 2 and whose common difference is $-2$.
20. The ninth element of an arithmetic sequence is 28 and the twenty-first element is 100. What is the 15th element?
21. The first three elements of an arithmetic sequence are 20, 16, and 12. Which element is $-96$?
22. In the arithmetic sequence whose first three elements are $\frac{1}{6}, \frac{1}{4}$, and $\frac{1}{3}$, which element is 4?
23. Insert four arithmetic means between 5 and 6.
24. Insert seven arithmetic means between 3 and 9.
25. Find the arithmetic mean of the following set of test scores: 72, 53, 85, 74, 62, 83.

In Exercises 26 through 31, find the sum of the given arithmetic series.

26. $\sum_{i=1}^{8} (3i - 1)$

27. $\sum_{i=1}^{18} \frac{2i - 1}{3}$

28. $\sum_{i=2}^{12} (8 - 2i)$

29. $\sum_{k=1}^{50} (2k - 1)$

30. $\sum_{j=1}^{20} (5j - 1)$

31. $\sum_{i=3}^{12} \left(\frac{1}{2}i - 5\right)$

32. Find the sum of all the positive integers less than 100.
33. Find the sum of all the positive even integers consisting of two digits.
34. Find the sum of all the integer multiples of 8 between 9 and 199.
35. The sum of $1000 is distributed among four people so that each person after the first receives $20 less than the preceding person. How much does each person receive?
36. To dig a well a company charges $10 for the first foot, $12.50 for the second foot, $15 for the third foot, and so on; the cost of each foot is $2.50 more than the cost of the preceding foot. What is the depth of a well that costs $2925 to dig?
37. A contractor who does not meet the deadline on the construction of a building is fined $100 per day for each of the first 10 days of extra time, and for each additional day thereafter the fine is increased by $20 each day. If the contractor is fined $2520, by how many extra days was the construction time extended?
38. It is desired to pile some logs in layers so that the top layer contains one log, the next layer contains two logs, the next layer contains three logs, and so on; each layer containing one more log than the layer on top of it. If there are 190 logs, determine if all the logs can be used in such a grouping, and if so how many logs are in the bottom layer?

## 8.4 Geometric Sequences and Series

The sequence

$$1, 2, 4, 8, 16, 32, 64, 128 \qquad (1)$$

is an example of a "geometric sequence." Each element, except the first, can be obtained by multiplying the preceding element by 2.

**8.4.1 DEFINITION** A *geometric sequence* is a sequence such that any element after the first can be obtained by multiplying the preceding element by a constant multiplier.

A geometric sequence is also called a *geometric progression*.

Because each element in a geometric sequence is obtained by multiplying the preceding element by a constant multiplier, there is a "common ratio" of two successive elements. For this reason the constant multiplier is called the *common ratio*, and we denote it by $r$. We may compute $r$ by dividing any term by the preceding one.

**ILLUSTRATION 1.** In geometric sequence (1), we have

$$\frac{2}{1} = 2; \quad \frac{4}{2} = 2; \quad \frac{8}{4} = 2; \quad \frac{16}{8} = 2; \quad \frac{32}{16} = 2; \quad \frac{64}{32} = 2; \quad \frac{128}{64} = 2$$

The common ratio $r$ is 2. ●

As with an arithmetic sequence, the number of elements in a geometric sequence is denoted by $N$, the first element is denoted by $a_1$, and the last element is denoted by $a_N$. In sequence (1), $N = 8$, $a_1 = 1$, and $a_8 = 128$.

A geometric sequence can be defined by giving the values of $a_1$ and $N$ and a recursive formula

$$a_{n+1} = a_n r \qquad (2)$$

from which every element after the first can be obtained from the preceding one.

**ILLUSTRATION 2.** If $a_1 = 128$, $N = 5$, and $a_{n+1} = a_n(-\frac{1}{4})$ the geometric sequence is

$$128, \; -32, \; 8, \; -2, \; \frac{1}{2}$$

●

The general geometric sequence, for which the first element is $a_1$, the common ratio is $r$, and the number of elements is $N$, can be obtained by applying formula (2). Starting with the element $a_1$, we obtain each successive element by multiplying the preceding element by $r$. Doing this, we have

$$a_1, \; a_1 r, \; a_1 r^2, \; a_1 r^3, \; a_1 r^4, \; \ldots, \; a_N \qquad (3)$$

In the first five elements we observe that each element is the product of $a_1$ and a power of $r$, where the exponent of $r$ is one less than the number of the element. Therefore, our intuition suggests that the $N$th (last) element is given by $a_N = a_1 r^{N-1}$.

**8.4.2 THEOREM** The $N$th element of a geometric sequence is given by

$$a_N = a_1 r^{N-1} \qquad (4)$$

The proof of Theorem 8.4.2 is by mathematical induction and is left as an exercise (see Exercise 41 in Exercises 8.4).

## EXAMPLE 1

Find the tenth element of the geometric sequence for which the first element is $\frac{5}{2}$ and the second element is $-5$.

**SOLUTION**

Let $a_{10}$ be the tenth element of the geometric sequence

$$\tfrac{5}{2}, -5, \ldots, a_{10}$$

Then

$$r = (-5) \div \frac{5}{2}$$
$$= (-5)\left(\frac{2}{5}\right)$$
$$= -2$$

From formula (4) with $a_1 = \frac{5}{2}$ and $N = 10$, we have

$$a_{10} = a_1 r^{10-1}$$
$$= \frac{5}{2}(-2)^9$$
$$= -5(2)^8$$
$$= -1280$$

## EXAMPLE 2

A city has a population of 100,000. If the population increases 10 per cent every 5 years, what will be the population at the end of 40 years?

**SOLUTION**

The population at the end of 5 years will be $100{,}000 + 0.10(100{,}000) = (1.10)(100{,}000)$. The population at the end of each successive 5-year period is 1.10 times the population at the end of the preceding 5-year period. Hence we have the geometric sequence of nine elements

$$100{,}000,\ (1.10)(100{,}000),\ (1.10)^2(100{,}000),\ \ldots,\ a_9$$

where $a_9$ is the population at the end of 40 years. From formula (4) with $N = 9$, $a_1 = 100{,}000$, and $r = 1.10$, we have

$$a_9 = a_1 r^{9-1}$$
$$= 100{,}000(1.10)^8 \qquad (5)$$

We perform the computation by using logarithms. Equating the common logarithms of each member of equation (5), we get

$$\log a_9 = \log [100{,}000(1.10)^8]$$
$$= \log 100{,}000 + \log (1.10)^8$$
$$= 5 + 8 \log 1.10$$

Using Table 2, we find $\log 1.10 = 0.0414$. Therefore,

$$\log a_9 = 5 + 8(0.0414)$$
$$= 5.3312$$

Thus

$$a_9 = \text{antilog } 5.3312$$
$$= (2.144)10^5$$

Hence, to four significant digits, the population at the end of 40 years will be 214,400.

**8.4.3 DEFINITION** In any geometric sequence the elements between the first and last elements are a set of *geometric means* between these two elements.

**ILLUSTRATION 3.** The sequence

$$2, 6, 18, 54, 162$$

is a geometric sequence with $r = 3$. Therefore, it follows from Definition 8.4.3, that the numbers 6, 18, and 54 form a set of three geometric means between 2 and 162.

Because the sequence

$$2, -6, 18, -54, 162$$

is also a geometric sequence ($r = -3$), the numbers $-6$, 18, and $-54$ form another set of three geometric means between 2 and 162. ●

**EXAMPLE 3**
Insert two geometric means between 1000 and 64.

**SOLUTION**
We consider the geometric sequence

$$1000, a_2, a_3, 64$$

where $a_2$ and $a_3$ are the required geometric means. In this geometric sequence, $N = 4$, $a_1 = 1000$, and $a_4 = 64$. From formula (4) we have

$$a_4 = a_1 r^{4-1}$$

Therefore,

$$64 = 1000 r^3$$

$$\frac{8}{125} = r^3$$

$$r = \sqrt[3]{\frac{8}{125}}$$

$$= \frac{2}{5}$$

Therefore, the geometric sequence is 1000, 400, 160, 64. Thus 400 and 160 are two geometric means between 1000 and 64.

If $m$ is a geometric mean between two numbers $x$ and $y$, then we have the geometric sequence

$$x, m, y$$

and hence

$$\frac{m}{x} = \frac{y}{m}$$

$$m^2 = xy$$

This equation has two solutions: $m = \sqrt{xy}$ and $m = -\sqrt{xy}$. Because we want the geometric mean to be between the numbers $x$ and $y$, we choose for the value of $m$ the number having the same sign as $x$ and $y$. We have then the following definition.

**8.4.4 DEFINITION**  The *geometric mean of the numbers $x$ and $y$* is

$$\sqrt{xy} \quad \text{if } x \text{ and } y \text{ are positive}$$

and

$$-\sqrt{xy} \quad \text{if } x \text{ and } y \text{ are negative}$$

**EXAMPLE 4**
Find the geometric mean of each of the following sets of numbers.

(a) 4 and 9   (b) $-\tfrac{3}{10}$ and $-\tfrac{5}{6}$

**SOLUTION**
In each part, we let $m$ be the geometric mean and we compute $m$ by applying Definition 8.4.4.

(a) $m = \sqrt{4 \cdot 9}$
$= \sqrt{36}$
$= 6$

(b) $m = -\sqrt{(-\tfrac{3}{10})(-\tfrac{5}{6})}$
$= -\sqrt{\tfrac{15}{60}}$
$= -\sqrt{\tfrac{1}{4}}$
$= -\tfrac{1}{2}$

By generalizing Definition 8.4.4, we define a *geometric mean* of a set of numbers $x_1, x_2, x_3, \ldots, x_n$ to be the number $\sqrt[n]{x_1 x_2 x_3 \cdots x_n}$.

**ILLUSTRATION 4.**  A geometric mean of the numbers 4, 10, and 25 is

$$\sqrt[3]{(4)(10)(25)} = \sqrt[3]{1000}$$
$$= 10 \qquad \bullet$$

With any geometric sequence we have an associated *geometric series*, which is the indicated sum of the elements of the geometric sequence.

**ILLUSTRATION 5.**  The geometric sequence of Illustration 2 is

$$128, -32, 8, -2, \frac{1}{2}$$

Associated with this sequence is the geometric series

$$128 - 32 + 8 - 2 + \frac{1}{2}$$

which can be written with the sigma notation as

$$\sum_{i=1}^{5} 128 \left(-\frac{1}{4}\right)^{i-1}$$

If $S_N$ is the sum of $N$ terms of a geometric series, we can obtain a formula for $S_N$ by first writing $S_N$ as the sum of the elements of sequence (3).

$$S_N = a_1 + a_1 r + a_1 r^2 + a_1 r^3 + \cdots + a_1 r^{N-2} + a_1 r^{N-1} \qquad (6)$$

If we multiply both members of equation (6) by $r$, we have

$$rS_N = a_1 r + a_1 r^2 + a_1 r^3 + a_1 r^4 + \cdots + a_1 r^{N-1} + a_1 r^N \qquad (7)$$

The sum of 1 times equation (6) and $-1$ times equation (7) gives

$$S_n - rS_N = a_1 - a_1 r^N$$
$$(1 - r)S_N = a_1 - a_1 r^N$$

If $r - 1 \neq 0$, we can divide each member of this equation by $(1 - r)$, and we obtain

$$S_N = \frac{a_1 - a_1 r^N}{1 - r} \qquad \text{if } r \neq 1 \qquad (8)$$

$$S_N = \frac{a_1(1 - r^N)}{1 - r} \qquad \text{if } r \neq 1 \qquad (9)$$

If $r = 1$, we have the rather trivial geometric series

$$a_1 + a_1 + a_1 + \cdots + a_1 \qquad (N \text{ terms of } a_1)$$

For this series, $S_N = Na_1$.

A formula for $S_N$ in terms of $a_1$, $r$, and $a_N$ can be obtained by first expressing $a_1 r^N$ as $r(a_1 r^{N-1})$ in formula (8). Doing this, we have

$$S_N = \frac{a_1 - r(a_1 r^{N-1})}{1 - r}$$

From formula (4) it follows that we can replace $a_1 r^{N-1}$ by $a_N$. Thus

$$S_N = \frac{a_1 - ra_N}{1 - r} \qquad \text{if } r \neq 1 \qquad (10)$$

## EXAMPLE 5
Find the sum of the geometric series

$$\sum_{i=1}^{5} 2 \left(\frac{1}{3}\right)^{i-1}$$

### SOLUTION
For the given series, $a_1 = 2$, $r = \frac{1}{3}$, and $N = 5$. Hence, from formula (9), we obtain

$$S_5 = \frac{a_1(1 - r^5)}{1 - r}$$

$$= \frac{2\left[1 - \left(\frac{1}{3}\right)^5\right]}{1 - \frac{1}{3}}$$

$$= \frac{2\left(1 - \frac{1}{243}\right)}{\frac{2}{3}}$$

$$= 3 - \frac{1}{81}$$

$$= 2\frac{80}{81}$$

## EXAMPLE 6
Find the sum of the geometric series associated with the sequence of Example 1.

### SOLUTION
The sequence of Example 1, consisting of 10 elements, is

$$\frac{5}{2}, -5, \ldots, -1280$$

Hence the associated geometric series is

$$\frac{5}{2} - 5 + 10 - \cdots - 1280 \quad (10 \text{ terms})$$

Using formula (10) with $a_{10} = -1280$, $r = -2$, and $a_1 = \frac{5}{2}$, we have

$$S_{10} = \frac{a_1 - ra_{10}}{1 - r}$$

$$= \frac{\frac{5}{2} - (-2)(-1280)}{1 - (-2)}$$

$$= \frac{\frac{5}{2} - 2560}{3}$$

$$= \frac{5 - 5120}{6}$$

$$= -852\frac{1}{2}$$

In the next example we use formula (4) of Section 4.6, which states that if $P$ dollars is invested at an interest rate of $100i$ per cent compounded $m$ times per year and if $A_n$ is the number of dollars in the amount of the investment at the end of $n$ interest periods, then

$$A_n = P\left(1 + \frac{i}{m}\right)^n \qquad (11)$$

### EXAMPLE 7

In order to create a sinking fund that will provide capital to build a new plant, a company deposits $25,000 into an account on January 1 every year for 10 years. If the account earns 6 per cent interest, compounded annually, how much is in the sinking fund immediately after the tenth deposit is made?

### SOLUTION

Immediately after the tenth deposit is made, the tenth payment has earned no interest; the ninth payment has earned interest for 1 year; the eighth payment has earned interest for 2 years; and so on; and the first payment has earned interest for 9 years. To find the number of dollars in the fund immediately after the tenth payment, we apply formula (11) (with $P = 25{,}000$, $i = 0.06$, and $m = 1$) to find the dollar amount of each payment. The results are as follows:

10th payment: $25{,}000$ (no interest)
9th payment: $25{,}000(1.06)^1$ (interest for 1 year; $n = 1$)
8th payment: $25{,}000(1.06)^2$ (interest for 2 years; $n = 2$)
$\vdots$
1st payment: $25{,}000(1.06)^9$ (interest for 9 years; $n = 9$)

If $x$ dollars is the total amount in the sinking fund immediately after the tenth deposit is made, then

$$x = 25{,}000 + 25{,}000(1.06)^1 + 25{,}000(1.06)^2 + \cdots + 25{,}000(1.06)^9$$

Therefore, $x$ is the sum of a geometric series for which $N = 10$, $r = 1.06$, and $a_1 = 25{,}000$. Applying formula (9), we have

$$x = \frac{25{,}000[1 - (1.06)^{10}]}{1 - 1.06} \qquad (12)$$

We compute $(1.06)^{10}$ by logarithms. We have

$$\begin{aligned}\log(1.06)^{10} &= 10 \log 1.06 \\ &= 10(0.0253) \\ &= 0.253\end{aligned}$$

Hence

$$\begin{aligned}(1.06)^{10} &= \text{antilog } 0.253 \\ &= 1.79\end{aligned}$$

Substituting this value into equation (12), we obtain

$$\begin{aligned}x &= \frac{25{,}000(1 - 1.79)}{-0.06} \\ &= (3.29)10^5\end{aligned}$$

Hence the amount in the sinking fund, immediately after the tenth deposit is made, is $329,000, to three significant digits.

## EXERCISES 8.4

In Exercises 1 through 8, write the first five elements of a geometric sequence whose first element is a and whose common ratio is r.

1. $a = 5; r = 3$
2. $a = 3; r = 2$
3. $a = 8; r = -\frac{1}{2}$
4. $a = 2; r = \sqrt{2}$
5. $a = -\frac{9}{16}; r = -\frac{2}{3}$
6. $a = -81; r = \frac{1}{3}$
7. $a = \frac{x}{y}; r = -\frac{y}{x}$
8. $a = \frac{s}{t}; r = \frac{1}{u}$

In Exercises 9 through 16, determine if the given elements form a geometric sequence. If they do, write the next two elements of the geometric sequence.

9. 1, 3, 9
10. 2, −4, 8
11. $\sqrt{2}, \sqrt{6}, 3\sqrt{2}$
12. $\frac{1}{2}, \frac{1}{3}, \frac{1}{4}$
13. 3.33, 2.22, 1.11
14. $3^{-2}, 3^0, 3^2$
15. $-6, 2, -\frac{2}{3}$
16. $\sqrt[3]{3}, \sqrt[6]{3}, 1$

17. Find the third element of a geometric sequence whose fifth element is 81 and whose ninth element is 16.
18. If the first element of a geometric sequence is $\frac{1}{8}$ and the eighth element is $-16$, find the sixth element.
19. Find the common ratio of a geometric sequence whose third element is $-2$ and whose sixth element is 54.
20. If the first element of a geometric sequence is 1 and the common ratio is 3, find the smallest four-digit numeral that represents an element of this geometric sequence.
21. In the geometric sequence whose first element is 0.0003 and whose common ratio is 10, which element is 3,000,000?
22. In the geometric sequence whose first three elements are 27, −18, and 12, which element is $-\frac{512}{729}$?
23. Insert five geometric means between 1 and 64.
24. Insert three geometric means between 162 and 2.
25. Insert two geometric means between $\sqrt{3}$ and 3.
26. Find the geometric mean of 16 and 25.
27. Find the geometric mean of $-\frac{2}{3}$ and $-6$.
28. Find a geometric mean of the numbers 9, 21, and 49.

In Exercises 29 through 34, find the sum of the given geometric series.

29. $\sum_{i=1}^{8} 2^i$
30. $\sum_{i=2}^{7} (-3)^i$
31. $\sum_{j=3}^{9} 5\left(\frac{1}{3}\right)^{j-3}$
32. $\sum_{k=1}^{6} \left(\frac{2}{3}\right)^k$
33. $\sum_{i=1}^{10} (1.02)^i$
34. $\sum_{i=1}^{12} (1.02)^{1-i}$

35. Find a sequence of four numbers, the first of which is 6 and the fourth of which is 16, if the first three numbers form an arithmetic sequence, and the last three numbers form a geometric sequence.
36. Three numbers form an arithmetic sequence having a common difference of 4. If the first number is increased by 2, the second number is increased by 3, and the third number is increased by 5, the resulting numbers form a geometric sequence. Find the original numbers.
37. From a barrel containing 1 gallon of wine, 1 pint is withdrawn and then the barrel is filled with water. If this procedure is followed six times, what fractional part of the original contents is in the barrel?

38. If a town having a population of 5000 in 1971 has a 20 per cent increase every 5 years, what is its expected population in the year 2001?
39. Payments of $1000 are deposited into a sinking fund every 6 months and the account earns 4 per cent interest, compounded semiannually. How much is in the fund immediately after the twentieth payment is made?
40. Three numbers, whose sum is 3, form an arithmetic sequence, and their squares form a geometric sequence. What are the numbers?
41. Prove Theorem 8.4.2.

## 8.5 Infinite Geometric Series

Our discussion of series has so far been restricted to those associated with finite sequences. The series associated with the infinite sequence

$$a_1, a_2, a_3, \ldots, a_n, \ldots$$

is denoted by

$$a_1 + a_2 + a_3 + \cdots + a_n + \cdots \tag{1}$$

and is called an *infinite series*. But what is the meaning of expression (1)? That is, what do we mean by the "sum" of an infinite number of terms, and under what circumstances does such a "sum" exist? As stated in Section 8.1, the answers to these questions depend upon the concept of "limit" which is studied in a course in the calculus. However, for some particular infinite series we can give an intuitive idea of the concept of "sum."

Consider an infinite arithmetic sequence of positive elements, for instance, the sequence

$$3, 5, 7, \ldots, [3 + (n - 1)(2)], \ldots$$

whose associated arithmetic series is

$$3 + 5 + 7 + \cdots + [3 + (n - 1)(2)] + \cdots \tag{2}$$

We can obtain a number that is greater than any preassigned positive number by finding the sum of a sufficient number of terms of series (2). Hence no finite number, representing the "sum" of infinite series (2), can exist. A similar argument holds for any infinite arithmetic series of positive terms.

Suppose that we have an infinite geometric series. If $|r| > 1$, then the absolute value of each term is greater than the absolute value of the preceding term, and therefore such an infinite series cannot have a finite "sum." Observe that this is the situation with the infinite geometric series

$$2 + 6 + 18 + 54 + \cdots + 2(3)^{n-1} + \cdots$$

for which $r = 3$, and the infinite geometric series

$$1 - 5 + 25 - 125 + \cdots + (-5)^{n-1} + \cdots$$

for which $|r| = 5$. Furthermore, an infinite geometric series for which $|r| = 1$ cannot have a finite "sum" because the absolute value of each term is the same, and hence the sum of the absolute values of the terms can again be made greater than any preassigned positive number by taking a sufficient number of terms of the series.

But an infinite geometric series for which $|r| < 1$ provides us with a different situation. For instance, consider the series

$$1 + \frac{1}{2} + \frac{1}{4} + \frac{1}{8} + \cdots + \frac{1}{2^{n-1}} + \cdots \qquad (3)$$

where $r = \frac{1}{2}$. From formula (9) of Section 8.4, if $S_N$ is the sum of $N$ terms of a geometric series,

$$S_N = \frac{a_1(1 - r^N)}{1 - r} \qquad (4)$$

Therefore, for a finite number $N$ of terms of series (3) with $a_1 = 1$ and $r = \frac{1}{2}$, we have

$$S_N = \frac{(1)\left[1 - \left(\frac{1}{2}\right)^N\right]}{1 - \frac{1}{2}}$$

$$= \frac{1 - \left(\frac{1}{2}\right)^N}{\frac{1}{2}}$$

or, equivalently,

$$S_N = 2\left[1 - \left(\frac{1}{2}\right)^N\right] \qquad (5)$$

Applying formula (5) to successive values of $N$, we obtain

$$S_1 = 2\left[1 - \left(\frac{1}{2}\right)^1\right] \qquad S_2 = 2\left[1 - \left(\frac{1}{2}\right)^2\right]$$
$$= 2\left(\frac{1}{2}\right) \qquad\qquad\quad = 2\left(\frac{3}{4}\right)$$
$$= 1 \qquad\qquad\qquad\quad\; = \frac{3}{2}$$
$$\qquad\qquad\qquad\qquad\qquad = 1\frac{1}{2}$$

$$S_3 = 2\left[1 - \left(\frac{1}{2}\right)^3\right] \qquad S_4 = 2\left[1 - \left(\frac{1}{2}\right)^4\right]$$
$$= 2\left(\frac{7}{8}\right) \qquad\qquad = 2\left(\frac{15}{16}\right)$$
$$= \frac{7}{4} \qquad\qquad = \frac{15}{8}$$
$$= 1\frac{3}{4} \qquad\qquad = 1\frac{7}{8}$$
$$S_5 = 2\left[1 - \left(\frac{1}{2}\right)^5\right] \qquad S_6 = 2\left[1 - \left(\frac{1}{2}\right)^6\right]$$
$$= 2\left(\frac{31}{32}\right) \qquad\qquad = 2\left(\frac{63}{64}\right)$$
$$= \frac{31}{16} \qquad\qquad = \frac{63}{32}$$
$$= 1\frac{15}{16} \qquad\qquad = 1\frac{31}{32}$$
$$\vdots$$
$$S_{10} = 2\left[1 - \left(\frac{1}{2}\right)^{10}\right]$$
$$= 2\left(\frac{1023}{1024}\right)$$
$$= \frac{1023}{512}$$
$$= 1\frac{511}{512}$$

and so on. From these values, it follows that

$$2 - S_1 = 1$$
$$2 - S_2 = \frac{1}{2}$$
$$2 - S_3 = \frac{1}{4}$$
$$2 - S_4 = \frac{1}{8}$$
$$2 - S_5 = \frac{1}{16}$$
$$2 - S_6 = \frac{1}{32}$$
$$\vdots$$
$$2 - S_{10} = \frac{1}{512}$$

and so on. We intuitively see that we can make the value of $S_N$ as close to 2 as we please by taking $N$ large enough. In other words, we can make the difference betwen 2 and $S_N$ as small as we please by taking $N$ sufficiently large. Therefore, we state that the limiting value of $S_N$, as $N$ increases without bound, is 2, and we write

$$\lim_{N \to \infty} S_N = 2 \tag{6}$$

Equation (6) is read "the limit of $S_N$ as $N$ increases without bound is 2." Observe that "$N$ increases without bound" is written "$N \to \infty$," where $\infty$ is the symbol for infinity. We emphasize again that equation (6) has not been defined precisely in this intuitive discussion.

Consider now the general infinite geometric series

$$a_1 + a_1 r + a_1 r^2 + a_1 r^3 + \cdots + a_1 r^{n-1} + \cdots \qquad |r| < 1$$

The sum of the first $N$ terms of this series is given by formula (4) or, equivalently, by

$$S_N = \frac{a_1}{1 - r}(1 - r^N) \tag{7}$$

Because $|r| < 1$, as $N$ increases the absolute value of $r^N$ decreases, and the absolute value of $r^N$ can be made as small as we please by taking $N$ sufficiently large. (Note that when $r = \frac{1}{2}$, we have $(\frac{1}{2})^1 = \frac{1}{2}$, $(\frac{1}{2})^2 = \frac{1}{4}$, $(\frac{1}{2})^3 = \frac{1}{8}$, $(\frac{1}{2})^4 = \frac{1}{16}$, $(\frac{1}{2})^5 = \frac{1}{32}$, $(\frac{1}{2})^6 = \frac{1}{64}$, $\ldots$, $(\frac{1}{2})^{10} = \frac{1}{1024}$, and so on.) Hence $\lim_{N \to \infty} r^N = 0$, and thus from formula (7), we have

$$\lim_{N \to \infty} S_N = \frac{a_1}{1 - r}$$

We therefore have the following definition.

**8.5.1 DEFINITION**   The *sum* $S$ of an infinite geometric series, for which $|r| < 1$, is given by

$$S = \frac{a_1}{1 - r}$$

**ILLUSTRATION 1.**  For series (3), $a_1 = 1$ and $r = \frac{1}{2}$. Therefore, from Definition 8.5.1, if $S$ is the sum of this series,

$$S = \frac{a_1}{1 - r}$$
$$= \frac{1}{1 - \frac{1}{2}}$$
$$= 2$$

This result agrees with the discussion leading to equation (6). ●

## EXAMPLE 1

Find the sum of the infinite geometric series

$$6 + 4 + \frac{8}{3} + \cdots + 6\left(\frac{2}{3}\right)^{n-1} + \cdots$$

**SOLUTION**
Applying Definition 8.5.1 with $a_1 = 6$ and $r = \frac{2}{3}$, we have

$$S = \frac{a_1}{1-r}$$

$$= \frac{6}{1 - \frac{2}{3}}$$

$$= \frac{6}{\frac{1}{3}}$$

$$= 18$$

An important application of Definition 8.5.1 is to express a nonterminating repeating decimal as a fraction, thus showing that such a decimal numeral is a representation of a rational number. To indicate a nonterminating repeating decimal, we write a bar over the repeated digits. Then $0.33\overline{3}$ indicates $0.3333\ldots$ and $4.024\overline{24}$ indicates $4.024242424\ldots$.

**ILLUSTRATION 2.** The nonterminating repeating decimal $0.33\overline{3}$ can be written as

$$0.3 + 0.03 + 0.003 + 0.0003 + \cdots$$

or, equivalently,

$$\frac{3}{10} + \frac{3}{100} + \frac{3}{1000} + \frac{3}{10{,}000} + \cdots + \frac{3}{10^n} + \cdots$$

which is an infinite geometric series with $a_1 = \frac{3}{10}$ and $r = \frac{1}{10}$. Denoting the sum of this series by $S$, we have from Definition 8.5.1,

$$S = \frac{a_1}{1-r}$$

$$= \frac{\frac{3}{10}}{1 - \frac{1}{10}}$$

$$= \frac{\frac{3}{10}}{\frac{9}{10}}$$

$$= \frac{1}{3}$$

Therefore, the nonterminating repeating decimal $0.33\overline{3}$ and the fraction $\frac{1}{3}$ are representations for the same rational number. •

**EXAMPLE 2**
Express the nonterminating repeating decimal $4.024\overline{24}$ as a fraction.

**SOLUTION**
The given decimal can be written as

$$4 + \left[\frac{24}{1000} + \frac{24}{100{,}000} + \frac{24}{10{,}000{,}000} + \cdots + \frac{24}{10^{2n+1}} + \cdots\right]$$

The series in brackets is an infinite geometric series with $a_1 = \frac{24}{1000}$ and $r = \frac{1}{100}$. If $S$ is the sum of this series, then

$$S = \frac{a_1}{1-r}$$

$$= \frac{\frac{24}{1000}}{1 - \frac{1}{100}}$$

$$= \frac{\frac{24}{1000}}{\frac{99}{100}}$$

$$= \frac{24}{990}$$

$$= \frac{4}{165}$$

Thus

$$4.024\overline{24} = 4 + \frac{4}{165}$$

$$= \frac{664}{165}$$

**EXAMPLE 3**
A ball is dropped from a height of 36 meters and each time it strikes the ground it rebounds to a height of two-thirds of the distance from which it fell. Find the total distance traveled by the ball before it comes to rest.

**SOLUTION**
Let $d$ meters be the total distance traveled by the ball. Then

$$d = 36 + \left[(36)\left(\frac{2}{3}\right) + (36)\left(\frac{2}{3}\right) + (36)\left(\frac{2}{3}\right)^2 + (36)\left(\frac{2}{3}\right)^2 + \cdots\right]$$

$$= 36 + 2\left[(36)\left(\frac{2}{3}\right) + (36)\left(\frac{2}{3}\right)^2 + \cdots + (36)\left(\frac{2}{3}\right)^n + \cdots\right]$$

The series in brackets is an infinite geometric series with $a_1 = (36)(\frac{2}{3})$ and

$r = \frac{2}{3}$. If $S$ is the sum of this series, then

$$S = \frac{a_1}{1-r}$$

$$= \frac{24}{1-\frac{2}{3}}$$

$$= \frac{24}{\frac{1}{3}}$$

$$= 72$$

Therefore,

$$d = 36 + 2(72)$$
$$= 180$$

Thus the ball travels 180 meters before coming to rest.

## EXERCISES 8.5

*In Exercises 1 through 10, find the sum of the given infinite geometric series.*

1. $16 + 4 + 1 + \cdots$
2. $\frac{1}{3} + \frac{1}{9} + \frac{1}{27} + \cdots$
3. $60 - 6 + 0.6 - \cdots$
4. $4 - 1.6 + 0.64 - \cdots$
5. $\frac{2}{3} + \frac{1}{9} + \frac{1}{54} + \cdots$
6. $1 + (1.04)^{-1} + (1.04)^{-2} + \cdots$
7. $\frac{4}{3} - 1 + \frac{3}{4} - \cdots$
8. $-2 - \frac{1}{2} - \frac{1}{8} - \cdots$
9. $3 + \sqrt{3} + 1 + \cdots$
10. $(2 + \sqrt{3}) + 1 + (2 - \sqrt{3}) + \cdots$

*In Exercises 11 through 22, express the given nonterminating repeating decimal as a fraction.*

11. $0.6\overline{66}$
12. $0.27\overline{27}$
13. $0.81\overline{81}$
14. $0.252\overline{252}$
15. $2.9\overline{99}$
16. $3.1416\overline{1416}$
17. $1.234\overline{234}$
18. $7.9\overline{99}$
19. $0.4653\overline{4653}$
20. $2.045\overline{045}$
21. $3.225\overline{44}$
22. $6.507\overline{11}$

23. Express the nonterminating repeating decimal $2.464\overline{646}$ as a fraction by two methods: (a) Consider $2.464\overline{646}$ as $2 + 0.46 + 0.0046 + 0.000046 + \cdots$; and (b) consider $2.464\overline{646}$ as $2.4 + 0.064 + 0.00064 + 0.0000064 + \cdots$.

24. Express the nonterminating repeating decimal $5.1696\overline{969}$ as a fraction by two methods: (a) Consider $5.1696\overline{969}$ as $5.1 + 0.069 + 0.00069 + 0.0000069 + \cdots$; and (b) consider $5.1696\overline{969}$ as $5.16 + 0.0096 + 0.000096 + 0.00000096 + \cdots$.

25. A ball is dropped from a height of 12 meters. Each time it strikes the ground it bounces back to a height of three-fourths of the distance from which it fell. Find the total distance traveled by the ball before it comes to rest.

26. What is the total distance traveled by a tennis ball

before coming to rest if it is dropped from a height of 100 meters and if, after each fall, it rebounds $\frac{11}{20}$ of the distance from which it fell?

27. The path of each swing, after the first, of a pendulum bob is 0.93 as long as the path of the previous swing (from one side to the other side). If the path of the first swing is 28 centimeters long, and air resistance eventually brings the pendulum to rest, how far does the bob travel before it comes to rest?

28. Find an infinite geometric series whose sum is 6 and such that each term is four times the sum of all the terms that follow it.

## REVIEW EXERCISES (CHAPTER 8)

1. Write the first six elements of the sequence whose general element is

$$a_n = (-1)^{n-1} \frac{2n-1}{3^n}$$

2. Write the series and find the sum of the series.

$$\sum_{i=2}^{7} \frac{i+1}{i-1}$$

3. Write the following series with sigma notation.

$$\frac{1}{2}x^2 - \frac{1}{4}x^4 + \frac{1}{8}x^6 - \frac{1}{16}x^8$$

In Exercises 4 through 7, find the sum of the given series.

4. $\sum_{j=1}^{10} 3\left(\frac{1}{2}\right)^j$

5. $\sum_{k=1}^{30} (3k+1)$

6. $\sum_{i=1}^{20} \left(\frac{1}{3}i + 3\right)$

7. $\sum_{i=1}^{10} (1.01)^i$

In Exercises 8 through 11, use mathematical induction to prove that the given formula is true for all positive-integer values of n.

8. $\sum_{i=1}^{n} (4i+1) = n(2n+3)$

9. $\sum_{i=1}^{n} 4^i = \frac{4(4^n-1)}{3}$

10. $\sum_{i=1}^{n} \frac{1}{(3i-2)(3i+1)} = \frac{n}{3n+1}$

11. $\sum_{i=1}^{n} i(2i+1) = \frac{4n^3 + 9n^2 + 5n}{6}$

12. Find $x$ so that the numbers 25, $x$, and 9 form a geometric sequence.

13. Find $x$ so that the numbers $\frac{1}{16}$, $\frac{1}{4}$, and $x$ form an arithmetic sequence.

14. Find the sum of the positive odd integers between 10 and 100.

15. Find the first element of a geometric sequence whose fourth element is $-3$ and whose eighth element is $-243$.

16. Insert five geometric means between 192 and 3.

17. In the arithmetic sequence whose first three elements are $-8$, $-5$, and $-2$, which element is 52?

18. Find the thirtieth element of the arithmetic sequence whose seventeenth element is 7 and whose forty-seventh element is 31.

19. Find the sum of the infinite geometric series:

$$0.4 + 0.02 + 0.001 + \cdots$$

*In Exercises 20 and 21, express the given nonterminating repeating decimal as a fraction.*

**20.** $0.727\overline{272}$

**21.** $4.6636\overline{363}$

**22.** Prove by mathematical induction that $(x - y)$ is a factor of $(x^{2n} - y^{2n})$ for all positive-integer values of $n$.

**23.** A pile of logs has 30 logs in the bottom layer, 29 logs in the next to bottom layer, and so on, and the top layer contains 5 logs; each layer except the last contains one less log than the layer beneath it. How many logs are in the pile?

**24.** In a certain culture, the number of bacteria increases 20 per cent every 30 minutes. If there are 1000 bacteria present initially, find a formula for determining the number of bacteria in the culture at the end of $t$ hours. How many bacteria are in the culture at the end of 5 hours?

**25.** How many ancestors, to the nearest thousand, did you have 20 generations ago under the assumption that each ancestor appears only once in your family tree? What is the total number of ancestors, to the nearest thousand, in all 20 generations?

**26.** A man borrows $20,000 and places a mortgage on his home. He agrees that at the end of each year for 10 years, he will repay $2000 of the principal together with interest at the rate of 10 per cent per year on the amount outstanding during the year. What is the total amount to be paid in 10 years?

**27.** Three numbers whose sum is 35 form a geometric sequence. If 1 is subtracted from the first number, 2 is subtracted from the second number, and 8 is subtracted from the third number, the resulting differences form an arithmetic sequence. What are the numbers?

# 9
# Permutations, Combinations, and the Binomial Theorem

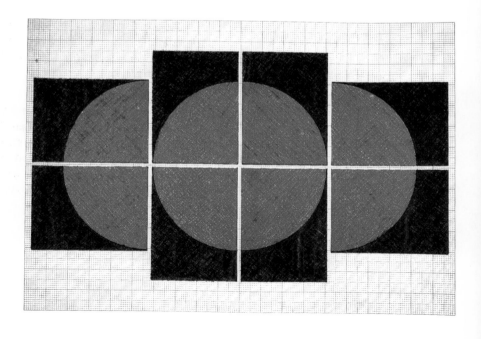

# 9.1 Counting

We wish to consider the number of ways a set of events can occur under certain conditions. For instance, how many positive integers of two different digits can be formed from the integers 1, 2, 3, and 4? In how many ways can a chairman and a secretary be chosen from a committee of five people? If there are seven doors giving access to a building, in how many ways can a person enter the building by one door and leave by a different door? The answers to these questions can be obtained by applying the following axiom.

**9.1.1 AXIOM** *Fundamental Principle of Counting.* If one event can occur in $m$ different ways, and if, after it has happened in one of these ways, a second event can occur in $n$ different ways, then both events can occur in the order stated, in $m \cdot n$ different ways.

**ILLUSTRATION 1.** To determine how many positive integers of two different digits can be formed from the integers 1, 2, 3, and 4, we apply Axiom 9.1.1 with $m = 4$ (the number of ways the tens' digit can be selected from among the four digits) and $n = 3$ (the number of ways the units' digit can be selected from among the three remaining digits after the tens' digit has been chosen). Hence there are $4 \cdot 3$, or 12, positive integers. To enumerate the twelve positive integers, we use a tree diagram as shown in Figure 9.1.1. The twelve positive integers are 12, 13, 14, 21, 23, 24, 31, 32, 34, 41, 42, and 43. ●

**ILLUSTRATION 2.** If in Illustration 1, repetition of digits is allowed, then the tens' digit can be chosen in four ways and the units' digit can also be chosen in four ways. Therefore, using Axiom 9.1.1 with $m = 4$ and $n = 4$, we have $4 \cdot 4$, or 16, possible positive integers. ●

**ILLUSTRATION 3.** Suppose that we wish to determine how many ways a chairman and a secretary can be chosen from a committee of five people. The number of ways that the chairman can be selected from among the five people on the committee is five and then, after the chairman has been chosen, there are four ways in which the secretary can be selected from the remaining four people. Hence, from Axiom 9.1.1, the two positions can be filled in $5 \cdot 4$, or 20, different ways. ●

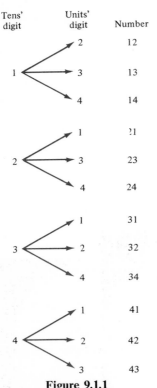

**Figure 9.1.1**

### EXAMPLE 1
If there are seven doors giving access to a building, in how many ways can a person enter the building by one door and leave by a different door?

### SOLUTION
The entry door can be chosen in any one of seven ways, and then the exit door can be selected from among the six remaining doors. Therefore, from Axiom 9.1.1, the number of ways that a person can enter the building by one door and leave by a different door is $7 \cdot 6$, or 42.

Two events are said to be *independent* of each other if neither one of them is influenced by the outcome of the other. For two independent events we have the following principle of counting, which is similar to Axiom 9.1.1.

**9.1.2 AXIOM** *Fundamental Principle of Counting.* If one of two independent events can occur in $m$ different ways and the other can occur in $n$ different ways, then both events can occur, without regard to order, in $m \cdot n$ different ways.

**ILLUSTRATION 4.** A lunch is to consist of a bowl of soup and a sandwich. If there are three different kinds of soup and five different kinds of sandwich, we wish to determine how many different lunches can be made. The choice of the soup and the choice of the sandwich are independent of each other, and so we apply Axiom 9.1.2. Therefore, there are $3 \cdot 5$, or 15, different lunches. •

**ILLUSTRATION 5.** In Illustration 1, we enumerated, by the use of a tree diagram, the twelve positive integers of two different digits that can be formed from the integers 1, 2, 3, and 4. Suppose now that we wish to find how many positive integers of three different digits can be formed with the same integers 1, 2, 3, and 4. In this case we have a tree diagram similar to that of Figure 9.1.1 but at the ends of each of the branches there are two additional branches. Such a tree diagram is shown in Figure 9.1.2. We see that there are 24 possible positive integers. Observe that 24 is the product of the number of ways, four, that the hundreds' digit can be chosen, the number of ways, three, that the tens' digit can be selected (after determining the hundreds' digit), and the number of possible choices, two, for the units' digit (after the tens' digit is chosen); that is, $24 = 4 \cdot 3 \cdot 2$. •

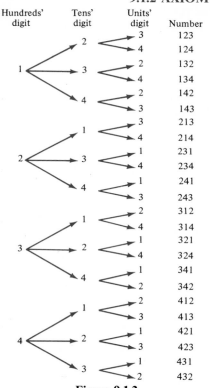

**Figure 9.1.2**

Illustration 5 shows an application of the following axiom.

**9.1.3 AXIOM** *Fundamental Principle of Counting.* If one event can occur in $n_1$ different ways, a second event can occur in $n_2$ different ways, a third event can occur in $n_3$ different ways, and so on, and a $k$th event can occur in $n_k$ different ways, then the total number of ways that the events may take place in the stated order is $n_1 n_2 n_3 \cdots n_k$.

**ILLUSTRATION 6.** We wish to determine how many different arrangements of six distinct books each can be made on a shelf with space for six books. We apply Axiom 9.1.3. In the first space we may place any one of the six books. In the next space we may place any one of the five remaining books; in the next space we may place any one of the four remaining books; and so on. Therefore, the number of different arrangements is $6 \cdot 5 \cdot 4 \cdot 3 \cdot 2 \cdot 1$, or 720. •

When applying Axiom 9.1.3, it is sometimes convenient to write a horizontal mark for each of the events that is to occur. Then, in the order in which the events occur, insert above the corresponding mark the number of ways that the event can happen. The number of ways the set of events can occur in the required order is equal to the product of the numbers above the horizontal marks. If there are any special conditions affecting the order of events, these conditions should be considered first. The following example demonstrates the procedure.

EXAMPLE 2

How many different arrangements, each consisting of four different letters, can be formed from the letters of the word "personal" if each arrangement is to begin and end with a vowel?

SOLUTION

There are three vowels and five consonants and four positions to be filled. We write four horizontal marks: $\_\ \_\ \_\ \_$. In the first position we may choose any one of the three vowels and for the fourth position we may choose any one of the two remaining vowels. The second position can then be filled by any one of the six remaining letters, and the third position by any one of the five remaining letters. We write the numbers of choices over the horizontal marks and multiply, and we have

$$\underline{3} \cdot \underline{6} \cdot \underline{5} \cdot \underline{2} = 180$$

Therefore, there are 180 different arrangements.

EXAMPLE 3

How many positive even integers of three different digits each, are less than 400?

SOLUTION

The three digits are selected from the digits 0, 1, 2, 3, 4, 5, 6, 7, 8, and 9. If a positive integer of three digits is less than 400, its hundreds' digit must be 1, 2, or 3. We consider separately the numbers for which the hundreds' digit is 2 and those for which it is 1 or 3.

If the hundreds' digit is 2, it can be chosen in only one way. Because the number is even, it must have a units' digit of 0, 4, 6, or 8; hence the units' digit can be chosen in one of four ways. The tens' digit can be any one of the eight digits not already used. Hence the number of positive even integers less than 400, for which the hundreds' digit is 2, is

$$\underline{1} \cdot \underline{8} \cdot \underline{4} = 32$$

If the hundreds' digit is either 1 or 3, it can be selected in two ways. The units' digit may be 0, 2, 4, 6, or 8, and thus it can be selected in five ways. The tens' digit can then be chosen from any one of the eight remaining digits. Thus the number of positive even integers less than 400 for which the hundreds' digit is either 1 or 3 is

$$\underline{2} \cdot \underline{8} \cdot \underline{5} = 80$$

We, therefore, conclude that the number of positive even integers less than 400, of three different digits, is 32 + 80, or 112.

## EXAMPLE 4

If three of the books in Illustration 6 are mathematics books and three are physics books, how many different arrangements of the six books can be made on the shelf if books on the same subject are to be kept together?

## SOLUTION

If the mathematics books are on the left, then the first book may be chosen in any one of three ways; the second book can be chosen from either of the two remaining books; and the third book must be the remaining mathematics book. The fourth book can be any one of the three physics books; the fifth book can be selected from either of the two remaining physics books; and the sixth book must be the remaining physics book. Therefore, if the mathematics books are on the left, the number of arrangements is

$$\underline{3} \cdot \underline{2} \cdot \underline{1} \cdot \underline{3} \cdot \underline{2} \cdot \underline{1} = 36$$

There are also 36 arrangements of the books for which the physics books are on the left. Therefore, the total number of arrangements of the six books, for which books on the same subject are kept together, is 36 + 36, or 72.

## EXAMPLE 5

In how many ways can the books in Example 4 be arranged on a shelf if the three mathematics books are to be kept together, but the three physics books can be placed anywhere?

## SOLUTION

We first consider the three mathematics books as a single element and each of the physics books as a single element. Then, we are arranging four elements in a row on a shelf. The first space can be occupied by any one of the four elements, the second space can be occupied by any one of the remaining three elements, and so on. The number of ways of arranging the four elements in order is

$$\underline{4} \cdot \underline{3} \cdot \underline{2} \cdot \underline{1} = 24$$

However, for each way of arranging the four elements, the three mathematics books can be arranged in order 3 · 2 · 1, or 6, ways. Hence the six books can be arranged in the desired order in 24 · 6, or 144 ways.

## EXERCISES 9.1

1. How many positive integers of three digits can be formed from the digits 1, 3, 5, 7, and 9 if repetitions (a) are not permitted, and (b) are permitted?
2. How many positive integers of four digits can be formed from the digits 1, 2, 3, 4, 5, and 6 if repetitions (a) are not permitted, and (b) are permitted?
3. How many positive even integers of four different digits can be formed from the digits of Exercise 2?
4. How many positive integers of four different digits, each greater than 3000, can be formed from the digits of Exercise 2?
5. How many positive integers of three different digits, each less than 400, can be formed from the digits of Exercise 2?
6. If two cubical dice, one white and one black, are thrown, in how many ways can they fall?
7. In how many ways can three different-colored cubical dice fall?
8. In a contest with twelve entries, in how many ways can a jury award first, second, and third prizes?
9. In a club of ten members, in how many ways can the offices of president, secretary, and treasurer be filled if no person is to hold more than one office?
10. In a manufacturing plant, raw material is processed by three different operations. If the first operation can be done on any one of eight machines, the second operation on any one of five machines, and the third on either of two machines,

over how many different runs can the raw material be processed?

11. If there are six airlines flying between Los Angeles and San Francisco and four bus lines going between San Francisco and Sonoma, in how many ways can a person go by plane from Los Angeles to San Francisco, and then by bus from San Francisco to Sonoma, and return, without using the same company twice?
12. In how many ways can nine books be arranged on a shelf if four particular books are to be kept together?
13. In how many ways can four French books, two Spanish books, and three German books be arranged on a shelf so that all books in the same language are together?
14. There are ten desks in a row. In how many ways can five students be seated at consecutive desks? (*Hint:* First determine the number of ways that the five consecutive desks can be chosen.)
15. If, in Exercise 14, the five students are seated for a test and there is to be exactly one empty desk between each two students, in how many ways can the students be seated in the row?
16. A man and woman invite four other men and four other women to dinner. After the host and hostess are seated at the ends of the table, in how many ways can the guests be seated so that no two women are seated next to each other?
17. How many different arrangements, each consisting of five different letters, can be formed from the letters of the word "equation," if each arrangement is to begin and end with a consonant?
18. How many arrangements, each consisting of five different letters, can be formed from the letters of the word "equation" if vowels and consonants alternate?
19. How many different arrangements can be formed of the eight letters of the word "equation" if all the vowels are to be kept together?
20. How many different arrangements can be formed of the eight letters of the word "equation" if in each arrangement the letter q is to be immediately followed by the letter u?
21. How many four-digit positive integers greater than 7500 can be made without repeating any digits?
22. How many of the positive integers in Exercise 21 are odd?

## 9.2 Permutations

In Illustration 1 of Section 9.1 we determined how many positive integers of two different digits can be formed from the integers 1, 2, 3, and 4. We obtained the following twelve positive integers.

$$12 \quad 13 \quad 14 \quad 21 \quad 23 \quad 24 \quad 31 \quad 32 \quad 34 \quad 41 \quad 42 \quad 43$$

Each of these orderings is called a "permutation" of the elements of the set $\{1, 2, 3, 4\}$ taken two at a time.

**9.2.1 DEFINITION** Let $S$ be a set containing $n$ elements, and suppose $r$ is a positive integer such that $r \leq n$. Then a *permutation* of $r$ elements of $S$ is an arrangement in a definite order, without repetitions, of $r$ elements of $S$.

We use the notation $_nP_r$ to denote the number of permutations of $n$ elements taken $r$ at a time. Other symbols used for this number are $P(n, r)$, $P_{n,r}$ and $P_r^n$.

**ILLUSTRATION 1.** If $S = \{1, 2, 3, 4\}$, we have seen that there are twelve permutations of these four elements taken two at a time. Thus

$$_4P_2 = 12$$

We wish to find a formula for computing $_nP_r$. The number of arrangements in a definite order of $n$ distinct elements taken $r$ at a time is the number of ways that $r$ positions can be filled from $n$ elements. The first position can be chosen in $n$ ways. Then there are $n - 1$ ways of choosing the second position, $n - 2$ ways of choosing the third position, and so on. The $r$th position can be selected from any of the remaining $n - (r - 1)$ (or, equivalently, $(n - r + 1)$) elements. Hence, from Axiom 9.1.3, we have

$$_nP_r = n(n - 1)(n - 2) \cdots (n - r + 1)$$

We have, therefore, proved the following theorem.

**9.2.2 THEOREM** The number of permutations of $n$ elements taken $r$ at a time is given by

$$_nP_r = n(n - 1)(n - 2) \cdots (n - r + 1) \qquad (1)$$

**ILLUSTRATION 2.** If in formula (1), $n = 4$ and $r = 2$, we have

$$_4P_2 = 4 \cdot 3$$
$$= 12$$

This result agrees with that of Illustration 1.

**ILLUSTRATION 3.** The number of positive integers of three different digits that can be formed with the integers 1, 2, 3, and 4 is the number of permutations of four elements taken three at a time. From formula (1)

$$_4P_3 = 4 \cdot 3 \cdot 2$$
$$= 24$$

This result agrees with that of Illustration 5 in Section 9.1.

**EXAMPLE 1**

A bus has seven vacant seats. If three additional passengers enter the bus, in how many different ways can they be seated?

**SOLUTION**

Each seating arrangement of the three passengers is a different arrangement of three seats from a set of seven seats. Therefore, the number of ways the passengers can be seated is $_7P_3$, and

$$_7P_3 = 7 \cdot 6 \cdot 5$$
$$= 210$$

To determine the number of permutations of $n$ elements taken $n$ (or all) at a time, we use formula (1) with $r = n$ and then $(n - r + 1)$ is $(n - n + 1)$, or

1, and so we have

$$_nP_n = n(n-1)(n-2) \cdots 3 \cdot 2 \cdot 1 \tag{2}$$

Thus $_nP_n$ is the product of the first $n$ positive integers.

We introduce the notation $n!$, read *"factorial n"* (or *"n factorial"*), to denote the product of the first $n$ positive integers; that is,

$$n! = n(n-1)(n-2) \cdots 3 \cdot 2 \cdot 1$$

Thus formula (2) can be written as

$$_nP_n = n! \tag{3}$$

**ILLUSTRATION 4**

$$1! = 1 \qquad 2! = 1 \cdot 2 \qquad 3! = 1 \cdot 2 \cdot 3 \qquad 4! = 1 \cdot 2 \cdot 3 \cdot 4$$
$$= 2 \qquad\qquad = 6 \qquad\qquad = 24$$

and so on. •

Because

$$n! = n(n-1)(n-2) \cdots 3 \cdot 2 \cdot 1$$

and

$$(n-1)! = (n-1)(n-2) \cdots 3 \cdot 2 \cdot 1$$

it follows that

$$n! = n(n-1)! \tag{4}$$

In particular,

$$5! = 5 \cdot 4! \qquad \text{and} \qquad 26! = 26 \cdot 25!$$

Furthermore, if we substitute 1 for $n$ in formula (4), we obtain

$$1! = 1(1-1)!$$

or, equivalently,

$$1! = 1 \cdot 0!$$

Therefore, we define

$$0! = 1$$

**ILLUSTRATION 5.** To simplify $\dfrac{10!}{7!}$, we first write $10! = 10 \cdot 9 \cdot 8 \cdot (7!)$ and then divide numerator and denominator by $7!$.

$$\frac{10!}{7!} = \frac{10 \cdot 9 \cdot 8(7!)}{7!}$$
$$= 10 \cdot 9 \cdot 8$$
$$= 720 \qquad \bullet$$

By a procedure similar to that used in Illustration 5 we can obtain an alternative formula for computing $_nP_r$ that involves the factorial notation. If $n$ and $r$ are positive integers and $r \leq n$, then

$$\frac{n!}{(n-r)!} = \frac{[n(n-1)(n-2) \cdots (n-r+1)] \cdot (n-r)!}{(n-r)!}$$

Thus

$$\frac{n!}{(n-r)!} = n(n-1)(n-2) \cdots (n-r+1) \qquad (5)$$

From formulas (1) and (5) it follows that

$$_nP_r = \frac{n!}{(n-r)!} \qquad (6)$$

Observe that if in formula (6) $r = n$, we have

$$_nP_n = \frac{n!}{(n-n)!}$$
$$= \frac{n!}{0!}$$

or, equivalently, because $0! = 1$.

$$_nP_n = n!$$

which is formula (3).

**EXAMPLE 2**

In how many ways can four boys and four girls be seated in a row containing eight seats if (a) a person may sit in any seat, and (b) boys and girls must sit in alternate seats?

**SOLUTION**

(a) The number of ways that eight people may be arranged in order eight at a time is $_8P_8$, and

$$_8P_8 = 8 \cdot 7 \cdot 6 \cdot 5 \cdot 4 \cdot 3 \cdot 2 \cdot 1$$
$$= 40{,}320$$

(b) If a boy is in the first seat on the left, then the first seat can be occupied by any one of the four boys, the second seat can be occupied by any one of the four girls, the third seat can be occupied by any one of the three remaining boys, the fourth seat can be occupied by any one of the three remaining girls, and so on. Hence, if a boy is in the first seat on the left, the number of ways the eight people can be seated is

$$\underline{4} \cdot \underline{4} \cdot \underline{3} \cdot \underline{3} \cdot \underline{2} \cdot \underline{2} \cdot \underline{1} \cdot \underline{1} = 576$$

There are also 576 arrangements for which a girl is in the first seat on the left. Hence the total number of ways that four boys and four girls can be seated in a row containing eight seats, if boys and girls must alternate, is $576 + 576 = 1152$.

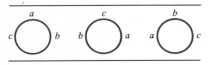

**Figure 9.2.1**

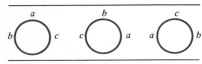

**Figure 9.2.2**

When elements are placed in order in a circle, we consider only their relative order as we go around the circle in a specific direction. For instance, the three circular arrangements of the elements $a$, $b$, and $c$ as shown in Figure 9.2.1 are considered the same. Furthermore, the three circular arrangements in Figure 9.2.2 are also considered the same. Thus there are only two different arrangements of three elements in a circle. Therefore, to determine the number of circular arrangements we first consider the position of one element as fixed and then calculate the number of permutations of the remaining elements as if they were in a straight line. Hence the number of ways that $n$ distinct elements can be arranged in a circle is $(n-1)!$.

**ILLUSTRATION 6.** If eight people are to occupy eight seats at a circular table, the number of different arrangements is

$$(8-1)! = 7!$$
$$= 5040 \qquad \bullet$$

In arranging keys on a ring (or different beads on a necklace), it is agreed that two arrangements are the same if one arrangement can be obtained from the other by turning over the ring (or the necklace). Thus the arrangements in Figures 9.2.1 and 9.2.2 are the same under such conditions. Consequently, there is only one arrangement of three different keys on a ring (or three different beads on a necklace); that is, the number of different arrangements is $\dfrac{(3-1)!}{2} = 1$. More generally, the number of different arrangements of $n$ keys on a ring (or $n$ different beads on a necklace) is $\dfrac{(n-1)!}{2}$.

EXAMPLE 3
How many different necklaces can be formed by stringing eight beads of different colors?

SOLUTION
The number of different necklaces is

$$\frac{(8-1)!}{2} = \frac{7!}{2}$$
$$= \frac{5040}{2}$$
$$= 2520$$

**ILLUSTRATION 7.** The number of permutations of the six distinct letters of the word PAINTS, taken six at a time is 6!, or 720. Now consider the letters of the word DEGREE. For this word, we also have six letters, but because there are three E's the letters are not distinct. Hence the number of *distinguishable* permutations of the six letters of the word DEGREE taken six at a time is not 720. To demonstrate this concept, we consider reordering the second, fifth, and sixth letters in each of the words DEGREE and PAINTS, without changing the order of the other letters. For the word DEGREE the three E's are distinguished by using subscripts and we write

$$D \ E_1 \ G \ R \ E_2 \ E_3$$

Rearranging the second, fifth, and sixth letters in each of the words gives the following permutations.

$$
\begin{array}{llllll}
D & E_1 & G & R & E_2 & E_3 \quad P \ A \ I \ N \ T \ S \\
D & E_1 & G & R & E_3 & E_2 \quad P \ A \ I \ N \ S \ T \\
D & E_2 & G & R & E_1 & E_3 \quad P \ T \ I \ N \ A \ S \\
D & E_2 & G & R & E_3 & E_1 \quad P \ T \ I \ N \ S \ A \\
D & E_3 & G & R & E_1 & E_2 \quad P \ S \ I \ N \ A \ T \\
D & E_3 & G & R & E_2 & E_1 \quad P \ S \ I \ N \ T \ A
\end{array}
$$

Observe that corresponding to the one permutation DEGREE there are 6, or 3!, permutations for the letters of the word PAINTS. Such a situation exists for each distinguishable permutation of the letters of the word DEGREE. Let $P$ be the number of distinguishable permutations of the six letters

$$D, E, G, R, E, E$$

Then, because for each of these permutations there are 3! ways in which the E's can be rearranged without changing the order of the other letters, we have

$$3! \cdot P = 6!$$

$$P = \frac{6!}{3!}$$

Hence $P = 120$. ●

**ILLUSTRATION 8.** To determine the number of different nine-digit numerals that can be formed from the digits 6, 6, 6, 6, 5, 5, 5, 4, and 3, we first consider one such numeral, for instance,

$$665566543 \tag{7}$$

With this ordering of the nine digits, there are 4! permutations of the digits 6 and 3! permutations of the digits 5 which have no effect on the numeral. Therefore, there are $4! \cdot 3!$ arrangements of digits in the numeral given by (7) that do not result in a distinguishable permutation of the given nine digits. Hence if $P$ is the number of distinguishable permutations of the nine digits,

$$4! \cdot 3! \cdot P = 9!$$

because 9! is the number of permutations of nine distinct elements taken nine at a time. Thus

$$P = \frac{9!}{4! \cdot 3!}$$
$$= \frac{9 \cdot 8 \cdot 7 \cdot 6 \cdot 5}{3 \cdot 2 \cdot 1} \cdot \frac{4!}{4!}$$
$$= 2520$$

Therefore, there are 2520 different nine-digit numerals that can be formed from the given digits. ●

By the procedure used in Illustrations 7 and 8, we can prove the following theorem. The details of the proof are omitted.

**9.2.3 THEOREM** If we are given $n$ elements, of which exactly $m_1$ are alike of one kind, exactly $m_2$ are alike of a second kind, ..., and exactly $m_k$ are alike of a $k$th kind, and if $n = m_1 + m_2 + \cdots + m_k$, then the number of distinguishable permutations that can be made of the $n$ elements taking them all at one time is

$$\frac{n!}{m_1! \cdot m_2! \cdots m_k!}$$

**EXAMPLE 4**
How many different signals, each consisting of eight flags hung one above the other, can be formed from a set of three indistinguishable red flags, two indistinguishable blue flags, two indistinguishable white flags, and one black flag?

**SOLUTION**
We have a set of eight elements in which three are alike of one kind (the red flags), two are alike of a second kind (the blue flags), two are alike of a third kind (the white flags), and one is of a fourth kind (the black flag). If $P$ is the number of different signals (the number of distinguishable permutations of these eight elements taken all at one time), it follows from Theorem 9.2.3 that

$$P = \frac{8!}{3! \cdot 2! \cdot 2! \cdot 1!}$$
$$= \frac{8 \cdot 7 \cdot 6 \cdot 5 \cdot 4}{2 \cdot 2 \cdot 1} \cdot \frac{3!}{3!}$$
$$= 1680$$

## EXERCISES 9.2

*In Exercises 1 through 6, simplify the given rational expression.*

1. $\dfrac{6!}{9!}$
2. $\dfrac{51!}{49!}$
3. $\dfrac{12!}{8!6!}$
4. $\dfrac{2!5!}{7!}$
5. $\dfrac{3! + 4!}{7!}$
6. $\dfrac{8!}{6! + 7!}$

*In Exercises 7 through 12, find the number.*

7. $_5P_3$
8. $_8P_4$
9. $_6P_6$
10. $_5P_5$
11. $_7P_1$
12. $_{20}P_2$

13. How many different signals, each consisting of three flags hung one above the other, can be made from seven different flags?
14. If a bookshelf has space for five books and there are ten different books available, how many different arrangements can be made of the five books on the shelf?
15. After a baseball coach has selected a team of nine members, in how many ways can the coach prepare a batting order?
16. In how many ways can the baseball coach of Exercise 15 prepare the batting order if the coach still has to select the team from among twelve players?
17. If the baseball coach of Exercise 15 wishes the pitcher to bat last and the best hitter to bat fourth, in how many ways can the batting order be prepared after the team has been selected?
18. In how many ways can a baseball coach assign positions to a team of nine men if only two men are qualified to be pitcher and only three other men are qualified to be catcher, but all of the men are qualified to play any other position?
19. A classroom has only eight vacant seats, three in the front row and five in the back row. In how many ways can five additional students be seated if two of them insist on sitting in the front row?
20. In how many ways can five students be seated in a row of eight seats if a certain two students insist on sitting next to each other?
21. In how many ways can five students be seated in a row of eight seats if a certain two students refuse to sit next to each other?
22. Six children join hands in a ring. In how many different orders can they be arranged?
23. Three boys and three girls join hands in a ring and boys and girls are to alternate. In how many different orders can they be arranged?
24. In how many ways can six children be seated at a round table if a certain two children refuse to sit next to each other?
25. In how many ways can six children be seated at a round table if a certain two children insist on sitting next to each other?
26. If five distinct keys are placed on a key ring, how many different orders are possible?
27. In how many different orders can nine different charms be placed on a charm bracelet?
28. From the letters of the word COLLEGE, how many different permutations can be formed if the letters are taken all at a time?
29. From the letters of the word BOOKWORM, how many different permutations can be formed if the letters are taken all at a time?
30. How many different permutations can be formed from the letters of the word MISSISSIPPI, taken all at a time?
31. How many different permutations can be formed from the letters of the word TENNESSEE, taken all at a time?
32. How many different eight-digit numerals can be formed from the digits 2, 2, 2, 4, 4, 6, 6, and 8?
33. How many different ten-digit numerals can be formed from the digits 3, 3, 3, 3, 1, 1, 1, 7, 7, and 5?
34. In a display window a grocer wishes to put a row of fifteen cans of soup consisting of five identical cans of tomato soup, four identical cans of mushroom soup, three identical cans of celery soup, and three identical cans of vegetable soup. How many different displays are possible?

**35.** How many of the displays in Exercise 34 have a can of tomato soup at each end?

**36.** How many of the displays in Exercise 34 have a can of the same kind of soup at each end?

*In Exercises 37 through 40, solve for n.*

**37.** $_nP_4 = 6(_nP_3)$    **38.** $_nP_5 = 9(_{n-1}P_4)$    **39.** $_nP_5 = 56(_nP_3)$    **40.** $_{n^2}P_2 = 600$

## 9.3 Combinations

We have been concerned with permutations, which are ordered arrangements of elements of a set. Our discussion now pertains to subsets of a set without regard to the relative order of the elements in the subsets.

**9.3.1 DEFINITION** Let $S$ be a set containing $n$ elements, and suppose $r$ is a positive integer such that $r \leq n$. Then a *combination* of $r$ elements of $S$ is a subset of $S$ containing $r$ distinct elements.

The notation we use for the number of combinations of $n$ elements taken $r$ at a time is $_nC_r$. Other symbols used for this number are $C(n, r)$, $C_{n,r}$, $C_r^n$, and $\binom{n}{r}$.

**ILLUSTRATION 1.** Suppose that $S = \{a, b, c, d\}$

(a) There is only one subset of the four letters taken all at a time. Hence
$$_4C_4 = 1$$

(b) There are four subsets of the four letters taken three at a time. They are
$$\{a, b, c\} \quad \{a, b, d\} \quad \{a, c, d\} \quad \{b, c, d\}$$
Therefore,
$$_4C_3 = 4$$

(c) There are six subsets of the four letters taken two at a time, and they are
$$\{a, b\} \quad \{a, c\} \quad \{a, d\} \quad \{b, c\} \quad \{b, d\} \quad \{c, d\}$$
Thus
$$_4C_2 = 6$$

(d) The subsets of the four letters taken one at a time are
$$\{a\} \quad \{b\} \quad \{c\} \quad \{d\}$$
Therefore,
$$_4C_1 = 4$$
●

The distinction between permutations and combinations is that changing the order of a set of elements gives a different permutation but the same combination. Observe in Illustration 1 that there are four subsets of $\{a, b, c, d\}$, taken three at a time, and therefore $_4C_3 = 4$. But the elements of each of these four subsets can be arranged in a definite order 3!, or 6, different ways. For instance, the elements of the subset $\{a, b, c\}$ can be arranged in order as

$$abc \quad acb \quad bac \quad bca \quad cab \quad cba$$

Therefore, $_4P_3 = 4 \cdot 6$, or 24, which agrees with the computation using formula (1) of Section 9.2: $_4P_3 = 4 \cdot 3 \cdot 2 \cdot 1$, or 24.

The formula for computing $_nC_r$ is given by the following theorem.

**9.3.2 THEOREM** The number of combinations of $n$ elements taken $r$ at a time is given by

$$_nC_r = \frac{_nP_r}{r!} \tag{1}$$

**Proof.** To find $_nC_r$, we must find the total number of subsets of $r$ elements each that can be obtained from a set of $n$ elements. The elements of each of the subsets can be arranged in order $r!$ different ways. Therefore, for each combination of $n$ elements taken $r$ at a time, there are $r!$ permutations; thus, for the total of all the possible combinations the number of different permutations is $r! \cdot {_nC_r}$. However, these are all the possible permutations of $n$ elements taken $r$ at a time, and therefore

$$r! \cdot {_nC_r} = {_nP_r}$$

$$_nC_r = \frac{_nP_r}{r!}$$

**ILLUSTRATION 2**

(a) $_4C_4 = \dfrac{_4P_4}{4!}$ 　　(b) $_4C_3 = \dfrac{_4P_3}{3!}$

$\phantom{(a) {_4C_4}} = \dfrac{4!}{4!}$ 　　$\phantom{(b) {_4C_3}} = \dfrac{4 \cdot 3 \cdot 2}{3 \cdot 2 \cdot 1}$

$\phantom{(a) {_4C_4}} = 1$ 　　$\phantom{(b) {_4C_3}} = 4$

(c) $_4C_2 = \dfrac{_4P_2}{2!}$ 　　(d) $_4C_1 = \dfrac{_4P_1}{1!}$

$\phantom{(c) {_4C_2}} = \dfrac{4 \cdot 3}{2 \cdot 1}$ 　　$\phantom{(d) {_4C_1}} = \dfrac{4}{1}$

$\phantom{(c) {_4C_2}} = 6$ 　　$\phantom{(d) {_4C_1}} = 4$

These results agree with those of Illustration 1. ●

## EXAMPLE 1

A football conference consists of eight teams. If each team plays every other team, how many conference games are played?

### SOLUTION

The number of conference games that are played is the number of two-element subsets of an eight-element set. This number is $_8C_2$, and

$$_8C_2 = \frac{_8P_2}{2!}$$
$$= \frac{8 \cdot 7}{2 \cdot 1}$$
$$= 28$$

Formula (6) in Section 9.2 is

$$_nP_r = \frac{n!}{(n-r)!}$$

Substituting from this formula into formula (1), we obtain the following alternate formula for computing $_nC_r$.

$$_nC_r = \frac{n!}{(n-r)!r!} \tag{2}$$

**ILLUSTRATION 3.** Using formula (2) to compute $_8C_2$ in Example 1, we have

$$_8C_2 = \frac{8!}{(8-2)!2!}$$
$$= \frac{8 \cdot 7}{2 \cdot 1} \cdot \frac{6!}{6!}$$
$$= 28 \qquad \bullet$$

### 9.3.3 THEOREM

$$\boxed{_nC_r = {_nC_{n-r}}}$$

**Proof.** From formula (2) we have

$$_nC_r = \frac{n!}{(n-r)!r!} \tag{3}$$

Using this formula with $r$ replaced by $n - r$, we get

$$_nC_{n-r} = \frac{n!}{[n-(n-r)]!(n-r)!}$$

$$_nC_{n-r} = \frac{n!}{r!(n-r)!} \tag{4}$$

Comparing equations (3) and (4), it follows that $_nC_r = {_nC_{n-r}}$.

Theorem 9.3.3 seems reasonable if you realize that each time a subset of $r$ elements is chosen from a set of $n$ elements, another set of $n - r$ elements remains unchosen.

## EXAMPLE 2
A student has ten posters to pin up on the walls of her room, but there is space for only seven. In how many ways can she choose the posters to be pinned up?

### SOLUTION
The number of ways she can choose the posters is the number of combinations of ten elements taken seven at a time, which is $_{10}C_7$. From Theorem 9.3.3, $_{10}C_7 = {_{10}C_3}$. Hence

$$_{10}C_7 = {_{10}C_3}$$
$$= \frac{10 \cdot 9 \cdot 8}{3 \cdot 2 \cdot 1}$$
$$= 120$$

## EXAMPLE 3
How many committees of five can be formed from eight sophomores and four freshmen if each committee is to have at least three sophomores?

### SOLUTION
If the committee is to have at least three sophomores, then it can have (a) three sophomores and two freshmen, (b) four sophomores and 1 freshman, or (c) five sophomores.

(a) The number of committees consisting of three sophomores and two freshmen is

$$_8C_3 \cdot {_4C_2} = \frac{8 \cdot 7 \cdot 6}{3 \cdot 2 \cdot 1} \cdot \frac{4 \cdot 3}{2 \cdot 1}$$
$$= 56 \cdot 6$$
$$= 336$$

(b) The number of committees consisting of four sophomores and one freshman is

$$_8C_4 \cdot {_4C_1} = \frac{8 \cdot 7 \cdot 6 \cdot 5}{4 \cdot 3 \cdot 2 \cdot 1} \cdot \frac{4}{1}$$
$$= 280$$

(c) The number of committees consisting of five sophomores is

$$_8C_5 = {_8C_3}$$
$$= \frac{8 \cdot 7 \cdot 6}{3 \cdot 2 \cdot 1}$$
$$= 56$$

Thus the total number of committees with at least three sophomores is $336 + 280 + 56 = 672$.

## EXAMPLE 4
(a) In how many ways can six tickets be divided among three people so that each person receives two tickets?

### SOLUTION
(a) The number of ways the first person can be allocated two tickets is $_6C_2$. With each of these ways, the number of ways that the second person can be allocated two tickets is $_4C_2$. The third person then receives the remaining two tickets in only one way, or $_2C_2$. Hence the number of ways

(b) In how many ways can six tickets be divided into three groups of two tickets each?

that six tickets can be divided among three people so that each person receives two tickets is

$$_6C_2 \cdot {_4C_2} \cdot {_2C_2} = \frac{6 \cdot 5}{2 \cdot 1} \cdot \frac{4 \cdot 3}{2 \cdot 1} \cdot 1$$
$$= 90$$

(b) The three groups, unlike the three people in part (a), are indistinguishable. When the three groups of two tickets each are established, these groups of tickets can be distributed among the three people in 3! different ways. Thus if $x$ is the number of ways that six tickets can be divided into three groups of two tickets each,

$$3! \cdot x = 90$$
$$x = \frac{90}{6}$$
$$x = 15$$

EXAMPLE 5
From six history books and eight economics books, in how many ways can a person select two history books and three economics books and arrange them on a shelf?

SOLUTION
The number of ways that two history books can be selected from six is $_6C_2$, and the number of ways that three economics books can be selected from eight is $_8C_3$. Then, after the five books are selected, the number of ways that they can be arranged on a shelf is $_5P_5$. Hence if $x$ is the number of ways that a person can select the books and arrange them on a shelf,

$$x = {_6C_2} \cdot {_8C_3} \cdot {_5P_5}$$
$$= \frac{6 \cdot 5}{2 \cdot 1} \cdot \frac{8 \cdot 7 \cdot 6}{3 \cdot 2 \cdot 1} \cdot 5 \cdot 4 \cdot 3 \cdot 2 \cdot 1$$
$$= 100{,}800$$

*EXERCISES 9.3*

In Exercises 1 through 6, find the number.

1. $_7C_3$  
2. $_9C_4$  
3. $_{12}C_9$  
4. $_{20}C_{17}$  
5. $_{50}C_{48}$  
6. $_{15}C_{15}$

In Exercises 7 through 10, solve for n.

7. $_nC_2 = 36$  
8. $2(_nC_4) = 3(_nC_3)$  
9. $_{n+1}C_4 = 6(_{n-1}C_2)$  
10. $_{n^2}C_2 = 30(_nC_3)$

11. If $_nP_4 = 5040$, find $_nC_4$.
12. If $_nC_3 = 165$, find $_nP_3$.
13. How many different committees of three persons each can be chosen from a group of twelve persons?
14. How many different hands of five cards each can be dealt from a standard deck of 52 cards?
15. If a student is to answer any six questions on a test

containing nine questions, in how many different ways can the student choose the questions?
16. If, in Exercise 15, the first three questions must be answered, in how many different ways can the student choose the six questions?
17. From seven points, no three of which are collinear, how many different lines are determined?
18. How many triangles are determined from the points in Exercise 17?
19. A committee of five is to be chosen from seven Republicans and four Democrats. In how many ways can the committee be chosen if it is to contain exactly three Republicans?
20. If, in Exercise 19, the committee is to contain at least three Republicans, in how many ways can it be chosen?
21. In how many ways can the committee in Exercise 19 be chosen if it is to contain at most three Republicans?
22. A bag contains four red balls, six white balls, and five blue balls. In how many ways can six balls be chosen if there are to be two balls of each color?
23. For the bag of Exercise 22, in how many ways can the six balls be chosen if there are to be exactly four white balls?
24. For the bag of Exercise 22, in how many ways can the six balls be chosen if there are to be no white balls?
25. For the bag of Exercise 22, in how many ways can four balls be chosen if all the balls are to be the same color?

*In Exercises 26 through 29, note that in a standard deck of cards there are 13 spades, 13 hearts, 13 diamonds, and 13 clubs.*

26. In how many different ways can a bridge hand of thirteen cards be dealt from a standard deck?
27. In how many different ways can a bridge hand containing six spades, four hearts, two diamonds, and one club be dealt from a standard deck?
28. In how many different ways can a bridge hand of thirteen cards, containing exactly eight spades, be dealt from a standard deck?
29. In how many different ways can a bridge hand of thirteen cards, containing at least ten spades, be dealt from a standard deck?
30. From eighteen boys, in how many different ways can two teams of nine players each be chosen to play baseball?
31. In how many ways can twelve persons be divided into four groups of three persons each?
32. From twelve different charms, how many different charm bracelets, containing eight charms each, can be formed?
33. Prove: $_{n-1}C_r + {}_{n-1}C_{r-1} = {}_nC_r$
34. Prove: $r({}_nC_r) = n({}_{n-1}C_{r-1})$

# 9.4 The Binomial Theorem

A power of a binomial is a polynomial, which is a special kind of series called a *binomial expansion*. For instance,

$$(a + b)^1 = a + b \tag{1}$$

and

$$(a + b)^2 = a^2 + 2ab + b^2 \tag{2}$$

$(a + b)^3 = (a + b)(a + b)^2$; thus $(a + b)^3 = (a + b)(a^2 + 2ab + b^2)$. We obtain

$$(a + b)^3 = a^3 + 3a^2b + 3ab^2 + b^3 \qquad (3)$$

In a similar way, because $(a + b)^4 = (a + b)(a + b)^3$ we have $(a + b)^4 = (a + b)(a^3 + 3a^2b + 3ab^2 + b^3)$. Therefore,

$$(a + b)^4 = a^4 + 4a^3b + 6a^2b^2 + 4ab^3 + b^4 \qquad (4)$$

Multiplying the expansion of $(a + b)^4$ by $(a + b)$, we obtain

$$(a + b)^5 = a^5 + 5a^4b + 10a^3b^2 + 10a^2b^3 + 5ab^4 + b^5 \qquad (5)$$

and multiplying the expansion of $(a + b)^5$ by $(a + b)$, we have

$$(a + b)^6 = a^6 + 6a^5b + 15a^4b^2 + 20a^3b^3 + 15a^2b^4 + 6ab^5 + b^6 \qquad (6)$$

Equations (1) through (6) give the binomial expansion of

$$(a + b)^n$$

for $n = 1, 2, 3, 4, 5,$ and 6. In each of these equations the first term on the right side can be written with a factor $b^0$ and the last term on the right side can be written with a factor $a^0$. Therefore, each term contains a nonnegative integer power of $a$ and a nonnegative integer power of $b$. Referring to the expansions (1) through (6), we note the following properties for these expressions.

1. There are $n + 1$ terms in the expansion.
2. The sum of the exponents of $a$ and $b$ in any term is $n$: the exponent of $a$ decreases by 1 and the exponent of $b$ increases by 1 from each term to the next term.
3. (i) The first term in the expansion is

$$a^n \qquad \text{or, equivalently,} \qquad {}_nC_0 a^n$$

(ii) The second term is

$$\frac{n}{1} a^{n-1}b \qquad \text{or, equivalently,} \qquad {}_nC_1 a^{n-1}b$$

(iii) The third term is

$$\frac{n(n-1)}{2 \cdot 1} a^{n-2}b^2 \qquad \text{or, equivalently,} \qquad {}_nC_2 a^{n-2}b^2$$

(iv) The fourth term is

$$\frac{n(n-1)(n-2)}{3 \cdot 2 \cdot 1} a^{n-3}b^3 \qquad \text{or, equivalently,} \qquad {}_nC_3 a^{n-3}b^3$$

(v) The fifth term is
$$\frac{n(n-1)(n-2)(n-3)}{4\cdot 3\cdot 2\cdot 1}a^{n-4}b^4 \quad \text{or, equivalently,} \quad {}_nC_4 a^{n-4}b^4$$

(vi) The term involving $b^r$ is
$$\frac{n(n-1)(n-2)\cdots(n-r+1)}{r!}a^{n-r}b^r \quad \text{or, equivalently,} \quad {}_nC_r a^{n-r}b^r$$

(vii) The last term is
$$b^n \quad \text{or, equivalently,} \quad {}_nC_n b^n$$

**ILLUSTRATION 1.** Equation (6) is
$$(a+b)^6 = a^6 + 6a^5b + 15a^4b^2 + 20a^3b^3 + 15a^2b^4 + 6ab^5 + b^6$$

We show that the preceding properties apply to this expansion.

1. There are seven terms in the expansion.
2. The sum of the exponents of $a$ and $b$ in any term is 6; the exponent of $a$ decreases by 1 and the exponent of $b$ increases by 1 from each term to the next term.
3. (i) The first term in the expansion is
$$a^6 = {}_6C_0 a^6$$

(ii) The second term is
$$\frac{6}{1}a^{6-1}b = {}_6C_1 a^5 b$$

(iii) The third term is
$$\frac{6\cdot 5}{2\cdot 1}a^{6-2}b^2 = {}_6C_2 a^4 b^2$$

(iv) The fourth term is
$$\frac{6\cdot 5\cdot 4}{3\cdot 2\cdot 1}a^{6-3}b^3 = {}_6C_3 a^3 b^3$$

(v) The fifth term is
$$\frac{6\cdot 5\cdot 4\cdot 3}{4\cdot 3\cdot 2\cdot 1}a^{6-4}b^4 = {}_6C_4 a^2 b^4$$

(vi) The sixth term is
$$\frac{6\cdot 5\cdot 4\cdot 3\cdot 2}{5\cdot 4\cdot 3\cdot 2\cdot 1}a^{6-5}b^5 = {}_6C_5 a b^5$$

(vii) The last term is
$$b^6 = {}_6C_6 b^6 \qquad \bullet$$

Illustration 1 and the discussion preceding it suggest a similar expression for the expansion of $(a + b)^n$, where $n$ is any positive integer. We state this in the binomial theorem.

**9.4.1 THEOREM** *The Binomial Theorem.* If $n$ is any positive integer, then

$$(a + b)^n = {}_nC_0 a^n + {}_nC_1 a^{n-1} b + {}_nC_2 a^{n-2} b^2 \\ + \cdots + {}_nC_r a^{n-r} b^r + \cdots + {}_nC_n b^n \qquad (7)$$

**Proof.** If $n$ is a positive integer, then $(a + b)^n$ represents the product of n equal factors of $(a + b)$; that is,

$$(a + b)^n = (a + b)(a + b)(a + b) \cdots (a + b) \qquad n \text{ factors of } (a + b)$$

When performing the multiplication, we select either $a$ or $b$ from each of the $n$ factors and multiply these $n$ numbers; we form every such product by considering all possible choices of either $a$ or $b$ from each of the $n$ factors. The term involving $a^{n-r} b^r$ is a sum of terms, each of which is a product of $r$ factors of $b$ (one factor from each of $r$ factors $(a + b)$) and $(n - r)$ factors of $a$ (one factor from each of the remaining $(n - r)$ factors $(a + b)$). The number of ways that the $r$ factors of $b$ can be chosen from $n$ factors of $(a + b)$ is ${}_nC_r$. Therefore, there are ${}_nC_r$ terms involving $a^{n-r} b^r$. Thus the coefficient of $a^{n-r} b^r$ is ${}_nC_r$. Because $r$ can be any integer from 0 through $n$, inclusive, formula (7) follows.

The binomial theorem can also be proved by mathematical induction; however, this proof is quite lengthy and is omitted.

**EXAMPLE 1**
Expand and simplify by the binomial theorem
$$(x^2 + 3y)^5$$

**SOLUTION**
Applying formula (7) where $a$ is $x^2$, $b$ is $3y$, and $n$ is 5, we have

$$(x^2 + 3y)^5 = {}_5C_0 (x^2)^5 + {}_5C_1 (x^2)^4 (3y)^1 + {}_5C_2 (x^2)^3 (3y)^2 \\ + {}_5C_3 (x^2)^2 (3y)^3 + {}_5C_4 (x^2)^1 (3y)^4 + {}_5C_5 (3y)^5$$

$$= 1 \cdot x^{10} + \frac{5}{1} x^8 (3y) + \frac{5 \cdot 4}{2 \cdot 1} x^6 (9y^2) + \frac{5 \cdot 4 \cdot 3}{3 \cdot 2 \cdot 1} x^4 (27y^3)$$

$$+ \frac{5 \cdot 4 \cdot 3 \cdot 2}{4 \cdot 3 \cdot 2 \cdot 1} x^2 (81 y^4) + 1 (243 y^5)$$

$$= x^{10} + 15 x^8 y + 90 x^6 y^2 + 270 x^4 y^3 + 405 x^2 y^4 + 243 y^5$$

Formula (2) of Section 9.3 is

$$_nC_r = \frac{n!}{(n-r)!r!}$$

or, equivalently,

$$_nC_r = \frac{n(n-1)(n-2) \cdots (n-r+1)}{r!} \cdot \frac{(n-r)!}{(n-r)!}$$

or, equivalently,

$$_nC_r = \frac{n(n-1)(n-2) \cdots (n-r+1)}{r!}$$

Therefore, formula (7) can be written as

$$(a+b)^n = a^n + \frac{n}{1!}a^{n-1}b + \frac{n(n-1)}{2!}a^{n-2}b^2 + \frac{n(n-1)(n-2)}{3!}a^{n-3}b^3$$
$$+ \cdots + \frac{n(n-1)(n-2)\cdots(n-r+1)}{r!}a^{n-r}b^r + \cdots + b^n \qquad (8)$$

Formula (8) is used in the next example.

**EXAMPLE 2**
Expand and simplify

$$\left(2\sqrt{t} - \frac{1}{t}\right)^4$$

**SOLUTION**
We use formula (8) where $a$ is $2\sqrt{t}$, $b$ is $-\frac{1}{t}$, and $n$ is 4.

$$\left(2\sqrt{t} - \frac{1}{t}\right)^4 = (2\sqrt{t})^4 + \frac{4}{1!}(2\sqrt{t})^3\left(-\frac{1}{t}\right)^1 + \frac{4 \cdot 3}{2!}(2\sqrt{t})^2\left(-\frac{1}{t}\right)^2$$
$$+ \frac{4 \cdot 3 \cdot 2}{3!}(2\sqrt{t})^1\left(-\frac{1}{t}\right)^3 + \frac{4 \cdot 3 \cdot 2 \cdot 1}{4!}\left(-\frac{1}{t}\right)^4$$
$$= 16t^2 - 32t^{1/2} + \frac{24}{t} - \frac{8}{t^{5/2}} + \frac{1}{t^4}$$

**EXAMPLE 3**
Find the value of $(1.02)^8$ to four significant digits.

**SOLUTION**
$(1.02)^8 = (1 + 0.02)^8$. We apply formula (8) where $a$ is 1, $b$ is 0.02, and $n$ is 8.

$$(1.02)^8 = 1^8 + \frac{8}{1!}1^7(0.02)^1 + \frac{8 \cdot 7}{2!}1^6(0.02)^2$$
$$+ \frac{8 \cdot 7 \cdot 6}{3!}1^5(0.02)^3 + \cdots + (0.02)^8$$
$$= 1 + 0.16 + 0.0112 + 0.000448 + \cdots + 0.0000000000000256$$
$$\approx 1.172$$

From formula (8), the term involving $b^r$ in the expansion of $(a + b)^n$ is

$$\frac{n(n-1)(n-2)\cdots(n-r+1)}{r!}a^{n-r}b^r \qquad (9)$$

Notice in expression (9) that the denominator is the product of the first $r$ positive integers and the numerator has $r$ factors beginning with $n$ and successively decreasing by 1; that is, the number of factors in both the numerator and denominator of the coefficient is the same as the exponent of $b$. Expression (9) is the $(r + 1)$th term of the binomial expansion of $(a + b)^n$. Hence the $r$th term of the expansion of $(a + b)^n$ is

$$\frac{n(n-1)(n-2)\cdots(n-r+2)}{(r-1)!}a^{n-r+1}b^{r-1} \qquad (10)$$

**EXAMPLE 4**
Find the seventh term of the expansion of $(2r^3 - \frac{1}{4}s^4)^{10}$.

**SOLUTION**
Applying expression (10), where $r$ is 7, $n$ is 10, $a$ is $2r^3$, and $b$ is $-\frac{1}{4}s^4$, we have

$$\frac{10\cdot 9\cdot 8\cdot 7\cdot 6\cdot 5}{6!}(2r^3)^4\left(-\frac{1}{4}s^4\right)^6 = \frac{105}{128}r^{12}s^{24}$$

**EXAMPLE 5**
Find the term involving $x^3$ in the expansion of $(x - 3x^{-1})^9$.

**SOLUTION**
From expression (10), where $a$ is $x$, $b$ is $-3x^{-1}$, and $n$ is 9, the $r$th term has the factors

$$x^{10-r}(-3x^{-1})^{r-1} = (-3)^{r-1}x^{11-2r}$$

The term involving $x^3$ is the one for which the exponent of $x$ is 3; hence we solve the equation

$$11 - 2r = 3$$
$$r = 4$$

Thus the fourth term is the desired term. It is

$$\frac{9\cdot 8\cdot 7}{3!}x^6(-3x^{-1})^3 = -2268x^3$$

The formula for the expansion of $(a + b)^n$ was proved when $n$ is a positive integer. There is a similar formula for the expansion of $(1 + b)^n$, when $n$ is any real number, and $|b| < 1$. However, in such a situation the formula gives an infinite series, called a *binomial series*.

**9.4.2 THEOREM**   *Formula for Binomial Series.*   If $n$ is any real number, and $|b| < 1$, then

$$(1+b)^n = 1 + nb + \frac{n(n-1)}{2!}b^2 + \frac{n(n-1)(n-2)}{3!}b^3 + \cdots$$
$$+ \frac{n(n-1)(n-2)\cdots(n-r+1)}{r!}b^r + \cdots \qquad (11)$$

The proof of Theorem 9.4.2 is omitted because it depends upon concepts studied in the calculus.

Observe that formula (11) is similar to formula (8), where $a$ is 1 and $n$ is any real number, except that formula (11) gives an infinite series while in formula (8) the expansion terminates.

**EXAMPLE 6**

Write a binomial series for
$$(1 + \sqrt{x})^{-1}$$
if $\sqrt{x} < 1$.

**SOLUTION**

From formula (11) with $n = -1$ and $b = \sqrt{x}$, we have

$$(1 + \sqrt{x})^{-1} = 1 + (-1)x^{1/2} + \frac{(-1)(-2)}{2!}(x^{1/2})^2 + \frac{(-1)(-2)(-3)}{3!}(x^{1/2})^3$$
$$+ \cdots + \frac{(-1)(-2)(-3)\cdots(-r)}{r!}(x^{1/2})^r + \cdots$$

or, equivalently,

$$(1 + x^{1/2})^{-1} = 1 - x^{1/2} + x - x^{3/2} + \cdots + (-1)^r x^{r/2} + \cdots \quad (12)$$

Note that the infinite series in the right member of equation (12) is an infinite geometric series where the first term is 1 and the common ratio is $-x^{1/2}$. Definition 8.5.1 states that the sum of an infinite geometric series, for which the first term is $a_1$ and the common ratio is $r$, with $|r| < 1$, is given by

$$\frac{a_1}{1-r}$$

If $a_1 = 1$ and $r = -x^{1/2}$, then

$$\frac{a_1}{1-r} = \frac{1}{1-(-x^{1/2})}$$
$$= \frac{1}{1+x^{1/2}}$$
$$= (1 + x^{1/2})^{-1}$$

Thus the result of Example 6 is in agreement with the discussion in Section 8.5.

**EXAMPLE 7**

Compute the value of $\sqrt[3]{25}$ accurate to three decimal places by using the binomial series for $(1 + x)^{1/3}$.

**SOLUTION**

From formula (11) we have if $|x| < 1$

$$(1 + x)^{1/3} = 1 + \frac{1}{3}x + \left(\frac{1}{3}\right)\left(-\frac{2}{3}\right)\frac{x^2}{2!} + \left(\frac{1}{3}\right)\left(-\frac{2}{3}\right)\left(-\frac{5}{3}\right)\frac{x^3}{3!} + \cdots \quad (13)$$

Because

$$\sqrt[3]{25} = \sqrt[3]{27} \sqrt[3]{\frac{25}{27}}$$
$$= 3\sqrt[3]{1 - \frac{2}{27}}$$

then
$$\sqrt[3]{25} = 3\left(1 - \frac{2}{27}\right)^{1/3} \tag{14}$$

We can use equation (13) with $x = -\frac{2}{27}$, and we have

$$\left(1 - \frac{2}{27}\right)^{1/3} = 1 + \frac{1}{3}\left(-\frac{2}{27}\right) - \frac{2}{3^2 \cdot 2!}\left(-\frac{2}{27}\right)^2$$
$$+ \frac{2 \cdot 5}{3^3 \cdot 3!}\left(-\frac{2}{27}\right)^3 + \cdots$$
$$= 1 - 0.0247 - 0.0006 - 0.00003 - \cdots \tag{15}$$

The fourth term has a zero in the fourth decimal place, and so has each successive term. It can be proved that no term after the third term affects the first four decimal places. Using the first three terms of series (15), we obtain

$$\left(1 - \frac{2}{27}\right)^{1/3} \approx 0.9747$$

Substituting into equation (14), we have

$$\sqrt[3]{25} \approx 3(0.9747)$$
$$= 2.9241$$

Rounding off to three decimal places gives $\sqrt[3]{25} \approx 2.924$.

## EXERCISES 9.4

*In Exercises 1 through 8, expand the given power of the binomial.*

1. $(x + 3y)^5$
2. $(2x - y)^6$
3. $(4 - ab)^6$
4. $(2t + s^2)^7$
5. $(u^{-1} - u^2)^5$
6. $(e^x - e^{-x})^9$
7. $(a^{1/2} + b^{1/2})^6$
8. $(xy^{-1} - x^{-1}y)^7$

*In Exercises 9 through 12, find the first four terms in the expansion of the given power of the binomial and simplify each term.*

9. $(x^2 + y^2)^{12}$
10. $(e^{x/2} - e^{-x/2})^{20}$
11. $(a^{1/3} - b^{1/3})^9$
12. $(\frac{2}{5}a^{2/3} + b^{3/2})^{11}$

*In Exercises 13 through 16, find to four significant digits the value of the given power by using a binomial expansion.*

13. $(1.06)^4$
14. $(0.49)^7$
15. $(99)^4$
16. $(51)^4$

17. Find the seventh term of the expansion of $(a + b)^{12}$.
18. Find the sixth term of the expansion of $(\frac{1}{2}a - b)^{13}$.
19. Find the sixth term of the expansion of $(2x - 3)^9$.
20. Find the tenth term of the expansion of $(\sqrt{t} - t^{-1/2})^{15}$.

21. Find the middle term of the expansion of $(1 - x^3 y^{-2})^{12}$.
22. Find the middle term of the expansion of $(\tfrac{1}{2}a + \sqrt{a})^{10}$.
23. Find the term involving $a^6$ in the expansion of $(\tfrac{1}{2} + a)^{12}$.
24. Find the term involving $x^{12}$ in the expansion of $(x^2 - \tfrac{1}{2})^{11}$.
25. Find the term that does not contain $x$ in the expansion of $(x^2 - 2x^{-2})^{10}$.
26. Find the term involving $t^{-9}$ in the expansion of $(t^{-1} - \tfrac{1}{5})^{13}$.

In Exercises 27 through 32, find the first four terms of the binomial series for the given expression.

27. $(1 + x^2)^{-1}$, $x^2 < 1$
28. $(1 - x)^{1/3}$, $|x| < 1$
29. $(1 - 2x)^{1/2}$, $|x| < \tfrac{1}{2}$
30. $(1 - x^2)^{-1/2}$, $x^2 < 1$
31. $(8 + x)^{1/3}$, $|x| < 8$
32. $(3 + x)^{-1}$, $|x| < 3$

In Exercises 33 through 38, compute the value of the given quantity accurate to three decimal places by using a binomial series.

33. $\sqrt{1.04}$
34. $\sqrt[3]{0.99}$
35. $\sqrt[3]{63}$
36. $\sqrt{38}$
37. $\sqrt[4]{620}$
38. $\dfrac{1}{\sqrt[4]{15}}$

## REVIEW EXERCISES (CHAPTER 9)

In Exercises 1 and 2, simplify the given rational expression.

1. $\dfrac{12!}{4!9!}$
2. $\dfrac{4! + 5!}{6!}$

In Exercises 3 through 6, find the number.

3. $_6P_4$
4. $_5P_2$
5. $_7C_3$
6. $_8C_6$

In Exercises 7 through 10, solve for n.

7. $_nP_3 = 3 \cdot _nP_2$
8. $_nP_4 = 2 \cdot _nP_2$
9. $5 \cdot _nC_5 = 8 \cdot _{n-1}C_3$
10. $15 \cdot _{n-1}C_2 = 2 \cdot _{n+1}C_4$

11. In how many ways can six students be seated in a row of six desks?
12. In how many ways can the six students of Exercise 11 be seated in the row of six desks if a certain two students insist on sitting next to each other?
13. In how many ways can the six students of Exercise 11 be seated in the row of six desks if a certain two students refuse to sit next to each other?
14. In how many ways can six students be seated at consecutive desks if there are nine desks in the row?
15. In how many ways can six students be seated in six chairs at a round table?
16. There are ten questions on an examination and a student is required to answer any five of them. How many different sets of five questions can be selected?
17. A committee of four is to be formed from eight sophomores and six freshmen. How many different committees can be chosen if the committee is to contain two sophomores and two freshmen?
18. If in Exercise 17, the committee is to contain at

least two sophomores, how many different committees can be chosen?
19. If in Exercise 17, the committee is to contain at most two sophomores, how many different committees can be chosen?
20. A basketball team consists of twelve players, of which only four play the center position. How many starting line-ups of five players are possible?
21. From the digits 0, 1, 2, 3, 4, 5, 6, 7, 8, 9, how many three-digit numerals can be formed?
22. How many of the numerals in Exercise 21 are odd?
23. How many of the numerals in Exercise 21 are even?
24. From the letters of the word PEOPLE, how many different permutations can be formed by taking all the letters at a time?
25. How many of the permutations in Exercise 24 begin and end with a vowel?
26. For an audition a tenor is to sing three operatic arias from a selection of ten. (a) In how many ways can he choose the three arias? (b) In how many different orders can he present three arias?
27. There are twelve people qualified to operate a machine that requires three persons at one time. (a) How many different groups of three people can operate the machine? (b) In how many of these groups does one particular person appear?
28. What is the binomial expansion of $(2a - b)^8$?
29. Write and simplify the first four terms of the binomial expansion of $(x + 2x^{-1})^{20}$.
30. Find the tenth term of the expansion of $(x^{1/2} - x^{-1/2})^{15}$.
31. Use a binomial expansion to find to four decimal places the value of $(0.95)^6$.
32. Use a binomial series to compute accurately to three decimal places the value of $\sqrt[4]{17}$.

# Appendix

Table 1. Powers and Roots

| $n$ | $n^2$ | $\sqrt{n}$ | $n^3$ | $\sqrt[3]{n}$ | $n$ | $n^2$ | $\sqrt{n}$ | $n^3$ | $\sqrt[3]{n}$ |
|---|---|---|---|---|---|---|---|---|---|
| 1 | 1 | 1.000 | 1 | 1.000 | 51 | 2,601 | 7.141 | 132,651 | 3.708 |
| 2 | 4 | 1.414 | 8 | 1.260 | 52 | 2,704 | 7.211 | 140,608 | 3.732 |
| 3 | 9 | 1.732 | 27 | 1.442 | 53 | 2,809 | 7.280 | 148,877 | 3.756 |
| 4 | 16 | 2.000 | 64 | 1.587 | 54 | 2,916 | 7.348 | 157,464 | 3.780 |
| 5 | 25 | 2.236 | 125 | 1.710 | 55 | 3,025 | 7.416 | 166,375 | 3.803 |
| 6 | 36 | 2.449 | 216 | 1.817 | 56 | 3,136 | 7.483 | 175,616 | 3.826 |
| 7 | 49 | 2.646 | 343 | 1.913 | 57 | 3,249 | 7.550 | 185,193 | 3.848 |
| 8 | 64 | 2.828 | 512 | 2.000 | 58 | 3,364 | 7.616 | 195,112 | 3.871 |
| 9 | 81 | 3.000 | 729 | 2.080 | 59 | 3,481 | 7.681 | 205,379 | 3.893 |
| 10 | 100 | 3.162 | 1,000 | 2.154 | 60 | 3,600 | 7.746 | 216,000 | 3.915 |
| 11 | 121 | 3.317 | 1,331 | 2.224 | 61 | 3,721 | 7.810 | 226,981 | 3.936 |
| 12 | 144 | 3.464 | 1,728 | 3.289 | 62 | 3,844 | 7.874 | 238,328 | 3.958 |
| 13 | 169 | 3.606 | 2,197 | 2.351 | 63 | 3,969 | 7.937 | 250,047 | 3.979 |
| 14 | 196 | 3.742 | 2,744 | 2.410 | 64 | 4,096 | 8.000 | 262,144 | 4.000 |
| 15 | 225 | 3.873 | 3,375 | 2.466 | 65 | 4,225 | 8.062 | 274,625 | 4.021 |
| 16 | 256 | 4.000 | 4,096 | 2.520 | 66 | 4,356 | 8.124 | 287,496 | 4.041 |
| 17 | 289 | 4.123 | 4,913 | 2.571 | 67 | 4,489 | 8.185 | 300,763 | 4.062 |
| 18 | 324 | 4.243 | 5,832 | 2.621 | 68 | 4,624 | 8.246 | 314,432 | 4.082 |
| 19 | 361 | 4.359 | 6,859 | 2.668 | 69 | 4,761 | 8.307 | 328,509 | 4.102 |
| 20 | 400 | 4.472 | 8,000 | 2.714 | 70 | 4,900 | 8.367 | 343,000 | 4.121 |
| 21 | 441 | 4.583 | 9,261 | 2.759 | 71 | 5,041 | 8.426 | 357,911 | 4.141 |
| 22 | 484 | 4.690 | 10,648 | 2.802 | 72 | 5,184 | 8.485 | 373,248 | 4.160 |
| 23 | 529 | 4.796 | 12,167 | 2.844 | 73 | 5,329 | 8.544 | 389,017 | 4.179 |
| 24 | 576 | 4.899 | 13,824 | 2.884 | 74 | 5,476 | 8.602 | 405,224 | 4.198 |
| 25 | 625 | 5.000 | 15,625 | 2.924 | 75 | 5,625 | 8.660 | 421,875 | 4.217 |
| 26 | 676 | 5.099 | 17,576 | 2.962 | 76 | 5,776 | 8.718 | 438,976 | 4.236 |
| 27 | 729 | 5.196 | 19,683 | 3.000 | 77 | 5,929 | 8.775 | 456,533 | 4.254 |
| 28 | 784 | 5.291 | 21,952 | 3.037 | 78 | 6,084 | 8.832 | 474,552 | 4.273 |
| 29 | 841 | 5.385 | 24,389 | 3.072 | 79 | 6,241 | 8.888 | 493,039 | 4.291 |
| 30 | 900 | 5.477 | 27,000 | 3.107 | 80 | 6,400 | 8.944 | 512,000 | 4.309 |
| 31 | 961 | 5.568 | 29,791 | 3.141 | 81 | 6,561 | 9.000 | 531,441 | 4.327 |
| 32 | 1,024 | 5.657 | 32,768 | 3.175 | 82 | 6,724 | 9.055 | 551,368 | 4.344 |
| 33 | 1,089 | 5.745 | 35,937 | 3.208 | 83 | 6,889 | 9.110 | 571,787 | 4.362 |
| 34 | 1,156 | 5.831 | 39,304 | 3.240 | 84 | 7,056 | 9.165 | 592,704 | 4.380 |
| 35 | 1,225 | 5.916 | 42,875 | 3.271 | 85 | 7,225 | 9.220 | 614,125 | 4.397 |
| 36 | 1,296 | 6.000 | 46,656 | 3.302 | 86 | 7,396 | 9.274 | 636,056 | 4.414 |
| 37 | 1,369 | 6.083 | 50,653 | 3.332 | 87 | 7,569 | 9.327 | 658,503 | 4.431 |
| 38 | 1,444 | 6.164 | 54,872 | 3.362 | 88 | 7,744 | 9.381 | 681,472 | 4.448 |
| 39 | 1,521 | 6.245 | 59,319 | 3.391 | 89 | 7,921 | 9.434 | 704,969 | 4.465 |
| 40 | 1,600 | 6.325 | 64,000 | 3.420 | 90 | 8,100 | 9.487 | 729,000 | 4.481 |
| 41 | 1,681 | 6.403 | 68,921 | 3.448 | 91 | 8,281 | 9.539 | 753,571 | 4.498 |
| 42 | 1,764 | 6.481 | 74,088 | 3.476 | 92 | 8,464 | 9.592 | 778,688 | 4.514 |
| 43 | 1,849 | 6.557 | 79,507 | 3.503 | 93 | 8,649 | 9.643 | 804,357 | 4.531 |
| 44 | 1,936 | 6.633 | 85,184 | 3.530 | 94 | 8,836 | 9.695 | 830,584 | 4.547 |
| 45 | 2,025 | 6.708 | 91,125 | 3.557 | 95 | 9,025 | 9.747 | 857,375 | 4.563 |
| 46 | 2,116 | 6.782 | 97,336 | 3.583 | 96 | 9,216 | 9.798 | 884,736 | 4.579 |
| 47 | 2,209 | 6.856 | 103,823 | 3.609 | 97 | 9,409 | 9.849 | 912,673 | 4.595 |
| 48 | 2,304 | 6.928 | 110,592 | 3.634 | 98 | 9,604 | 9.899 | 941,192 | 4.610 |
| 49 | 2,401 | 7.000 | 117,649 | 3.659 | 99 | 9,801 | 9.950 | 970,299 | 4.626 |
| 50 | 2,500 | 7.071 | 125,000 | 3.684 | 100 | 10,000 | 10.000 | 1,000,000 | 4.642 |

Table 2. Common Logarithms

| N | 0 | 1 | 2 | 3 | 4 | 5 | 6 | 7 | 8 | 9 |
|---|---|---|---|---|---|---|---|---|---|---|
| 10 | 0000 | 0043 | 0086 | 0128 | 0170 | 0212 | 0253 | 0294 | 0334 | 0374 |
| 11 | 0414 | 0453 | 0492 | 0531 | 0569 | 0607 | 0645 | 0682 | 0719 | 0755 |
| 12 | 0792 | 0828 | 0864 | 0899 | 0934 | 0969 | 1004 | 1038 | 1072 | 1106 |
| 13 | 1139 | 1173 | 1206 | 1239 | 1271 | 1303 | 1335 | 1367 | 1399 | 1430 |
| 14 | 1461 | 1492 | 1523 | 1553 | 1584 | 1614 | 1644 | 1673 | 1703 | 1732 |
| 15 | 1761 | 1790 | 1818 | 1847 | 1875 | 1903 | 1931 | 1959 | 1987 | 2014 |
| 16 | 2041 | 2068 | 2095 | 2122 | 2148 | 2175 | 2201 | 2227 | 2253 | 2279 |
| 17 | 2304 | 2330 | 2355 | 2380 | 2405 | 2430 | 2455 | 2480 | 2504 | 2529 |
| 18 | 2553 | 2577 | 2601 | 2625 | 2648 | 2672 | 2695 | 2718 | 2742 | 2765 |
| 19 | 2788 | 2810 | 2833 | 2856 | 2878 | 2900 | 2923 | 2945 | 2967 | 2989 |
| 20 | 3010 | 3032 | 3054 | 3075 | 3096 | 3118 | 3139 | 3160 | 3181 | 3201 |
| 21 | 3222 | 3243 | 3263 | 3284 | 3304 | 3324 | 3345 | 3365 | 3385 | 3404 |
| 22 | 3424 | 3444 | 3464 | 3483 | 3502 | 3522 | 3541 | 3560 | 3579 | 3598 |
| 23 | 3617 | 3636 | 3655 | 3674 | 3692 | 3711 | 3729 | 3747 | 3766 | 3784 |
| 24 | 3802 | 3820 | 3838 | 3856 | 3874 | 3892 | 3909 | 3927 | 3945 | 3962 |
| 25 | 3979 | 3997 | 4014 | 4031 | 4048 | 4065 | 4082 | 4099 | 4116 | 4133 |
| 26 | 4150 | 4166 | 4183 | 4200 | 4216 | 4232 | 4249 | 4265 | 4281 | 4298 |
| 27 | 4314 | 4330 | 4346 | 4362 | 4378 | 4393 | 4409 | 4425 | 4440 | 4456 |
| 28 | 4472 | 4487 | 4502 | 4518 | 4533 | 4548 | 4564 | 4579 | 4594 | 4609 |
| 29 | 4624 | 4639 | 4654 | 4669 | 4683 | 4698 | 4713 | 4728 | 4742 | 4757 |
| 30 | 4771 | 4786 | 4800 | 4814 | 4829 | 4843 | 4857 | 4871 | 4886 | 4900 |
| 31 | 4914 | 4928 | 4942 | 4955 | 4969 | 4983 | 4997 | 5011 | 5024 | 5038 |
| 32 | 5051 | 5065 | 5079 | 5092 | 5105 | 5119 | 5132 | 5145 | 5159 | 5172 |
| 33 | 5185 | 5198 | 5211 | 5224 | 5237 | 5250 | 5263 | 5276 | 5289 | 5302 |
| 34 | 5315 | 5328 | 5340 | 5353 | 5366 | 5378 | 5391 | 5403 | 5416 | 5428 |
| 35 | 5441 | 5453 | 5465 | 5478 | 5490 | 5502 | 5514 | 5527 | 5539 | 5551 |
| 36 | 5563 | 5575 | 5587 | 5599 | 5611 | 5623 | 5635 | 5647 | 5658 | 5670 |
| 37 | 5682 | 5694 | 5705 | 5717 | 5729 | 5740 | 5752 | 5763 | 5775 | 5786 |
| 38 | 5798 | 5809 | 5821 | 5832 | 5843 | 5855 | 5866 | 5877 | 5888 | 5899 |
| 39 | 5911 | 5922 | 5933 | 5944 | 5955 | 5966 | 5977 | 5988 | 5999 | 6010 |
| 40 | 6021 | 6031 | 6042 | 6053 | 6064 | 6075 | 6085 | 6096 | 6107 | 6117 |
| 41 | 6128 | 6138 | 6149 | 6160 | 6170 | 6180 | 6191 | 6201 | 6212 | 6222 |
| 42 | 6232 | 6243 | 6253 | 6263 | 6274 | 6284 | 6294 | 6304 | 6314 | 6325 |
| 43 | 6335 | 6345 | 6355 | 6365 | 6375 | 6385 | 6395 | 6405 | 6415 | 6425 |
| 44 | 6435 | 6444 | 6454 | 6464 | 6474 | 6484 | 6493 | 6503 | 6513 | 6522 |
| 45 | 6532 | 6542 | 6551 | 6561 | 6571 | 6580 | 6590 | 6599 | 6609 | 6618 |
| 46 | 6628 | 6637 | 6646 | 6656 | 6665 | 6675 | 6684 | 6693 | 6702 | 6712 |
| 47 | 6721 | 6730 | 6739 | 6749 | 6758 | 6767 | 6776 | 6785 | 6794 | 6803 |
| 48 | 6812 | 6821 | 6830 | 6839 | 6848 | 6857 | 6866 | 6875 | 6884 | 6893 |
| 49 | 6902 | 6911 | 6920 | 6928 | 6937 | 6946 | 6955 | 6964 | 6972 | 6981 |
| 50 | 6990 | 6998 | 7007 | 7016 | 7024 | 7033 | 7042 | 7050 | 7059 | 7067 |
| 51 | 7076 | 7084 | 7093 | 7101 | 7110 | 7118 | 7126 | 7135 | 7143 | 7152 |
| 52 | 7160 | 7168 | 7177 | 7185 | 7193 | 7202 | 7210 | 7218 | 7226 | 7235 |
| 53 | 7243 | 7251 | 7259 | 7267 | 7275 | 7284 | 7292 | 7300 | 7308 | 7316 |
| 54 | 7324 | 7332 | 7340 | 7348 | 7356 | 7364 | 7372 | 7380 | 7388 | 7396 |

**Table 2. Common Logarithms (cont.)**

| N | 0 | 1 | 2 | 3 | 4 | 5 | 6 | 7 | 8 | 9 |
|---|---|---|---|---|---|---|---|---|---|---|
| 55 | 7404 | 7412 | 7419 | 7427 | 7435 | 7443 | 7451 | 7459 | 7466 | 7474 |
| 56 | 7482 | 7490 | 7497 | 7505 | 7513 | 7520 | 7528 | 7536 | 7543 | 7551 |
| 57 | 7559 | 7566 | 7574 | 7582 | 7589 | 7597 | 7604 | 7612 | 7619 | 7627 |
| 58 | 7634 | 7642 | 7649 | 7657 | 7664 | 7672 | 7679 | 7686 | 7694 | 7701 |
| 59 | 7709 | 7716 | 7723 | 7731 | 7738 | 7745 | 7752 | 7760 | 7767 | 7774 |
| 60 | 7782 | 7789 | 7796 | 7803 | 7810 | 7818 | 7825 | 7832 | 7839 | 7846 |
| 61 | 7853 | 7860 | 7868 | 7875 | 7882 | 7889 | 7896 | 7903 | 7910 | 7917 |
| 62 | 7924 | 7931 | 7938 | 7945 | 7952 | 7959 | 7966 | 7973 | 7980 | 7987 |
| 63 | 7993 | 8000 | 8007 | 8014 | 8021 | 8028 | 8035 | 8041 | 8048 | 8055 |
| 64 | 8062 | 8069 | 8075 | 8082 | 8089 | 8096 | 8102 | 8109 | 8116 | 8122 |
| 65 | 8129 | 8136 | 8142 | 8149 | 8156 | 8162 | 8169 | 8176 | 8182 | 8189 |
| 66 | 8195 | 8202 | 8209 | 8215 | 8222 | 8228 | 8235 | 8241 | 8248 | 8254 |
| 67 | 8261 | 8267 | 8274 | 8280 | 8287 | 8293 | 8299 | 8306 | 8312 | 8319 |
| 68 | 8325 | 8331 | 8338 | 8344 | 8351 | 8357 | 8363 | 8370 | 8376 | 8382 |
| 69 | 8388 | 8395 | 8401 | 8407 | 8414 | 8420 | 8426 | 8432 | 8439 | 8445 |
| 70 | 8451 | 8457 | 8463 | 8470 | 8476 | 8482 | 8488 | 8494 | 8500 | 8506 |
| 71 | 8513 | 8519 | 8525 | 8531 | 8537 | 8543 | 8549 | 8555 | 8561 | 8567 |
| 72 | 8573 | 8579 | 8585 | 8591 | 8597 | 8603 | 8609 | 8615 | 8621 | 8627 |
| 73 | 8633 | 8639 | 8645 | 8651 | 8657 | 8663 | 8669 | 8675 | 8681 | 8686 |
| 74 | 8692 | 8698 | 8704 | 8710 | 8716 | 8722 | 8727 | 8733 | 8739 | 8745 |
| 75 | 8751 | 8756 | 8762 | 8768 | 8774 | 8779 | 8785 | 8791 | 8797 | 8802 |
| 76 | 8808 | 8814 | 8820 | 8825 | 8831 | 8837 | 8842 | 8848 | 8854 | 8859 |
| 77 | 8865 | 8871 | 8876 | 8882 | 8887 | 8893 | 8899 | 8904 | 8910 | 8915 |
| 78 | 8921 | 8927 | 8932 | 8938 | 8943 | 8949 | 8954 | 8960 | 8965 | 8971 |
| 79 | 8976 | 8982 | 8987 | 8993 | 8998 | 9004 | 9009 | 9015 | 9020 | 9025 |
| 80 | 9031 | 9036 | 9042 | 9047 | 9053 | 9058 | 9063 | 9069 | 9074 | 9079 |
| 81 | 9085 | 9090 | 9096 | 9101 | 9106 | 9112 | 9117 | 9122 | 9128 | 9133 |
| 82 | 9138 | 9143 | 9149 | 9154 | 9159 | 9165 | 9170 | 9175 | 9180 | 9186 |
| 83 | 9191 | 9196 | 9201 | 9206 | 9212 | 9217 | 9222 | 9227 | 9232 | 9238 |
| 84 | 9243 | 9248 | 9253 | 9258 | 9263 | 9269 | 9274 | 9279 | 9284 | 9289 |
| 85 | 9294 | 9299 | 9304 | 9309 | 9315 | 9320 | 9325 | 9330 | 9335 | 9340 |
| 86 | 9345 | 9350 | 9355 | 9360 | 9365 | 9370 | 9375 | 9380 | 9385 | 9390 |
| 87 | 9395 | 9400 | 9405 | 9410 | 9415 | 9420 | 9425 | 9430 | 9435 | 9440 |
| 88 | 9445 | 9450 | 9455 | 9460 | 9465 | 9469 | 9474 | 9479 | 9484 | 9489 |
| 89 | 9494 | 9499 | 9504 | 9509 | 9513 | 9518 | 9523 | 9528 | 9533 | 9538 |
| 90 | 9542 | 9547 | 9552 | 9557 | 9562 | 9566 | 9571 | 9576 | 9581 | 9586 |
| 91 | 9590 | 9595 | 9600 | 9605 | 9609 | 9614 | 9619 | 9624 | 9628 | 9633 |
| 92 | 9638 | 9643 | 9647 | 9652 | 9657 | 9661 | 9666 | 9671 | 9675 | 9680 |
| 93 | 9685 | 9689 | 9694 | 9699 | 9703 | 9708 | 9713 | 9717 | 9722 | 9727 |
| 94 | 9731 | 9736 | 9741 | 9745 | 9750 | 9754 | 9759 | 9763 | 9768 | 9773 |
| 95 | 9777 | 9782 | 9786 | 9791 | 9795 | 9800 | 9805 | 9809 | 9814 | 9818 |
| 96 | 9823 | 9827 | 9832 | 9836 | 9841 | 9845 | 9850 | 9854 | 9859 | 9863 |
| 97 | 9868 | 9872 | 9877 | 9881 | 9886 | 9890 | 9894 | 9899 | 9903 | 9908 |
| 98 | 9912 | 9917 | 9921 | 9926 | 9930 | 9934 | 9939 | 9943 | 9948 | 9952 |
| 99 | 9956 | 9961 | 9965 | 9969 | 9974 | 9978 | 9983 | 9987 | 9991 | 9996 |

## Table 3. Exponential Functions

| $x$ | $e^x$ | $e^{-x}$ |
|---|---|---|
| 0 | 1.0000 | 1.0000 |
| 0.1 | 1.1052 | 0.90484 |
| 0.2 | 1.2214 | 0.81873 |
| 0.3 | 1.3499 | 0.74082 |
| 0.4 | 1.4918 | 0.67032 |
| 0.5 | 1.6487 | 0.60653 |
| 0.6 | 1.8221 | 0.54881 |
| 0.7 | 2.0138 | 0.49659 |
| 0.8 | 2.2255 | 0.44933 |
| 0.9 | 2.4596 | 0.40657 |
| 1.0 | 2.7183 | 0.36788 |
| 1.1 | 3.0042 | 0.33287 |
| 1.2 | 3.3201 | 0.30119 |
| 1.3 | 3.6693 | 0.27253 |
| 1.4 | 4.0552 | 0.24660 |
| 1.5 | 4.4817 | 0.22313 |
| 1.6 | 4.9530 | 0.20190 |
| 1.7 | 5.4739 | 0.18268 |
| 1.8 | 6.0496 | 0.16530 |
| 1.9 | 6.6859 | 0.14957 |
| 2.0 | 7.3891 | 0.13534 |
| 2.1 | 8.1662 | 0.12246 |
| 2.2 | 9.0250 | 0.11080 |
| 2.3 | 9.9742 | 0.10026 |
| 2.4 | 11.023 | 0.09072 |
| 2.5 | 12.182 | 0.08208 |
| 2.6 | 13.464 | 0.07427 |
| 2.7 | 14.880 | 0.06721 |
| 2.8 | 16.445 | 0.06081 |
| 2.9 | 18.174 | 0.05502 |
| 3.0 | 20.086 | 0.04979 |
| 3.1 | 22.198 | 0.04505 |
| 3.2 | 24.533 | 0.04076 |
| 3.3 | 27.113 | 0.03688 |
| 3.4 | 29.964 | 0.03337 |
| 3.5 | 33.115 | 0.03020 |
| 3.6 | 36.598 | 0.02732 |
| 3.7 | 40.447 | 0.02472 |
| 3.8 | 44.701 | 0.02237 |
| 3.9 | 49.402 | 0.02024 |
| 4.0 | 54.598 | 0.01832 |
| 4.1 | 60.340 | 0.01657 |
| 4.2 | 66.686 | 0.01500 |
| 4.3 | 73.700 | 0.01357 |
| 4.4 | 81.451 | 0.01228 |
| 4.5 | 90.017 | 0.01111 |
| 4.6 | 99.484 | 0.01005 |
| 4.7 | 109.95 | 0.00910 |
| 4.8 | 121.51 | 0.00823 |
| 4.9 | 134.29 | 0.00745 |
| 5.0 | 148.41 | 0.00674 |

## Table 4. Natural Logarithms

| $n$ | $\ln n$ | $n$ | $\ln n$ | $n$ | $\ln n$ |
|---|---|---|---|---|---|
|  | * | 4.5 | 1.5041 | 9.0 | 2.1972 |
| 0.1 | 7.6974 | 4.6 | 1.5261 | 9.1 | 2.2083 |
| 0.2 | 8.3906 | 4.7 | 1.5476 | 9.2 | 2.2192 |
| 0.3 | 8.7960 | 4.8 | 1.5686 | 9.3 | 2.2300 |
| 0.4 | 9.0837 | 4.9 | 1.5892 | 9.4 | 2.2407 |
| 0.5 | 9.3069 | 5.0 | 1.6094 | 9.5 | 2.2513 |
| 0.6 | 9.4892 | 5.1 | 1.6292 | 9.6 | 2.2618 |
| 0.7 | 9.6433 | 5.2 | 1.6487 | 9.7 | 2.2721 |
| 0.8 | 9.7769 | 5.3 | 1.6677 | 9.8 | 2.2824 |
| 0.9 | 9.8946 | 5.4 | 1.6864 | 9.9 | 2.2925 |
| 1.0 | 0.0000 | 5.5 | 1.7047 | 10 | 2.3026 |
| 1.1 | 0.0953 | 5.6 | 1.7228 | 11 | 2.3979 |
| 1.2 | 0.1823 | 5.7 | 1.7405 | 12 | 2.4849 |
| 1.3 | 0.2624 | 5.8 | 1.7579 | 13 | 2.5649 |
| 1.4 | 0.3365 | 5.9 | 1.7750 | 14 | 2.6391 |
| 1.5 | 0.4055 | 6.0 | 1.7918 | 15 | 2.7081 |
| 1.6 | 0.4700 | 6.1 | 1.8083 | 16 | 2.7726 |
| 1.7 | 0.5306 | 6.2 | 1.8245 | 17 | 2.8332 |
| 1.8 | 0.5878 | 6.3 | 1.8405 | 18 | 2.8904 |
| 1.9 | 0.6419 | 6.4 | 1.8563 | 19 | 2.9444 |
| 2.0 | 0.6931 | 6.5 | 1.8718 | 20 | 2.9957 |
| 2.1 | 0.7419 | 6.6 | 1.8871 | 25 | 3.2189 |
| 2.2 | 0.7885 | 6.7 | 1.9021 | 30 | 3.4012 |
| 2.3 | 0.8329 | 6.8 | 1.9169 | 35 | 3.5553 |
| 2.4 | 0.8755 | 6.9 | 1.9315 | 40 | 3.6889 |
| 2.5 | 0.9163 | 7.0 | 1.9459 | 45 | 3.8067 |
| 2.6 | 0.9555 | 7.1 | 1.9601 | 50 | 3.9120 |
| 2.7 | 0.9933 | 7.2 | 1.9741 | 55 | 4.0073 |
| 2.8 | 1.0296 | 7.3 | 1.9879 | 60 | 4.0943 |
| 2.9 | 1.0647 | 7.4 | 2.0015 | 65 | 4.1744 |
| 3.0 | 1.0986 | 7.5 | 2.0149 | 70 | 4.2485 |
| 3.1 | 1.1314 | 7.6 | 2.0281 | 75 | 4.3175 |
| 3.2 | 1.1632 | 7.7 | 2.0412 | 80 | 4.3820 |
| 3.3 | 1.1939 | 7.8 | 2.0541 | 85 | 4.4427 |
| 3.4 | 1.2238 | 7.9 | 2.0669 | 90 | 4.4998 |
| 3.5 | 1.2528 | 8.0 | 2.0794 | 100 | 4.6052 |
| 3.6 | 1.2809 | 8.1 | 2.0919 | 110 | 4.7005 |
| 3.7 | 1.3083 | 8.2 | 2.1041 | 120 | 4.7875 |
| 3.8 | 1.3350 | 8.3 | 2.1163 | 130 | 4.8676 |
| 3.9 | 1.3610 | 8.4 | 2.1282 | 140 | 4.9416 |
| 4.0 | 1.3863 | 8.5 | 2.1401 | 150 | 5.0106 |
| 4.1 | 1.4110 | 8.6 | 2.1518 | 160 | 5.0752 |
| 4.2 | 1.4351 | 8.7 | 2.1633 | 170 | 5.1358 |
| 4.3 | 1.4586 | 8.8 | 2.1748 | 180 | 5.1930 |
| 4.4 | 1.4816 | 8.9 | 2.1861 | 190 | 5.2470 |

*Subtract 10 for $n < 1$. Thus $\ln 0.1 = 7.6974 - 10 = -2.3026$.

# Answers to Odd-numbered Exercises

**Exercises 1.1**

1. $\{2, 4, 6, 8\}$   3. $\{8, 16, 24, 32, 40, 48, 56, 64, 72, 80, 88, 96\}$
5. $\{w, e, d, n, s, a, y\}$   7. $\{x \mid x$ is a natural number less than $10\}$
9. $\{x \mid x$ is one of the first four letters of the English alphabet$\}$
11. Infinite   13. Finite   15. $\{0, 1, 2\}; \{0, 1\}; \{0, 2\}; \{1, 2\}; \{0\}; \{1\}; \{2\};$ $\emptyset$   17. $\subseteq$   19. $\not\subseteq$   21. $\not\subseteq$   23. $\{1, 2, 3, 4, 5, 6, 7, 8, 9\}$
25. $\{1, 5, 9\}$   27. $\{1, 2, 4, 6, 8\}$   29. $\emptyset$   31. $\{2, 4, 8\}$   33. $\{6\}$
35. $\{1, 2, 4, 8\}$   37. $\emptyset$   39. $\{6, 9\}$   41. $\{1, 2, 3, 4, 6, 8, 9\}$   43. $\in$
45. $\notin$   47. $\in$   49. $\notin$   51. $N \subseteq W$   53. $\bar{N} \subseteq R^-$   55. $H \subseteq R$
57. $Q$   59. $R$   61. $N$   63. $\{-38, 0, 12, 571\}$   65. $\{\sqrt{7}, -\sqrt{2}, \pi\}$
67. $\{-38\}$

**Exercises 1.2**

1. Axiom 1.2.2   3. Axiom 1.2.5   5. Axiom 1.2.2   7. Axiom 1.2.3
9. Axiom 1.2.6   11. Axiom 1.2.2   13. Axiom 1.2.7   15. Axiom 1.2.2
17. Axiom 1.2.5   19. Axiom 1.2.5   21. Axiom 1.2.4   23. Axiom 1.2.3
25. $-10, -7, -\sqrt{5}, -2, -\frac{7}{4}, -\frac{5}{3}, -1, 0, \frac{2}{3}, \frac{3}{4}, \sqrt{2}, 3, 5, 21$
27. $8 > -9; -9 < 8$   29. $-12 < -3; -3 > -12$
31. $4x - 5 < 0; 0 > 4x - 5$   33. $3t + 7 \geq 0; 0 \leq 3t + 7;$
35. $2 < r < 8; 8 > r > 2$   37. $-5 < a - 2 \leq 7; 7 \geq a - 2 > -5$

**Exercises 1.3**

1. Binomial; second degree in $x$; first degree in $y$; second-degree expression
3. Monomial; first degree in $x$; third degree in $y$; second degree in $z$; sixth-degree expression
5. Trinomial; second degree in $u$; third degree in $v$; fourth-degree expression
7. Binomial; first degree in $x$; third degree in $y$; third degree in $z$; sixth-degree expression
9. $-148$   11. $-30$   13. $-2$   15. $-\frac{3}{2}$   17. 9   19. $-4$
21. $-0.1$   23. $\frac{2}{5}$   25. 2   27. $\frac{6}{5}$   29. 3   31. $7i$   33. $6\sqrt{3}i$
35. $4 + 3i$   37. $-4 + 2i$   39. $4\sqrt{3} - 4\sqrt{3}i$   41. $2 - \frac{5}{4}i$
43. $5\sqrt{2} + 10\sqrt{2}i$

**Exercises 1.4**

1. $\{3\}$   3. $\{2\}$   5. $\{3\}$   7. $\emptyset$   9. $\{-\frac{13}{49}\}$   11. $\{-33\}$   13. $\{2\}$
15. $\{2\}$   17. $\{2\}$   19. $\{x \mid x \neq 2\}$   21. $\{-1\}$   23. $\{3\}$   25. $\{\frac{1}{2}\}$
27. $\{\frac{4}{3}\}$   29. $x = \frac{3}{4}b; a \neq 0$   31. $y = a + 3b; a - 2b \neq 0$

33. $x = 3a - 5b$; $a + b \neq 0$   35. $h = \dfrac{2A}{a+b}$   37. $r = \dfrac{E - IR}{I}$

39. $p = \dfrac{fq}{q - f}$   41. $r = \dfrac{a - S}{1 - S}$; $1 \neq S$

### Exercises 1.5

1. 12 cm; 8 cm   3. $12,000 at 14 per cent and $18,000 at 10 per cent
5. $8\frac{3}{4}$   7. 150 ounces of $26 perfume and 120 ounces of $12.50 perfume
9. 1 hour   11. 9 meters per second; $9\frac{1}{3}$ meters per second
13. 18 minutes   15. Newer press, $4\frac{1}{2}$ hours; older press, 9 hours
17. Fundamentals, 650; composition, 590
19. $2500 in camera A; $2500 in camera B; $1500 in camera C
21. 6 liters   23. 36 seconds   25. 54 minutes

### Exercises 1.6

1. $\{x \mid x \geq 5\}$   3. $\{x \mid x < \frac{7}{2}\}$   5. $\{x \mid x > -5\}$   7. $\{x \mid x \leq \frac{17}{2}\}$
9. $\{x \mid x < -\frac{17}{3}\}$   11. $\{x \mid x > -8\}$   13. $\{x \mid 4 \leq x < 8\}$
15. $\{x \mid -4 < x < 1\}$   17. $\{x \mid -6 \leq x \leq -4\}$   19. $\{x \mid -\frac{5}{3} < x \leq \frac{4}{3}\}$
21. $\{x \mid x < 0\} \cup \{x \mid x > \frac{20}{3}\}$   23. $\{x \mid x \leq -\frac{1}{2}\} \cup \{x \mid x > 0\}$
25. $\{x \mid 4 < x \leq 24\}$   27. $\{x \mid x < -\frac{23}{9}\} \cup \{x \mid x > -\frac{5}{3}\}$
37. At least 14 grams and at most 20 grams.   39. $6000   41. 49

### Exercises 1.7

1. (a) 10; (b) 6; (c) 8; (d) 0   3. (a) 24; (b) 24; (c) $-24$; (d) $-24$; (e) 24
5. (a) $-2$; (b) 5; (c) 2; (d) $\frac{2}{9}$   7. $\{1, 9\}$   9. $\{\frac{4}{3}, 4\}$   11. $\{-5, \frac{5}{2}\}$
13. $\{-1, 8\}$   15. $\{1, 3\}$   17. $\{\frac{9}{4}, 4\}$   19. $\{x \mid -5 \leq x \leq 5\}$
21. $\{x \mid x < -6\} \cup \{x \mid x > 8\}$   23. $\{x \mid 2 \leq x \leq 8\}$
25. $\{x \mid -1 < x < 8\}$   27. $\{x \mid x < -1\} \cup \{x \mid x > 8\}$
29. $\{x \mid x < -\frac{16}{3}\} \cup \{x \mid x > -\frac{8}{3}\}$   31. $\{x \mid x \leq -\frac{13}{3}\} \cup \{x \mid x \geq -2\}$
33. $\{x \mid x < -3\} \cup \{x \mid -3 < x < \frac{9}{11}\} \cup \{x \mid x > \frac{5}{3}\}$

### Exercises 1.8

9. (a) 7; (b) $-7$   11. (a) $-4$; (b) 4   13. (a) $-10$; (b) 6   15. 5
17. 13   19. $3\sqrt{2}$   21. $5\sqrt{5}$   23. $|\overline{AB}| = 10$; $|\overline{AC}| = 13$; $|\overline{BC}| = \sqrt{17}$
25. 3   27. $-\frac{3}{4}$   29. 0   41. 8

### Exercises 1.9

17. $3x - 7y = 0$   19. $x - y + 1 = 0$   21. $x = 4$
23. $2x - y - 10 = 0$   25. $2x - 3y + 19 = 0$   27. $y + 3 = 0$
29. $y = 2x + 5$; $m = 2$; $b = 5$   31. $y = 2x - \frac{2}{3}$; $m = 2$; $b = -\frac{2}{3}$
33. $y = -\frac{8}{3}$; $m = 0$; $b = -\frac{8}{3}$   35. $4x - 3y - 15 = 0$
37. $5x + y - 14 = 0$   39. $6x + y - 30 = 0$   41. (a) $y = 0$; (b) $x = 0$

### Review Exercises (Chapter 1)

1. $\{1, 2, 4, 6, 8, 9\}$   3. $\{4\}$   5. $\{2, 17\}$   7. $\{-4, 17, \frac{3}{4}, -5, 0, -\frac{1}{3}, 2\}$
9. $\{2, 4, 6, 8\}$; $\{2, 4, 8\}$; $\{4, 6, 8\}$; $\{4, 8\}$
11. Commutative law for multiplication (Axiom 1.2.2)
13. Identity element for addition (Axiom 1.2.5)
15. Associative law for multiplication (Axiom 1.2.3)
17. Existence of additive inverse (Axiom 1.2.6)
19. 16   21. $\frac{5}{9}$   23. 3   25. $8 - 8i$   27. $\{\frac{4}{3}\}$   29. $\{\frac{4}{3}\}$   31. $\{\frac{1}{2}, 2\}$
33. $x = \dfrac{a}{a - b}$; $a + b \neq 0$; $a - b \neq 0$   35. $y = -\dfrac{A}{B}x - \dfrac{C}{B}$

37. $\{x \mid x \le 4\}$  39. $\{x \mid x \ge 12\}$  41. $\{x \mid 3 < x \le 5\}$
43. $\{x \mid -1 < x < 6\}$  45. $4\sqrt{10}$  51. parallelogram  53. rectangle
55. $3x - 11y - 1 = 0$  57. $2x - 3y + 13 = 0$
59. \$10,000 at 8 per cent; \$20,000 at 11 per cent  61. \$9, \$6
63. 26 minutes and 40 seconds  65. $1\frac{1}{3}$ liters  67. at least 95

### Exercises 2.1

1. $\{-7, 7\}$  3. $\{-3i, 3i\}$  5. $\{-\frac{2}{5}\sqrt{15}, \frac{2}{5}\sqrt{15}\}$  7. $\{0, \frac{1}{4}\}$  9. $\{3, 5\}$
11. $\{-\frac{3}{2}, \frac{1}{4}\}$  13. $\{-\frac{3}{4}, \frac{5}{8}\}$  15. $\{-\frac{6}{7}\}$  17. $\{-\frac{5}{3}, -\frac{3}{2}\}$  19. $\{\frac{1}{3}a, \frac{2}{3}a\}$
21. $\left\{-\dfrac{b}{2a}, \dfrac{7b}{5a}\right\}$  23. $\{-3, 7\}$  25. $\{-\frac{1}{3}, 2\}$  27. $\{-2, -\frac{5}{3}\}$
29. $x^2 - 3x - 10 = 0$  31. $12x^2 - 17x + 6 = 0$  33. $9x^2 + x = 0$
35. $abx^2 - (a^2 + b^2)x + ab = 0$  37. $x^2 + 16 = 0$  39. 8, 10
41. 27 centimeters by 12 centimeters
43. 30 feet per second and 25 feet per second, or 25 feet per second and 20 feet per second.
45. 6 kilometers per hour, or 48 kilometers per hour

### Exercises 2.2

1. $x^2 + 6x + 9; (x + 3)^2$  3. $x^2 - 11x + \frac{121}{4}; (x - \frac{11}{2})^2$
5. $w^2 + \frac{1}{2}w + \frac{1}{16}; (w + \frac{1}{4})^2$  7. $x^2 - \frac{4}{3}ax + \frac{4}{9}a^2; (x - \frac{2}{3}a)^2$  9. $\{-5, -2\}$
11. $\{-3, \frac{4}{3}\}$  13. $\left\{\dfrac{1 - \sqrt{5}}{2}, \dfrac{1 + \sqrt{5}}{2}\right\}$  15. $\left\{\dfrac{-2 - i\sqrt{2}}{3}, \dfrac{-2 + i\sqrt{2}}{3}\right\}$
17. $\{-1, 4\}$  19. $\{-3, \frac{5}{2}\}$  21. $\{1 - \sqrt{3}, 1 + \sqrt{3}\}$
23. $\{2 - i\sqrt{3}, 2 + i\sqrt{3}\}$  25. $\left\{\dfrac{2 - i\sqrt{3}}{5}, \dfrac{2 + i\sqrt{3}}{5}\right\}$
27. 361; roots are unequal rational numbers
29. 52; roots are unequal irrational numbers
31. $-31$; roots are unequal imaginary numbers
33. 0; roots are equal rational numbers
35. 225; roots are unequal irrational numbers
37. $-60, 60$  39. $-1$  41. $\frac{3}{8}, \frac{3}{2}$  43. $x = \dfrac{3 \pm \sqrt{9 + 40a^2}}{10a}; a \ne 0$
45. $x = \dfrac{b}{1 + a}, x = \dfrac{b}{1 - a}; a \ne -1, a \ne 1$  47. $x = 5 - 2y; x = \dfrac{y - 3}{2}$
49. $x = \dfrac{-y \pm y\sqrt{3}}{2}$  51. $x = y + 2 \pm 2\sqrt{y^2 + y + 1}$  55. $\frac{1}{3}; -\frac{1}{2}$
57. $-\dfrac{3}{m}; -\dfrac{n}{m}$  59. $x = \dfrac{-b \pm \sqrt{b^2 - 4ac + 4ay}}{2a}$  61. 15.51 meters

### Exercises 2.3

1. $\{64\}$  3. $\emptyset$  5. $\{\frac{7}{2}\}$  7. $\{4\}$  9. $\{3\}$  11. $\{4\}$  13. $\emptyset$
15. $\{0, 3\}$  17. $\{7\}$  19. $\{-2\}$  21. $\{\frac{5}{3}, 3\}$  23. $\{-7\}$  25. $\{1, 2\}$
27. $\{-3, 0\}$  29. $\{-2, -1, 1, 2\}$  31. $\{-\sqrt{3}, -\sqrt{2}, \sqrt{2}, \sqrt{3}\}$
33. $\left\{-\dfrac{\sqrt{5}}{2}, \dfrac{\sqrt{5}}{2}, -\dfrac{\sqrt{2}}{2}i, \dfrac{\sqrt{2}}{2}i\right\}$
35. $\left\{2, 3, \dfrac{-3 - 3i\sqrt{3}}{2}, \dfrac{-3 + 3i\sqrt{3}}{2}, -1 - i\sqrt{3}, -1 + i\sqrt{3}\right\}$  37. $\{1\}$
39. $\{-5, -1, 3\}$  41. $\emptyset$

## Exercises 2.4

**31.** $xy = 0$

## Exercises 2.5

**1.** $(0, \frac{1}{4}); y = -\frac{1}{4}$   **3.** $(0, 2); y = -2$   **5.** $(0, -4); y = 4$   **7.** $(0, -\frac{1}{2}); y = \frac{1}{2}$
**9.** $(0, 3); x = 0; (0, \frac{11}{4}); y = \frac{13}{4}$   **11.** $(3, -9); x = 3; (3, -\frac{35}{4}); y = -\frac{37}{4}$
**13.** $(-2, -6); x = -2; (-2, -\frac{23}{4}); y = -\frac{25}{4}$
**15.** $(-2, 3); x = -2; (-2, \frac{23}{8}); y = \frac{25}{8}$   **17.** $(2, -2); x = 2; (2, 0); y = -4$
**19.** $x^2 = -12y$   **21.** $x^2 = 3y$   **23.** $x^2 + 4x + 60 = 8y$
**27.** $(-\frac{1}{32}, 0); x = \frac{1}{32}$   **29.** $(\frac{5}{8}, 0); x = -\frac{5}{8}$

## Exercises 2.6

**1.** Circle   **3.** Circle   **5.** Ellipse   **7.** Hyperbola
**9.** Two distinct lines through the origin   **11.** Point: the origin
**13.** Hyperbola   **15.** Ellipse   **17.** No graph
**19.** (a) Ellipse; (b) hyperbola; (c) point: the origin; (d) two distinct lines through the origin
**21.** (a) Ellipse; (b) hyperbola; (c) no graph; (d) two distinct lines through the origin
**23.** $x^2 + y^2 - 6x + 10y + 18 = 0$   **25.** $(5, 6); 7$

## Exercises 2.7

**1.** $\{x \mid 1 < x < 3\}$   **3.** $\{x \mid x < -4\} \cup \{x \mid x > 2\}$   **5.** $\{x \mid -2 \leq x \leq \frac{3}{2}\}$
**7.** $\{x \mid x < -3\} \cup \{x \mid x > 3\}$   **9.** $\{x \mid -3 < x < 4\}$
**11.** $\{x \mid x < 1\} \cup \{x \mid x > 3\}$   **13.** $\{x \mid -4 \leq x \leq 1\}$
**15.** $\left\{x \mid \frac{3 - \sqrt{3}}{2} < x < \frac{3 + \sqrt{3}}{2}\right\}$   **17.** $\{x \mid x < 1 - \sqrt{2}\} \cup \{x \mid x > 1 + \sqrt{2}\}$
**19.** $\{x \mid x < -\frac{5}{3}\} \cup \{x \mid \frac{4}{3} < x < 2\}$   **21.** $\{x \mid x < -\frac{23}{9}\} \cup \{x \mid x > -\frac{5}{3}\}$
**23.** $\{x \mid \frac{5}{4} < x < 2\}$   **25.** More than 10 and less than 150

## Exercises 2.8

**17.** $2y < 12 - 3x$   **19.** $2x - 7y + 3 \leq 0$

## Review Exercises (Chapter 2)

**1.** $\{-\frac{6}{5}, \frac{1}{2}\}$   **3.** $\left\{\frac{2 - \sqrt{14}}{2}, \frac{2 + \sqrt{14}}{2}\right\}$   **5.** $\left\{\frac{1 - i\sqrt{5}}{3}, \frac{1 + i\sqrt{5}}{3}\right\}$
**7.** $\{1 - \sqrt{13}, 1 + \sqrt{13}\}$   **9.** $\{\frac{1}{3}\}$   **11.** $\{-2\}$   **13.** $\{-\frac{1}{2}, -\frac{1}{3}, \frac{1}{3}, \frac{1}{2}\}$
**15.** $x = \frac{y - 1}{3}, x = \frac{1}{2}$   **29.** Center at $(2, -5)$, radius 4
**31.** $\{x \mid x < -2\} \cup \{x \mid x > \frac{5}{2}\}$   **33.** $\{x \mid -5 < x < 5\}$
**35.** $\{x \mid x \leq -5\} \cup \{x \mid x \geq \frac{3}{2}\}$   **37.** $\{x \mid \frac{1}{2} \leq x < 3\} \cup \{x \mid x \geq 4\}$
**43.** $(0, \frac{3}{2}); y = -\frac{3}{2}$   **45.** $x^2 = -16y$   **47.** 70 kilometers per hour
**49.** $4\sqrt{21}$ kilometers per hour

## Exercises 3.1

**1.** Domain: $\{-3, -1, 1, 3\}$; range: $\{0, 2, 4, 6\}$; function
**3.** Domain: $\{0, 2, 4, 6\}$; range: $\{0, 2, 4\}$; function
**5.** Domain: $\{-1, 1, 3\}$; range: $\{1, 3, 4, 5\}$; not a function
**7.** Domain: $\{-2, 0, 2\}$; range: $\{-1\}$; function
**9.** Domain $\{2, 3, 4\}$; range: $\{-2, 0, 1, 3, 5\}$; not a function   **11.** Function
**13.** Function   **15.** Function   **17.** Not a function
**19.** $S$ and $T$ are functions; $R$ is not a function   **21.** Domain: $R$; range: $R$
**23.** Domain: $R$; range: $\{y \mid y \geq -4\}$
**25.** Domain: $\{x \mid x \geq 4\}$; range: $\{y \mid y \geq 0\}$

27. Domain: $\{x\,|-6\leq x\leq 6\}$; range: $\{y\,|\,0\leq y\leq 6\}$
29. Domain: $R$; range: $\{-5\}$   31. Domain: $R$; range: $\{-3,3\}$
33. Domain: $\{x\,|\,x\neq 3\}$; range: $\{y\,|\,y\neq 6\}$   35. Domain: $R$; range: $\{y\,|\,y\neq 6\}$
37. Domain: $\{x\,|\,x\neq -5,1\}$; range: $\{y\,|\,y\neq -9,-3\}$
39. Domain: $R$; range: $\{y\,|\,y\geq -4\}$
41. Domain: $R$; range: $\{y\,|\,y\leq -2\}\cup\{y\,|\,0<y\leq 4\}$
43. Domain: $\{x\,|\,x\leq -1\}\cup\{x\,|\,x\geq 5\}$; range: $\{y\,|\,y\geq 0\}$

**Exercises 3.2**

1. 0   3. 30   5. 3   7. $\frac{5}{3}$   9. 13   11. 16   13. $-11$
15. (a) $8h-10$; (b) $8h-5$; (c) $8x-5$; (d) $4x+3$; (e) $4x^2-5$
17. (a) $4h^2+2$; (b) $8h^2+1$; (c) $8x^2+1$; (d) $2x^2+8x+9$; (e) $2x^4+1$
19. (a) $6-10h-4h^2$; (b) $3-10h-8h^2$; (c) $3-10x-8x^2$;
    (d) $2x^2+3x+1$; (e) $3-5x^2+2x^4$
21. (a) $\dfrac{4}{1+h}$; (b) $\dfrac{2}{1+2h}$; (c) $\dfrac{2}{1+2x}$; (d) $\dfrac{2}{x+3}$; (e) $\dfrac{2}{1+x^2}$
23. 6   25. $-2x-h$   27. $6x+3h+4$   29. $3x^2+3hx+h^2+3$
31. $\dfrac{-11}{(3+x)(3+x+h)}$
33. (a) even; (b) odd; (c) neither; (d) even; (e) neither; (f) even; (g) odd;
    (h) neither

**Exercises 3.3**

1. (a) $2x+5$, domain: $R$; (b) $-9$, domain: $R$; (c) $x^2+5x-14$, domain: $R$;
   (d) $\dfrac{x-2}{x+7}$, domain: $\{x\,|\,x\neq -7\}$; (e) $\dfrac{x+7}{x-2}$, domain: $\{x\,|\,x\neq 2\}$;
   (f) $x^2-4x+4$, domain: $R$
3. (a) $6+4x-x^2$, domain: $R$; (b) $x^2+4x-2$, domain: $R$;
   (c) $8+16x-2x^2-4x^3$, domain: $R$; (d) $\dfrac{2+4x}{4-x^2}$, domain: $\{x\,|\,x\neq -2,2\}$;
   (e) $\dfrac{4-x^2}{2+4x}$, domain: $\{x\,|\,x\neq -\tfrac{1}{2}\}$; (f) $16x^2+16x+4$, domain: $R$
5. (a) $5x^2-9x-18$, domain: $R$; (b) $-x^2+7x-12$, domain: $R$;
   (c) $6x^4-19x^3-43x^2-123x+45$, domain: $R$; (d) $\dfrac{2x+5}{3x+1}$, domain:
   $\{x\,|\,x\neq -\tfrac{1}{3},3\}$; (e) $\dfrac{3x+1}{2x+5}$, domain: $\{x\,|\,x\neq -\tfrac{5}{2},3\}$;
   (f) $4x^4-4x^3-59x^2+30x+225$, domain: $R$
7. (a) $\dfrac{1}{x}+\sqrt{x}$, domain: $\{x\,|\,x>0\}$; (b) $\dfrac{1}{x}-\sqrt{x}$, domain: $\{x\,|\,x>0\}$; (c) $\dfrac{1}{\sqrt{x}}$,
   domain: $\{x\,|\,x>0\}$; (d) $\dfrac{1}{x\sqrt{x}}$; domain: $\{x\,|\,x>0\}$; (e) $x\sqrt{x}$, domain:
   $\{x\,|\,x>0\}$; (f) $\dfrac{1}{x^2}$; domain: $\{x\,|\,x\neq 0\}$
9. (a) $\dfrac{x^3+x^2+x-1}{x^2-x}$, domain: $\{x\,|\,x\neq 0,1\}$; (b) $\dfrac{-x^3-x^2+x-1}{x^2-x}$,
   domain: $\{x\,|\,x\neq 0,1\}$; (c) $\dfrac{x+1}{x-1}$, domain: $\{x\,|\,x\neq 0,1\}$; (d) $\dfrac{x-1}{x^3+x^2}$,
   domain: $\{x\,|\,x\neq -1,0,1\}$; (e) $\dfrac{x^3+x^2}{x-1}$, domain: $\{x\,|\,x\neq 0,1\}$; (f) $\dfrac{1}{x^2}$,
   domain: $\{x\,|\,x\neq 0\}$

11. (a) $x + 5$, domain: $R$; (b) $x + 5$, domain: $R$; (c) $x - 4$, domain: $R$;
(d) $x + 14$, domain: $R$
13. (a) $x^2 - 1$, domain: $R$; (b) $(x - 1)^2$, domain: $R$; (c) $x - 2$, domain: $R$;
(d) $x^4$, domain: $R$
15. (a) $\sqrt{x^2 - 4}$, domain: $\{x \mid x \leq -2\} \cup \{x \mid x \geq 2\}$; (b) $x - 4$, domain: $\{x \mid x \geq 2\}$; (c) $\sqrt{\sqrt{x - 2} - 2}$, domain: $\{x \mid x \geq 6\}$; (d) $x^4 - 4x^2 + 2$, domain: $R$
17. (a) $\dfrac{1}{\sqrt{x}}$, domain: $\{x \mid x > 0\}$; (b) $\dfrac{1}{\sqrt{x}}$, domain: $\{x \mid x > 0\}$; (c) $x$, domain: $\{x \mid x \neq 0\}$; (d) $\sqrt[4]{x}$, domain: $\{x \mid x \geq 0\}$
19. (a) $|x + 2|$, domain: $R$; (b) $||x| + 2|$, domain: $R$; (c) $|x|$, domain: $R$;
(d) $||x + 2| + 2|$, domain: $R$
23. $g(x) = x - 3$

## Exercises 3.4

1. (a)  3. (a)  5. (c)  7. (b)  9. (c)  11. $-1, 3$  13. $\dfrac{1 \pm \sqrt{3}}{2}$

15. $-1$ is a minimum value  17. 4 is a maximum value
19. $-\frac{1}{3}$ is a maximum value  21. 5 and 5  23. 144 feet; 3 seconds
25. 50 yards by 50 yards  27. 130; $84,500  29. 40; $900

## Exercises 3.5

1. (a) domain: $\{x \mid x \neq 0\}$; (b) $x = 0, y = 0$; (c) none
3. (a) domain: $\{x \mid x \neq 0\}$; (b) $x = 0, y = 0$; (c) $y < 0$
5. (a) domain: $\{x \mid x \neq 4\}$; (b) $x = 4, y = 1$; (c) none
7. (a) domain: $\{x \mid x \neq 1\}$; (b) $x = 1$; (c) $0 < y < 4$
9. (a) domain: $\{x \mid x \neq \pm 2\}$; (b) $x = -2, x = 2, y = 1$; (c) $0 < y < 1$
11. (a) domain: $R$; (b) $y = 0$; (c) $y < -1, y > 1$
13. (a) domain: $\{x \mid x \neq \pm 2\}$; (b) $x = -2, x = 2, y = 0$; (c) none
15. (a) domain: $\{x \mid x \neq \pm 1\}$; (b) $x = -1, x = 1, y = 1$; (c) $-1 < y < 1$
17. (a) domain: $\{x \mid x \neq 1, 2\}$; (b) $x = 1, x = 2, y = 1$; (c) $-8 < y < 0$
19. (a) domain: $\{x \mid x \neq -\frac{5}{2}, \frac{1}{2}\}$; (b) $x = -\frac{5}{2}, x = \frac{1}{2}, y = 2$; (c) none

## Exercises 3.6

1. $S^{-1} = \{(3, -4), (-3, -2), (-1, 0), (1, 2), (0, 4)\}$; $S^{-1}$ is a function
3. $S^{-1} = \{(5, -1), (-2, 0), (-1, 1), (2, 1), (4, 2)\}$; $S^{-1}$ is a function
5. $S^{-1} = \{(x, y) \mid 2y^2 + x - 4 = 0\}$; $S^{-1}$ is not a function
7. $S^{-1} = \{(x, y) \mid x^2 + y^2 = 25\}$; $S^{-1}$ is not a function
9. $S^{-1} = \{(x, y) \mid x = \sqrt{9 - y^2}\}$; $S^{-1}$ is not a function
11. $S^{-1} = \{(x, y) \mid xy = 6\}$; $S^{-1}$ is a function
13. $f^{-1} = \{(7, -5), (4, -2), (1, 1), (-2, 4)\}$; $f^{-1}$ is a function
15. $f^{-1} = \{(x, y) \mid 2x + 5y = 10\}$; $f^{-1}$ is a function
17. $f^{-1} = \{(x, y) \mid x = |y + 1|\}$; $f^{-1}$ is not a function
19. $f^{-1} = \{(x, y) \mid x = -\sqrt{y + 4}\}$; $f^{-1}$ is a function
21. $f^{-1} = \{(x, y) \mid y^2 - 6y - 6x + 15 = 0\}$; $f^{-1}$ is not a function
23. $f^{-1} = \{(x, y) \mid y = \sqrt[3]{x} + 2\}$; $f^{-1}$ is a function
25. (b) $f^{-1}(x) = \dfrac{x + 3}{4}$; (c) $-\frac{1}{2}$  27. (b) $f^{-1}(x) = \sqrt[3]{x - 2}$; (c) $-2$

29. (a) Range of $f = \{y\,|\,y \geq -5\}$; (b) $f^{-1} = \{(x,y)\,|\,y = \sqrt{x+5}\}$, domain of $f^{-1} = \{x\,|\,x \geq -5\}$; (c) $f^{-1}$ is a function
31. (a) Range of $f = \{y\,|\,y \geq -4\}$; (b) $f^{-1} = \{(x,y)\,|\,y = 2 - \sqrt{3x+12}\}$, domain of $f^{-1} = \{x\,|\,x \geq -4\}$; (c) $f^{-1}$ is a function
33. (a) Range of $f = \{y\,|\,0 \leq y \leq 4\}$; (b) $f^{-1} = \{x,y\}\,|\,x = \sqrt{16-y^2}\}$, domain of $f^{-1} = \{x\,|\,0 \leq x \leq 4\}$; (c) $f^{-1}$ is not a function
35. (a) $F^{-1} = \{(x,y)\,|\,y = \sqrt{x}+2\}$    37. (a) $F^{-1} = \{(x,y)\,|\,y = 2 - \sqrt[3]{x}\}$

## Exercises 3.7

1. (a) $y = 3x$; (b) 9    3. (a) $v = \frac{5}{8}u^2$; (b) 20    5. (a) $p = \dfrac{6\sqrt{2}}{\sqrt{q}}$; (b) 2

7. (a) $z = \dfrac{9x^3}{2y^2}$; (b) 9    9. (a) $r = \dfrac{5st^3}{2u^2}$; (b) 3    11. $\frac{625}{16}$ candlepower

13. $2\frac{1}{2}$ seconds    15. $41\frac{2}{3}$ amperes    17. 200 cubic meters
19. $z$ is multiplied by $\frac{4}{3}$    21. 50 per cent    23. 212.5 per cent

## Review Exercises (Chapter 3)

1. Function    3. Not a function    5. Function
7. Domain: $R$; range: $\{y\,|\,y \geq -4\}$
9. Domain: $\{x\,|\,x \leq -2\} \cup \{x\,|\,x \geq 2\}$; range: $\{y\,|\,y \geq 0\}$
11. Domain: $\{x\,|\,x \neq -4\}$; range: $\{y\,|\,y \neq -8\}$
13. Domain: $R$; range: $\{y\,|\,y \geq 1\}$    15. 0    17. $-2$    19. (a) 9; (b) 57
21. (a) $48x^2 - 8x + 1$; (b) $12x^2 - 8x + 4$
23. (a) $3x^2 - 14x + 17$; (b) $3x^2 - 2x - 8$
25. (a) $x^2 + x - 6$, domain: $R$; (b) $-x^2 + x - 4$, domain: $R$;

  (c) $x^3 - 5x^2 - x + 5$, domain: $R$; (d) $\dfrac{x-5}{x^2-1}$, domain: $\{x\,|\,x \neq \pm 1\}$;

  (e) $\dfrac{x^2-1}{x-5}$, domain: $\{x\,|\,x \neq 5\}$; (f) $x^2 - 10x + 25$, domain: $R$

27. (a) $\sqrt{x} + \dfrac{1}{x^2}$, domain: $\{x\,|\,x > 0\}$; (b) $\sqrt{x} - \dfrac{1}{x^2}$, domain: $\{x\,|\,x > 0\}$;

  (c) $\dfrac{\sqrt{x}}{x^2}$, domain: $\{x\,|\,x > 0\}$; (d) $x^2\sqrt{x}$, domain: $\{x\,|\,x > 0\}$;

  (e) $\dfrac{1}{x^2\sqrt{x}}$, domain: $\{x\,|\,x > 0\}$; (f) $x$, domain: $\{x\,|\,x \geq 0\}$

29. (a) $\sqrt{x^2+1}$, domain: $R$; (b) $x + 1$, domain: $\{x\,|\,x \geq 0\}$;
  (c) $\sqrt[4]{x}$, domain: $\{x\,|\,x \geq 0\}$; (d) $x^4 + 2x^2 + 2$, domain: $R$
31. (a)    33. (b)    35. $-3$ is a minimum value
37. (a) domain: $\{x\,|\,x \neq 3\}$; (b) $x = 3$; (c) $0 < y < 12$
39. (a) domain: $\{x\,|\,x \neq \pm 5\}$; (b) $x = -5$, $x = 5$, $y = -1$; (c) $-1 < y < 0$
41. $f^{-1} = \{(x,y)\,|\,x = 3y\}$; $f^{-1}$ is a function
43. $f^{-1} = \{(x,y)\,|\,x = \sqrt{y-4}\}$; $f^{-1}$ is a function
45. $f^{-1} = \{(x,y)\,|\,x = 9 - y^2, y \geq 0\}$; $f^{-1}$ is a function
47. 9 and 9
49. 24 or 25; $375
51. (a) 80 amperes; (b) 0.25 ohm

## Exercises 4.1

1. $-\frac{1}{125}$  3. $\frac{1}{36}$  5. 6  7. 9  9. 4  11. $-0.064$  13. $\frac{1}{56}$
15. $-15\frac{15}{16}$  17. $\frac{8}{3}$  19. $\frac{14}{51}$  21. 9  23. $\frac{1}{x}$  25. $t^6$  27. $a^{5/12}$
29. $\frac{1}{x^{1/4}}$  31. $\frac{y^2}{x^4}$  33. $x^{1/4}$  35. $\frac{2}{3}x^4y^4z^4$  37. $\frac{b-a}{ab}$  39. $\frac{2y}{x-2y}$
41. $-\frac{1}{a^2b+ab^2}$  43. $\frac{y^2+x^2}{y^2-x^2}$  45. $\frac{z}{x^{1/2}y^{1/3}}$  47. $y^{9/4}+y^{1/8}$
49. $x + 2x^{1/2}y^{1/2} + y$  51. $\frac{a + 2a^{1/2}b^{1/2} + b}{ab}$
53. $a + 3a^{2/3}b^{1/3} + 3a^{1/3}b^{2/3} + b$  55. $x^2$  57. $x^{n+16}$  59. $\frac{1}{a^{2n-2}}$
61. 100  63. $2|x|y^2$  65. $25(x-5)^2$  67. $4(x-2)^2|2-y|$
69. (a) $\begin{cases} -6 \text{ if } x \leq -3 \\ 2x \text{ if } -3 < x < 3 \\ 6 \text{ if } x \geq 3 \end{cases}$; (b) $x \geq 3$

## Exercises 4.2

1. 9  3. $-0.1$  5. $\frac{4}{25}$  7. $\frac{6}{5}$  9. 5  11. $4\sqrt{3}$  13. $3\sqrt[3]{2}$
15. $2c^2\sqrt[3]{c^2}$  17. $-2x^5y^2\sqrt[5]{3y^2}$  19. $|b|$  21. $10\sqrt{3}$  23. $-24\sqrt[3]{2}$
25. $-3s^2t^2\sqrt[3]{2s}$  27. $6x^2y^2\sqrt[3]{15x^2y}$  29. $7\sqrt{7}$  31. $27\sqrt[3]{3}$
33. $5a\sqrt{2a}$  35. $\frac{3-s}{3}\sqrt{st}$  37. $\frac{2b^3}{c^2}\sqrt{c}$  39. $6\sqrt{3} - 6\sqrt{6}$
41. $\sqrt{12} - 10$  43. $30 + 12\sqrt{6}$  45. $1 - 2\sqrt[3]{6} + \sqrt[3]{36}$  47. $5a + 13b$
49. $\frac{\sqrt{35}}{7}$  51. $\frac{2}{5}\sqrt[3]{15}$  53. $\frac{1}{5}\sqrt[3]{10}$  55. $-\frac{3s}{2t}\sqrt{2t}$  57. $-\frac{1}{6y}\sqrt[3]{6y}$
59. $\frac{2a\sqrt[4]{12a^2b^2}}{3b}$  61. $-\frac{3}{2x}\sqrt[3]{28x^2}$  63. $\frac{14}{3}\sqrt{3} - \frac{21}{2}\sqrt{2}$
65. $\frac{4}{7}(3 + \sqrt{2})$  67. $1 + \sqrt{5}$  69. $\frac{13\sqrt{14} - 54}{10}$

## Exercises 4.3

11. $3^{6\sqrt{2}}$  13. $5^{3\sqrt{10}}$  15. $2^{5\sqrt{2}}$  17. 3033  19. 22.3 grams

## Exercises 4.4

1. $\log_3 81 = 4$  3. $\log_5 125 = 3$  5. $\log_{10} 0.001 = -3$  7. $\log_8 4 = \frac{2}{3}$
9. $\log_{625} \frac{1}{125} = -\frac{3}{4}$  11. $8^2 = 64$  13. $3^4 = 81$  15. $7^0 = 1$
17. $(\frac{1}{3})^{-2} = 9$  19. $9^{-1/2} = \frac{1}{3}$  21. 2  23. $\frac{2}{3}$  25. $-3$  27. $-\frac{1}{3}$
29. $-\frac{4}{3}$  31. $\frac{1}{3}$  33. 343  35. 81  37. 12  39. 216  41. $\frac{1}{128}$
43. 0  45. 1  47. 3  49. 0

## Exercises 4.5

1. $\log_b 5 + \log_b x + \log_b y$  3. $\log_b y - \log_b z$  5. $\log_b x - \log_b y - \log_b z$
7. $\log_b x + 5\log_b y$  9. $\frac{1}{2}\log_b x + \frac{1}{2}\log_b y$  11. $\frac{1}{3}\log_b x + 3\log_b z$
13. $\log_b x + \frac{1}{2}\log_b y - 4\log_b z$  15. $\frac{2}{3}\log_b x - \frac{1}{3}\log_b y - \frac{2}{3}\log_b z$
17. $\frac{2}{3}\log_b x + \frac{1}{2}\log_b y + \frac{1}{2}\log_b z$  19. $\log_{10} x^4\sqrt{y}$  21. $\log_b \frac{\sqrt[4]{x^3}}{y^6\sqrt[5]{z^4}}$
23. $\ln \frac{1}{3}\pi r^2 h$  25. 1.1461  27. 1.1761  29. 1.7993  31. 3.1461
33. 0.3404  35. $-2.7744$  37. 0.6309  39. $\{125\}$  41. $\{3\}$  43. $\{7\}$

Answers to Odd-numbered Exercises  A-13

**Exercises 4.6**
1. $(5.260)10^1$   3. $(6.1)10^{-3}$   5. $(1.72)10^5$   7. $(3.960)10^{-2}$
9. $(8.0022)10^{-6}$   11. 2.5611   13. 1.7143   15. $9.4314 - 10$
17. $8.9605 - 10$   19. 5.5416   21. 252   23. 8.43   25. 0.0298
27. 0.000195   29. 581,000   31. 0.4400   33. 3.6617   35. $7.9667 - 10$
37. $5.5228 - 10$   39. 5.8920   41. 3.525   43. 9311   45. 0.5116
47. 0.001446   49. 7,396,000   51. 73.2   53. 0.0002269   55. 2.384
57. 0.1501   59. 0.2258   61. 4.73   63. 4.03   65. 1.64 seconds
67. 3.259 feet   69. $10,940

**Exercises 4.7**
1. $\{1.403\}$   3. $\{0.4307\}$   5. $\{4.299\}$   7. $\{8.632\}$   9. $\{32.2\}$
11. $\{1.015\}$   13. 2.262   15. 4.169   17. 5.043   19. 0.9376
21. Between $18\frac{1}{2}$ and 19 years   23. 68.4 years   25. 2.5 per cent

**Review Exercises (Chapter 4)**
1. 16   3. $\frac{1}{32}$   5. $\frac{y^8}{x^6}$   7. $\frac{y-x}{x}$   9. $2y^2|xz^3|$   11. $4\sqrt{3}$
13. $2\sqrt[3]{2} - \sqrt[3]{4}$   15. $6x - 2x\sqrt{3y} - 12xy$   17. $\sqrt{x} + \sqrt{y}$   21. 625
23. $\frac{4}{3}$   25. 64   27. $\log_b x + \frac{1}{3}\log_b y - 4\log_b z$   29. $\log_b \frac{4}{3}\pi r^2 h$   31. $\{\frac{1}{2}\}$
33. $\{1.113\}$   35. 0.4584   37. 8.82   39. 0.9358   41. 47.63
43. 11.9 years   45. 57.65 mg.

**Exercises 5.1**
1. $12 + 3i$   3. $-6$   5. $1 + 3i$   7. $8 + 26i$   9. $-9 + \sqrt{5}i$
11. $-15$   13. $-4$   15. $-72i$   17. $8 - 18\sqrt{2}$   19. 53
21. $-24 + 5\sqrt{3}i$   23. $-18 + 18\sqrt{3}i$   25. $5i$   27. $-\frac{3}{13} - \frac{2}{13}i$
29. $\frac{4}{5} + \frac{7}{5}i$   31. $\frac{1}{2} - \frac{3}{4}i$   33. $\frac{3}{11} - \frac{\sqrt{2}}{11}i$   35. $\frac{5}{169} - \frac{12}{169}i$   37. $-i$
39. $i$   41. $-i$   43. $i$   45. $16 - 16i$   47. 0   49. 0

**Exercises 5.2**
19. $\sqrt{29}$   21. $\sqrt{5}$   23. 2   25. 10

**Exercises 5.3**
1. $\{-3i, 3i\}$   3. $\left\{-\frac{\sqrt{6}}{3}i, \frac{\sqrt{6}}{3}i\right\}$   5. $\{-1 - i\sqrt{5}, -1 + i\sqrt{5}\}$
7. $\{\frac{1}{2} - i, \frac{1}{2} + i\}$   9. $\{-\frac{1}{3} - \frac{1}{3}i\sqrt{2}, -\frac{1}{3} + \frac{1}{3}i\sqrt{2}\}$   11. $x^2 - 10x + 26 = 0$
13. $25x^2 - 10\sqrt{3}x + 19 = 0$   15. $36x^3 - 36x^2 + 25x = 0$
17. $x^4 + 4x^2 + 3 = 0$   19. $3x^4 - 4x^3 - 12x^2 + 20x - 15 = 0$
21. $\{-2i\}$   23. $\{\frac{1}{17} - \frac{4}{17}i\}$   25. $\{\frac{4}{5} + \frac{3}{5}i\}$   27. $\{-\frac{1}{10} + \frac{7}{10}i\}$
29. $\{-i, \frac{5}{2}i\}$   31. $\{-i, 4i\}$   33. $\left\{-\frac{\sqrt{11}}{5} + \frac{2}{5}i, \frac{\sqrt{11}}{5} + \frac{2}{5}i\right\}$
35. $1, \frac{-1 \pm i\sqrt{3}}{2}$   37. $-3, \frac{-3 \pm 3i\sqrt{3}}{2}$   39. $-5, 5, -5i, 5i$
41. $-1 \pm i, 1 \pm i$

**Exercises 5.4**
1. 20   3. 5   5. 1   7. $-8$   9. Yes   11. No   13. Yes   15. Yes
17. $-7$   19. 1   21. 1 and $-3$   23. All positive integer values
25. Positive odd integer values
27. Positive even integer values not divisible by 4

Exercises 5.5

1. The quotient is $(2x^2 + 7x + 31)$, and the remainder is 135
3. The quotient is $(2x^3 - 3x^2 + 12x - 50)$, and the remainder is 199
5. The quotient is $(3x^4 + 7x^3 + 14x^2 + 24x + 48)$, and the remainder is 103
7. The quotient is $(x^4 - x^3 - 4x^2 - x - 4)$, and the remainder is $-16$
9. The quotient is $(6x^2 - 3x + 3)$, and the remainder is 1
11. The quotient is $(x^6 + x^5 + x^4 + x^3 + x^2 + x + 1)$, and the remainder is 0
13. The quotient is $(x^3 + 2ix^2 + i)$, and the remainder is $-2$
15. The quotient is $[x^2 + (1 + i)x - 2]$, and the remainder is $(4 - 2i)$
17. $8; -157$   19. $-7; 126$   21. $4; \frac{19}{2}$   23. $3 - 4i; 3 - 4i$
25. $-19 - 2i; -19 + 2i$   35. No   37. Yes

Exercises 5.6

1. 4, multiplicity two; $-2$; 2
3. 0, multiplicity two; $-1$, multiplicity two; $i$; $-i$
5. $-7$, multiplicity three; $\sqrt{7}$, multiplicity two; $-\sqrt{7}$, multiplicity two
7. $-\frac{4}{3}$, multiplicity three; $\frac{3}{2}$, multiplicity two; $-\frac{3}{2}$, multiplicity two; $\frac{3}{2}i$; $-\frac{3}{2}i$
9. $4, -1$   11. $-\frac{1}{2} \pm \frac{\sqrt{5}}{2}$   13. $\pm i\sqrt{2}$   15. $\frac{1}{2}, -1$   17. $\frac{2}{5}, 1$
19. $-2 \pm \sqrt{3}$   21. $1, 2$   23. $\frac{\sqrt{11}}{2} + \frac{1}{2}i, -\frac{\sqrt{11}}{2} + \frac{1}{2}i$
25. $x^2 - 8x + 25$   27. $x^3 - 7x + 36$   29. $x^4 + 18x^2 + 81$
31. $x^5 - 9x^4 + 31x^3 - 51x^2 + 58x - 66$   33. $\left\{\frac{1 \pm \sqrt{31}}{3}, -2i, 2i\right\}$
35. $\left\{\frac{1 \pm i}{2}, -2 - 4i, -2 + 4i\right\}$

Exercises 5.7

1. $-1, 1, 3$   3. $-2, -1, 3$   5. $-5, 3, \frac{-1 \pm i\sqrt{3}}{2}$   7. $\frac{1}{3}, \frac{-3 \pm \sqrt{5}}{2}$
9. $\frac{1}{2}, \frac{5}{3}, 2 \pm \sqrt{3}$   11. 2 is the only rational zero   13. $-4, -1, 1$
15. $\frac{3}{2}, 2, 3$   17. $-4, -2, 0, 1, 3$   19. $-\frac{1}{4}, \frac{2}{3}, \pm\sqrt{3}$   21. $-6, \frac{1 \pm i\sqrt{11}}{6}$
23. $-2$ is the only rational root   29. 2 cm or $(4 - \sqrt{11})$ cm   31. 6 cm

Exercises 5.9

1. 1 positive, 0 negative, 2 imaginary
3. 2 positive, 1 negative, and 0 imaginary; or 0 positive, 1 negative, and 2 imaginary
5. 3 positive, 0 negative, and 0 imaginary; or 1 positive, 0 negative, and 2 imaginary
7. 1 positive, 1 negative, and 2 imaginary
9. 0 positive, 1 negative, and 2 imaginary; one root is 0
11. 3 positive, 0 negative, and 2 imaginary; or 1 positive, 0 negative, and 4 imaginary
13. One negative irrational root between $-3$ and $-2$; one positive irrational root between 0 and 1; one positive irrational root between 2 and 3
15. One negative irrational root between $-1$ and 0; one positive irrational root between 0 and 1; two imaginary roots

Answers to Odd-numbered Exercises  A-15

17. $-1$ is a root; one positive irrational root between 1 and 2; two imaginary roots
19. 4 is a root; one negative irrational root between $-1$ and 0; one positive irrational root between 0 and 1; one positive irrational root between 2 and 3
21. 2.65   23. 0.51   25. $-1.12$

**Review Exercises (Chapter 5)**

1. $18 + i$   3. $\frac{1}{4} - \frac{2}{3}i$   5. $-21$   7. $10 - 10i$   9. $-\frac{14}{25} + \frac{23}{25}i$   13. 5
15. $x^4 - x^3 + 10x^2 - 16x - 96$   17. $-\frac{7}{2}i, i$   19. $-2, 1 \pm i\sqrt{3}$
21. $-7$   23. Yes   25. $5i$
27. The quotient is $2x^3 + x^2 - 3x + 5$, and the remainder is $-10$.
29. The quotient is $x^5 + 2x^4 + 4x^3 + 8x^2 + 16x + 32$, and the remainder is 0.
31. 17; 57   33. $-2i$, multiplicity two; $2i$, multiplicity two; $-3$; $-\frac{5}{3}$; $\frac{3}{2}$; 3
35. $-1 - i$; $\dfrac{-1 \pm \sqrt{5}}{2}$   37. $x^4 - 4x^3 + 10x^2 - 12x + 8$
39. $-3, 2, -i\sqrt{2}, i\sqrt{2}$   41. 3, $\dfrac{-1 \pm i\sqrt{11}}{2}$   43. $-\frac{1}{3}, \frac{1}{2}, 2 \pm \sqrt{5}$
49. 2 positive, 1 negative, and 0 imaginary; or 0 positive, 1 negative, and 2 imaginary
51. 2 is a root; one negative irrational root between $-4$ and $-3$; one positive irrational root between 0 and 1; one positive irrational root between 3 and 4
53. $-1.58$

**Exercises 6.1**

1. $\{(3, -5)\}$   3. $\{(\frac{9}{4}, \frac{3}{2})\}$   5. Inconsistent   7. Consistent and dependent
9. $\{(-1.5, -1.3)\}$   11. $\{(1, 0)\}$   13. $\{(0, -2)\}$   15. $\{(\frac{1}{2}, \frac{1}{6})\}$
17. $\{(\frac{2}{3}, \frac{1}{2})\}$   19. $\{(-4, 7)\}$   21. $\{(-\frac{12}{17}, \frac{42}{17})\}$   23. $\{(-\frac{15}{4}, -\frac{15}{2})\}$
25. $\{(\frac{3}{4}, -\frac{1}{5})\}$

**Exercises 6.2**

1. $\{(3, 1, 0)\}$   3. $\{(1, 2, 1)\}$   5. $\{(-3, 2, -4)\}$   7. Inconsistent
9. $\{(-3, -5, 0)\}$   11. $\{(\frac{1}{4}, \frac{1}{2}, 1)\}$
13. $\{(t - 3, 2t - 4, t)\}$; $(-3, -4, 0), (-2, -2, 1), (-1, 0, 2), (-4, -6, -1), (-5, -8, -2)$
15. Consistent; $\{(3, 2, -4)\}$   17. Inconsistent

**Exercises 6.3**

1. India tea, $2.56 per pound; China tea, $2.32 per pound   3. 20; $120
5. 600 kilometers per hour, 840 kilometers
7. Any fraction of the form $\dfrac{t}{2t - 4}$, where $t \neq 2$ and $t \neq 0$
9. $6\frac{1}{4}$ miles per hour; 12 miles per hour
11. Girl, 20 minutes; brother, 25 minutes
13. Twenty grams of first alloy, forty grams of second alloy, and ten grams of third alloy
15. $y = 4x^2 - 8x + 3$
17. Two large-size cards, seven medium-size cards, and one small-size card; three large-size cards, five medium-size cards, and two small-size cards; four large-size cards, three medium-size cards, and three small-size cards; or five large-size cards, one medium-size card, and four small-size cards

**Exercises 6.4**

1. $\{(-4, -3), (3, 4)\}$   3. $\{(-1, 0), (\frac{3}{2}, \frac{5}{4})\}$   5. $\{(\frac{17}{5}, \frac{8}{5})\}$
7. $\{(\frac{1 + \sqrt{5}}{2}, \frac{-5 + \sqrt{5}}{2}), (\frac{1 - \sqrt{5}}{2}, \frac{-5 - \sqrt{5}}{2})\}$
9. $\{(-2 + 2\sqrt{3}, 2 - \sqrt{3}), (-2 - 2\sqrt{3}, 2 + \sqrt{3})\}$
11. $\{(2, 0), (-2, 0), (0, 2)\}$; $(0, 2)$ is a double root
13. $\{(4, 1), (-4, -1), (i, -4i), (-i, 4i)\}$
15. $\{(2\sqrt{3}, \sqrt{13}), (-2\sqrt{3}, \sqrt{13}), (2\sqrt{3}, -\sqrt{13}), (-2\sqrt{3}, -\sqrt{13})\}$
17. $\{(-3, 0), (5, 4), (5, -4)\}$; $(-3, 0)$ is a double root
19. $\{(2, 4), (-2, -4)\}$; $(2, 4)$ and $(-2, -4)$ are double roots
21. $\{(-1, 2), (1, -2), (\frac{2}{3}\sqrt{6}, \frac{1}{3}\sqrt{6}), (-\frac{2}{3}\sqrt{6}, -\frac{1}{3}\sqrt{6})\}$
23. $\{(\frac{1}{2}, i), (-\frac{1}{2}, i), (\frac{1}{2}, -i), (-\frac{1}{2}, -i)\}$   25. 6 and 10   27. 20; $7.50
29. 24 miles; 6 miles per hour

**Exercises 6.5**

19. $3\frac{11}{16}$ at $(\frac{25}{8}, \frac{17}{8})$   21. 8 units of product A and 16 units of product B.
23. 3 from $W_1$ to $R_1$; 4 from $W_2$ to $R_2$; 2 from $W_1$ to R3

**Review Exercises (Chapter 6)**

1. $(i)$; $\{(3, -2)\}$   3. $(ii)$; $\{(t, 2t - 3)\}$   5. $(i)$; $\{(2, -5)\}$   7. $(iii)$
9. $(i)$; $\{(3, 1, -2)\}$   11. $(ii)$; $\{(3t + 1, 2t - 1, t)\}$
13. Consistent; $\{(4, \frac{1}{2}, -\frac{2}{3})\}$
15. $\{(-4t - 5, -2t - 3, t)\}$; $(-5, -3, 0)$, $(-9, -5, 1)$, $(-13, -7, 2)$, $(-1, -1, -1)$, $(3, 1, -2)$
17. $\{(1, \sqrt{2}), (1, -\sqrt{2}), (-1, \sqrt{2}), (-1, -\sqrt{2})\}$
19. $\{(1, -2), (-1, 2), (\frac{1}{2}, -\frac{5}{2}), (-\frac{1}{2}, \frac{5}{2})\}$
25. 60 kilometers per hour; 540 kilometers per hour
27. $12,500; 6 per cent

**Exercises 7.1**

1. $\{(\frac{1}{2}, -1)\}$   3. $\{(4, 6)\}$   5. $\{(3, -1, 2)\}$   7. $\{(2, -3, 8)\}$
9. $\{(-10 - 7t, 8 + 6t, t)\}$   11. $\{(-2, 2, -3, 1)\}$   13. $\emptyset$

**Exercises 7.2**

1. (a) $2 \times 4$; (b) $3 \times 1$; (c) $-2$; (d) $\begin{bmatrix} -3 & 2 & -7 & 0 \\ -4 & -5 & 1 & -8 \end{bmatrix}$;
(e) $\begin{bmatrix} 6 \\ -2 \\ -5 \end{bmatrix}$; (f) $\begin{bmatrix} 9 & -6 & 21 & 0 \\ 12 & 15 & -3 & 24 \end{bmatrix}$; (g) $\begin{bmatrix} 12 \\ -4 \\ -10 \end{bmatrix}$

3. (a) $4 \times 3$; (b) $2 \times 2$; (c) 0; (d) $\begin{bmatrix} 0 & 0 & 0 \\ 0 & 0 & 0 \\ 0 & 0 & 0 \\ 0 & 0 & 0 \end{bmatrix}$;
(e) $\begin{bmatrix} 0 & 0 \\ 0 & 0 \end{bmatrix}$; (f) $\begin{bmatrix} -28 & -8 & 4 \\ 12 & 16 & 0 \\ 8 & -4 & 20 \\ 0 & 16 & 4 \end{bmatrix}$; (g) $\begin{bmatrix} -18 & 30 \\ 6 & -12 \end{bmatrix}$

5. $\begin{bmatrix} -1 & -4 \\ -9 & 8 \end{bmatrix}$   7. $\begin{bmatrix} -11 & 5 & -6 \\ 13 & 5 & -8 \end{bmatrix}$   9. $\begin{bmatrix} -5 & 0 & 3 \\ 5 & -9 & 0 \\ -5 & -5 & -1 \end{bmatrix}$

11. $a = 13, b = -7, c = -3, d = -1$   13. $\begin{bmatrix} -4 & 8 \\ 12 & 20 \\ -24 & 4 \end{bmatrix}$

Answers to Odd-numbered Exercises   A-17

**15.** $\begin{bmatrix} 6 & -12 & 2 \\ -8 & 4 & 0 \\ 10 & -2 & 16 \end{bmatrix}$    **17.** $\begin{bmatrix} 4 & 7 \\ -2 & 16 \end{bmatrix}$    **19.** $\begin{bmatrix} 2 & 2 \\ 7 & 3 \end{bmatrix}$    **21.** $[7]$

**23.** $\begin{bmatrix} 2 & 4 & -6 \\ 1 & 2 & -3 \\ -1 & -2 & 3 \end{bmatrix}$    **37.** $\begin{bmatrix} 4 & 2 \\ 10 & 7 \end{bmatrix}$    **39.** $\begin{bmatrix} -2 & -1 \\ -4 & -1 \\ 1 & 2 \end{bmatrix}$

### Exercises 7.3

**1.** $M_{11} = \begin{vmatrix} 2 & -5 \\ 3 & 6 \end{vmatrix}$; $M_{23} = \begin{vmatrix} -3 & 4 \\ -1 & 3 \end{vmatrix}$; $M_{32} = \begin{vmatrix} -3 & 0 \\ 1 & -5 \end{vmatrix}$

**3.** $A_{21} = -\begin{vmatrix} 4 & 0 \\ 3 & 6 \end{vmatrix}$; $A_{31} = \begin{vmatrix} 4 & 0 \\ 2 & -5 \end{vmatrix}$; $A_{13} = \begin{vmatrix} 1 & 2 \\ -1 & 3 \end{vmatrix}$

**5.** $M_{12} = \begin{vmatrix} -1 & -1 & 2 \\ 3 & 6 & 0 \\ 1 & 4 & 1 \end{vmatrix}$; $M_{22} = \begin{vmatrix} 5 & 0 & 3 \\ 3 & 6 & 0 \\ 1 & 4 & 1 \end{vmatrix}$; $M_{32} = \begin{vmatrix} 5 & 0 & 3 \\ -1 & -1 & 2 \\ 1 & 4 & 1 \end{vmatrix}$;

$M_{42} = \begin{vmatrix} 5 & 0 & 3 \\ -1 & -1 & 2 \\ 3 & 6 & 0 \end{vmatrix}$

**7.** $A_{31} = \begin{vmatrix} -2 & 0 & 3 \\ 4 & -1 & 2 \\ 3 & 4 & 1 \end{vmatrix}$; $A_{32} = -\begin{vmatrix} 5 & 0 & 3 \\ -1 & -1 & 2 \\ 1 & 4 & 1 \end{vmatrix}$; $A_{33} = \begin{vmatrix} 5 & -2 & 3 \\ -1 & 4 & 2 \\ 1 & 3 & 1 \end{vmatrix}$;

$A_{34} = -\begin{vmatrix} 5 & -2 & 0 \\ -1 & 4 & -1 \\ 1 & 3 & 4 \end{vmatrix}$

**9.** 22    **11.** $-6$    **13.** 11    **15.** 112    **17.** 1    **19.** $-108$    **21.** 1
**23.** $-2, 2$    **25.** $-2$    **27.** $-\frac{3}{2}, 1$
**31.** $-a_{14}a_{21}a_{32}a_{43}a_{55}$, $a_{14}a_{23}a_{32}a_{41}a_{55}$
**33.** $-a_{12}a_{23}a_{34}a_{41}a_{55}$, $a_{12}a_{23}a_{35}a_{41}a_{54}$, $a_{14}a_{23}a_{32}a_{41}a_{55}$, $-a_{14}a_{23}a_{35}a_{41}a_{52}$,
$-a_{15}a_{23}a_{32}a_{41}a_{54}$, $a_{15}a_{23}a_{34}a_{41}a_{52}$

### Exercises 7.4

**1.** Theorem 7.4.1    **3.** Theorem 7.4.5    **5.** Theorem 7.4.2
**7.** Theorem 7.4.3    **9.** Theorem 7.4.3    **11.** Theorem 7.4.4
**13.** Theorem 7.4.4    **15.** $-1$    **17.** $-1$    **19.** 50    **21.** $-11$    **23.** $-12$
**27.** $9x + 5y = 7$

### Exercises 7.5

**1.** $\begin{bmatrix} 2 & -3 \\ -4 & 5 \\ 1 & 0 \end{bmatrix}$    **3.** $\begin{bmatrix} -5 & 3 & 0 \\ 0 & -6 & -1 \\ 2 & 8 & 4 \end{bmatrix}$    **5.** $\begin{bmatrix} 1 & -2 \\ -2 & 5 \end{bmatrix}$    **7.** Singular

**9.** $\begin{bmatrix} -\frac{2}{3} & \frac{1}{3} \\ -\frac{5}{3} & \frac{4}{3} \end{bmatrix}$    **11.** $\begin{bmatrix} 1 & 0 & -1 \\ 1 & 1 & -2 \\ -5 & -3 & 9 \end{bmatrix}$    **13.** $\begin{bmatrix} \frac{1}{3} & \frac{1}{3} & 0 \\ -\frac{1}{3} & -\frac{1}{3} & 1 \\ 0 & -1 & 2 \end{bmatrix}$    **15.** $\begin{bmatrix} 3 & 4 \\ 2 & 3 \end{bmatrix}$

**29.** $\{(1, 4)\}$    **31.** $\{(-3, 4)\}$    **33.** $\{(-3, 1, 4)\}$    **35.** $\{(1, 1, 2)\}$

### Exercises 7.6

**1.** $\{(-\frac{1}{7}, -\frac{10}{7})\}$    **3.** $\{(-\frac{1}{2}, \frac{2}{3})\}$    **5.** $\{(2, 3, 2)\}$
**7.** $\{(-\frac{1}{2}t, -t, t)\}$; dependent    **9.** $\{(-\frac{5}{12}, -\frac{1}{6}, -\frac{4}{3})\}$    **11.** $\emptyset$; inconsistent
**13.** $\{(-1, 2, 1, 0)\}$    **15.** $\{(\frac{3}{4}, \frac{1}{4}, \frac{1}{2}, -\frac{1}{4})\}$    **17.** $\{(2t, 1 - t, 1 + t, t)\}$; dependent

**Review Exercises (Chapter 7)**   1. $\{(5, 3)\}$   3. $\{(-4, -2, 3)\}$   5. $\{(\frac{1}{2}, 3, -\frac{1}{3}, 4)\}$   7. $\begin{bmatrix} -3 & 3 \\ -10 & 6 \end{bmatrix}$

9. $\begin{bmatrix} -2 & -6 \\ -7 & -21 \\ 19 & 5 \end{bmatrix}$   11. $[2]$   13. $\begin{bmatrix} 2 & -1 & 3 \\ -2 & -4 & -3 \end{bmatrix}$   15. $\begin{bmatrix} a & c \\ b & d \end{bmatrix}$

17. $\begin{bmatrix} 0 & 0 & \frac{1}{2} \\ \frac{1}{3} & 0 & -\frac{1}{6} \\ -\frac{1}{3} & 1 & \frac{1}{6} \end{bmatrix}$   19. $[2 \ \ -3 \ \ 2]$   21. $\begin{bmatrix} -1 & 3 \\ 0 & 2 \end{bmatrix}$   23. $\begin{bmatrix} 2 \\ -1 \\ -3 \end{bmatrix}$

25. $1$   27. $-35$   29. $34$   33. $\begin{bmatrix} -\frac{3}{2} & -\frac{1}{2} \\ -2 & -1 \end{bmatrix}$   35. $\{(-1, 2, 1)\}$

37. $\{(\frac{2}{3}, -1)\}$   39. $\{(\frac{1}{6}, \frac{1}{2}, -\frac{2}{3})\}$

**Exercises 8.1**   1. $5, 7, 9, 11, 13, 15$   3. $\frac{2}{1}, \frac{5}{2}, \frac{10}{3}, \frac{17}{4}, \frac{26}{5}, \frac{37}{6}$   5. $\frac{2}{1}, -\frac{3}{3}, \frac{4}{5}, -\frac{5}{7}, \frac{6}{9}, -\frac{7}{11}$

7. $-\frac{2}{3}, \frac{4}{5}, -\frac{8}{9}, \frac{16}{17}, -\frac{32}{33}, \frac{64}{65}$   9. $\frac{1}{3}x, -\frac{1}{4}x^2, \frac{1}{5}x^3, -\frac{1}{6}x^4, \frac{1}{7}x^5, -\frac{1}{8}x^6$

11. $0, 4, 0, 8, 0, 12$   13. $1, 2, \frac{1}{2}, 2, \frac{1}{3}, 2$   15. $1, 1, 2, 2, 3, 3$

17. $1 + 5 + 9 + 13 + 17; 45$   19. $\frac{2}{1} + \frac{3}{2} + \frac{4}{3} + \frac{5}{4} + \frac{6}{5}; \frac{437}{60}$

21. $5 + 5 + 5 + \cdots + 5$, (100 terms); $500$   23. $1 + \frac{1}{2} + \frac{1}{4} + \frac{1}{8} + \frac{1}{16} + \frac{1}{32}; \frac{63}{32}$

25. $x - x^3 + x^5 - x^7 + x^9 - x^{11} + x^{13} - x^{15}$

27. $f(x_1) + f(x_2) + f(x_3) + \cdots + f(x_n)$   29. $\sum_{i=1}^{6}(2i - 1)$

31. $\sum_{i=1}^{5}(-1)^{i-1}(3i + 1)$   33. $\sum_{i=1}^{6}\frac{2i-1}{i^2}$   35. $\sum_{i=1}^{5}(-1)^{i=1}\frac{x^{2i-2}}{2i}$

37. $1, 3, 5, 7, 9, 11, \ldots, 2n - 1, \ldots$;

$1, 3, 5, 5, 9, 7, \ldots, a_n, \ldots$, where $a_n = \begin{cases} 2n - 1, \text{ if } n \text{ is odd} \\ n + 1, \text{ if } n \text{ is even} \end{cases}$;

$1, 3, 5, 13, 33, 71, \ldots, 2n - 1 + (n - 1)(n - 2)(n - 3), \ldots$

**Exercises 8.3**   1. $5, 8, 11, 14, 17$   3. $10, 6, 2, -2, -6$   5. $-5, -12, -19, -26, -33$
7. $x, x + 2y, x + 4y, x + 6y, x + 8y$
9. An arithmetic sequence; $-13, -17$   11. Not an arithmetic sequence
13. An arithmetic sequence; $0, -\frac{1}{12}$
15. An arithmetic sequence; $4x + 3y, 5x + 4y$   17. $35$   19. $16$
21. Thirtieth   23. $5\frac{1}{5}, 5\frac{2}{5}, 5\frac{3}{5}, 5\frac{4}{5}$   25. $71\frac{1}{2}$   27. $108$   29. $2500$
31. $-\frac{25}{2}$   33. $2430$   35. $\$280, \$260, \$240, \$220$   37. $18$ days

**Exercises 8.4**   1. $5, 15, 45, 135, 405$   3. $8, -4, 2, -1, \frac{1}{2}$   5. $-\frac{9}{16}, \frac{3}{8}, -\frac{1}{4}, \frac{1}{6}, -\frac{1}{9}$
7. $\frac{x}{y}, -1, \frac{y}{x}, -\frac{y^2}{x^2}, \frac{y^3}{x^3}$   9. A geometric sequence; $27, 81$
11. A geometric sequence; $3\sqrt{6}, 9\sqrt{2}$   13. Not a geometric sequence
15. A geometric sequence; $\frac{2}{9}, -\frac{2}{27}$   17. $\frac{729}{4}$   19. $-3$   21. Eleventh
23. $2, 4, 8, 16, 32$   25. $\sqrt[3]{9}, \sqrt[6]{243}$   27. $-2$   29. $510$   31. $\frac{5465}{729}$
33. $11.17$   35. $6, 9, 12, 16$; or, $6, 1, -4, 16$   37. $\frac{117,649}{262,144}$   39. $\$24,300$

**Exercises 8.5**  1. $\frac{64}{3}$  3. $\frac{600}{11}$  5. $\frac{4}{5}$  7. $\frac{16}{21}$  9. $\dfrac{9+3\sqrt{3}}{2}$  11. $\frac{2}{3}$  13. $\frac{9}{11}$  15. 3  17. $\frac{137}{111}$  19. $\frac{47}{101}$  21. $\frac{29{,}029}{9{,}000}$  23. $\frac{244}{99}$  25. 84 meters  27. 400 centimeters

**Review Exercises (Chapter 8)**  1. $\frac{1}{3}, -\frac{3}{9}, \frac{5}{27}, -\frac{7}{81}, \frac{9}{243}, -\frac{11}{729}$  3. $\sum_{i=1}^{4}(-1)^{i-1}\dfrac{x^{2i}}{2^{i}}$  5. 1425  7. 10.56  13. $\frac{7}{16}$  15. $\frac{1}{9}$ or $-\frac{1}{9}$  17. Twenty-first  19. $\frac{8}{19}$  21. $\frac{513}{110}$  23. 455  25. 1,049,000; 2,098,000  27. 5, 10, 20; or 20, 10, 5

**Exercises 9.1**  1. (a) 60; (b) 125  3. 180  5. 60  7. 216  9. 720  11. 360  13. 1728  15. 240  17. 720  19. 2880  21. 1232

**Exercises 9.2**  1. $\frac{1}{504}$  3. $\frac{33}{2}$  5. $\frac{1}{168}$  7. 60  9. 720  11. 7  13. 210  15. 362,880  17. 5040  19. 720  21. 5040  23. 12  25. 48  27. 20,160  29. 6720  31. 3780  33. 12,600  35. 1,201,200  37. 9  39. 11

**Exercises 9.3**  1. 35  3. 220  5. 1225  7. 9  9. 8  11. 210  13. 220  15. 84  17. 21  19. 210  21. 301  23. 540  25. 21  27. 1,244,117,160  29. 2,672,060  31. 15,400

**Exercises 9.4**  1. $x^5 + 15x^4y + 90x^3y^2 + 270x^2y^3 + 405xy^4 + 243y^5$  3. $4096 - 6144ab + 3840a^2b^2 - 1280a^3b^3 + 240a^4b^4 - 24a^5b^5 + a^6b^6$  5. $u^{-5} - 5u^{-2} + 10u - 10u^4 + 5u^7 - u^{10}$  7. $a^3 + 6a^{5/2}b^{1/2} + 15a^2b + 20a^{3/2}b^{3/2} + 15ab^2 + 6a^{1/2}b^{5/2} + b^3$  9. $x^{24} + 12x^{22}y^2 + 66x^{20}y^4 + 220x^{18}y^6$  11. $a^3 - 9a^{8/3}b^{1/3} + 36a^{7/3}b^{2/3} - 84a^2b$  13. 1.262  15. 96,060,000  17. $924a^6b^6$  19. $-489{,}888x^4$  21. $\dfrac{924x^{18}}{y^{12}}$  23. $\frac{231}{16}a^6$  25. $-8064$  27. $1 - x^2 + x^4 - x^6$  29. $1 - x - \frac{1}{2}x^2 - \frac{1}{2}x^3$  31. $2 + \frac{1}{12}x - \frac{1}{288}x^2 + \frac{5}{10{,}736}x^3$  33. 1.020  35. 3.979  37. 4.990

**Review Exercises (Chapter 9)**  1. 55  3. 360  5. 35  7. 5  9. 8  11. 720  13. 480  15. 120  17. 420  19. 595  21. 648  23. 328  25. 72  27. (a) 220; (b) 55  29. $x^{20} + 40x^{18} + 760x^{16} + 9120x^{14}$  31. 0.7351

# Index

## A

Abscissa, 64
Absolute inequality, 49
Absolute value, 55, 278
Absolute value function, 165
Addition
  of complex numbers, 267
  of functions, 167
  of radicals, 219
  of real numbers, 11
Additive identity
  for complex numbers, 268
  for real numbers, 11
Additive inverse
  for complex numbers, 268
  for real numbers, 11
Additive-inverse axiom, 11
Algebraic equation, 28
Algebraic expression, 20
Algebraic function, 164
Antilogarithm, 250
Arithmetic means, 447, 448
Arithmetic progression, 444
Arithmetic sequence
  common difference of, 444
  definition of, 444
  $N$th element of, 445
Arithmetic series, 448
  sum of, 449

Associative law
  for addition of real numbers, 11
  for multiplication of real numbers, 11
Asymptotes, 120, 132
  horizontal, 180
  vertical, 179
Augmented matrix, 377
Axiom(s)
  of addition, 11
  meaning of, 11
  of multiplication, 11
  of order, 13
Axis
  of a coordinate system, 15, 64
  of a parabola, 122
  of pure imaginary numbers, 276
  of real numbers, 276

## B

Base
  exponential function with, 226
  of a logarithm, 231
  logarithmic function with, 231
  of a power, 20
Binomial
  definition of, 21
  expansion, 487

A-21

Binomial (*cont.*)
  series, 492
  theorem, 490

## C

Cartesian coordinate system, rectangular, 65
Cartesian product, 63
Center of a circle, 126
Characteristic of a logarithm, 246
Circle
  definition of, 126
  equation of, 127
  exterior of, 146
  interior of, 146
Closed half plane, 144
Closure
  law for addition of real numbers, 11
  law of multiplication of real numbers, 11
Coefficient, 21
Coefficient matrix, 377
Cofactor, 395
Collinear, 72
Column of a matrix, 376
Combinations, 482
Common logarithm, 245
Commutative law
  for addition of real numbers, 11
  for multiplication of real numbers, 11
Complete ordered field, 16
Completeness property, 14
Completing a square, 99
Complex number, 24, 266
Complex plane, 276
Component of an ordered pair, 63
Composite function, 169
Compound interest, 254
Conclusion of a theorem, 13
Conditional equation, 29
Conditional inequality, 50
Conic section, 134

Conjugate, 222, 269
Consistent equations, 329, 339
Constant, 20
Constant function, 162
Constant of proportionality, 196
Constant of variation, 196
Constraints, 369
Convex region, 368
Coordinate(s), 16, 65
Coordinate axis, 15, 64
Coordinate system, rectangular Cartesian, 65
Counting numbers, 2
Cramer's rule, 423, 425
Cubic function, 163
Curve, 76

## D

Decimal(s)
  nonterminating, 8
  repeating, 8
  terminating, 8
Decreasing function, 192
Degree
  of equation, 29
  of monomial, 21
  of polynomial, 22
Denominator, 12
  rationalizing a, 220
Dependent equations, 329, 344
Dependent variable, 155
Descartes' rule of signs, 319
Determinant(s)
  in Cramer's rule, 423, 425
  definition of, 393, 394
  evaluation of, 393, 394, 401
  $n$th order, 401
  properties of, 403
  second-order, 393
  third-order, 394
Difference
  of functions, 167
  of real numbers, 12

Direct variation, 196
Directed distance, 65
Directly proportional, 196
Directrix, 121
Discriminant, 102
Disjoint sets, 6
Distance formula, 68
Distinguishable permutations, 479
Distributive law of multiplication over addition, for real numbers, 11
Dividend, 12
Division
  of complex numbers, 268
  of functions, 167
  of real numbers, 12
  by zero, 13
Divisor, 12
Domain
  of a function, 153
  of a relation, 152
Double root, 92

## E

$e$, 228
  table of powers of, A-4
Elements(s)
  of a matrix, 376
  of a sequence, 433
  of a set, 2
Elementary row operations for a matrix, 377
Ellipse, 130
Empty set, 3
Equality
  of complex numbers, 266
  of sets, 4
Equation(s)
  algebraic, 28
  of circle, 127
  conditional, 29
  consistent, 329, 339
  degree of, 29
  dependent, 329, 344

  equivalent, 30
  exponential, 256
  first-degree in one variable, 32
  first-degree in two variables, 75
  graph of, 76
  of graph, 80
  inconsistent, 329, 343
  independent, 329, 339
  involving radicals, 107
  involving rational expressions, 31
  of line
    point-slope form, 81
    slope-intercept form, 82
    standard form, 81
    two-point form, 81
  linear, 77
  literal, 34
  logarithmic, 230
  of parabola, 121, 123
  polynomial, 29, 286
  quadratic, 90
  quadratic in form, 112
  root of, 29
  second-degree in one variable, 90
  solution of, 29
  solution set of, 29
  solving, 30
  systems of linear, 328, 338
  systems involving quadratic, 355
Equivalent equations, 30
Equivalent sets, 5
Equivalent systems of equations, 330
Evaluation of determinants, 393, 394, 401
Even function, 162
Expansion of a binomial, 487
Exponent(s)
  irrational, 225
  laws of, 206
  negative integer, 208
  positive integer, 20, 206
  rational, 211
  zero, 209
Exponential equation, 256
Exponential form of an equation, 239

Exponential function, 225
Expression, algebraic, 20
Extraneous solution, 108

## F

Factor(s)
  literal, 21
  numerical, 21
Factor theorem, 288
Factorial notation, 476
Field, 12
Field axioms, 12
Finite sequence, 432
Finite sequence function, 432
Finite set, 5
First-degree equation(s)
  in one variable, 32
  in two variables, 75
Focus of a parabola, 121
Formula
  distance, 68
  quadratic, 101
Fractions, 12
Function(s)
  absolute value, 165
  algebraic, 164
  composite, 169
  constant, 162
  cubic, 163
  decreasing, 192
  definition of, 153
  domain of, 153
  even, 162
  exponential, 225
  greatest integer, 165
  identity, 164
  increasing, 192
  inverse of, 187
  linear, 163
  logarithmic, 230
  odd, 162
  polynomial, 163, 286
  quadratic, 163, 171
  range of, 153
  rational, 164
  sequence, 432
  transcendental, 165
  value, 160
  zeros of, 172
Fundamental principle of counting, 470, 471
Fundamental theorem of algebra, 295

## G

General element of a sequence, 433
General term of a series, 436
Generator of a cone, 134
Geometric mean(s), 454, 455
Geometric progression, 452
Geometric sequence(s)
  common ratio of, 452
  definition of, 451
  $N$th element of, 452
Geometric series, 455
  infinite, 460
  sum of finite, 456
  sum of infinite, 463
Graph(s)
  of a complex number, 275
  equation of, 80
  of equation, 76
  of inequality, 51, 141
  of an ordered pair, 65
  of polynomial functions, 312
  of rational functions, 177
  of a real number, 16
  of a set of ordered pairs, 75
  sketch of, 77
Greater than, 14
Greatest integer function, 165

## H

Half-plane, 142
Horizontal asymptote, 180
Hyperbola, 120, 131
Hypothesis of a theorem, 13

## I

*i*
  definition of, 24
  powers of, 271
Identity, 29
Identity element
  for addition of real numbers, 11
  for multiplication of real numbers, 11
Identity function, 164
If and only if, meaning of, 3
Imaginary number, 25, 266
Imaginary part of complex number, 25, 266
Inconsistent equations, 329, 343
Increasing function, 192
Independent equations, 329, 339
Independent events, 471
Independent variable, 155
Index of radical, 22
Index of summation, 436
Inequalities
  absolute, 49
  conditional, 50
  equivalent, 50
  first-degree in one variable, 46
  graphs of, 51, 141
  involving absolute values, 58
  quadratic, 136
  solution of, 49
  solution set of, 49
  systems of, 363
    in two variables, 141
Infinite geometric series, 460
Infinite sequence, 432

Infinite sequence function, 432
Infinite series, 460
Infinite set, 5
Interest, compound, 254
Interpolation, linear, 251
Intersection of sets, 6
Inverse
  additive, for real numbers, 11
  of a function, 187
  multiplicative, for real numbers, 12
  of a relation, 187
  of a square matrix, 414
Inverse variation, 198
Inversely proportional, 198
Investment problem, 37
Irrational numbers
  as exponents, 225
  set of, 8

## J

Joint variation, 199

## L

Latus rectum of a parabola, 122
Laws of exponents, 206
Length of a line segment, 68
Less than, 14
Linear equation(s), 77
  graph of, 77
  point-slope form of, 81
  slope-intercept form of, 82
  standard form of, 81
  systems of
    in two variables, 328
    in three variables, 338
  two-point form of, 81
Linear function, 163
Linear interpolation, 251
Linear programming, 369
Literal equation, 34

Literal factors, 21
Logarithm(s), 231
　base of, 231
　changing base of, 258
　characteristic of, 246
　common, 245
　computations with, 253
　mantissa of, 247
　natural, 236
　properties of, 237
　table of common, A-2
　table of natural, A-4
Logarithmic equation, 243
Logarithmic form of an equation, 237
Logarithmic function, 230
Lower bound, 307
Lower limit of sum, 436

## M

Main diagonal of a matrix, 412
Mantissa of a logarithm, 247
Mathematical induction, 438
Mathematical programming, 369
Matrix, 376
Maximum function value, 174
Minimum function value, 173
Minor, 395
Mixture problem, 37
Modulus, 278
Monomial
　definition of, 20
　degree of, 21
Multiple roots, 92
Multiplication
　of complex numbers, 267
　of functions, 167
　of radicals, 217
　of real numbers, 11
Multiplicative identity
　for complex numbers, 260
　for matrices, 412
　for real numbers, 11
Multiplicative inverse
　for complex numbers, 270
　for real numbers, 12
Multiplicative-inverse axiom, 12

## N

$n$th element of a sequence, 432
$n$th power, 20, 206
$n$th root, 22
Natural logarithm, 236
Natural logarithmic function, 236
Natural numbers, set of, 2
Negative integers, set of, 7
Negative numbers, set of, 7
Nonnegative integers, 7
Nonnegative numbers, 17
Nonpositive integers, 7
Nonpositive numbers, 17
Nonsingular matrix, 415
Nonterminating decimal, 8
Notation
　factorial, 476
　function value, 160
　scientific, 245
　set, 2
　set-builder, 2
　sigma, 436
　summation, 436
Null set, 3
Number(s)
　absolute value, 55, 278
　complex, 24, 266
　counting, 2
　graph of a complex, 275
　graph of a real, 16
　imaginary, 25, 266
　integers, 7
　irrational, 8
　natural, 8
　negative, 7
　nonnegative, 17
　nonpositive, 7

ordered pairs of real, 63
positive, 13
rational, 7
real, 8
whole, 6
Numerator, 12
Numerical coefficient, 21
Numerical factors, 21
Number line, real, 16

## O

Odd function, 162
One-to-one correspondence, 4
Opposite of a real number, 11
Order
  axiom of, 13
  of a determinant, 394
  of a matrix, 376
  of a radical, 22
Ordered field, 14
Ordered pairs of real numbers, 63
  components of, 63
  graph of, 65
  as solutions of equations, 75
Ordered triple, 339
Ordinate, 64
Origin, 15, 64

## P

Pair of real numbers, 63
Parabola, 114, 121
Permutation, 474
Point-slope form of equation of line, 81
Polynomial(s)
  definition of, 20
  degree of, 22
  equation, 29, 286
  function, 163, 286
Positive integers, set of, 7
Positive numbers, set of, 13

Power(s)
  of binomials, 487
  definition of, 20
  of $i$, 271
Primary variables, 369
Principal $n$th root of a real number, 22
Principal square root of a negative number, 26
Principle of mathematical induction, 438
Product(s)
  Cartesian, 63
  of complex numbers, 267
  of functions, 167
  involving radicals, 217
  of real numbers, 11
Progression(s)
  arithmetic, 444
  geometric, 452
Proof of a theorem, 13
Proper subset, 3
Proportional
  directly, 196
  inversely, 198
Pure imaginary number, 25, 266
Pythagorean theorem, 16, 68

## Q

Quadrant, 65
Quadratic equation(s)
  definition of, 90
  discriminant of, 102
  formula for solving, 101
  solution of
    by completing a square, 99
    by factoring, 91
    by formula, 101
  standard form of, 90
  systems of, 355
  in two variables, 114
Quadratic formula, 101
Quadratic function, 163, 171

Quadratic inequality, 136
Quotient(s)
  of complex numbers, 268
  definition of, 12
  of functions, 167
  of real numbers, 12

## R

Radical(s), 22
  addition of, 219
  definition of, 22
  division of, 220
  equations involving, 107
  index of, 22
  multiplication of, 217
  order of, 22
Radical sign, 22
Radicand, 22
Radius of circle, 126
Range
  of a function, 153
  of a relation, 152
Ratio of a geometric sequence, 452
Rational function(s), 164
  graphs of, 177
Rational numbers, set of, 7
Rationalizing the denominator, 220
Real number(s), 8
  axioms for, 11
  set of, 8
Real number line, 16
Real part of complex number, 25, 266
Real plane, 64
Reciprocal, 12
Rectangular Cartesian coordinate system, 65
Rectangular matrix, 377
Recursive formula, 445
Relation
  definition of, 152
  domain of, 152
  inverse of, 187
  range of, 152
Remainder theorem, 287
Repeating decimal, 8
Replacement set, 49
Rise of a line, 69
Root(s)
  of an equation, 29
  $n$th, of real numbers, 22
Row of a matrix, 376
Run of a line, 69

## S

Scalar, 386
Scientific notation, 245
Second-degree equations (see quadratic equations)
Sequence(s)
  arithmetic, 444
  definition of, 432
  finite, 432
  function, 432
  general element of, 433
  geometric, 451
  infinite, 432
  $n$th element of, 432
Series
  arithmetic, 448
  definition of, 435
  geometric, 455
  infinite, 460
Set(s)
  of complex numbers, 25, 266
  of counting numbers, 2
  disjoint, 6
  elements of, 2
  empty, 3
  equality of, 4
  finite, 5
  infinite, 5
  of integers, 7
  intersection of, 6
  of irrational numbers, 8

of natural numbers, 2
of negative integers, 7
of negative real numbers, 9
null, 3
of positive integers, 7
of positive real numbers, 9
of rational numbers, 7
of real numbers, 8
replacement, 2
solution, 29, 75, 339
union of, 5
of whole numbers, 6
Set-builder notation, 2
Sigma notation, 436
Singular matrix, 415
Significant digits, 250
Slope, 70
Slope-intercept form of equation of line, 82
Solution(s)
   of an equation, 29, 75
   extraneous, 108
   of inequality, 49
   ordered pairs as, 75
   ordered triples as, 339
   of systems of equations, 328, 338
Solution set
   of an equation, 29, 75, 328, 338
   of an inequality, 49, 328, 338
   of a system of equations, 328, 338
Square matrix, 377
Square root of a real number, 23
   table of, A-1
Standard form
   of a linear equation, 81
   of a quadratic equation, 90
Subset, 3
Substitution, solution of system of equations by, 330
Subtraction
   of functions, 167
   of real numbers, 12
Sum
   of an arithmetic series, 449

of complex numbers, 267
of functions, 167
of a geometric series, 456
of an infinite geometric series, 463
of radicals, 219
of real numbers, 11
Summation
   index of, 436
   notation, 436
Symmetry
   of a graph, 117, 119
   with respect to a line, 117
   with respect to a point, 119
Synthetic division, 290
System(s)
   of inequalities, 363
   of linear equations
     in two variables, 328
     in three variables, 338
   of quadratic equations, 355
   solution set of, 328, 338

## T

Tables, numerical
   common logarithms, A-2
   exponential functions, A-4
   natural logarithms, A-4
   powers and roots, A-1
Terms(s)
   of an algebraic expression, 20
   general, of a series, 436
   of a series, 436
Terminating decimals, 8
Theorem
   conclusion of, 13
   definition of, 13
   hypothesis of, 13
Transcendental functions, 165
Transitive property of order, 46
Transpose of a matrix, 412
Triangular form, 342
Trinomial, 21
Two-point form of equation of line, 81

## U

Undirected distance, 67
Uniform motion problem, 40
Union of sets, 5
Unit points, 15
Unknown, variable as, 28
Upper bound, 307
Upper limit of sum, 436

## V

Value, absolute, 55, 495
Value of a function, 160
Variable(s), 2
   definition of, 2
   dependent, 155
   independent, 155
   replacement set of, 2
Variation
   constant of, 196
   direct, 196
   inverse, 198
   joint, 199
Vertex of a cone, 134
Vertex of a parabola, 122
Vertical asymptote, 179

## W

Well-defined set, 2
Whole numbers, set of, 6
Word problems, 37, 346
Work problem, 42

## X

$x$ axis, 64
$x$ coordinate, 64
$x$ intercept, 78

## Y

$y$ axis, 64
$y$ coordinate, 64
$y$ intercept, 78

## Z

Zero(s)
   division by, 13
   of a function, 172
Zero-matrix, 384